MATHEMATICS RESEARCH DEVELOPMENTS

DYNAMICAL SYSTEMS

THEORY, APPLICATIONS AND FUTURE DIRECTIONS

MATHEMATICS RESEARCH DEVELOPMENTS

MATHEMATICS RESEARCH DEVELOPMENTS

DYNAMICAL SYSTEMS

THEORY, APPLICATIONS AND FUTURE DIRECTIONS

JEAN MICHEL TCHUENCHE
EDITOR

nova publishers
New York

For permission to use material from this book please contact us:
Telephone 631-231-7269; Fax 631-231-8175
Web Site: http://www.novapublishers.com

LIBRARY OF CONGRESS CATALOGING-IN-PUBLICATION DATA

ISBN: 978-1-62808-001-8

Published by Nova Science Publishers, Inc. † New York

Dedication

To Talia and Jordi Ryan Tchuenche

CONTENTS

PREFACE

Virtually any advance in diseases dynamics requires a sophisticated mathematical approach in order to map out the parameters that drive the system. This book exploits some of the fundamental theory of Dynamical Systems and applies them to biological and social sciences. Topics includes: The asymptotic stability of almost periodic systems, modeling and vibration-based system identification and damage inspection of flexible structures, the stability and persistence of dynamical systems using Lyapunov functions, Krylov and Schur subspaces in construction of prediction error filter - its computational complexity and performance comparison, parametric resonance and homoclinic chaos in a Bullard-type dynamo; birhythmicity, synchonization and chaos in an enzyme-substrate reaction with a Ferroelectric behavior. Biological applications include the local stability of an age-structured Hepatitis B model, the dynamics of a predator-prey model, avian influenza, tuberculosis, malaria and HIV-TB co-dynamics. A possible application of nonlinear dynamical systems in social sciences is also investigated.

Mathematical models of scientific systems often lead to differential equations in which the independent variable is time. Such systems arise in astronomy, biology, chemistry, economics, engineering, physics and other disciplines. A dynamical system is a concept in mathematics where a fixed rule describes the time dependence of a point in a geometrical space. Examples include the mathematical models that describe the swinging of a clock pendulum, the flow of water in a pipe, and the number of fish each springtime in a lake.

Virtually any advance in diseases dynamics requires a sophisticated mathematical approach in order to map out the parameters that drive the system. In this book, the authors explore the asymptotic stability of almost periodic systems, modeling and vibration-based system identification and damage inspection of flexible structures, the stability and persistence of dynamical systems using Lyapunov functions, the computational complexity and performance comparison of Krylov and Schur subspaces in construction of prediction error filter, parametric resonance and homoclinic chaos in a Bullard-type dynamo; birhythmicity, synchonization and chaos in an enzyme-substrate reaction with a Ferroelectric behavior. Biological applications include the local stability of an age-structured Hepatitis B model, the dynamics of a predator-prey model, avian influenza, tuberculosis, malaria and HIV-TB co-dynamics. A possible application of nonlinear dynamical systems in social sciences is also investigated.

Anyone with interest in the field of dynamical systems and its applications in biology and social sciences is invited to explore the richness of this collection.

Jean M. Tchuenche
Atlanta, Georgia, US
January, 2013

In: Dynamical Systems
Editor: Jean Michel Tchuenche, pp. 1-20
ISBN: 978-1-62808-001-8

Chapter 1

MARKUS-SELL'S THEOREM FOR ASYMPTOTICALLY ALMOST PERIODIC SYSTEMS

***David Cheban*[1,*] *and Cristiana Mammana*[2,†]**
[1]State University of Moldova,
Department of Mathematics and Informatics
Chişinău, Moldova
[2]Dept. of Economic and Financial Institutions
University of Macerata, Macerata, Italy

Abstract

This chapter is dedicated to the study of asymptotic stability of asymptotically almost periodic systems. The authors formulate and prove the analog of Markus-Sell's theorem for asymptotically almost periodic systems (both finite and infinite dimensional cases). The authors study this problem in the framework of general non-autonomous dynamical systems. The obtained general results they apply to different classes of non-autonomous evolution equations: Ordinary Differential Equations, Difference Equations, Functional Differential Equations and Semi-Linear Parabolic Equations.

Keywords: Global attractor; non-autonomous dynamical system; asymptotic stability, almost periodic motions

1. Introduction

Denote by $\mathbb{R} := (-\infty, +\infty)$, $\mathbb{R}^n$ is a product space of n copies of $\mathbb{R}$, $F(t,x) := f(x) + p(t,x) \in C(\mathbb{R}_+ \times \mathbb{R}^n, \mathbb{R}^n)$ the right hand side of system (1), where $C(\mathbb{R}_+ \times \mathbb{R}^n, \mathbb{R}^n)$ is the space of all continuous functions $F : \mathbb{R}_+ \times \mathbb{R}^n \mapsto \mathbb{R}^n$ equipped with the compact open topology.

*E-mail address: cheban@usm.md
†E-mail address: mammana@unimc.it

A system of differential equation

$$x' = f(x) + p(t,x) \tag{1}$$

is said to be asymptotically autonomous, if the function $p \in C(\mathbb{R}_+ \times \mathbb{R}^n, \mathbb{R}^n)$ satisfies the following condition

$$\lim_{t\to\infty} |p(t,x)| = 0 \tag{2}$$

uniformly in x on every compact subset from $\mathbb{R}^n$, where $|\cdot|$ is a norm on $\mathbb{R}^n$. Autonomous system

$$x' = f(x) \tag{3}$$

is called a limiting system for (1).

Example 1.1. *(***Bessel's equation***) Consider the equation*

$$t^2x'' + tx' + (t^2 - \alpha^2)x = 0,$$

or equivalently

$$\begin{cases} x' = y \\ y' = -\frac{1}{t}y + (\frac{\alpha^2}{t^2} - 1)x, \end{cases}$$

with limiting system

$$\begin{cases} x' = y \\ y' = -x\ . \end{cases}$$

Denote by $C^1(\mathbb{R}^n, \mathbb{R}^n)$ the space of all continuously differentiable functions $f : \mathbb{R}^n \mapsto \mathbb{R}^n$.

Theorem 1.2. *(L. Markus [10]) Let $f \in C^1(\mathbb{R}^n, \mathbb{R}^n)$, $O = (0,0)$ be a critical point of limiting system (3), i.e, $f(0) = 0$. Assume that the variational system of (3) based on origin O have characteristic values with negative real parts. Then there exists a neighborhood U of O and a time T such that $\lim_{t\to\infty} |x(t)| = 0$ for any solution of equation (1) intersecting U no later than T, i.e., the origin is an attracting point for (1).*

Let $(C(\mathbb{R}_+ \times \mathbb{R}^n, \mathbb{R}^n), \mathbb{R}_+, \sigma)$ be the shift dynamical system [11, 12] (or Bebutov's dynamical system) on $C(\mathbb{R}_+ \times \mathbb{R}^n, \mathbb{R}^n)$. For every function $F \in C(\mathbb{R}_+ \times \mathbb{R}^n, \mathbb{R}^n)$ we denote by $H^+(F) := \overline{\{F_\tau : \tau \in \mathbb{R}_+\}}$ the closure of all positive translations of the function F and by Ω_F its ω-limit set, i.e., $\Omega_F := \{G :$ there exists a sequence $\tau_n \to +\infty$ such that $F_{\tau_n} \to G\}$, where F_τ is τ-shift of the function F, i.e., $F_\tau(t,x) := F(t+\tau, x)$ for all $(t,x) \in \mathbb{R}_+ \times \mathbb{R}^n$.

Let $F \in C(\mathbb{R}_+ \times \mathbb{R}^n, \mathbb{R}^n)$ be an arbitrary function. Consider the equation

$$x' = F(t,x). \tag{4}$$

Along with equation (4) we consider its H^+-class, i.e., the following family of equations

$$y' = G(t,y) \quad (G \in H^+(F)).$$

Example 1.3. *1. Let $F \in C(\mathbb{R}_+ \times \mathbb{R}^n; \mathbb{R}^n)$ be asymptotically autonomous, i.e., $F(t,x) = f(x) + p(t,x)$ and p satisfies condition (2). In this case $\Omega_F = \{f\}$, i.e., its ω-limit set contains a single function.*

2. Let $F \in C(\mathbb{R}_+ \times \mathbb{R}^n; \mathbb{R}^n)$ be asymptotically T periodic, i.e., $F(t,x) = f(t,x) + p(t,x)$, $f(t+T,x) = f(t,x)$ for all $(t,x) \in \mathbb{R} \times \mathbb{R}^n$ and p satisfies condition (2). In this case $\Omega_F = \{f_\tau : \tau \in [0,T)\}$, i.e., its ω-limit set contains a continuum functions and it is homeomorphic to unitary circle.

3. If $F \in C(\mathbb{R}_+ \times \mathbb{R}^n; \mathbb{R}^n)$ is asymptotically quasi periodic, i.e., $F(t,x) = f(t,x) + p(t,x)$, where $f(t,x)$ is a quasi periodic function with the spectrum of frequency $\nu_1, \nu_2, \ldots, \nu_m$ and p satisfies condition (2). In this case its ω-limit set is homeomorphic to an m-torus.

Theorem 1.4. *(G. Sell [11, Ch.VIII]) Let $F \in C(\mathbb{R}^n, \mathbb{R}^n)$ be regular, asymptotically autonomous and $O \in \mathbb{R}^n$ be the null solution equation (4), i.e, $F(t,0) = 0$ for all $t \in \mathbb{R}_+$. Assume that the null solution of limiting equation (3) is uniformly asymptotically stable. Then the null solution of equation (4) is uniformly asymptotically stable.*

Remark 1.5. *1. Note that Theorem 1.4 generalizes Theorem of L. Markus in the following directions:*

a. right hand side f of the limiting equation is only continuous (L. Markus, $f \in C^1(\mathbb{R}^n, \mathbb{R}^n)$);

b. the null solution of limiting equation (3) is only uniformly asymptotically stable (L. Markus, O is uniformly exponentially stable (In fact, $Re\lambda_i < 0$ $(i = 1, \ldots, n)$, $\lambda_1, \ldots, \lambda_n$ are characteristic values of the origin for the variational equation for (3)).

2. It is easy to see that there are examples with uniformly asymptotically stable origin which is not uniformly exponentially stable. For example $x' = -x^3$ $(n = 1)$.

Consider a differential equation

$$x' = f(t,x) \quad (f \in C(\mathbb{R} \times W, \mathbb{R}^n)), \tag{5}$$

where W is an open subset from $\mathbb{R}^n$ containing the origin (i.e., $0 \in W$), $C(\mathbb{R} \times W, \mathbb{R}^n)$ is the space of all continuous functions $f : \mathbb{R} \times W \mapsto \mathbb{R}^n$ equipped with compact open topology. Denote by $(C(\mathbb{R} \times W, \mathbb{R}^n), \mathbb{R}, \sigma)$ the *shift dynamical system* [5, 11] on the space $C(\mathbb{R} \times W, \mathbb{R}^n)$ (*dynamical system of translations* or *Bebutov's dynamical system*), i.e., $\sigma(\tau, f) := f_\tau$ for all $\tau \in \mathbb{R}$ and $f \in C(\mathbb{R} \times W, \mathbb{R}^n)$, where $f_\tau(t,x) := f(t+\tau, x)$ for all $(t,x) \in \mathbb{R} \times W$.

Below we will use the following conditions:

(A): for all $(t_0, x_0) \in \mathbb{R}_+ \times W$ the equation (5) admits a unique solution $x(t; t_0, x_0)$ with initial data (t_0, x_0) and defined on $\mathbb{R}_+ := [0, +\infty)$, i.e., $x(t_0; t_0, x_0) = x_0$;

(B): the hand right side f *is positively compact*, if the set $\Sigma_f^+ := \{f_\tau : \tau \in \mathbb{R}_+\}$ is a relatively compact subset of $C(\mathbb{R} \times W, \mathbb{R}^n)$;

(C): the equation

$$y' = g(t, y), \quad (g \in \Omega_f) \tag{6}$$

is called a *limiting equation* for (5), where Ω_f is the ω-limit set of f with respect to the shift dynamical system $(C(\mathbb{R} \times W, \mathbb{R}^n), \mathbb{R}, \sigma)$, i.e., $\Omega_f := \{g :$ there exists a sequence $\{\tau_k\} \to +\infty$ such that $f_{\tau_k} \to g$ as $k \to \infty\}$;

(D): *equation* (5) (or its hand right side f) *is regular* , if for all $p \in H^+(f)$ the equation

$$y' = p(t, y)$$

admits a unique solution $\varphi(t, x_0, p)$ defined on $\mathbb{R}_+$ with initial condition $\varphi(0, x_0, p) = x_0$ for all $x_0 \in W$, where $H^+(f) = \overline{\{f_\tau : \tau \in \mathbb{R}_+\}}$ and by bar is denoted the closure in the space $C(\mathbb{R} \times W, \mathbb{R}^n)$;

(E): equation (5) admits a null (trivial) solution, i.e., $f(t, 0) = 0$ for all $t \in \mathbb{R}_+$.

The trivial solution of equation (5) is said to be:

(i) *uniformly stable*, if for all positive number ε there exists a number $\delta = \delta(\varepsilon)$ $(\delta \in (0, \varepsilon))$ such that $|x| < \delta$ implies $|\varphi(t, x, f_\tau)| < \varepsilon$ for all $t, \tau \in \mathbb{R}_+$;

(ii) *uniformly attracting*, if there exists a positive number a such that

$$\lim_{t \to +\infty} |\varphi(t, x, f_\tau)| = 0$$

uniformly in $|x| \leq a$ and $\tau \in \mathbb{R}_+$;

(iii) *uniformly asymptotically stable*, if it is uniformly stable and uniformly attracting.

In connection with Theorems 1.2 and 1.4 by G. Sell was formulated the following problem.

G. Sell's conjecture ([11, Ch.VIII,p.134]). Let $f \in C(\mathbb{R} \times W, \mathbb{R}^n)$ be a regular function and f be positively pre-compact. Assume that W contains the origin 0 and $f(t, 0) = 0$ for all $t \in \mathbb{R}_+$. Assume further that there exists a positive number a such that the equality

$$\lim_{t \to +\infty} |\varphi(t, x, g)| = 0$$

takes place uniformly in $|x| \leq a$ and $g \in \Omega_f$. Then the trivial solution of (5) is uniformly asymptotically stable.

The positive solution of G. Sell's conjecture was obtained by Z. Artstein [1] and Bondi P. et al. [2].

Remark 1.6. *1. Bondi P. et al. [2] proved this conjecture under the additional assumption that the function f is locally Lipschitzian.*

2. Artstein Z. [1] proved this statement without Lipschitzian condition. In reality he proved a more general statement. Namely, he supposed that only limiting equations for (5) are regular, but the function f is not obligatory regular.

3. By D. Cheban [7] was formulated G. Sell's conjecture for abstract NDSs (the both with continuous and discrete time). In [7] it is given a positive answer to this conjecture and also are presented some applications of this result to different classes of evolution equations: infinite-dimensional differential equations, functional-differential equations and semi-linear parabolic equations .

In the paper [13] it was published the following false [3] statement.

Theorem 1.7. *Let f be a regular function with $f(t,0) = 0$ for all $t \geq 0$. If there exits a function $g \in \Omega_f$ such that the null solution of equation (6) is uniformly asymptotically stable, then the null solution of equation (5) is uniformly asymptotically stable.*

Bondi P. at al. [3] give the following counterexample to Theorem 1.7.

$$ax'' + bx' + cx = x \sin\sqrt{t} \ \ (x \in \mathbb{R},\ t \in \mathbb{R}_+,\ a, b > 0,\ c \in (0,1)). \tag{7}$$

For every $\mu \in [-1, 1]$

$$ax'' + bx' + cx = \mu x \tag{8}$$

is a limiting equation for (7). For $\mu \in [-1, c)$ the null solution of equation (8) is uniformly asymptotically stable, but the null solution of equation (7) is not uniformly stable.

Recall that a function $f \in C(\mathbb{R} \times W, E)$ is called:

- almost periodic (respectively, almost recurrent) in $t \in \mathbb{R}$ uniformly in u on every compact subset K from W, if for an arbitrary number $\varepsilon > 0$ and compact subset $K \subseteq W$ there exists a positive number $L = L(K, \varepsilon)$ such that on every segment $[a, a+L]$ $(a \in \mathbb{R})$ of the length L there exists at least one number τ such that

  $$\max_{u \in K,\ |t| \leq 1/\varepsilon} |f(t+s+\tau, u) - f(t+s)| < \varepsilon$$

  (respectively,

  $$\max_{u \in K,\ |t| \leq 1/\varepsilon} |f(t+\tau, u) - f(t, u)| < \varepsilon)$$

  for all $s \in \mathbb{R}$;

- recurrent (in $t \in \mathbb{R}$ uniformly in u on every compact subset K from W, if the function $f \in C(\mathbb{R} \times W, E)$ is almost recurrent and $H(f) := \overline{\{f_\tau : \tau \in \mathbb{R}\}}$ is compact;

- asymptotically recurrent (respectively, almost periodic) in $t \in \mathbb{R}$ uniformly in u on every compact subset K from W, if there exits two functions $P, R \in C(\mathbb{R} \times W, E)$ such that:

  (a)
  $$f(t, u) = P(t, u) + R(t, x) \tag{9}$$
  for all $(t, u) \in \mathbb{R} \times E$;

  (b) the function P is recurrent (respectively, almost periodic) in $t \in \mathbb{R}$ uniformly in u on every compact subset K from W and

  $$\lim_{t \to +\infty} \max_{u \in K} |R(t, u)| = 0 \tag{10}$$

  for every compact subset $K \subseteq W$.

From the main result of this paper (Theorems 3.7 and 3.8) it follows that Theorem 1.7 is true if the function f is asymptotically recurrent (in particularly, asymptotically almost periodic) in $t \in \mathbb{R}$ uniformly in u on every compact subset Q from $\mathbb{R}^n$.

The aim of this paper is investigation the problem of asymptotic stability of trivial solution for asymptotically almost periodic (respectively, asymptotically recurrent) systems. We study this problem in the framework of general *non-autonomous dynamical systems* (NDS). We formulate and prove the analog of Theorem 1.7 for abstract non-autonomous dynamical systems. The obtained result we apply to different classes of evolution equations: Ordinary Differential Equations (both finite and infinite-dimensional cases), Difference Equations, Functional Differential Equations, Semi-Linear Parabolic Equations .

The paper is organized as follows.

In Section 2, we collect some notions (global attractor, stability, asymptotic stability, uniform asymptotic stability, minimal set, point/compact dissipativity, recurrence, shift dynamical systems, etc) and facts from the theory of dynamical systems which will be necessary in this paper.

Section 3 is devoted to the study of asymptotic stability of NDS with asymptotically recurrent base. The main result of paper (Theorem 3.7 and Theorem 3.8) contain some tests of asymptotic stability for asymptotically recurrent (in particularly, asymptotically almost periodic) NDS.

Finally, Section 4 contains a series of applications of our general results from Sections 2-3 for Ordinary Differential Equations (Theorem 4.1 and Theorem 4.2), Functional-Differential Equations with finite delay (Theorem 4.7) and Semi-Linear Parabolic Equations (Theorem 4.9).

2. Asymptotic Stability of Dynamical Systems

In this section we collect some facts about stability and asymptotic stability of dynamical systems (both autonomous and non-autonomous) which use in this paper.

2.1. Compact Global Attractors of Dynamical Systems

Let X be a topological space, $\mathbb{R}$ $(\mathbb{Z})$ be a group of real (integer) numbers, $\mathbb{R}_+$ $(\mathbb{Z}_+)$ be a semi-group of the nonnegative real (integer) numbers, $\mathbb{S}$ be one of the two sets $\mathbb{R}$ or $\mathbb{Z}$ and $\mathbb{T} \subseteq \mathbb{S}$ be one of the sub-semigroups $\mathbb{R}_+$ (respectively, $\mathbb{Z}_+$) or $\mathbb{R}$ (respectively, $\mathbb{Z}$).

Triplet $(X, \mathbb{T}, \pi)$, where $\pi : \mathbb{T} \times X \to X$ is a continuous mapping satisfying the following conditions: $\pi(0, x) = x$ and $\pi(s, \pi(t, x)) = \pi(s + t, x)$ $(\forall t, s \in \mathbb{T})$ is called a *dynamical system*. If $\mathbb{T} = \mathbb{R}$ $(\mathbb{R}_+)$ or $\mathbb{Z}$ $(\mathbb{Z}_+)$, then $(X, \mathbb{T}, \pi)$ is called a *group (semi-group) dynamical system*. In the case, when $\mathbb{T} = \mathbb{R}_+$ or $\mathbb{R}$ the dynamical system $(X, \mathbb{T}, \pi)$ is called a *flow*, but if $\mathbb{T} \subseteq \mathbb{Z}$, then $(X, \mathbb{T}, \pi)$ is called a *cascade (discrete flow)*.

Below X will be a complete metric space with the distance ρ.

The function $\pi(\cdot, x) : \mathbb{T} \to X$ is called a *motion* passing through the point x at the moment $t = 0$ and the set $\Sigma_x := \pi(\mathbb{T}, x)$ is called a *trajectory* of this motion.

A nonempty set $M \subseteq X$ is called *positively invariant* (*negatively invariant*, *invariant*) with respect to dynamical system $(X, \mathbb{T}, \pi)$ or, simple, positively invariant (negatively invariant, invariant), if $\pi(t,M) \subseteq M$ ($M \subseteq \pi(t,M), \pi(t,M) = M$) for every $t \in \mathbb{T}_+ := \{t \in \mathbb{T} : t \geq 0\}$.

A closed positively invariant set (respectively, invariant set), which does not contain own closed positively invariant (respectively, invariant) subset, is called *minimal*.

Let $M \subseteq X$. The set

$$\omega(M) := \bigcap_{t \geq 0} \overline{\bigcup_{\tau \geq t} \pi(\tau, M)}$$

is called ω-limit for M.

The set $W^s(\Lambda)$, defined by equality

$$W^s(\Lambda) := \{x \in X | \lim_{t \to +\infty} \rho(\pi(t,x), \Lambda) = 0\}$$

is called a *stable manifold* of the set $\Lambda \subseteq X$.

The set M is called:

- *orbital stable*, if for every $\varepsilon > 0$ there exists $\delta = \delta(\varepsilon) > 0$ such that $\rho(x, M) < \delta$ implies $\rho(\pi(t,x), M) < \varepsilon$ for all $t \geq 0$;
- *attracting*, if there exists $\gamma > 0$ such that $B(M,\gamma) \subset W^s(M)$, where $B(M,\gamma) := \{x \in X : \rho(x,M) < \gamma\}$;
- *asymptotic stable*, if it is orbital stable and attracting;
- *uniform attracting*, if there exists $\gamma > 0$ such that

$$\lim_{t \to +\infty} \sup_{x \in B(M,\gamma)} \rho(\pi(t,x), M) = 0.$$

The system $(X, \mathbb{T}, \pi)$ is called:

– *compactly dissipative* if there exists a nonempty compact subset $K \subseteq X$ such that

$$\lim_{t \to +\infty} \beta(\pi(t,M), K) = 0 \tag{11}$$

for all compact subset $M \subseteq X$, where $\beta(A,B) := \sup_{a \in A} \rho(a,B)$ and $\rho(a,B) := \inf_{b \in B} \rho(a,b)$;

– *local completely continuous* (*compact*) if for all point $p \in X$ there are two positive numbers δ_p and l_p such that the set $\pi(l_p, B(p,\delta_p))$ is relatively compact.

Let $(X, \mathbb{T}, \pi)$ be compactly dissipative and K be a compact set attracting every compact subset from X. Let us set $J = \omega(K)$. It can be shown [5, Ch.I] that the set J doesn't depends on the choice of the attractor K, but is characterized only by the properties of the dynamical system $(X, \mathbb{T}, \pi)$ itself. The set J is called a *Levinson center* of the compactly dissipative dynamical system $(X, \mathbb{T}, \pi)$.

Lemma 2.1. *[7] Let $(X, \mathbb{T}, \pi)$ be a dynamical system and $x \in X$ be a point with relatively compact semi-trajectory $\Sigma_x^+ := \{\pi(t,x) : t \geq 0\}$. Then the following statements hold:*

(i) *the dynamical system* $(X, \mathbb{T}, \pi)$ *induces on the* $H^+(x) := \overline{\Sigma_x^+}$ *a dynamical system* $(H^+(x), \mathbb{T}_+, \pi)$, *where by bar is denoted the closure of* Σ_x^+ *in the space* X;

(ii) *the dynamical system* $(H^+(x), \mathbb{T}_+, \pi)$ *is compactly dissipative;*

(iii) *Levinson center* $J_{H^+(x)}$ *of* $(H^+(x), \mathbb{T}_+, \pi)$ *coincides with* ω*-limit set* ω_x *of the point* x.

2.2. G. Sell's Conjecture for Non-Autonomous Dynamical Systems

Let $\mathbb{T}_1 \subseteq \mathbb{T}_2 \subseteq \mathbb{S}$ be two sub-semigroups of $\mathbb{S}$ and $(Y, \mathbb{T}_2, \sigma)$ be a dynamical system on metric space Y. Recall that a triplet $\langle W, \varphi, (Y, T_2, \sigma)\rangle$ (or shortly φ), where W is a metric space and φ is a mapping from $\mathbb{T}_1 \times W \times Y$ into W, is said to be a *cocycle* over $(Y, \mathbb{T}_2, \sigma)$ with the fiber W, if the following conditions are fulfilled:

(i) $\varphi(0, u, y) = u$ for all $u \in W$ and $y \in Y$;

(ii) $\varphi(t + \tau, u, y) = \varphi(t, \varphi(\tau, u, y), \sigma(\tau, y))$ for all $t, \tau \in \mathbb{T}_1$, $u \in W$ and $y \in Y$;

(iii) the mapping $\varphi : \mathbb{T}_1 \times W \times Y \mapsto W$ is continuous.

Example 2.2. (**Bebutov's dynamical system**) Let X, W be two metric space. Denote by $C(\mathbb{T} \times W, X)$ the space of all continuous mappings $f : \mathbb{T} \times W \mapsto X$ equipped with the compact-open topology and σ be the mapping from $\mathbb{T} \times C(\mathbb{T} \times W, X)$ into $C(\mathbb{T} \times W, X)$ defined by the equality $\sigma(\tau, f) := f_\tau$ for all $\tau \in \mathbb{T}$ and $f \in C(\mathbb{T} \times W, X)$, where f_τ is the τ-translation (shift) of f in t, i.e., $f_\tau(t, x) = f(t + \tau, x)$ for all $(t, x) \in \mathbb{T} \times W$. Then [5, Ch.I] the triplet $(C(\mathbb{T} \times W, X), \mathbb{T}, \sigma)$ is a dynamical system on $C(\mathbb{T} \times W, X)$ which is called a *shift dynamical system* (*dynamical system of translations* or *Bebutov's dynamical system*).

A function $f \in C(\mathbb{T} \times W, X)$ is said to be *recurrent* (respectively, *almost periodic*) in $t \in \mathbb{T}$ uniformly in $x \in W$ on every compact subset from W, if $f \in C(\mathbb{T} \times W, X)$ is a recurrent (respectively, almost periodic) point of the Bebutov's dynamical system $(C(\mathbb{T} \times W, X), \mathbb{T}, \sigma)$.

Example 2.3. Consider differential equation (5) with regular second right hand side $f \in C(\mathbb{R} \times W, \mathbb{R}^n)$, where $W \subseteq \mathbb{R}^n$. Denote by $(H^+(f), \mathbb{R}_+, \sigma)$ a semi-group shift dynamical system on $H^+(f)$ induced by Bebutov's dynamical system $(C(\mathbb{R} \times W, \mathbb{R}^n), \mathbb{R}, \sigma)$, where $H^+(f) := \overline{\{f_\tau : \tau \in \mathbb{R}_+\}}$. Let $\varphi(t, u, g)$ a unique solution of equation

$$y' = g(t, y), \quad (g \in H^+(f)),$$

then from the general properties of the solutions of non-autonomous equations it follows that the following statements hold:

(i) $\varphi(0, u, g) = u$ for all $u \in W$ and $g \in H^+(f)$;

(ii) $\varphi(t + \tau, u, g) = \varphi(t, \varphi(\tau, u, g), g_\tau)$ for all $t, \tau \in \mathbb{R}_+$, $u \in W$ and $g \in H^+(f)$;

(iii) the mapping $\varphi : \mathbb{R}_+ \times W \times H^+(f) \mapsto W$ is continuous.

From above it follows that the triplet $\langle W, \varphi, (H^+(f), \mathbb{R}_+, \sigma)\rangle$ is a cocycle over $(H^+(f), \mathbb{R}_+, \sigma)$ with the fiber $W \subseteq \mathbb{R}^n$. Thus, every non-autonomous equation (5) with regular f naturally generates a cocycle which plays a very important role in the qualitative study of equation (5).

Suppose that $W \subseteq E$, where E is a Banach space with the norm $|\cdot|$, $0 \in W$ (0 is the null element of E) and the cocycle $\langle W, \varphi, (Y, T_2, \sigma)\rangle$ admits a trivial (null) motion/solution, i.e., $\varphi(t, 0, y) = 0$ for all $t \in \mathbb{T}_1$ and $y \in Y$.

The trivial motion/solution of cocycle φ is said to be:

(i) *uniformly stable*, if for all positive number ε there exists a number $\delta = \delta(\varepsilon)$ ($\delta \in (0, \varepsilon)$) such that $|u| < \delta$ implies $|\varphi(t, u, y)| < \varepsilon$ for all $t \geq 0$ and $y \in Y$;

(ii) *uniformly attracting*, if there exists a positive number a such that

$$\lim_{t \to +\infty} |\varphi(t, u, y)| = 0 \tag{12}$$

uniformly with respect to $|u| \leq a$ and $y \in Y$;

(iii) *uniformly asymptotically stable*, if it is uniformly stable and uniformly attracting.

Theorem 2.4. *[7] Let $\langle (X, \mathbb{T}_1, \pi), (Y, \mathbb{T}_2, \sigma), h\rangle$ be a NDS. Suppose that the following conditions are fulfilled:*

(i) Y is compact;

(ii) the dynamical system $(X, \mathbb{T}_1, \pi)$ is locally compact;

(iii) the trivial section Θ of (X, h, Y) is positively invariant;

(iv) the trivial section $\tilde{\Theta}$ of NDS $\langle (\tilde{X}, \mathbb{T}_1, \pi), (J_Y, \mathbb{T}_2, \sigma), h\rangle$ is uniformly attracting.

Then the trivial section Θ of non-autonomous dynamical system $\langle (X, \mathbb{T}_1, \pi), (Y, \mathbb{T}_2, \sigma), h\rangle$ is uniformly stable.

Corollary 2.5. *[7] Under the conditions of Theorem 2.4 the trivial section Θ of NDS $\langle (X, \mathbb{T}_1, \pi), (Y, \mathbb{T}_2, \sigma), h\rangle$ is uniformly asymptotically stable.*

Remark 2.6. *Note that Corollary 2.5 gives a positive answer to G.Sell's conjecture for local-compact NDS [7].*

3. Asymptotic Stability of NDS with Asymptotically Recurrent Base

Section 3 is devoted to the study of asymptotic stability of NDS with asymptotically recurrent base. The main result of paper (Theorem 3.7 and Theorem 3.8) contain some tests of asymptotic stability for asymptotically recurrent (in particularly, asymptotically almost periodic) NDS.

Remark 3.1. *If the bundle (X, h, Y) is locally trivial, then for all $y \in Y$, $x \in X_y$ and $y_n \to y$ there exists a sequences $x_n \in X_{y_n}$ such that $x_n \to x$.*

Let $M \subseteq Y$. Denote by $\Theta_M := h^{-1}(M) \bigcap \Theta$, where Θ is the null section of the vectorial bundle (X, h, Y).

Definition 3.2. *The set Θ_M is said to be:*

(i) uniform stable, *if for all $\varepsilon > 0$ there exists a number $\delta = \delta(\varepsilon, M) > 0$ such that $|x| < \delta$ $(x \in h^{-1}(M))$ implies $|\pi(t, x)| < \varepsilon$ for all $t \geq 0$;*

(ii) uniform attracting, *if there exists a positive number a such that*

$$\lim_{t \to \infty} \sup_{|x| \leq a,\ x \in h^{-1}(M)} |\pi(t, x)| = 0; \tag{13}$$

(iii) uniform asymptotic stable, *if it is uniform stable and uniform attracting.*

Lemma 3.3. *Suppose that the set Θ_M $(M \subseteq Y)$ is uniformly stable (respectively, uniformly attracting or uniformly asymptotically stable), then the set $\Theta_{\overline{M}}$ is also uniformly stable (respectively, uniformly attracting or uniformly asymptotically stable), where by bar is denoted the closure of M in Y.*

Proof. Let Θ_M be uniformly stable, $\varepsilon > 0$ and $\delta = \delta(\varepsilon, M) > 0$ a positive number figuring in the definition of uniform stability of Θ_M. Denote by $\nu(\varepsilon) := \frac{\delta(\varepsilon/2, M)}{2}$. Let now $x \in h^{-1}(\overline{M})$ with $|x| < \nu(\varepsilon)$, $y := h(x) \in \overline{M}$ and $y_n \in M$ such that $y_n \to y$. By Remark 3.1 there exist sequences $y_n \to y$ $(y_n \in M)$ and $x_n \in X_{y_n}$ such that $x_n \to x$. According to choose of the number $\nu(\varepsilon)$ there exists $n_0 \in \mathbb{N}$ such that $|x_n| \leq \nu(\varepsilon) < \frac{\delta(\varepsilon/2, M)}{2}$ and, consequently,

$$|\pi(t, x_n)| < \varepsilon/2 \tag{14}$$

for all $t \geq 0$. Passing into limit in (14) as n goes to ∞ we obtain $|\pi(t, x)| \leq \varepsilon/2 < \varepsilon$ for all $t \geq 0$.

Let now Θ_M be uniformly attracting, $a > 0$ be a positive number figuring in (13), and ε be an arbitrary positive number. According to equality (13) there exists a positive number $L = L(\varepsilon/2, M)$ such that

$$|\pi(t, x)| < \varepsilon/2 \tag{15}$$

for all $t \geq L$, $|x| \leq a$ with $x \in h^{-1}(M)$. Let now $x \in h^{-1}(\overline{M})$ with $|x| \leq a/2$, $y_n \to y$ $(y_n \in M)$ and $x_n \to x$ $(x_n \in X_{y_n})$, then there exists a number $n_0 \in \mathbb{N}$ such that $|x_n| \leq a$ for all $n \geq n_0$. From (15) we obtain

$$|\pi(t, x_n)| \leq \varepsilon/2 \tag{16}$$

for all $t \geq L$ and $n \geq n_0$. Passing into limit in (16) as n tends to ∞ we obtain $|\pi(t, x)| < \varepsilon$ for all $t \geq L$ and $x \in h^{-1}(\overline{M})$ with $|x| \leq b := a/2$. This means that

$$\lim_{t \to \infty} \sup_{|x| \leq b,\ x \in h^{-1}(\overline{M})} |\pi(t, x)| = 0,$$

i.e., the set $\Theta_{\overline{M}}$ is uniformly attracting.

Since the set $\Theta_{\overline{M}}$ is uniformly stable and uniformly attracting, then it is uniformly asymptotically stable. The lemma is completely proved. □

Definition 3.4. *Let $y_0 \in Y$, the null element $\theta_{y_0} \in X_{y_0}$ is said to be* uniformly stable *(respectively,* uniformly attracting *or* uniformly asymptotically stable*), if the set $\Theta_{\Sigma^+_{y_0}}$ is so, where $\Sigma^+_{y_0} := \{\sigma(t, y_0) : t \geq 0\}$ is the positive semi-trajectory of the point y_0.*

Corollary 3.5. *Let $y_0 \in Y$ and θ_{y_0} be uniformly stable (respectively, uniformly attracting or uniformly asymptotically stable), then the set $\Theta_{H^+(y_0)}$ is also uniformly stable (respectively, uniformly attracting or uniformly asymptotically stable).*

Proof. This statement follows directly from Lemma 3.3 if we apply it to the set $M = \Sigma^+_{y_0}$. □

Let $y \in Y$ be an asymptotically recurrent point, i.e., there exists a recurrent point $q \in Y$ such that

$$\lim_{t \to +\infty} \rho(\sigma(t, y), \sigma(t, q)) = 0. \tag{17}$$

Denote by $P_y := \{q \in Y : q \text{ is a recurrent point such that (17) holds}\}$.

Remark 3.6. *1. If the point $y \in Y$ is asymptotically recurrent, then its ω-limit set ω_y is a compact and minimal set of dynamical system $(Y, \mathbb{T}, \sigma)$.*

2. There exist points $y \in Y$ (see, for example, [6, Ch.I,p.13] Example 1.42) with the properties that Σ^+_y is relatively compact and ω_y is compact and minimal, but y is not asymptotically recurrent.

Theorem 3.7. *Let $\langle (X, \mathbb{T}_1, \pi), (Y, \mathbb{T}_2, \sigma), h \rangle$ be a NDS. Suppose that the following conditions are fulfilled:*

(i) the dynamical system $(X, \mathbb{T}_1, \pi)$ is locally compact;

(ii) the space Y is compact;

(iii) Levinson center J_Y of the dynamical system $(Y, \mathbb{T}_2, \sigma)$ is minimal;

(iv) the null section Θ of the bundle (X, h, Y) is positively invariant;

(v) there exists a point $q \in J_Y$ such that $\theta_q \in \Theta$ is uniformly attracting.

Then the trivial section Θ of $\langle (X, \mathbb{T}_1, \pi), (Y, \mathbb{T}_2, \sigma), h \rangle$ is uniformly asymptotically stable.

Proof. According to Corollary 2.5 the set $\Theta_{H^+(q)}$is uniformly attracting. Since q is recurrent and the set J_Y is minimal we obtain $J_Y = H^+(q)$. Thus the trivial section $\tilde{\Theta}$ of the NDS $\langle (\tilde{X}, \mathbb{T}_1, \pi), (J_Y, \mathbb{T}_2, \sigma), h \rangle$ is uniformly attracting. Now to finish the proof of Theorem 3.7 it is sufficient to apply Theorem 2.4 and Corollary 2.5. □

Theorem 3.8. *Let $\langle (X, \mathbb{T}_1, \pi), (Y, \mathbb{T}_2, \sigma), h \rangle$ be a NDS. Suppose that the following conditions are fulfilled:*

(i) the dynamical system $(X, \mathbb{T}_1, \pi)$ is locally compact;

(ii) there exists an asymptotically recurrent point $y \in Y$ such that $Y = H^+(y)$;

(iii) the null section Θ of the bundle (X, h, Y)) is positively invariant;

(iv) there exists a recurrent point $q \in P_y$ such that $\theta_q \in \Theta$ is uniformly attracting.

Then the trivial section Θ of $\langle (X, \mathbb{T}_1, \pi), (Y, \mathbb{T}_2, \sigma), h \rangle$ is uniformly asymptotically stable.

Proof. Since the point $y \in Y$ is asymptotically recurrent, $q \in P_y$ and $Y = H^+(y)$, then Y is compact, the dynamical system $(Y, \mathbb{T}_2, \sigma)$ is compactly dissipative and its Levinson center J_Y coincides with $\omega_y = H^+(q)$ and, consequently, it is minimal. Now to finish the proof of Theorem 3.8 it is sufficient to apply Theorem 3.7. □

Recall that a subset $M \subseteq X$ is called transitive, if there exists a point $p \in X$ such that $M = H(q) := \overline{\{\pi(t,p) : t \in \mathbb{R}\}}$, where by bar is denoted a closure in the space X.

A point $x \in X$ is said to be:

- Poisson stable (in the positive direction), if $x \in \omega_x$;

- asymptotically Poisson stable, if there exists a Poisson stable point $p \in X$ such that

$$\lim_{t \to +\infty} \rho(\pi(t,x), \pi(t,p)) = 0. \tag{18}$$

Lemma 3.9. *If the point $x \in X$ is asymptotically Poisson stable, then its ω-limit set ω_x is transitive.*

Proof. Let x be an asymptotically Poisson stable point and ω_x be its ω-limit set. Then there exists a Poisson stable point $p \in X$ such that (18) holds. From (18) it follows the equality $\omega_x = \omega_p$. On the other hand from the Poisson stability of x we obtain $\omega_p = H(p)$ and, consequently, $\omega_x = H(p)$. Lemma is proved. □

Remark 3.10. *All results of Section 3 remain true, if we replace the minimality of J_Y by its transitivity.*

Remark 3.11. *All results of Sections 2–3 remain true if*

1. *we replace the positive invariance of the trivial section Θ by the following condition: there exists a compact positively invariant set $M \subseteq X$ such that $M_y := \{x \in M : h(x) = y\}$ consists a single point for all $y \in Y$;*

2. *the compact metric space Y we replace by an arbitrary compact regular topological space.*

4. Some Applications

Section 4 contains a series of applications of our general results from Sections 2-3 for Ordinary Differential Equations, Functional-Differential Equations with finite delay and Semi-Linear Parabolic Equations.

4.1. Ordinary Differential Equations

Denote by $C(\mathbb{S}\times W,E)$ the space of all continuous mappings $f:\mathbb{S}\times W\mapsto E$ equipped with the compact open topology. On the space $C(\mathbb{S}\times W,E)$ it is defined a shift dynamical system [5, ChI] (dynamical system of translations or Bebutov's dynamical system) $(C(\mathbb{S}\times W,E),\mathbb{S},\sigma)$, where σ is a mapping from $\mathbb{S}\times C(\mathbb{S}\times W,E)$ onto $C(\mathbb{S}\times W,E)$ defined as follow $\sigma(\tau,f):=f_\tau$ for all $(\tau,f)\in\mathbb{S}\times C(\mathbb{S}\times W,E)$, where f_τ is the τ-translation of f in t, i.e., $f_\tau(t,x):=f(t+\tau,x)$ for all $(t,x)\in\mathbb{S}\times W$. Consider a differential equation

$$u'=f(t,u), \tag{19}$$

where $f\in C(\mathbb{R}\times W,E)$.

If the function f is regular, then equation (19) naturally defines a cocycle $\langle W,\varphi,(H^+(f),\mathbb{R}_+,\sigma)\rangle$, where $(H^+(f),\mathbb{R}_+,\sigma)$ is a (semi-group) dynamical system on $H^+(f)$ induced by Bebutov's dynamical system.

Applying the general results from Sections 2-3 we will obtain a series of results for equation (19). Below we formulate some of them.

Denote by $\Omega_f:=\{g\in H^+(f):$ there exists a sequence$\tau_n\to+\infty$ such that $g=\lim\limits_{n\to\infty}f_{\tau_n}\}$ the ω-limit set of f.

Theorem 4.1. *Assume that the following conditions are fulfilled:*

(i) the function f is regular;

(ii) the set $H^+(f)$ is compact;

(iii) the ω-limit set Ω_f of function f is a compact minimal set of Bebutov's dynamical system $(C(\mathbb{R}\times W,E),\mathbb{R},\sigma)$;

(iv) $f(t,0)=0$ for all $t\in\mathbb{R}_+$;

(v) there exists a neighborhood U of the origin 0 and a positive number l such that the set $\varphi(l,U,H^+(f))$ is relatively compact;

(vi) there exists a function $P\in\Omega_f$ such that the trivial solution of equation

$$x'=P(t,x) \tag{20}$$

is uniformly attraction, i.e., there exists a positive number a such that

$$\lim_{t\to+\infty}\sup_{|v|\le a,\tau\ge 0}|\varphi(t,v,P_\tau)|=0. \tag{21}$$

Then the null solution of equation (19) is uniformly asymptotically stable.

Proof. Consider the dynamical system $(H^+(f),\mathbb{R}_+,\sigma)$. Since the space $H^+(f)$ is compact, then $(H^+(f),\mathbb{R}_+,\sigma)$ is compactly dissipative and by Lemma 2.1 its Levinson center (maximal compact invariant set) $J_{H^+(f)}$ coincides with ω-limit set Ω_f of f. Let $Y:=H^+(f)$ and $(Y,\mathbb{R}_+,\sigma)$ be the shift dynamical system on Y. Denote by $X:=W\times Y$ and $(X,\mathbb{R}_+,\pi)$ the skew-product dynamical system generates by $(Y,\mathbb{R}_+,\sigma)$ and cocycle φ, i.e., $\pi(t,(v,g)):=(\varphi(t,v,g),\sigma(t,g))$ for all $t\in\mathbb{R}_+$ and $(v,g)\in X$. Now consider a non-autonomous dynamical system $\langle(X,\mathbb{R}_+,\pi),(Y,\mathbb{R}_+,\pi),h\rangle$ $(h:=pr_2)$ associated by equation (19). It is easy to verify that this NDS possess the following properties:

(i) by Lemma 2.1 the dynamical system $(Y, \mathbb{R}_+, \sigma)$ is compactly dissipative and its Levinson center J_Y coincides with Ω_f;

(ii) the null section Θ of $\langle (X, \mathbb{R}_+, \pi), (Y, \mathbb{R}_+, \pi), h\rangle$ coincides with $\{0\} \times Y$;

(iii) Θ is a positively invariant subset of $(X, \mathbb{R}_+, \pi)$;

(iv) according to (21) the null section $\tilde{\Theta}$ of NDS $\langle(\tilde{X}, \mathbb{R}_+, \pi), (J_Y, \mathbb{R}_+, \sigma), h\rangle$ is uniformly attracting (see Lemma 3.3) because $|\pi(t, x)| = |\varphi(t, v, g)|$ for all $t \in \mathbb{R}_+$ and $x := (v, g) \in X$.

Now to finish the proof it is sufficient to apply Corollary 2.5. □

Theorem 4.2. *Assume that the following conditions are fulfilled:*

(i) the function f is regular;

(ii) f is asymptotically recurrent in t uniformly in x on every compact subset from W;

(iii) $f(t, 0) = 0$ for all $t \in \mathbb{R}_+$;

(iv) there exists a neighborhood U of the origin 0 and a positive number l such that the set $\varphi(l, U, H^+(f))$ is relatively compact;

(v) the trivial solution of equation (20) is uniformly attracting, i.e., there exists a positive number a such that

$$\lim_{t \to +\infty} \sup_{|v| \leq a, \tau \geq 0} |\varphi(t, v, P_\tau)| = 0.$$

Then the null solution of equation (19) is uniformly asymptotically stable.

Proof. This statement it follows directly from Lemma 3.3 and Corollary 3.5 using the same arguments as in the proof of Theorem 4.1. □

Remark 4.3. *1. Note that the compactness and minimality of the set Ω_f in Theorem 4.1 are essential. In fact, in the work [3] there is an example of equation of type (19), where all conditions of Theorem 4.1 are fulfilled with the exception of minimality of Ω_f and for which equation the trivial solution is not uniformly stable.*

2. In the case, when the function f is asymptotically stationary (i.e., the function P figuring in (22) does not depend on $t \in \mathbb{R}$) and E is finite-dimensional Theorem 4.2 coincides with one result of G. Sell (see [11, Ch.VIII, p.135] Theorem 10) and L. Markus [10].

3. Note that assumption of the minimality of Ω_f is not necessary in this Subsection (see Remark 3.10 and Lemma 3.9). All the results from Subsection 4.1 remain true, if we replace the minimality of Ω_f by its transitivity (this means that there exists a function $P \in \Omega_f$ such that $\Omega_f = H(P) := \overline{\{P_\tau : \tau \in \mathbb{R}\}}$).

4.2. Difference Equations

Consider a difference equation

$$u(t+1) = f(t, u(t)), \tag{22}$$

where $f \in C(\mathbb{Z} \times W, E)$.

Along with equation (22) we consider the family of equations

$$v(t+1) = g(t, v(t)), \tag{23}$$

where $g \in H^+(f) := \overline{\{f_\tau : \tau \in \mathbb{Z}_+\}}$ and by bar is denoted the closure in the space $C(\mathbb{Z} \times W, E)$. Let $\varphi(t, v, g)$ be a unique solution of equation (23) with initial data $\varphi(0, v, g) = v$. Denote by $(H^+(f), \mathbb{Z}_+, \sigma)$ the shift dynamical system on $H^+(f)$, then the triplet $\langle W, \varphi, (H^+(f), \mathbb{Z}_+, \sigma)\rangle$ is a cocycle (with discrete time) over $(H^+(f), \mathbb{Z}_+, \sigma)$ with the fibre W.

Applying the results from Sections 2-3 we will obtain the following result for difference equation (22).

Theorem 4.4. *Assume that the following conditions are fulfilled:*

(i) the function $f \in C(\mathbb{Z} \times W, E)$ is asymptotically recurrent in t uniformly in x on every compact subset from W;

(ii) $f(t, 0) = 0$ for all $t \in \mathbb{Z}_+$;

(iii) there exists a neighborhood U of the origin 0 and a positive number l such that the set $\varphi(l, U, H^+(f))$ is relatively compact;

(iv) the trivial solution of equation

$$x(t+1) = P(t, x)$$

is uniformly attracting, i.e., there exists a positive number a such that

$$\lim_{t \to +\infty} \sup_{|v| \le a, \tau \ge 0} |\varphi(t, v, P_\tau)| = 0.$$

Then the null solution of equation (22) is uniformly asymptotically stable.

4.3. Functional Differential-Equations (FDEs) with Finite Delay

We will apply now the abstract theory developed in the previous Sections to the analysis of a class of functional differential equations.

Let us first recall some notions and notations from [8]. Let $r > 0$, $C([a, b], \mathbb{R}^n)$ be the Banach space of all continuous functions $\varphi : [a, b] \to \mathbb{R}^n$ equipped with the sup–norm. If $[a, b] = [-r, 0]$, then we set $\mathcal{C} := C([-r, 0], \mathbb{R}^n)$. Let $\sigma \in \mathbb{R}$, $A \ge 0$ and $u \in C([\sigma - r, \sigma + A], \mathbb{R}^n)$. We will define $u_t \in \mathcal{C}$ for all $t \in [\sigma, \sigma + A]$ by the equality $u_t(\theta) := u(t + \theta)$, $-r \le \theta \le 0$. Consider a functional differential equation

$$\dot{u} = f(t, u_t), \tag{24}$$

where $f : \mathbb{R} \times \mathcal{C} \to \mathbb{R}^n$ is continuous.

Denote by $C(\mathbb{R} \times \mathcal{C}, \mathbb{R}^n)$ the space of all continuous mappings $f : \mathbb{R} \times \mathcal{C} \mapsto \mathbb{R}^n$ equipped with the compact open topology. On the space $C(\mathbb{R} \times \mathcal{C}, \mathbb{R}^n)$ is defined (see, for example, [5, ChI]) a shift dynamical system $(C(\mathbb{R} \times \mathcal{C}, \mathbb{R}^n), \mathbb{R}, \sigma)$, where $\sigma(\tau, f) := f_\tau$ for all $f \in C(\mathbb{R} \times \mathcal{C}, \mathbb{R}^n)$ and $\tau \in \mathbb{R}$ and f_τ is τ-translation of f, i.e., $f_\tau(t, \phi) := f(t + \tau, \phi)$ for all $(t, \phi) \in \mathbb{R} \times \mathcal{C}$.

Let us set $H^+(f) := \overline{\{f_s : s \in \mathbb{R}_+\}}$, where by bar we denote the closure in $C(\mathbb{R} \times \mathcal{C}, \mathbb{R}^n)$.

Along with the equation (24) let us consider the family of equations

$$\dot{v} = g(t, v_t), \tag{25}$$

where $g \in H^+(f)$.

Below, in this subsection, we suppose that equation (24) is regular.

Remark 4.5. 1. Denote by $\tilde{\varphi}(t, u, f)$ the solution of equation (24) defined on $\mathbb{R}_+$ (respectively, on $\mathbb{R}$) with the initial condition $\varphi(0, u, f) = u \in \mathcal{C}$, i.e., $\varphi(s, u, f) = u(s)$ for all $s \in [-r, 0]$. By $\varphi(t, u, f)$ we will denote below the trajectory of equation (24), corresponding to the solution $\tilde{\varphi}(t, u, f)$, i.e., the mapping from $\mathbb{R}_+$ (respectively, $\mathbb{R}$) into $\mathcal{C}$, defined by $\varphi(t, u, f)(s) := \tilde{\varphi}(t + s, u, f)$ for all $t \in \mathbb{R}_+$ (respectively, $t \in \mathbb{R}$) and $s \in [-r, 0]$.

2. Due to item 1. of this remark, below we will use the notions of "solution" and "trajectory" for equation (24) as synonym concepts.

It is well known [4, 11] that the mapping $\varphi : \mathbb{R}_+ \times \mathcal{C} \times H^+(f) \mapsto \mathbb{R}^n$ possesses the following properties:

(i) $\varphi(0, v, g) = u$ for all $v \in \mathcal{C}$ and $g \in H^+(f)$;

(ii) $\varphi(t + \tau, v, g) = \varphi(t, \varphi(\tau, v, g), \sigma(\tau, g))$ for all $t, \tau \in \mathbb{R}_+$, $v \in \mathcal{C}$ and $g \in H^+(f)$;

(iii) the mapping φ is continuous.

Thus, a triplet $\langle \mathcal{C}, \varphi, (H^+(f), \mathbb{R}_+, \sigma) \rangle$ is a cocycle which is associated to equation (24). Applying the results from Sections 2-3 we will obtain certain results for functional differential equation (24).

A function $f \in C(\mathbb{R} \times W, \mathcal{C})$ is said to be completely continuous, if the set $f(\mathbb{R}_+ \times A)$ is bounded for all bounded subset $A \subseteq \mathcal{C}$.

Lemma 4.6. *[7] Suppose that the following conditions hold:*

(i) the function $f \in C(\mathbb{R} \times W, \mathcal{C})$ is regular and completely continuous;

(ii) the set $H^+(f)$ is compact.

Then the cocycle φ associated by (24) is completely continuous, i.e., for all bounded subset $A \subseteq W$ there exists a positive number $l = l(A)$ such that the set $\varphi(l, A, H^+(f))$ is relatively compact in $\mathcal{C}$.

Theorem 4.7. *Assume that the following conditions are fulfilled:*

a. the function $f \in C(\mathbb{R} \times \mathcal{C}, \mathcal{C})$ is regular and completely continuous;

b. f is asymptotically recurrent in t uniformly in x on every compact subset from $\mathcal{C}$;

c. $f(t, 0) = 0$ for all $t \in \mathbb{R}_+$;

d. the trivial solution of equation

$$x' = P(t, x_t)$$

is uniformly attraction.

Then the null solution of equation (24) is uniformly asymptotically stable.

Proof. Consider the dynamical system $(H^+(f), \mathbb{R}_+, \sigma)$. Since the space $H^+(f)$ is compact, then $(H^+(f), \mathbb{R}_+, \sigma)$ is compactly dissipative and by Lemma 2.1 its Levinson center $J_{H^+(f)}$ coincides with ω-limit set Ω_f of f. Let $Y := H^+(f)$ and $(Y, \mathbb{R}_+, \sigma)$ be the shift dynamical system on Y. Denote by $X := \mathcal{C} \times Y$ and $(X, \mathbb{R}_+, \pi)$ the skew-product dynamical system generates by $(Y, \mathbb{R}_+, \sigma)$ and cocycle φ. Now consider a NDS $\langle (X, \mathbb{R}_+, \pi), (Y, \mathbb{R}_+, \pi), h \rangle$ ($h := pr_2$) associated by equation (24). It easy to verify this NDS possesses the following properties:

(i) the dynamical system $(Y, \mathbb{R}_+, \sigma)$ is compactly dissipative and its Levinson center J_Y coincides with Ω_f;

(ii) the null section Θ of $\langle (X, \mathbb{R}_+, \pi), (Y, \mathbb{R}_+, \pi), h \rangle$ coincides with $\{0\} \times Y$;

(iii) Θ is a positively invariant subset of $(X, \mathbb{R}_+, \pi)$;

(iv) according to item e. the null section $\tilde{\Theta}$ of NDS $\langle (\tilde{X}, \mathbb{R}_+, \pi), (J_Y, \mathbb{R}_+, \sigma), h \rangle$ is uniformly attracting because $|\pi(t, x)| = |\varphi(t, v, g)|$ for all $t \in \mathbb{R}_+$ and $x := (v, g) \in X$;

(v) according to Lemma 4.6 the dynamical system $(X, \mathbb{R}_+, \pi)$ is completely continuous.

Now to finish the proof it is sufficient to apply Lemma 3.3 and Corollary 3.5. □

4.4. Semi-Linear Parabolic Equations

Let E be a Banach space, and let $A : D(A) \to E$ be a linear closed operator with the dense domain $D(A) \subseteq E$.

An operator A is called [9] sectorial if for some $\varphi \in (0, \pi/2)$, some $M \geq 1$, and some real a, the sector

$$S_{a,\varphi} := \{\lambda : \varphi \leq |\arg(\lambda - a)| \leq \pi, \lambda \neq a\}$$

lies in the resolvent set $\rho(A)$ and $\|(I\lambda - A)^{-1}\| \leq M|\lambda - a|^{-1}$ for all $\lambda \in S_{a,\varphi}$.

If A is a sectorial operator, then there exists an $a_1 \geq 0$ such that *Re* $\sigma(A + a_1 I) > 0$ $(\sigma(A) := \mathbb{C} \setminus \rho(A))$. Let $A_1 = A + a_1 I$. For $0 < \alpha < 1$, one defines the operator [9]

$$A_1^{-\alpha} := \frac{\sin \pi\alpha}{\pi} \int_0^{+\infty} \lambda^{-\alpha}(\lambda I + A_1)^{-1} d\lambda,$$

which is linear, bounded, and one-to-one. Set $E^\alpha := D(A_1^\alpha)$, and let us equip the space E^α with the graph norm $|x|_\alpha := |A_1^\alpha x|$ $(x \in E)$, $E^0 := E$, and $E^1 := D(A)$. Then E^α is a Banach space with the norm $|\cdot|_\alpha$ and is densely and continuously embedded in E.

Consider differential equation

$$x' + Ax = f(t,x), \tag{26}$$

where $f \in C(\mathbb{R} \times E^\alpha, E)$ and $C(\mathbb{R} \times E^\alpha, E)$ is the space of all the continuous functions equipped with compact open topology.

Along with equation (26), consider family of equations

$$y' + Ay = g(t,y), \tag{27}$$

where $g \in H^+(f) := \overline{\{f_\tau : \tau \in \mathbb{R}_+\}}$.

Recall that a function f is said to be regular, if for every $(v,g) \in E^\alpha \times H^+(f)$ equation (27) admits a unique solution [9, Ch.III] $\varphi(t,v,g)$ with initial data $\varphi(0,v,g) = v$ and the mapping $\varphi : \mathbb{R}_+ \times E^\alpha \times H^+(F) \mapsto E^\alpha$ is continuous.

Regularity conditions for f are given in Theorems 3.3.3, 3.3.4, 3.3.6, and 3.4.1 in [9, Ch.III].

Assuming that f is regular, a non-autonomous dynamical system can be associated in a natural way with equation (26). Namely, we set $Y := H^+(f)$ and by $(Y, \mathbb{R}_+, \sigma)$ denote the dynamical system of translations on Y. Further, let $X := E^\alpha \times Y$, and let $(X, \mathbb{R}_+, \pi)$ be the dynamical system on X defined by the relation $\pi^\tau(v,g) = \langle \varphi(\tau, v, g), g_\tau \rangle$. Finally, by setting $h = pr_2 : X \to Y$, we obtain the non-autonomous system $\langle (X, \mathbb{R}_+, \pi), (Y, \mathbb{R}_+, \sigma), h \rangle$ determined by equation (26).

Recall that a function $f \in C(\mathbb{R} \times E^\alpha, E)$ is said to be locally Hölder continuous in t and locally Lipschitz in x, if for every $(t_0, x_0) \in \mathbb{R} \times E^\alpha$ there exists a neighborhood V $((t_0, x_0) \in V)$ and positive numbers L and θ such that

$$|f(t_1, x_1) - f(t_2, x_2)| \le L(|t_1 - t_2|^\theta + |x_1 - x_2|_\alpha)$$

for all $(t_i, x_i) \in V$ $(i = 1, 2)$.

Lemma 4.8. *Suppose that the following conditions are fulfilled:*

(i) *A is a sectorial operator;*

(ii) *the resolvent of operator A is compact;*

(iii) $0 \le \alpha < 1$ *and* $f \in C(\mathbb{R} \times E^\alpha, E)$;

(iv) *the function f is locally Hölder continuous in t and locally Lipschitz in x;*

(v) *the set $f(\mathbb{R}_+ \times B)$ is bounded in E for all bounded subset B from E^α.*

Under the conditions listed above, if the function f is regular and the set $H^+(f)$ is compact, then the cocycle φ associated by equation (26) is completely continuous.

Proof. This statement can be proved with the slight modification of the proof of Theorem 3.3.6 [9, Ch.III]. □

Applying results from Sections 3-4 we obtain the following result for evolution equation (26).

Theorem 4.9. *Assume that the following conditions hold:*

a. the function $f \in C(\mathbb{R} \times E^{\alpha}, E)$ is asymptotically recurrent in t uniformly in x on every compact subset from E^{α};

b. $f(t, 0) = 0$ for all $t \in \mathbb{R}_+$;

c. the set $f(\mathbb{R}_+ \times B)$ is bounded in E for all bounded subset B from E^{α};

d. the trivial solution of equation

$$x' = Ax + P(t, x)$$

is uniformly attracting.

Then the null solution of equation (26) is uniformly asymptotically stable.

Proof. Consider the dynamical system $(H^+(f), \mathbb{R}_+, \sigma)$. Since the space $H^+(f)$ is compact, then according to Lemma 2.1 $(H^+(f), \mathbb{R}_+, \sigma)$ is compactly dissipative and its Levinson center $J_{H^+(f)}$ coincides with ω-limit set Ω_f of f. Let $Y := H^+(f)$ and $(Y, \mathbb{R}_+, \sigma)$ be the shift dynamical system on Y. Denote by $X := E^{\alpha} \times Y$ and $(X, \mathbb{R}_+, \pi)$ the skew-product dynamical system generates by $(Y, \mathbb{R}_+, \sigma)$ and cocycle φ. Consider a non-autonomous dynamical system $\langle (X, \mathbb{R}_+, \pi), (Y, \mathbb{R}_+, \pi), h \rangle$ $(h := pr_2)$ associated by equation (26). It easy to verify that this NDS possesses the following properties:

(i) the dynamical system $(Y, \mathbb{R}_+, \sigma)$ is compact dissipative and by Lemma 2.1 its Levinson center J_Y coincides with Ω_f;

(ii) the null section Θ of $\langle (X, \mathbb{R}_+, \pi), (Y, \mathbb{R}_+, \pi), h \rangle$ coincides with $\{0\} \times Y$;

(iii) Θ is a positively invariant subset of $(X, \mathbb{R}_+, \pi)$;

(iv) according to item b. the null section $\tilde{\Theta}$ of NDS $\langle (\tilde{X}, \mathbb{R}_+, \pi), (J_Y, \mathbb{R}_+, \sigma), h \rangle$ is uniformly attracting because $|\pi(t, x)| = |\varphi(t, v, g)|$ for all $t \in \mathbb{R}_+$ and $x := (v, g) \in X$;

(v) by Lemma 4.8 the cocycle φ and, consequently, the skew-product dynamical system $(X, \mathbb{R}_+, \pi)$ too, is completely continuous.

Now to finish the proof it is sufficient to apply Lemma 3.3 and Theorem 3.8. □

Remark 4.10. *Theorem 4.9 for asymptotically autonomous equations it was established in [9] (Chapter IV, Theorem 4.3.7).*

References

[1] Artstein Zvi, Uniform asymptotic stability via the limiting equations. *Journal of Differential Equations,* 27(2), 172-189, 1978.

[2] Bondi P., Moaurd V. and Visentin F., Limiting equations in the stability problem. *Nonlinear Analisys,* v.1, no.2, 123–128, 1977.

[3] Bondi P., Moaurd V. and Visentin F., Addendum to th Paper "Limiting equations in the stability problem". *Nonlinear Analysis,* v.1, no.6, 1977. pp.701.

[4] Bronsteyn I. U., *Extensions of Minimal Transformation Group,* Noordhoff, 1979.

[5] Cheban D. N., Global Attractors of Non-Autonomous Dissipstive Dynamical Systems. *Interdisciplinary Mathematical Sciences* 1. River Edge, NJ: World Scientific, 2004, xxiii+502 pp.

[6] Cheban D. N. *Asymptotically Almost Periodic Solutions of Differential Equations.* Hindawi Publishing Corporation, New York, 2009, ix+186 pp.

[7] Cheban D. N., Sell's conjecture for Non-Autonomous Dynamical Systems. *Nonlinear Analysis: TMA,* Vol.75, no.7, 2012, pp.3393-3406.

[8] Hale J. K., *Theory of Functional-Differential Equations.* Springer-Verlag, New York-Heidelberg-Berlin, 1977. [Russian translation: Theory of Functional-Differential Equations. Mir, Moscow, 1984.]

[9] Henry D., Geometric Theory of Semilinear Parabolic Equations. *Lecture Notes in Mathematics*, No. 840, Springer-Verlag, New York 1981.

[10] Markus L., *Contributions to Nonlinear Oscilations*, Vol.3 Princeton, Univ. Press, Princeton, N. J., 1956, pp.17-29.

[11] Sell G. R., *Topological Dynamics and Ordinary Differential Equations.* Van Nostrand-Reinhold, London, 1971.

[12] Shcherbakov B. A., Topologic Dynamics and Poisson Stability of Solutions of Differential Equations, *Ştiinţa*, Chişinău, 1972. (In Russian)

[13] S. M. Shamim Imdadi and M. Rama Mohama Rao, A note on topological dynamics and limiting equations. *Proceeding of The American Mathematical Society,* Vol.40, No.1, 1973, pp.154-158.

In: Dynamical Systems
Editor: Jean Michel Tchuenche, pp. 21-50
ISBN: 978-1-62808-001-8

Chapter 2

A Survey of the Method of Lyapunov Functions of Two Variables in the Study of Stability and Persistence of Dynamical Systems

***Jorge Lewowicz*[1,*], *Jorge Groisman*[1,†] *and Juan Tolosa*[2,‡]**
[1]IMERL, Facultad de Ingeniería, Universidad de la República
Montevideo, Uruguay
[2]Department of Natural Sciences and Mathematics
Richard Stockton College of New Jersey, Galloway, US

Abstract

In this chapter, the authors describe Lewowicz's method of Lyapunov functions of two variables, for the study of structural stability of dynamical systems, and related concepts, such as topological stability and persistence. The authors include a wide scope of applications of this method, among others, to perturbations of hyperbolic maps on the plane, Anosov diffeomorphisms, expansive homeomorphisms on surfaces and on the plane, geodesic flows on compact manifolds, and billiards.

Keywords: Quadratic forms, Lyapunov functions, persistence, structural stability, topological stability, hyperbolicity, expansive homeomorphisms, billiards, Anosov diffeomorphisms, pseudo-Anosov maps

1. Introduction

The (usual) Lyapunov functions were introduced in [27], to study stability of equilibria of systems of differential equations using *Lyapunov functions*. These are real-valued positive definite functions, whose derivative along the flow is negative. The wide applicability of the study of stability via Lyapunov functions resides, on the one hand, in the fact that this

*E-mail address: lew@fing.edu.uy
†E-mail address: jorgeg@fing.edu.uy
‡E-mail address: juan.tolosa@stockton.edu

derivative can be computed without actually solving the system; and on the other hand, in the fact that this method allows for the application of powerful geometric methods, for example, to the analysis of level surfaces of such functions.

If the system is linear, $\dot{x} = Ax$, with a constant matrix A such that all its eigenvalues have negative real part, we can even pick a *quadratic* Lyapunov function (see, for example, Lemma 4.2 in [36], p. 255), which guarantees the stability of the equilibrium $x = 0$. If the system is nonlinear, say, with $x = 0$ as an equilibrium point, the existence of a Lyapunov function, not necessarily quadratic, will still guarantee stability of the equilibrium.

In [18] J. Lewowicz proposed to use Lyapunov functions of two variables to study structural stability of homeomorphisms on compact manifolds, and similar concepts, such as topological stability and persistence. Since then, the method has been successfully applied to a wide spectrum of dynamical systems, such as hyperbolic and almost hyperbolic diffeomorphisms on manifolds, geodesic flows, pseudo-Anosov maps, billiards, expansive homeomorphisms on compact manifolds, in particular on surfaces, and on expansive homeomorphisms of the plane.

Much as in the case of Lyapunov's second method, for all the mentioned systems the hyperbolic prototypes—Anosov diffeomorphisms, geodesic flows on surfaces with negative curvature, and billiards with dispersing boundaries—are characterized by quadratic Lyapunov functions. These can then be used to prove structural stability for the corresponding hyperbolic systems. Further, the existence of a "good" Lyapunov function, not necessarily quadratic, provides a class of almost hyperbolic dynamical systems. Typically, they are no longer structurally stable, but they still are topologically stable.

In this chapter we showcase Lewowicz's method and its main applications. The structure of the chapter is as follows. We start in Section 2 with linear hyperbolic systems in $\mathbb{R}^n$, and its perturbations. Although Lewowicz's method was first introduced for the study of homeomorphisms on compact manifolds, linear hyperbolic systems in $\mathbb{R}^n$ are a "canonical case," in which many of the main ideas of the method are more transparent, since additional technicalities are reduced to a minimum. After the linear hyperbolic case, we discuss systems in $\mathbb{R}^n$ which, while no longer hyperbolic, still possess a "nice" Lyapunov function. These systems have persistence properties and in some cases are conjugate to their perturbations, provided sufficiently high jets are preserved at equilibrium. In particular, we show a global stability result: linear hyperbolic systems are "structurally stable" relative to C^1-perturbations that impose a strong proximity condition away from the origin. As a consequence, we obtain a new proof of Hartman's classical theorem ([13], p. 244), that an autonomous system is locally conjugate to its linear part near a hyperbolic equilibrium.

In Section 3 we describe the use of quadratic forms to characterize hyperbolic behavior. In all the situations reviewed—Anosov diffeomorphisms, geodesic flows, and non-uniformly hyperbolic maps (in particular, billiards)—the existence of an appropriate quadratic form turns out to be a necessary and sufficient condition for hyperbolicity. These systems are structurally stable in a suitable sense.

Section 4 is devoted to the corresponding generalizations, the *almost hyperbolic systems*. Those are systems that still admit a Lyapunov function, not necessarily quadratic. Although in general they are no longer structurally stable, they still have nice properties, such as topological stability, or persistence.

We start by discussing almost-hyperbolic diffeomorphisms on compact manifolds, which are shown to be topologically stable. We outline the proof of this result, and mention examples which pose interesting open questions. Next we review perturbations of geodesic flows for surfaces of non-positive curvature; we show how Lyapunov functions can be constructed to prove persistence results.

Section 5 contains the important case of expansive homeomorphisms in compact manifolds, in particular surfaces, and in the plane. We show that the existence of a Lyapunov function is a necessary and sufficient condition for a homeomorphism on a compact manifold to be expansive. Lyapunov functions are then used to derive important topological properties of expansive maps. For the case when the manifold is a compact surface, a detailed use of the topology of the plane allows for a complete classification of expansive homeomorphisms. Next, expansive homeomorphisms in the plane are considered, which either have a single equilibrium, or no equilibria. Many of the results of homeomorphisms on compact surfaces are generalized, and examples are provided, which show the possible structure of these homeomorphisms.

Section 6 contains other persistence results. In particular, we show that pseudo-Anosov maps are persistent, and indicate how one can obtain from here smooth, and even analytic, models of pseudo-Anosov maps. The section ends with reference to results on semi-persistence.

Most proofs in this chapter are either omitted or only the main ideas are indicated, with references to proofs in all cases. A little more detail has been given to linear hyperbolic systems in $\mathbb{R}^n$ and its perturbations, in Section 2.

This paper is close to our survey [36]. On the one hand, we have reduced, or omitted, several proofs; on the other hand, we have updated many results, especially regarding expansive homeomorphisms. Finally, the topics have been reorganized, in order to start with the presentation of several key ideas in a setting that requires a minimal background.

The remainder of this Introduction is devoted to definitions of several terms that will be used throughout the paper. In order to minimize clutter, many definitions have been deferred to the sections where they are needed.

Let $f\colon M \to M$ be a homeomorphism of a compact metric space M, and let d be the distance function.

Definition 1.1. *We say that f is persistent at $x \in M$ ([19], [22]) if for every $\varepsilon > 0$ there is an $r > 0$ such that if $g\colon M \to M$ is another homeomorphism of M satisfying $d(f(z), g(z)) < r$ for all $z \in M$, then there is a $y \in M$ such that $d(f^n(x), g^n(y)) < \varepsilon$ for all integers n.*

In other words, f is persistent at x if the f-trajectory through x has a close counterpart for all homeomorphisms g that are C^0-close to f.

Definition 1.2. *We say that f is persistent on a set $K \subset M$ if it is uniformly persistent at all $z \in K$, that is, the $r > 0$ depends on K and ε, but not on the particular x. In this situation we also say that the set K is persistent (with respect to f). It is not hard to show that if a set is persistent so is its closure, hence it suffices to consider closed persistent sets. If the set K is the entire space M, we will say that f is (plain) persistent.*

If in the definition of persistence we restrict ourselves to non-negative integers (non-positive integers), we get the notion of *persistent in the future* (*persistent in the past*). Notice, by the way, that f might be both persistent in the future and in the past at a point x without being persistent at x [22].

Definition 1.3. *For a fixed $x \in M$ and a given $\varepsilon > 0$ we define the (local) ε-stable set $S_\varepsilon(x)$ as*

$$S_\varepsilon(x) = \{y \in M \mid d(f^n(x), f^n(y)) \leq \varepsilon \quad \textit{for every } n \geq 0\},$$

and the ε-unstable set $U_\varepsilon(x)$ as

$$U_\varepsilon(x) = \{y \in M \mid d(f^n(x), f^n(y)) \leq \varepsilon \quad \textit{for every } n \leq 0\}.$$

Lyapunov functions of two variables will be defined slightly differently in different contexts. To convey the idea, here we introduce two of the definitions used, one for the case of a flow, namely, the suspension flow, and the other for the case of a homeomorphism.

First of all, let us recall the definition of the suspension manifold (see, for example, [15]).

Definition 1.4. *Given a homeomorphism $f\colon M \to M$ of a compact differentiable manifold M, the suspension manifold $\hat{M} = \hat{M}_f$ is obtained from $M \times \mathbb{R}$ by identifying $(x, t+1) \sim (f(x), t)$ for every $x \in M$ and every $t \in \mathbb{R}$. The suspension flow $\phi(x,t)$ on $\hat{M}$ is then defined as translation along the t-axis in $M \times \mathbb{R}$, under this identification.*

If f is a diffeomorphism, then $\hat{M}$ is also a differentiable manifold. Moreover, there is a natural projection $\pi\colon M \times \mathbb{R} \to \hat{M}$; we will denote $M_t = \pi(M \times \{t\})$. Often M itself is identified with $M_0 \subset \hat{M}$.

Definition 1.5. *Given a differentiable manifold M and a diffeomorphism $f\colon M \to M$, call $\hat{M}$ the suspension flow. Assume a continuous flow $\psi_t(x) = \psi(x,t)$ is defined on $\hat{M}$, such that $\psi_s M_t = M_{t+s}$. Then a Lyapunov function for ψ is a continuous real-valued function V defined on a neighborhood of the diagonal of $\bigcup_{t\in\mathbb{R}} M_t \times M_t$, such that $V(x,x) = 0$ and the derivative $\dot{V}$ of V along the flow,*

$$\dot{V}(x,y) = \lim_{t\to 0} \frac{1}{t} \left(V(\psi(x,t), \psi(y,t) - V(x,y)\right), \tag{1}$$

exists and is positive for $x \neq y$.

For the case of a homeomorphism $f\colon M \to M$, the derivative is replaced by the difference $\Delta V = f^\sharp V - V$, where $f^\sharp V(x,y) = V(f(x), f(y))$. Namely, a real-valued function V, defined on a neighborhood of the diagonal of $M \times M$, is a Lyapunov function for f, if $V(x,x) = 0$ and

$$\Delta V(x,y) = (f^\sharp V - V)(x,y) = V(f(x), f(y)) - V(x,y) > 0 \qquad \textit{when } x \neq y. \tag{2}$$

In Section 5 and in Section 4.1, Lyapunov functions will be defined somewhat differently, requiring additional properties.

Anosov diffeomorphisms on compact manifolds will be defined in Section 3.1. The following is a generalization of this idea, for two-dimensional manifolds.

Definition 1.6. *A homeomorphism $f\colon M \to M$ of a compact oriented surface M is pseudo-Anosov, if it preserves a transverse pair of foliations with measure, with finitely many singularitites, and there is a $\lambda > 1$ such that one measure is expanded by λ and the other is contracted by λ^{-1}.*

Pseudo-Anosov maps were introduced by Thurston [35] in his work of classification of surface homeomorphisms; see also the two volumes [10]. Typically they are smooth, except at the finitely many singularities, at which the foliations may have three or more "prongs;" that is why they are sometimes called "pseudo-Anosov diffeomorphisms."

Definition 1.7. *A diffeomorphism $f\colon M \to M$ of a compact manifold M is structurally stable, if there is a C^1-neighborhood of f such that for every g in that neighborhood there is a homeomorphism $h\colon M \to M$ such that $h \circ g = f \circ h$.*

Definition 1.8. *A homeomorphism $f\colon M \to M$ of a compact metric space M is topologically stable [38] if there is a $\delta > 0$ such that if $d(f(z), g(z)) < \delta$ for all $x \in M$ then there is a continuous onto map $\phi\colon M \to M$, close to the identity map, such that $\phi \circ g = f \circ \phi$.*

Thus, if f is topologically stable then it locally (in the C^0-sense) has the "simplest" dynamics. An equivalent definition of topological stability is given in [18], p. 192.

Notice that topological stability involves C^0-neighborhoods of f, whereas structural stability deals with C^1-neighborhoods.

2. Linear Hyperbolic Systems in $\mathbb{R}^n$ and Their Perturbations

As was mentioned in the Introduction, the method of Lyapunov functions of two variables was originally introduced in [18], for the case of Anosov diffeomorphisms on compact manifolds, and their generalizations, almost hyperbolic diffeomorphisms; this will be described in Sections 3 and 4. However, some key ideas can perhaps be better understood in the case of linear hyperbolic systems in Euclidean space and their quasi-hyperbolic generalizations, for which technicalities are reduced to a minimum. Hence, we will start with that case, and in subsequent sections we will show how the method can be successfully applied to more general situations. Most of the material in this section is due to direct ideas discussed by Lewowicz.

2.1. Linear Hyperbolic Systems and Quadratic Lyapunov Functions

We will consider a linear hyperbolic system in $\mathbb{R}^n$, for which a quadratic Lyapunov function can be adapted. Detailed proofs of all results in this Section are in [36], Section 4.

Let us consider the linear system

$$\dot{x} = Ax, \tag{3}$$

where $x \in \mathbb{R}^n$, and A is a *hyperbolic* $n \times n$ real-valued matrix, that is, all its eigenvalues have nonzero real parts. The matrix A defines a direct sum splitting

$$\mathbb{R}^n = \mathcal{S} \oplus \mathcal{U}, \tag{4}$$

where both $\mathcal{S}$ and $\mathcal{U}$ are A-invariant, the restriction $A|_{\mathcal{S}}$ is stable (all the eigenvalues have negative real parts), and $A|_{\mathcal{U}}$ is unstable (all the eigenvalues have positive real parts). Let us denote by $P_{\mathcal{S}}$ and $P_{\mathcal{U}}$ are the projections onto $\mathcal{S}$ and $\mathcal{U}$ respectively, associated with this splitting.

Let us call $L^t x$ the trajectory of (3) passing through x when $t = 0$.

Given a symmetric $n \times n$ matrix V, we construct the quadratic form

$$\mathbf{V}(x) = (Vx, x), \tag{5}$$

and denote $\dot{\mathbf{V}}$ the derivative (1) of $\mathbf{V}$ along the linear flow $L^t x$:

$$\dot{\mathbf{V}}(x) = (V\dot{x}, x) + (Vx, \dot{x}) = (VAx, x) + (Vx, Ax) = (Wx, x),$$

where

$$W = VA + A^T V.$$

This defines a map

$$\phi_A \colon V \mapsto W$$

from the space of all symmetric $n \times n$ matrices onto itself. It is shown that if A is stable, or unstable, then ϕ is an isomorphism. We now carry out the following construction. First, we pick (arbitrary) positive definite symmetric matrices $W_{\mathcal{S}}$, acting on $\mathcal{S}$, and $W_{\mathcal{U}}$, acting on $\mathcal{U}$, and define W acting on $\mathbb{R}^n$ as

$$W(x) = W_{\mathcal{S}}(P_{\mathcal{S}}x) + W_{\mathcal{U}}(P_{\mathcal{U}}x).$$

Using the fact that ϕ is an isomorphism, plus the splitting (4), we construct symmetric matrices U and V such that $\dot{\mathbf{U}} = \mathbf{V}$, $\dot{\mathbf{V}} = \mathbf{W}$, where $\mathbf{U}(x) = (Ux, x)$, $\mathbf{V}(x) = (Vx, x)$ and, moreover,

$$U_{\mathcal{S}} > 0, \qquad V_{\mathcal{S}} < 0, \qquad W_{\mathcal{S}} > 0;$$

$$U_{\mathcal{U}} > 0, \qquad V_{\mathcal{U}} > 0, \qquad W_{\mathcal{U}} > 0.$$

We have thus constructed three quadratic forms $\mathbf{U}$, $\mathbf{V}$, and $\mathbf{W}$ such that both $\mathbf{U}$ and $\mathbf{W}$ are positive definite, while $\mathbf{V}$ is positive definite on $\mathcal{U}$ and negative definite on $\mathcal{S}$. We will call $\mathbf{U}(x)$, with these properties, a (quadratic) *Lyapunov function* for (3).

We will now consider a non-linear system

$$\dot{y} = f(y), \tag{6}$$

where $y \in \mathbb{R}^n$, and $f \colon \mathbb{R}^n \to \mathbb{R}^n$ is a C^1-function. Let us denote by $N^t y$ the solution of (6) satisfying $N^0 y = y$.

We will say that (6) is a δ-*perturbation* of the linear system (3), if the following conditions hold for every $x \in \mathbb{R}^n$:

$$\|f(x) - Ax\| < \delta, \tag{$H1$}$$

$$\|f'(x)f(x) - Af(x)\| < \delta, \tag{$H2$}$$

$$\|f'(x) - A\| < \delta. \tag{$H3$}$$

In (H1) and (H2), $\|\cdot\|$ denotes the usual norm in $\mathbb{R}^n$, whereas in (H3) it denotes any norm in the space of $n \times n$-matrices that satisfies $\|Mx\| \leq \|M\| \cdot \|x\|$; $f'(x)$ is the Jacobian matrix of $f(x)$. If δ is small, these conditions are quite strong for large $\|x\|$, but near zero they are satisfied by the usual assumptions in local stability theorems, such as Hartman's classical result [13] (see Theorem 2.3 below).

In order to estimate how close to each other are the trajectories of (3) and (6), one can estimate the $\mathbb{R}^n$-norm of the difference $\|N^t y - L^t x\|$. Equivalently, one may use $\mathbf{U}(N^t y - L^t x)$, where $\mathbf{U}$ is the positive definite Lyapunov function constructed earlier. This corresponds to replacing the usual norm in $\mathbb{R}^n$ by the equivalent norm

$$|x| = \sqrt{(\mathbf{U}x, x)}\,. \tag{7}$$

As in [18], to track the variation of this function, it therefore makes sense to replace the derivative (1) along the linear flow $L^t x$ by the *mixed derivative* $\mathbf{U}^*$, defined as

$$\mathbf{U}^*(y, x) = \frac{d}{dt}\,\mathbf{U}(N^t y - L^t x)\big|_{t=0}\,. \tag{8}$$

Using this mixed derivative, plus assumptions (H1) and (H2), one can prove *persistence* of linear hyperbolic systems, as stated in the following main result of this section.

Theorem 2.1. *If A is hyperbolic then for every $\varepsilon > 0$ there exists a $\delta > 0$ such that if systems (3) and (6) satisfy (H1) and (H2), then for every $x \in \mathbb{R}^n$ there is a $y \in \mathbb{R}^n$ such that*

$$|N^t y - L^t x| \leq \varepsilon$$

for all $t \in \mathbb{R}$.

In other words, the linear system (3) is persistent, if δ-closeness is gauged by conditions (H1) and (H2).

Proof. We will sketch the proof, since it is the heart of the argument in many applications of the method of Lyapunov functions of two variables.

Arguing by contradiction, assume there is a positive ε such that for all positive δ there is an $x \in \mathbb{R}^n$ such that no trajectory $N^t y$ stays ε-close to $L^t x$ for all t. This means that for every $y \in \mathbb{R}^n$ there is a t for which $|N^t y - L^t x| > \varepsilon$. Given these ε and x, let fix a small $\delta > 0$, and define

$$K_t = \{y \in \mathbb{R}^n : \ |y - L^t x| \leq \varepsilon\}.$$

We can think of the union of all $K_t \times \{t\}$ as a "tube" with axis $L^t x$. Call D the disk

$$D = \{u \in \mathbb{R}^n : \ |u| \leq \varepsilon\}$$

with boundary ∂D, and write $y = x + u$, so that $y \in K_0$ will be equivalent to $u \in D$.

By our contradiction assumption, for every $u \in D$ the trajectory $N^t(x + u)$ will leave the tube for some t. This allows us to define a non-negative real-valued function $t(u)$ on D, where $t(u)$ is characterized by the following two properties (see Figure 1):

1. $N^t(x + u) \in K_t$ for every $|t| \leq t(u)$;

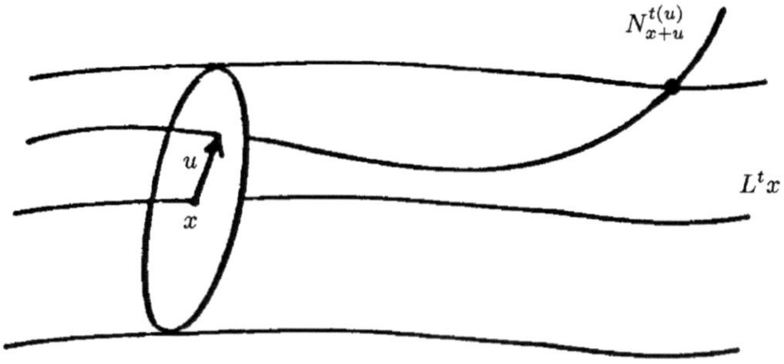

Figure 1. Trajectory leaving tube.

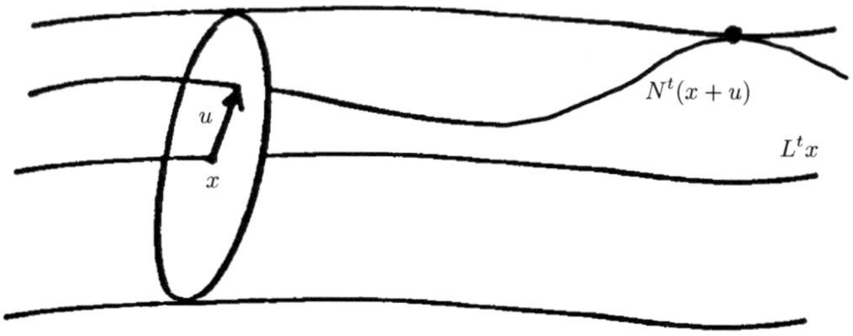

Figure 2. Inner tangency.

2. for every positive σ we have $N^t(x+u) \notin K_t$ either for some t in $[t(u), t(u)+\sigma)$ or for some t in $(-t(u)-\sigma, -t(u)]$.

In other words, $t(u)$ is the (absolute value of) the first escape time, either in the future or in the past.

One now proves that if δ is small enough then $t(u)$ satisfies the following two important properties.

- $t(u) = 0$ for every u in the boundary ∂D of the disk D.
- $t(u)$ is a continuous function for $u \in D$.

Both properties basically follow from the fact that, once a trajectory hits the boundary ∂D, it must leave the tube, that is, there are no "inner tangencies" (Figure 2). This, in turn, follows from the properties of the constructed Lyapunov functions $\mathbf{U}$, $\mathbf{V}$, $\mathbf{W}$—especially the fact that $\mathbf{W} = \ddot{\mathbf{U}}$ is positive definite—and the closeness conditions (H1) and (H2).

The next step is to define a map $\rho\colon D \to \mathbb{R}^n$ as

$$\rho(u) = P_{\mathcal{U}}(N^{t(u)}(x+u) - L^{t(u)}x) + P_{\mathcal{S}}(N^{-t(u)}(x+u) - L^{-t(u)}x). \tag{9}$$

The continuity of $t(u)$ implies that $\rho(u)$ is a continuous function. Moreover, using again the properties of the Lyapunov functions $\mathbf{U}$ and $\mathbf{V}$, one shows that, for small δ, if the trajectory $N^t(x+u)$ leaves the tube in the future, then the first term of (9) is nonzero, whereas if the trajectory leaves in the past, then the second term is nonzero. Thus, $\rho(u)$ is never zero on D. Finally, if u lies in the boundary ∂D, then $t(u) = 0$, which implies that $\rho(u) = u$. But then the function

$$u \mapsto \varepsilon \frac{\rho(u)}{|\rho(u)|}$$

is a retraction of D onto ∂D, which is impossible. □

Theorem 2.1 proves persistence of system (3); in other words, it shows that every trajectory $L^t x$ of (3) has a counterpart trajectory $N^t y$ of (6) that stays ε-close for all t, if δ is small enough. Now we want to show that this shadowing trajectory is unique. This follows from the fact that if (H3) holds for small δ, then system (6) is expansive, with any expansivity constant. Namely, if for some arbitrary $\varepsilon > 0$ we have $\|N^t y_1 - N^t y_2\| \le \varepsilon$ for every t, then necessarily $y_1 = y_2$.

Next one shows that the map $h\colon x \mapsto y$, where y is the unique element such that $|L^t x - N^t y| \le \varepsilon$, is in fact a homeomorphism of $\mathbb{R}^n$ onto itself, which is close to the identity map. This leads to the following result.

Theorem 2.2. *Let A be hyperbolic, and assume that $f\colon \mathbb{R}^n \to \mathbb{R}^n$ is C^1. For every $\varepsilon > 0$ there exists a $\delta > 0$ such that if conditions (H1), (H2), and (H3) hold, then there is a homeomorphism $h\colon \mathbb{R}^n \to \mathbb{R}^n$ satisfying*

$$h(L^t x) = N^t h(x) \qquad \text{for every } x \in \mathbb{R}^n \text{ and every } t; \tag{10}$$

moreover,

$$\|h(x) - x\| \le \varepsilon \qquad \text{for all } x \in \mathbb{R}^n. \tag{11}$$

As a consequence of this global conjugacy result, we get a new proof of Hartman's local conjugacy theorem [13].

Theorem 2.3 (Hartman). *Let $f\colon \mathbb{R}^n \to \mathbb{R}^n$ be C^1 with $f(0) = 0$, and assume that $f'(0) = A$ is hyperbolic. Then there are neighborhoods $\mathcal{N}_1$ and $\mathcal{N}_2$ of 0 in $\mathbb{R}^n$ and a homeomorphism h from $\mathcal{N}_1$ onto $\mathcal{N}_2$ such that $h(L^t x) = N^t h(x)$ for all t and all x as long as $L^t x \in \mathcal{N}_1$. Here $N^t x$ is the solution of $x' = f(x)$ satisfying $N^0 x = x$, while $L^t x$ is the solution of the linearized system $x' = Ax$ with $L^0 x = x$.*

The result is obtained from Theorem 2.2, if one replaces $f(x)$ by a C^1-function equal to $f(x)$ inside a small neighborhood of 0, and to Ax outside a slightly larger neighborhood.

2.2. Perturbations of Non-hyperbolic Systems

In [24] the ideas in 2.1 are applied to quasi-hyperbolic systems in $\mathbb{R}^n$. A C^1-vector field $X\colon \mathbb{R}^n \to \mathbb{R}^n$, $X(0) = 0$, is *quasi-hyperbolic*, if it admits a smooth Lyapunov function $U\colon \mathbb{R}^n \times \mathbb{R}^n \to \mathbb{R}$, similar to the function $\mathbf{U}(x, y)$ constructed in Section 2.1, but not necessarily quadratic. Namely, we will say that $U\colon \mathbb{R}^n \times \mathbb{R}^n \to \mathbb{R}$ is a *Lyapunov function* for the vector field X, if U and its second derivative $\ddot{U}$ along the flow of X are positive definite (that is, they are non-negative and are zero only on the diagonal), and there is a splitting $\mathbb{R}^n = \mathcal{S} \oplus \mathcal{U}$ with the property that $\dot{U}(x, x+y) < 0$ for any nonzero $y \in \mathcal{S}$ and $\dot{U}(x, x+y) > 0$ for $y \in \mathcal{U}$, $y \neq 0$ (plus a few technical assumptions). We show that a quasi-hyperbolic system is persistent, in the sense that if Y is a C^1-field close to X, where closeness is measured by conditions similar to (H1) and (H2) in Section 2, then every trajectory of X has a counterpart in Y, as in Theorem 2.1.

In particular, the planar system

$$\begin{cases} \dot{x}_1 = x_1^3, \\ \dot{x}_2 = -x_2^3, \end{cases} \tag{12}$$

is quasi-hyperbolic, as the Lyapunov function $U(x, y) = \|x - y\|^2$ shows. In this case persistence is all one can hope for, since for $\varepsilon > 0$ small, the system

$$\begin{cases} \dot{y}_1 = y_1^3 - \varepsilon^2 y_1, \\ \dot{y}_2 = -y_2^3 \end{cases}$$

is a small perturbation of (12) containing a whole segment $[-\varepsilon, \varepsilon]$ on the y_1-axis consisting of trajectories close to the equilibrium of (12). However, we show that if the perturbation preserves the 3-jet at the origin, then the two systems are conjugate.

This method can be generalized to higher-dimensional systems [24].

3. Quadratic Forms and Hyperbolic Behavior

As we have shown in Section 2, hyperbolic behavior in $\mathbb{R}^n$ is connected with the existence of a quadratic Lyapunov function. In this section we will discuss other important examples of hyperbolic behavior, starting with Anosov diffeomorphisms on compact manifolds. It is in the context of Anosov diffeomorphisms and their generalizations, discussed in Section 4, that the method of Lyapunov functions of two variables was first introduced [18]. It is interesting to observe that, as Theorem 3.2 below shows, the existence of a quadratic Lyapunov form is in fact *equivalent* to a diffeomorphism being Anosov.

3.1. Anosov Diffeomorphisms

We will assume that $f \colon M \to M$ is a diffeomorphism of a compact Riemannian manifold M. Let us recall the definition of Anosov diffeomorphism ([1] and [29]).

Definition 3.1. *We say that f is Anosov, if the tangent bundle TM can be split as $TM = S \oplus U$, where S and U are continuous and f'-invariant sub-bundles such that there exist positive constants K and λ, $0 < \lambda < 1$, with the property that for all $n > 0$, $\|(f^n)'(s)\| \leq K\lambda^n \|s\|$ for every $s \in S$ and $\|(f^{-n})'(u)\| \leq K\lambda^n \|u\|$ for all $u \in U$.*

An *Anosov flow* $\phi \colon \mathbb{R} \times M \to M$ is defined much as an Anosov diffeomorphism, except that the tangent bundle is split into three continuous invariant sub-bundles rather than two; apart from the previous two (one contracting and one expanding), there is an additional one-dimensional bundle, tangent to the flow, which is neither expanding nor contracting; see [29] and [1].

If the flow is generated by a continuous nonvanishing vector field X on M, we will say that X is an *Anosov vector field* if the corresponding flow is Anosov.

We define a *quadratic form* on TM as a real-valued function B such that $B_x = B|_{T_xM}$ is a quadratic form on T_xM for all x. In a similar way we define non-degenerate and positive quadratic forms B. Finally, we define $f^{\#}(B)$ as $f^{\#}(B)_x(v) = B_{f(x)}(f'(x)(v))$ for $v \in T_xM$. The following result in [18] completely characterizes Anosov diffeomorphisms in terms of quadratic forms.

Theorem 3.2. *Let f be a C^r diffeomorphism, $r \geq 1$. Then f is Anosov if, and only if, there exists a continuous non-degenerate quadratic form B on TM such that $f^{\#}(B) - B > 0$.*

In the proof of the existence of such B for an Anosov diffeomorphism, an important ingredient is the following result, which has independent interest [18].

Proposition 3.3. *If f is an Anosov diffeomorphism, there exists a positive integer m such that for every nonzero $v \in TM$ we have $\|(f^n)'(v)\| > 2\|v\|$ either for $n = m$ or for $n = -m$.*

Assume f is Anosov, and let B be as in Theorem 3.2. Since the set of diffeomorphisms $g\colon M \to M$ satisfying $g^{\#}(B) - B > 0$ is open in the C^1 topology, this result provides a new proof that the set of Anosov diffeomorphisms has the same property. In fact, we get the stronger result that there is a C^1-neighborhood $\mathcal{N}$ of f such that any finite composition of elements of $\mathcal{N}$ is also Anosov.

Anosov diffeomorphisms are also topologically stable. This was proved by Walters [38]; we obtain a new proof of this fact, as a consequence of Theorem 4.1 concerning almost-hyperbolic diffeomorphisms (see Section 4.1).

3.2. Quasi-Anosov Diffeomorphisms

In [28] Mañé defines a *Quasi- Anosov diffeomorphism* $f\colon M \to M$ of a smooth manifold M as a map such that the set $\{\|(f')^n(v)\| \mid n \in \mathbb{Z}\}$ is unbounded for every nonzero vector $v \in TM$. This amounts to stating that the tangent map $f'\colon TM \to TM$ is expansive (see Section 5). Therefore, using the results in [18] one can prove a result similar to Theorem 3.2, except that the corresponding quadratic form need no longer be non-degenerate. In fact, M. Cerminara [6] established the following result.

Theorem 3.4. *Let $f\colon M \to M$ be a diffeomorphism of a compact Riemannian manifold. Then f is quasi-Anosov if, and only if, there exists a continuous quadratic form B on TM such that $f^{\#}(B) - B > 0$.*

3.3. Geodesic Flows

Geodesic flows on manifolds with negative curvature are Anosov and hence structurally stable. This was proved in Anosov's celebrated paper [1]; they are also topologically stable [14].

The application of the method of quadratic forms to geodesic flows is discussed in [20]. The first result shown in [20] is a version of Theorem 3.2 for flows.

Theorem 3.5. *Let M be a compact smooth Riemannian manifold with a given C^1-vector field X, and assume that there is a continuous vector sub-bundle $W \subset T(M)$ transversal to X and invariant under the flow generated by X. Then X is an Anosov vector field if, and only if, there exists a non-degenerate continuous quadratic form $B\colon W \to \mathbb{R}$ such that $\mathcal{L}_X(B)$ is positive definite. Here $\mathcal{L}_X$ stands for the Lie derivative.*

The previous result can be applied to the geodesic flow ϕ on a two-dimensional oriented compact connected Riemannian manifold M with non-positive curvature. Let us denote by $S(M)$ the bundle of M consisting of unit tangent vectors. The *geodesic flow* ϕ is defined on $S(M)$ as follows (see, for example, [16] and [15]). If $(x, u) \in S(M)$ and if γ is the unique geodesic with $\gamma(0) = x$ and $\dot\gamma(0) = u$, we set $\phi((x, u), t) = (\gamma(t), \dot\gamma(t))$. We will now define a suitable quadratic form using differential forms. There is a unique connection form θ on $S(M)$ (the *Riemannian connection* form [16]) such that

$$d\theta = -(K \circ \pi)\, \omega \wedge \omega^{\perp},$$

$$d\omega = \theta \wedge \omega^{\perp},$$

$$d\omega^{\perp} = -\theta \wedge \omega.$$

Here ω is defined as $\omega_{(x,u)}(\nu) = (\pi'\nu, u)$, where $\pi\colon S(M) \to M$ is the natural projection and $\nu \in T(S(M))$; $\omega^{\perp}$ is given by ${\omega^{\perp}}_{(x,u)}(\nu) = (\pi'\nu, iu)$; iu is the unit perpendicular vector to u such that the orientation of $\{u, iu\}$ agrees with that of M; K is the curvature function of M. Finally, $(\cdot,\cdot)$ is the dot product given by the Riemannian metric.

Let us define the non-degenerate quadratic form

$$B(\nu) = \theta(\nu) \cdot \omega^{\perp}(\nu), \qquad \text{for } \nu \in T(S(M)).$$

If we call X the vector field generated by the geodesic flow ϕ, and $\mathcal{L}_X$ the corresponding Lie derivative, it is shown [20] that the bundle

$$W = \{\nu \in T(S(M)) \mid \omega(\nu) = 0\}$$

is invariant under ϕ and transversal to X and, moreover,

$$\mathcal{L}_X(B) = -(K \circ \pi)(\omega^{\perp})^2 + \theta^2.$$

Hence, if $K(x) < 0$ for all $x \in M$ this form is positive definite on W. Therefore, an application of Theorem 3.5 yields another proof of Anosov's classical result.

Corollary 3.6. *If the curvature K of M is negative everywhere, then the geodesic flow is Anosov.*

To obtain a more general result, let us fix a $t_0 > 0$ and consider the following integral version of the previous quadratic form:

$$\hat{B} = \int_0^{t_0} {\phi_s}^{\sharp}(B)\, ds, \quad \text{where} \quad (\phi^{\sharp}(B))_{\xi}(\nu) = B_{\phi(\xi,s)}(\phi_s'(\xi)\nu),$$

for $\nu \in T_\xi S(M)$. Then

$$\mathcal{L}_X(\hat{B}) = \int_0^{t_0} \phi_s^\sharp(\mathcal{L}_X(B))\, ds.$$

This form is still positive definite on W in the more general case $K \leq 0$, provided that for each $\xi \in S(M)$ there exists a $t \in [0, t_0]$ for which $K \circ \pi(\phi(\xi, t)) < 0$. It turns out that $\hat{B}$ is non-degenerate [20]. Hence, again by Theorem 3.5, we have the following result.

Theorem 3.7. *If M has non-positive curvature, then the geodesic flow ϕ is Anosov, unless $K^{-1}(\{0\})$ contains a whole geodesic.*

On the other hand, if $K^{-1}(\{0\})$ contains a whole geodesic, the flow is *never* Anosov [20].

3.4. Non-uniformly Hyperbolic Maps

The results in 3.1 can be extended to non-uniformly hyperbolic maps, which need not even be defined on the whole manifold (smooth maps with singularities). Namely, assume M is a smooth compact Riemannian manifold of dimension $d \geq 2$, with a probability measure ν absolutely continuous with respect to the Riemannian volume, and let N be a subset of M with zero ν-measure. Let us call $H = M \setminus N$ and let $f\colon H \to H$ be a C^r ν-invariant diffeomorphism. Let us denote $\Sigma(f)$ the Pesin region, that is, the set of points of H for which the Lyapunov exponents of f exist and are nonzero. If $\nu(\Sigma(f)) = 1$, we say that f is *non-uniformly hyperbolic.*

One can show that the existence of a bounded non-degenerate quadratic form B on H such that $f^\sharp B - B$ is positive implies that f is non-uniformly hyperbolic; see [23], p. 390. It turns out that this condition is also necessary for non-uniform hyperbolicity, as proved by Markarian in [31].

In order to provide a precise statement, for a quadratic form B on TM let us denote

$$S_x = \{x \in T_xM \mid B((f^n)'u) < 0 \text{ for all } n \geq 0\};$$

$$U_x = \{x \in T_xM \mid B((f^n)'u) > 0 \text{ for all } n \leq 0\}.$$

Then the following results hold ([23], [31]).

Theorem 3.8. *Let f be a C^1 diffeomorphism. Assume there exists a quadratic form B on TM with the following properties.*

1. *For every $x \in H$ we have $T_xM = L_x \oplus K_x$ such that if $u \neq 0$ then $B(u) < 0$ when $u \in L_y$ and $B(u) > 0$ for $u \in K_y$;*
2. $0 < \dim L_x < d$ *and* $\dim L_{f^n(x)} = \dim L_x$ *for every $x \in H$ and every integer n;*
3. *B_x is non-degenerate in H and depends measurably on x;*
4. *$\Delta B_x = (f^\sharp B - B)_x$ is positive for all $x \in H$.*

Then $\nu(\Sigma(f)) = 1$ and $S_x = E_x^s$, $U_x = E_x^u$ for every $x \in \Sigma(f)$, where E_x^s (resp. E_x^u) is the sum of all eigenspaces corresponding to negative (resp. positive) Lyapunov exponents.

Note. Conditions (1) and (2) hold if either $d = 2$ or if B is continuous, non-degenerate, and M is connected (see [30]).

The idea to prove the main part of Theorem 3.8 is as follows ([23], [31]). One picks a nonzero u in U_x, and shows that $B((f^n)'u)$ grows exponentially for x in a subset of H of full measure. Then an application of Birkhoff's ergodic theorem yields a positive estimate from below for $\liminf \frac{1}{n} \log \|(f^n)'u\|$; this allows one to conclude that the upper Lyapunov exponent $\lambda^+(x) = \min\{\lambda_i(x) \mid \lambda_i > 0\}$ is positive on a set of full measure. A similar argument is used for $u \in S_x$.

Next, if f is at least C^2, then the converse result is also true.

Theorem 3.9. *Let f be C^2. If $\nu(\Sigma(f)) = 1$, then there exists a quadratic form B on TM satisfying conditions (1)–(4) of Theorem 3.8.*

Application to Billiards

In a sense, billiards can be regarded as a generalization of geodesic flows. (For more on billiards, see [7], [17] part V, [34], and [8].) Let us consider the simplest case of a *plane billiard.* This is the free motion of a point mass inside a bounded, connected domain $Q \subset \mathbb{R}^2$ with elastic reflections at the boundary. We assume the boundary ∂Q consists of finitely many smooth curves ∂Q_i with bounded curvature K. The *regular components* $\partial\tilde{Q}_i$ are the ∂Q_i with the "corners" removed.

We parametrize the boundary using arc length s on each component and in such a way that if $q(s)$ is the position vector, then on regular components the unit inward normal at q is $n = iq'(s)$, the rotation of the tangent vector $q'(s)$ by a positive angle $\pi/2$. The regular components may have positive curvature (*focusing components*), negative curvature (*dispersing components*) or zero curvature (*neutral components*).

To define the domain of the billiard map, we first consider the set

$$M = \{(q, v) \mid q \in \partial\tilde{Q}_i,\ v \in \mathbb{R}^2,\ \|v\| = 1,\ v \cdot n(q) > 0\}$$

and for $x = (q, v) \in M$ we define $T(x) = (\bar{q}, \bar{v})$ as the point obtained moving forward from q in the direction v until the next intersection $\bar{q}$ with ∂Q, and $\bar{v} = v - 2[n(\bar{q}) \cdot v]\, n(\bar{q})$. (The "$\cdot$" means dot product.) Finally, we set $H = M \setminus N$, where N is the set at which T^k is either not defined or not continuous for some $k \in \mathbb{N}$ (for example, when the trajectory hits a corner, or is tangent to a component $\partial\tilde{Q}_i$). Then $T\colon H \to H$ is a smooth map with singularities. If we call θ the angle from n to v and parametrize $x = (q, v)$ using s and θ, then T preserves the measure $d\nu = \cos\theta\, ds\, d\theta$. An element in the tangent space T_xM is obtained by considering a smooth curve (a *wave front*) $(q(s), v(s))$ in M with $(q(0), v(0)) = (q, v) = x$; then $(q', v') \overset{\text{def}}{=} (q'(0), v'(0)) \in T_xM$.

The following 1-forms are defined in [31]:

$$\alpha = \alpha(q') = q' \cdot iv \qquad \text{and} \qquad V = V(v') = v' \cdot iv.$$

Then the non-degenerate quadratic form $B_x(q', v') = \alpha(q')V(v')$ is the analogue of the form considered in [20] for the geodesic flow (see Section 3.3). And, indeed, the form

$\Delta B = T^\sharp B - B$ turns out to be equal to

$$(\Delta B)_x(q',v') = \sigma'^2 \left[-2K_2 \cos\theta_2 + t_1(K_2 - \theta_2')^2\right],$$

which is positive if the regular components of the boundary are dispersing. Here $x = (q_1, v_1)$ with parameters s_1, θ_1, $T(x) = (q_2, v_2)$ with parameters s_2, θ_2; t_1 is the time (distance) from x to $T(x)$, K_2 is the curvature at $T(x)$, $\theta_2' = d\theta_2/ds_2$, and $\sigma' = ds_2/ds_1$. Finally, as in Section 3.1, $T^\sharp B$ is defined as $T^\sharp B_x(q',v') = B_{T(x)}(T_x'(q',v'))$.

Applying now Theorem 3.8 we conclude that the billiard map T is non-uniformly hyperbolic if all the regular components are dispersing (Sinai's result [33]).

To address the case of billiards with focusing components (positive curvature), instead of αV one considers a more general quadratic form of the type

$$B_x(q',v') = a\,\alpha^2 + 2b\,\alpha V + c\,V^2,$$

where a, b, and c are measurable functions of x with $ac - b^2 \neq 0$. This guarantees that condition (3) of Theorem 3.8 holds. One can then choose a, b, and c so that the other conditions be also satisfied. By considering suitable particular cases, Markarian ([30], [31]) shows non-uniform hyperbolicity of wide classes of billiards, which generalize those in [33] and [39]. In particular, $\nu(\Sigma(T)) = 1$ for billiards with finitely many dispersing and neutral components, plus smooth focusing components satisfying certain restrictions. It is also shown in [31] that Bunimovich's billiards ([4], [5]) admit C^4 perturbations that preserve non-uniform hyperbolicity; see also [9].

Other Applications

Recently, Araujo and Salgado [2] applied Lewowicz's idea of using quadratic forms (or *infinitesimal Lyapunov functions*) to the study of non-uniformly hyperbolic models. They studied a singular flow X_t on a manifold M (possibly with boundary), and obtained a useful characterization of sectional hyperbolic sets, expressed in terms of non-degenerate quadratic forms. If there are no singularities, their results allow one to characterize uniform hyperbolicity. In particular, when X_t is generated by an incompressible vector field X and the quadratic forms have codimension two, they show that X_t is an Anosov flow.

4. Almost Hyperbolic Systems

When a system admits a "good" Lyapunov function which, however, is not necessarily quadratic, in general it is no longer structurally stable, but still shares nice stability and persistence properties, which are present in the hyperbolic class. This is the class of *almost hyperbolic systems*.

4.1. Almost-Hyperbolic Diffeomorphisms

Let $f\colon M \to M$ be a diffeomorphism of a compact connected Riemannian manifold M. In Section 3.1 we saw that f is Anosov (i.e., hyperbolic) if, and only if, it admits a non-degenerate quadratic form B such that ΔB is positive definite. Generalizing this idea, we will say that f is *almost hyperbolic* [18] if it admits a smooth function $V\colon M \times M \to \mathbb{R}$ (a *Lyapunov function*) with the following properties.

1. There is a positive $\rho > 0$ and a continuous splitting $T_xM = S_x \oplus U_x$ for every $x \in M$ such that if $v \in T_xM$, $0 < \|v\| \le \rho$, then $V(x, \exp_x v) < 0$ when $v \in S_x$ and $V(x, \exp_x v) > 0$ for $v \in U_x$.

2. There exists an $\alpha > 0$ such that $\Delta V(x,y) = V(f(x), f(y)) - V(x,y) > 0$ for every $x, y \in M$, $0 < d(x,y) \le \alpha$.

If the splitting is not f'-invariant, then an additional technical condition is imposed on V [18].

Unlike Anosov (i.e., hyperbolic) diffeomorphisms, almost hyperbolic diffeomorphisms in general are not structurally stable. However, as in the hyperbolic case, they are topologically stable.

Theorem 4.1 ([18]). *If f is almost hyperbolic, then it is topologically stable.*

Some central ideas in the proof of this important result have been outlined in Section 2 in simplified form, for the case of a linear hyperbolic system in $\mathbb{R}^n$.

Proof. Here is an outline of the proof. One starts by considering the suspension manifold $\hat{M} = \hat{M}_f$ and the corresponding suspension flow $\phi(x,t)$ of the diffeomorphism f in M (Definition 1.4). As before, let us call $M_t = \pi(M \times \{t\})$, where $\pi\colon M \times \mathbb{R} \to \hat{M}$ is the natural projection.

Next, one constructs smooth functions U, V, W on a neighborhood of the diagonal of $M_t \times M_t$, with U and V positive for $x \ne y$ and with V satisfying condition (1) above, for the natural extension to T_xM_t of the splitting on T_xM assumed in (1); the functions are such that $\dot{U} = V$ and $\dot{V} = W$, where the dot means differentiation along the suspension flow ϕ.

The next step is to show persistence of the flow ϕ among all continuous flows ψ on $\hat{M}$ such that $\psi_t M_s = M_{s+t}$. Here one argues by contradiction and applies the ideas contained in the proof of Theorem 2.1 of Section 2 to the tube $\bigcup_{t\in\mathbb{R}} K_t$, with sections

$$K_t(x) = \{y \in M_t \mid U(\phi(x,t), y) \le k\},$$

where $k > 0$ is small.

The argument is now more involved, since the splitting is not necessarily invariant, and vectors need to be moved from one tangent space to another. The function $\rho(v)$, analogous to (9), which was used in the proof of Theorem 2.1, is now defined on

$$B_x = \{v \in T_xM \mid V(x, \exp_x v) \le k\}$$

as $\rho(v) = \alpha(v) + \beta(v)$, where

$$\alpha(v) = P_{t(v)}\phi'_{-t(v)} \exp^{-1}_{\phi(x,t(v))} \psi(\exp_x v, t(v))$$

and

$$\beta(v) = Q_{t(v)}\phi'_{t(v)} \exp^{-1}_{\phi(x,-t(v))} \psi(\exp_x v, -t(v)).$$

Here P_t is the projection of TM_t onto $\phi'_t U_{\phi(x,-t)}$ along $\phi'_{-t} S_{\phi(x,t)}$, and Q_t is the complementary projection. As in Section 2, $t(v)$ is the “escape time”. This function is proved to be continuous, which is done by showing that there are no "inner tangencies", using the positivity of $\ddot{U} = W$ as in the proof of Theorem 2.1. In turn, this implies that both $\alpha(v)$ and $\beta(v)$ are continuous

As in the proof of Theorem 2.1, using the assumptions on $\dot{U} = V$ one shows that $\alpha(u)$ and $\beta(u)$ cannot vanish simultaneously. This yields a nonzero vector field $\gamma(v) = \alpha(v) + \beta(v)$ on B_x, which coincides with the identity on ∂B_x. From here then one constructs a continuous retraction of a larger ball B, $B_x \subset B \subset T_xM$, onto its boundary, which is impossible.

Once persistence of ϕ is established, to show topological stability one needs a symmetric result, namely, that if ψ is a continuous flow close to ϕ that preserves the foliation M_t, then for every $y \in M_s$ there is an $x \in M_s$ such that $U(\phi(x,t), \psi(y,t)) \leq k$ for all t. To do this, one uses a similar argument to the previous one, but using now a tube with sections

$$L_t(y) = \{x \in M_t \mid U(x, \psi(y,t)) \leq k\},$$

with "axis" on $\psi(y,t))$ instead of $\phi(x,t)$. The continuous onto map, figuring in the Definition 1.8 of topological stability, is now established by sending every $y \in M_s$ to the $x \in M_s$ that satisfies $U(\phi(x,t), \psi(y,t)) \leq k$ for all t. □

4.1.1. Examples of Almost Hyperbolic Diffeomorphisms

Consider the family of diffeomorphisms $F\colon T^2 \to T^2$ defined by

$$F(x,y) = (2x + y - g(x), x + y - g(x)),$$

where $g\colon \mathbb{R} \to \mathbb{R}$ is a C^1-function of period 1 such that $g'(1) < 1$, except perhaps at finitely many points on each period interval. The Lyapunov function

$$V((x_1,y_1),(x_2,y_2)) = -(y_2 - y_1)(y_2 - x_2 - (y_1 - x_1))$$

is non-degenerate and such that ΔV is positive definite, proving that F is almost hyperbolic; it is also measure-preserving. If $g(0) = 0$ and $g'(0) = 1$, then $(0,0)$ is a non-hyperbolic fixed point, so F is not Anosov. Moreover, there is no F'-invariant splitting of the tangent space at this point.

A particular case is the *Lewowicz map*

$$f(x,y) = \left(2x + y - \frac{\sin(2\pi x)}{2\pi}, x + y - \frac{\sin(2\pi x)}{2\pi}\right),$$

which is the object of active research (see, for example, [40] and [26]). When embedded in the one-parameter family

$$f_c(x,y) = \left(2x + y - c\frac{\sin(2\pi x)}{2\pi}, x + y - c\frac{\sin(2\pi x)}{2\pi}\right),$$

one can show that for $0 \leq c < 1$ the map f_c is Anosov; this implies [18] that $f = f_1$ is conjugate to the linear Anosov automorphism f_0. For $c > 1$ small, $(0,0)$ becomes an elliptic point. The coexistence of elliptic and hyperbolic behavior for the Lewowicz map is an open problem; in particular, it would be interesting to obtain estimates for the Lyapunov exponents.

In [18] there are also several other interesting examples of almost hyperbolic diffeomorphisms.

4.2. Perturbations of Hyperbolic Geodesic Flows

Using quadratic forms, in section 3.3 it was shown that if a compact oriented surface M has non-negative curvature, then the geodesic flow ϕ is Anosov, unless $K^{-1}(\{0\})$ contains a whole geodesic [20].

To address the case when $(K \circ \pi)^{-1}(\{0\})$ contains a non-empty invariant set of ϕ, one considers Lyapunov functions that are not necessarily quadratic forms. To create such a function, one starts by defining $p_\xi : \mathcal{U} \to S(M)$, where $\mathcal{U}$ is a suitable neighborhood of 0 in $\mathbb{R}^3$, as follows. For every $(m, n, \alpha) \in \mathcal{U}$ we set

$$p_\xi(m, n, \alpha) = e^{-\left(\frac{\pi}{2}-\alpha\right)i}\phi(i\phi(\xi, m), n).$$

We have $p_\xi(0, 0, 0) = \xi$, and also

$$p_\xi(m, n, \alpha) = p^0_{\phi(\xi,m)}(n, \alpha), \text{ where } p^0_\xi(n, \alpha) = p_\xi(0, n, \alpha).$$

One finds

$$p'_\xi(0)\frac{\partial}{\partial m} = X_\xi, \quad p'_\xi(0)\frac{\partial}{\partial n} = iX_\xi, \quad p'_\xi(0)\frac{\partial}{\partial \alpha} = V_\xi.$$

This implies that p_ξ is a diffeomorphism from $\mathcal{U}$ onto a neighborhood of ξ in $S(M)$.

Next, let $N = \{n, \alpha\}$ be an open ball centered at $0 \in \mathbb{R}^2$ so that $p^0_\xi|_N$ is an embedding and $p^0_\xi(N)$ is transversal to X for every $\xi \in S(M)$. Then the sets $p^0_\xi(N)$ for $\xi \in S(M)$ will constitute a family of local sections of the geodesic flow ϕ. However, in general this family is not invariant under ϕ.

The domain of the Lyapunov function will be the set

$$\{(\xi, \eta) \in S(M) \times S(M) \,|\, p^0_\xi(n, \alpha) = \eta,\ (n, \alpha) \in \frac{1}{2}N\},$$

and the function $\mathcal{V}$ itself will be defined as

$$\mathcal{V}(\xi, \eta) = \sin^2 \alpha + n^2.$$

A little care has to be taken as to how to define the derivative $\dot{\mathcal{V}}$ of $\mathcal{V}$ along the flow ϕ since, as was observed, the family of local sections $p^0_\xi(N)$ is not invariant under ϕ. We define

$$\dot{\mathcal{V}}(\xi, \eta) = \lim_{t \to 0} \frac{1}{t}\left[\mathcal{V}(\phi_\tau(\xi), \phi_t(\eta)) - \mathcal{V}(\xi, \eta)\right],$$

where $\tau = \tau_\eta(t)$ is the unique real number close to 0 such that $\phi_t(\eta) \in p^0_{\phi(\xi,\tau)}(N)$. The second derivative $\ddot{\mathcal{V}} = \left(\dot{\mathcal{V}}\right)^{\cdot}$ is defined using the same procedure.

Finally, we pick a small $T > 0$ and define

$$\mathcal{U}(\xi, \eta) = \int_0^T \mathcal{V}(\phi(\xi, \tau_\eta(s)), \phi(\eta, s))\, ds.$$

Let us assume that M has non-positive curvature and $K^{-1}(\{0\})$ has empty interior. Then one can prove that there is a $\rho > 0$ such that $\mathcal{U}$ is defined and smooth on the set

$$\{(\xi, \eta) \in S(M) \times S(M) \,|\, p^0_\xi(n, \alpha) = \eta, \|n\, iX_\xi + \alpha V_\xi\| < \rho\}.$$

Moreover, $\mathcal{U} > 0$ and $\ddot{\mathcal{U}} > 0$ if $\xi \neq \eta$, and there exists a continuous splitting $W_\xi = S_\xi \oplus U_\xi$ such that $\dot{\mathcal{U}}(\xi, \eta) < 0$ (> 0) for nonzero $n\, iX_\xi + \alpha V_\xi \in S_\xi$ (resp. $n\, iX_\xi + \alpha V_\xi \in U_\xi$).

Thus, $\mathcal{U}$ is a Lyapunov function for the geodesic flow, with properties similar to those of the **U** defined in Section 2 or the U constructed in Section 4.1. Therefore, we can apply the techniques used in the proof of Theorem 2.1, with suitable modifications, to the tube with sections

$$H_t(\xi) = \{\eta \in p^0_{\phi(\xi,t)}(N) \mid \mathcal{U}(\phi(\xi,t), \eta) \leq h\}$$

and prove the following persistence result.

Theorem 4.2 ([20]). *Let M be a compact oriented surface, and let X be the vector field generated by the geodesic flow ϕ on $S(M)$. If Y is a vector field on $S(M)$ that is C^0-close to X then for each $\xi \in S(M)$ there exists $\eta \in S(M)$ and a smooth increasing onto function $\tau_\eta \colon \mathbb{R} \to \mathbb{R}$, $\tau_\eta(0) = 0$, such that $\psi(\eta, t) \in H_{\tau_\eta(t)}(\xi)$ for every $t \in \mathbb{R}$, where ψ is the flow generated by Y.*

Next, as in [18], one proves a symmetric result, switching places η and ξ. Putting together both results, the following is proved.

Theorem 4.3 ([20]). *If M has non-positive curvature K, and $K^{-1}(\{0\})$ has empty interior, then the geodesic flow is topologically stable.*

Corollary 4.4 ([20]). *With the same assumptions on K, the geodesic flow is C^1-structurally stable in the class of vector fields Y on $S(M)$ producing geodesic flows on M with similar properties on K.*

5. Expansive Homeomorphisms

Lyapunov functions provide a very convenient characterization of expansive homeomorphisms. Moreover, for the important case of expansive homeomorphisms on compact surfaces, a combination of this method with in-depth arguments involving the topology of surfaces, allows for their complete characterization [20]. The results have recently been extended to expansive homeomorphisms on the plane ([11], [12]).

5.1. The Compact Case

Let us recall the definition of an expansive homeomorphism in a compact metric space. Assume $f \colon M \to M$ is a homeomorphism of a compact metric space M, with distance function d.

Definition 5.1. *We say that f is expansive if there is a positive number α such that $d(f^n(x), f^n(y)) < \alpha$ for every integer n implies that $x = y$. Such α is called an expansivity constant for f.*

Here is the definition of a Lyapunov function, which will be used in this setting.

Definition 5.2. *Let $f \colon M \to M$ be a homeomorphism of a compact manifold M. We will say that $U \colon M \times M \to \mathbb{R}$ is a Lyapunov function for f if it is continuous, vanishes* *on the diagonal, and is such that both $U(x,y)$ and $\Delta\Delta U(x,y)$ are positive for (x,y) on a neighborhood of the diagonal, whenever $x \neq y$. (Here $\Delta U(x,y) = U(f(x), f(y)) - U(x,y)$.)*

Let $\hat{M} = \hat{M}_f$ be the suspension manifold, with suspension flow ϕ (Definition 1.4); as before, call $M_t = \pi(M \times \{t\})$. Let U be a differentiable, real valued function defined on a neighborhood of the diagonal of $\bigcup_{t\in\mathbb{R}} M_t \times M_t$. We will say that U is a Lyapunov function for ϕ, if $U(x,x) = 0$, and both $U(x,y)$ and $\ddot{U}(x,y)$ are positive for $x \neq y$. Here $\dot{U}$ is the derivative of U along the flow ϕ, as in (1), and $\ddot{U} = (\dot{U})^{\cdot}$.

The following result characterizes expansive homeomorphisms in terms of Lyapunov functions.

Theorem 5.3. *Let f be a homeomorphism of a compact manifold M. The following conditions are equivalent.*

1. *f is expansive;*
2. *there exists a Lyapunov function for f;*
3. *there exists a Lyapunov function for the suspension flow of f.*

The proof of this result for diffeomorphisms f is in [18]; see Section 4 and Lemma 3.3 of that paper. Additional arguments required for the case of a homeomorphism are discussed in [21], Section 1.

Note. If we compare Definition 5.2 with the weaker Definition 1.5 of a Lyapunov function, given in the Introduction, we can see that $V = \Delta U$ for the discrete case, or $V = \dot{U}$ for the continuous case, are Lyapunov functions in the sense of Definition 1.5. The existence of such a V suffices to prove expansivity, as is easily seen (cf. Lemma 3.3 in [18]). Thus, the existence of a "strong" Lyapunov function U is equivalent to the existence of a "weak" Lyapunov function V. Notice, however, that given a "weak" V, generally speaking the corresponding "strong" U will be such that ΔU, or $\dot{U}$, does not necessarily coincide with the original V.

Topological Consequences of Expansivity

Let us start by mentioning the interesting fact [19] that expansive homeomorphisms with a fixed expansivity constant $\alpha > 0$ form an open set in the C^0-topology, if we identify "indistinguishable points."

Proposition 5.4. *Let $f\colon M \to M$ be an expansive homeomorphism of a compact metric space M, with expansivity constant α. For every $\delta > 0$ there exists a C^0-neighborhood $\mathcal{N}$ of f such that if $x, y \in M$, $g \in \mathcal{N}$ and $d(g^n(x), g^n(y)) \leq \alpha$ for all $n \in \mathbb{Z}$ then also $d(g^n(x), d(g^n(y)) < \delta$ for all $n \in \mathbb{Z}$.*

Assume now that $f\colon M \to M$ is a C^1-expansive homeomorphism of a compact connected Riemannian manifold M. Using the existence of Lyapunov functions established in Theorem 5.3, one proves the following important result. Before we state it, recall that a point $x \in M$ is called *stable* (or *Lyapunov stable*) if for every $\varepsilon > 0$ there is a $\delta > 0$ such that if $d(x, y) < \delta$ then $d(f^n(x), f^n(y)) \leq \varepsilon$ for every $n \geq 0$. An *unstable* point $x \in M$ is defined analogously, replacing "$n \geq 0$" by "$n \leq 0$."

Theorem 5.5 ([19]). *If f is expansive, there are no stable points or unstable points.*

The assumption that M is here a compact connected manifold is crucial. If the space is not locally connected, the result may fail to hold. For example, on subshifts of finite type there are expansive homeomorphisms with stable points.

A second important result is that both the local stable and the local unstable sets (Definition 1.3) of expansive homeomorphisms contain "reasonably large" connected sets. (cf. [21] and [25]; see also a proof in [36], Theorem 3.5). Namely, let U be a (strong) Lyapunov function for f, defined on a neighborhood of the diagonal of $M \times M$. Let $\varepsilon > 0$ be so small that $U(x, y)$ is positive whenever $d(x, y) < \varepsilon$ and $x \neq y$; pick $\sigma > 0$ such that $U(x, y) < \sigma$ implies $d(x, y) < \varepsilon$, and finally let $\delta > 0$ be such that $d(x, y) < \delta$ implies $U(x, y) < \sigma$.

Theorem 5.6. *Both the stable set $S_\varepsilon(x)$ and the unstable set $U_\varepsilon(x)$ contain nontrivial connected sets joining x to a point at distance δ from x.*

Expansive homeomorphisms on surfaces. In the case when M is two-dimensional (that is, a compact surface), applying the previous techniques, combined with an elaborate use of topological properties of the plane, one can obtain deeper topological properties of expansive homeomorphisms on M.

Let f be an expansive homeomorphism of a compact surface M. We say that a point $x \in M$ has *local product structure* if there is a homeomorphism of $\mathbb{R}^2$ onto an open neighborhood of x that maps horizontal (vertical) lines onto open subsets of local stable (respectively unstable) sets.

Combining Theorem 5.6 with the topology of the plane, the following result is obtained [21], which reveals the fine structure of expansive homeomorphisms on surfaces.

Theorem 5.7. *Except for a finite number of points, which we shall call singular, every $x \in M$ has a local product structure. Any singular point $y \in M$ is periodic, and its local stable (unstable) set consists of the union of r arcs, $r \geq 3$, that meet only at y. Moreover, the stable (unstable) arcs separate unstable (respectively stable) sectors.*

In turn, this result leads to the following classification theorem for expansive homeomorphisms on surfaces [21].

Theorem 5.8. *Let $f\colon M \to M$ be a homeomorphism of a compact oriented surface M.*

1. *If $M = S^2$, then f is not expansive, i. e. there are no expansive homeomorphisms on the sphere.*

2. *If f is expansive and M is the torus T^2, then f is conjugate to an Anosov diffeomorphism.*

3. *If f is expansive and M has genus ≥ 2, then f is conjugate to a pseudo-Anosov map.*

It then follows that expansive homeomorphisms are topologically stable on T^2 and persistent on surfaces of higher genus.

The three-dimensional case. For an expansive homeomorphism on a surface M, as shown in Theorem 5.7, both the local stable and the local unstable sets (locally) separate M. For a three-dimensional compact oriented manifold M this need no longer be the case, which significantly complicates the study and classification of expansive homeomorphisms. On the other hand, the three dimensional case is special. As Vieitez showed [37], for an expansive diffeomorphism f on M satisfying $\Omega(f) = M$, then one of these sets locally separates M, while the other does not. Using Lyapunov functions, plus this asymmetry, he was able to show the existence of a local product structure in M. This allowed him to prove that M must in fact be the three-torus, and f is conjugate to a linear Anosov diffeomorphism.

For expansive homeomorphisms in compact manifolds M of dimension higher than three, in general one can no longer guarantee the existence of a local product structure. However, Potrie, Artigue and Brum [3] proved that if one assumes that the set of topologically hyperbolic points of f is dense, then, in any dimension, one can show the existence of a local product structure in an open and dense subset of M (in fact, in a neighborhood of every periodic point). This allowed them to prove that if, in addition, some hyperbolic point has codimension one, then f is conjugate to a linear automorphism on a torus.

5.2. Non-compact Case: Expansive Homeomorphisms of the Plane

The methods and results in Section 5.1 have recently been generalized to address the case of expansive homeomorphisms of the plane with one fixed point, or with no fixed points ([11], [12]). Here, the notion of *U-expansivity* is used, for a suitable Lyapunov function U.

Namely, let $f\colon \mathbb{R}^2 \to \mathbb{R}^2$ be a homeomorphism of the plane, and let $U\colon \mathbb{R}^2 \times \mathbb{R}^2 \to \mathbb{R}$ be a continuous real-valued function. We define the difference ΔU as $\Delta U(x, y) = U(f(x), f(y)) - U(x, y)$.

Definition 5.9. *We say that U is a metric Lyapunov function for f if U is a metric in $\mathbb{R}^2$ that induces the usual topology in the plane, and the second difference $W = \Delta\Delta U(x, y)$ is positive for all $x \neq y$.*

In particular, both $U(x, y)$ and $W(x, y)$ are positive definite, that is, they are positive whenever $x \neq y$, and are zero on the diagonal $x = y$.

Next, we define the concept of (infinite) U-expansivity.

Definition 5.10. *Let U be a metric Lyapunov function for the homeomorphism f. We say that f is U-expansive if for any two points x, y in the plane and any $k > 0$ there exists an integer $n \in \mathbb{Z}$ such that $U(f^n(x), f^n(y)) > k$.*

Thus, f is assumed to be expansive with respect to the metric U, for any expansivity constant k.

Given a metric Lyapunov function U for f, and given $x \in \mathbb{R}^2$, we shall call

$$B_k(x) = \text{connected component of the set } \{y \in \mathbb{R}^2 \mid U(x, y) \leq k\} \text{ that contains } x. \quad (13)$$

The following important assumption refers to the behavior of U at infinity.

Definition 5.11. *Let f be a homeomorphism of the plane, and let U be a metric Lyapunov function for f. Call $V = \Delta U$ and $W = \Delta V = \Delta\Delta U$. We say that U satisfies Condition* **HP**, *if for any compact set $C \subset \mathbb{R}^2$ the following properties hold:*

- *there exists a $k > 0$ such that $|V(x,y) - V(x,z)| \leq k$ for all $y, z \in C$ and all $x \in \mathbb{R}^2$;*
- *$W(x,y)$ tends to infinity as $\|x\| \to \infty$, uniformly with respect to $y \in C$.*

Finally, we will need the following concept, for a metric Lyapunov function U for f.

Condition 5.12. Existence of both signs for the first difference of U. *For every $x \in \mathbb{R}^2$ and every $k > 0$, there exist points y and z on the boundary of $B_k(x)$ such that $V(x,y) > 0$ and $V(x,z) < 0$, respectively; here $V = \Delta U$.*

It is shown [12] that a homeomorphism of the plane that admits a Lyapunov metric function is U-expansive.

Moreover, the previous conditions on U guarantee the same detailed local behavior of U-expansive homeomorphisms in the plane as for the case of expansive homeomorphisms on compact surfaces. In fact, in [11] the following analogue of Theorem 5.7 is proved.

Theorem 5.13. *Let f be a homeomorphism of $\mathbb{R}^2$ that has a metric Lyapunov function U, satisfying Condition 5.12. Then, except for a finite number of points, which we shall call singular, every $x \in \mathbb{R}^2$ has a local product structure. Any singular point $y \in \mathbb{R}^2$ is periodic, and its local stable (unstable) set consists of the union of r arcs, $r \geq 3$, that meet only at y. Moreover, the stable (unstable) arcs separate unstable (respectively stable) sectors.*

In principle, the neighborhoods in which a local product structure exists, may become arbitrarily small. However, in [11] it is shown that, for every $x \in \mathbb{R}^2$, these stable and unstable arcs may be extended to *stable and unstable curves*, denoted by $W^s(x)$ and $W^u(x)$ respectively; moreover, each stable curve, as well as each unstable curve, separates the plane.

Using this result, plus a detailed analysis of the behavior of stable and unstable curves, the following characterization result is obtained in [11] and [12].

Theorem 5.14. *Let f be a homeomorphism of $\mathbb{R}^2$ with no singularities, which has a metric Lyapunov function U, satisfying Condition 5.12. Then*

- *if f has a single fixed point, then f is conjugate to a linear hyperbolic automorphism if, and only if, U satisfies Condition* **HP**;
- *if f has no fixed points, then f is topologically conjugate to a translation of the plane if, and only if, U satisfies Condition* **HP**.

In particular, from this result it follows that if f is a homeomorphism satisfying the assumptions of the theorem, then it is the time-one map of a flow in $\mathbb{R}^2$.

For the case of a homeomorphism f or $\mathbb{R}^2$ with a single fixed point, if Condition **HP** fails to hold, different behaviors may appear. In particular, one can no longer guarantee that every stable curve will intersect every unstable curve. In [11], necessary and sufficient conditions (weaker than Condition **HP**) are obtained for the restriction of f to each of the quadrants determined by the stable and unstable curves of the fixed point, to be either conjugate to a linear hyperbolic automorphism, or conjugate to the restriction of such automorphism to certain invariant domain. Since the behavior in each quadrant is independent of what happens in the other quadrants, various combinations can be obtained, for which in some quadrants every stable curve intersects every unstable curve, while in other quadrants this does not happen. Some possibilities are shown in Figure 3. Two detailed examples in [11] illustrate this kind of behavior.

For the case of a homeomorphism with no fixed points, the example of the translation T of the plane defined by $T(x, y) = (x + 1, y + 1)$ is considered. Using a cleverly defined Riemannian metric on the plane, in [12] a metric Lyapunov function U is constructed for f, which satisfies Condition **HP** and Condition 5.12. In particular, the topology generated by the metric U is the same as the usual topology in the plane, but the corresponding metric spaces are not equivalent. Indeed, T is U-expansive (since U satisfies Condition 5.12), but is not expansive in the usual metric of $\mathbb{R}^2$.

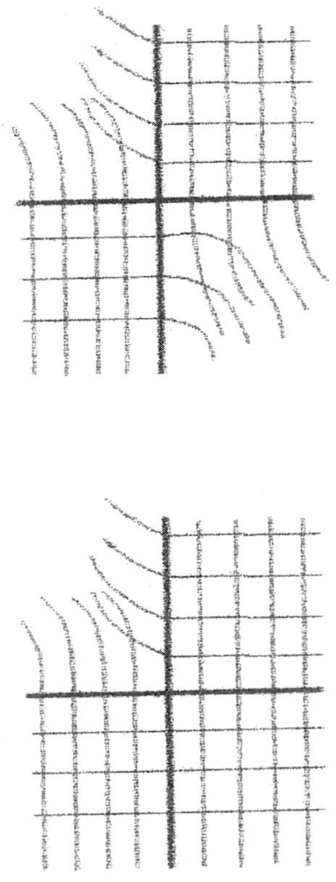

Figure 3. Two possible behaviors.

5.3. Open Questions

Let us mention several interesting open questions and work in process, regarding expansive homeomorphisms.

1. Homeomorphisms of the plane that admit a Lyapunov metric defined in $\mathbb{R}^2 \times \mathbb{R}^2$. The existence of such function implies (infinite) expansivity with respect to such metric. Let us assume that the homeomorphism has a fixed point, which is (Lyapunov) stable. We believe that then the homeomorphism is topologically conjugate to a similarity; in other words, the equilibrium is globally asymptotically stable. Jorge Groisman has completed a preprint on this topic.

2. It would be good to extend the characterization of an expansive homeomorphism f of the plane, carried out in [11] and [12], to the case when f admits singularities. The set of singularities is now discrete, but not necessarily finite. Can the use of metric Lyapunov functions allow one to recognize when f is conjugate to a lift of a pseudo-Anosov map? This is a sensible question, since, among other things, it is known that such liftings admit a Lyapunov metric satisfying Condition **HP**.

3. We are interested in exploring the connection between Condition **HP** and the completeness of the Lyapunov metric.

4. Even for the case of the plane, it would be nice to be able to describe the finitely expansive homeomorphisms. One possible way to do this would be to use Lyapunov functions that are defined only in a neighborhood of the diagonal of $\mathbb{R}^2 \times \mathbb{R}^2$.

5. Can any of the results about expansive homeomorphisms in the plane be generalized to higher dimensions?

6. Expansivity, Persistence and Pseudo-Anosov Maps

In [19] persistence is shown for expansive maps satisfying additional conditions.

Theorem 6.1. *Let $f\colon M \to M$ be an expansive diffeomorphism of a two-dimensional compact connected Riemannian manifold M satisfying the following additional assumptions.*

1. *There is an f-invariant set P such that every $x \in P$ belongs to both its α- and its ω-limit sets.*

2. *At each $x \in P$ there are transversal smooth local stable and unstable manifolds S_x, U_x, and a number $r_x > 0$, such that if $y = f^k(x)$ for some k and $d(x, y) < r_x$ then S_y and U_x are transversal and have nonempty intersection.*

3. *For every $x \in P$, either S_x or U_x is one-dimensional.*

Then f is persistent on the closure of P.

The proof uses results from [18] and the techniques shown in Sections 2 and 4.1, taking advantage of the fact that for dimension 1 the retraction arguments can be carried out using connectedness. Additional complications arise due to the fact that the existence of the local manifolds S_x and U_x is assumed only for $x \in P$ (and not in all of M), and they are not assumed to depend continuously on x.

The following propositions are consequences of Theorem 6.1.

Corollary 6.2. *If $f\colon M \to M$ is expansive and the omega-limit set Ω of f contains a dense set of hyperbolic points, then f is persistent on Ω.*

Corollary 6.3. *If f is expansive and has an invariant measure given by a volume form, then the closure of the set for which all Lyapunov exponents is nonzero is persistent under f.*

From Corollary 6.2 we get an important consequence for pseudo-Anosov maps.

Theorem 6.4 ([19]). *Pseudo-Anosov maps are persistent. Moreover, if f is pseudo-Anosov, there is a C^1-neighborhood $\mathcal{U}$ of f such that every homeomorphism $g \in \mathcal{U}$ that coincides with f at singular points is conjugate to f.*

This result shows that pseudo-Anosov maps display a kind of structural stability.

In order to derive this fact from the previous results, one needs to construct a non-degenerate Lyapunov function U such that not only $f^\sharp U - U$ is positive definite, but also $g^\sharp U - U$ is positive definite, where g is a nearby homeomorphism satisfying the above assumptions.[1] In brief, this is accomplished as follows. First, one considers an iterate of f such that all its singularities are fixed points. Next, using the fact that at singular points there are coordinate neighborhoods such that f coincides with the time 1 map of the flow defined by the differential equation

$$\dot{z} = \frac{\bar{z}^p}{z^{p-1}} \qquad p = \frac{n}{2}, n = 2, 3, \dots$$

one shows that $U = \mathrm{Re}[(z - w)(z^{2p-1} - w^{2p-1})]$ satisfies the requirements near singular points. Outside singular points for nearby x, y one defines $U(x, y) = \mu^2(x, y) - \hat{\mu}^2(x, y)$. Here we call z the intersection of the stable leaf through y with the unstable leaf through x and $\hat{z}$, the intersection of the unstable leaf through y with the stable leaf through x.

Finally, $\mu(x, y)$ is the transversal measure of the segment $[x, z]$, and $\hat{\mu}(x, y)$ is the transversal measure of $[x, \hat{z}]$.

Using similar techniques, one proves [19] that the homeomorphisms cited in [32] are persistent. These are constructed by lifting an Anosov diffeomorphism $h\colon T^2 \to T^2$ on the torus. Here the construction near singular points is as before, and outside a neighborhood of these one takes the lift of a non-degenerate quadratic Lyapunov function for h (see Section 3.1).

[1] As before, $f^\sharp U$ is defined as $(f^\sharp U)(x, y) = U(f(x), f(y))$.

Smooth and Analytic Models of Pseudo-Anosov Maps

The persistence of the pseudo-Anosov maps is used in [23] to construct smooth models, and even analytic models, of these maps. For a given pseudo-Anosov map f, one constructs a family of smooth diffeomorphisms g arbitrarily close to f such that they are all expanding with a fixed expansivity constant $\rho > 0$. This implies that for every $x \in M$, the $y \in M$ (guaranteed by persistence) such that the orbits of $f^n(x)$ and $g^n(y)$ remain close is in fact unique, and the map sending x to y provides a conjugacy between f and the smooth diffeomorphism g.

To get such a family, one uses a Lyapunov function that is a suitable modification of the function U described earlier. Moreover, since this function also guarantees that $h^\sharp(U) - U$ is positive definite for diffeomorphisms h that are C^1-close to g which agree with g at singular points together with a finite set of jets, this construction also leads to analytic models of pseudo-Anosov maps [23].

Finally, if we call B the quadratic part of U, one proves that $h^\sharp(B) - B$ is a positive definite quadratic form. Hence, using Theorem 3.8 in Section 3.4, we conclude that if h preserves a normalized measure μ with smooth positive density, then the Pesin region of h has full μ-measure.

As a consequence, one gets analytic homeomorphisms h that are conjugate to f and Bernoulli with respect to μ, where μ is any probability measure of M with positive analytic density.

Semi-persistence

In [22] Lyapunov functions are used to study one-sided persistence of a homeomorphism f at a given point x of a compact manifold M. We will address persistence in the future; analogous results can be obtained for persistence in the past.

Let $\hat{M} = \hat{M}_f$ be the suspension manifold, with suspension flow ϕ, and identify M with M_0 (recall Definition 1.4). We define persistence for ϕ similarly to what was done in Section 1 for homeomorphisms.

Definition 6.5. *We say that the flow ϕ is persistent (persistent in the future) at $x \in M$ if for every $\varepsilon > 0$ there is an $r > 0$ such that if ψ is a flow on $\hat{M}$ satisfying $\psi_s(M_t) = M_{s+t}$ and $d(\phi(x,t), \psi(x,t)) < r$, $0 \le t \le 1$, then there is a $y \in M_0$ such that $d(\phi(x,t), \psi(y,t)) < \varepsilon$ for all t (respectively for all $t \ge 0$).*

Let us assume that f is expansive. Then, by Theorem 5.3, there is a Lyapunov function U for the suspension flow, such that both U and $\ddot{U}$ are positive definite. To state the following result, let us call $\dot{U} = V$. (Compare with Theorem 4.1.)

Theorem 6.6. *Let $x \in M = M_0$. Assume there is a $\rho > 0$ such that*

1. *there exists a subspace $\mathcal{U} \subset T_x M_0$ such that $V(x, \exp_x v) > 0$ whenever $v \in \mathcal{U}$, $0 < \|v\| \le \rho$;*

2. *for $t \ge 0$ there exist subspaces $E_{\phi(x,t)} \subset T_{\phi(x,t)} M_t$, which depend continuously on t, such that $V(\phi(x,t), \exp_{\phi(x,t)} v) < 0$ for $v \in E_{\phi(x,t)}$, $0 < \|v\| \le \rho$;*

3. $\mathcal{U} \oplus \phi'_{-t} E_{\phi(x,t)} = T_x M_0$ *for all* $t \geq 0$.

Then the flow ϕ *is persistent in the future at* x.

If the point $x \in M$ in question lies on a minimal set, the assumptions on the Lyapunov function U can be localized (see [22]).

Applying Theorem 6.6, one obtains geometric conditions for persistence of f itself in the future at a given x [22].

Conclusion

In this chapter, we have introduced the concept of a *Lyapunov function of two variables*, which is a powerful method to analyze structural stability of homeomorphisms, diffeomorphisms, and flows, on compact manifolds and in $\mathbb{R}^n$. This concept generalizes the idea of a classical Lyapunov function, which has been used, also very successfully, to investigate the (usual, or *Lyapunov*) stability of equilibria as well as of single trajectories of a dynamical system.

We have shown that, similar to the classical case, the existence of a a quadratic Lyapunov function for a dynamical system characterizes hyperbolic behavior, which has strong stability properties of the system (such as structural stability), whereas the existence of a "nice" Lyapunov function, not necessarily quadratic, means the system—which we called *almost hyperbolic*—still has some nice stability properties (such as topological stability, or persistence).

We have applied the method to a wide range of systems, both hyperbolic and almost hyperbolic: systems in $\mathbb{R}^n$, diffeomorphisms on compact manifolds, geodesic flows and its perturbations, billiards, expansive homeomorphisms, and others. In all cases, important results were obtained concerning not only stability and related concepts of the dynamical systems, but also, in several cases, the deep local, and global, structure of these systems.

References

[1] D. V. Anosov, Geodesic flows on closed Riemannian manifolds with negative curvature, *Proc. Steklov Inst. Math.* 90, 674–685. *Amer. Math. Soc.*, Providence, RI, 1967.

[2] V. Araujo and L Salgado, *Infinitesimal Lyapunov functions for singular flows*, preprint (2012).

[3] A. Artigue, J. Brum, and R. Potrie, Local product structure for expansive homeomorphisms, *Topology and its Applications* 156 (2009),

[4] L. A. Bunimovich, On billiards close to dispersing, *Math. USSR Sb.* 23 no. 1 (1974), 45–67.

[5] L. A. Bunimovich, On ergodic properties of nowhere dispersing billiards, *Commun. Math. Phys.* 65 (1979), 295–312.

[6] M. Cerminara, *Quasi-Anosov diffeomorphisms and quadratic forms*, preprint.

[7] I. P. Cornfeld, S. V. Fomin, and Ya. G. Sinai, *Ergodic Theory*, Springer (1982), Chapter 6. (Transl. of Russian ed. 1980.)

[8] N. Chernov and R. Markarian, *Chaotic Billiards*, American Mathematical Society, 2006.

[9] V. Donnay, Using integrability to produce chaos: billiards with positive entropy, *Commun. Math. Phys.* 141 (1991), 225–257.

[10] A. Fathi, F. Laudenbach & V. Poénaru, Travaux de Thurston sur les surfaces, Seminaire Orsay, *Astérisque* 66–67 (1979).

[11] J. Groisman, Expansive homeomorphisms of the plane, *Discrete and Continuous Dynamical Systems* 29 (2011), 213–239.

[12] J. Groisman, Expansive and fixed point free homeomorphisms of the plane, *Discrete and Continuous Dynamical Systems* 32 (2012), 1709–1721.

[13] P. Hartman, *Ordinary differential equations*, Baltimore, 1973.

[14] K. Kato and A. Morimoto, Topological stability of Anosov flows and their centralizers, *Topology* 12 (1973), 255–273.

[15] A. Katok and B. Hasselblatt, *Introduction to the modern theory of dynamical systems*, Cambridge University Press, 1995.

[16] S. Kobayashi and K. Nomizu, *Foundations of Differential Geometry*, Vol. I, Interscience, New York, 1969.

[17] A. Katok and J.-M. Strelcyn, Invariant Manifolds, Entropy and Billiards; Smooth maps with singularities, *Lecture Notes in Math.* 1222, Springer-Verlag, 1986

[18] J. Lewowicz, Lyapunov functions and topological stability, *J. Differential Equations* 38 (1980), 192–209.

[19] J. Lewowicz, Persistence in expansive systems, *Ergodic Th. & Dynamical Syst.* 3 (1983), 567–578.

[20] J. Lewowicz, Lyapunov functions and stability of geodesic flows *Lecture Notes in Math.* 2007, Geometric Dynamics, 1983, 463–479.

[21] J. Lewowicz, Expansive homeomorphisms of surfaces, *Bol. Soc. Bras. Mat.* 20 (1989)113–133.

[22] J. Lewowicz, Persistence of semi-trajectories, to appear in *Journal of Dynamics and Differential Equations.*

[23] J. Lewowicz and E. Lima de Sa, Analytic models of pseudo-Anosov maps, *Ergod. Th. & Dynam. Sys.* 6 (1986), 385–392.

[24] J. Lewowicz and J. Tolosa, Local conjugacy for quasi-hyperbolic systems, in *Differential equations: qualitative theory*, Vol. II, Colloquia Mathematica Societatis János Bolyai, 47, North-Holland, Hungary, 1987, 687–704.

[25] J. Lewowicz and J. Tolosa, Genericity of homeomorphisms with connected stable and unstable sets, *Contemporary Mathematics* 182 (1993), 203–213.

[26] C. Liverani, Birth of an elliptic island in a chaotic sea, *Mathematical Physics Electronic Journal* 10 (2004), 1–13.

[27] A. M. Lyapunov, Problème général de la stabilité du mouvement, *Ann. Fac. Sci. Univ. Toulouse* 9 (1907), 203–475; reproduced in *Ann. Math. Study* 17, Princeton, 1947. [Original Russian paper published in 1892.]

[28] R. Mañé, Quasi-Anosov diffeomorphisms and hyperbolic manifolds, *Trans. Amer. Math. Soc.* 229 (1977), 351–370.

[29] R. Mañé, *Ergodic Theory and Differentiable Dynamics*, Springer-Verlag, New York, 1983.

[30] R. Markarian, Billiards with Pesin region of measure one, *Commun. Math. Phys.* 118 (1988), 87–97.

[31] R. Markarian, Non-uniformly hyperbolic billiards, *Annales de la faculté des sciences de Toulouse* 3 (1994), 223–257.

[32] T. O'Brien and W. Reddi, Each compact orientable surface of positive genus admits an expansive homeomorphism, *Pacific J. Math.* (3) 35 no. 1 (1970), 737–741.

[33] Ya. G. Sinai, Dynamical systems with elastic reflections, *Russian Math. Surveys* 25 (1970), 137–189.

[34] S. Tabachnikov, Billiards, *Société Math. de France*, 1995. (In English.)

[35] W. Thurston, On the geometry and dynamics of diffeomorphisms of surfaces, *Bull. Amer. Math. Soc.* 19 (1988), 417–431.

[36] J. Tolosa, The method of Lyapunov functions of two variables, *Contemporary Mathematics* 440 (2007), 243–271.

[37] J. L. Vieitez, Lyapunov functions and expansive diffeomorphisms on 3D-manifolds, *Ergod. Th. & Dynam. Sys.* 22 (2002), 601-632.

[38] P. Walters, Anosov diffeomorphisms are topologically stable, *Topology* 9 (1970), 71–78.

[39] M. Wojtkowski, Principles for the design of billiards with nonvanishing Lyapunov exponents, *Commun. Math. Phys.* 105 (1986), 391–414.

[40] G. M. Zaslavsky and M. Edelman, *Stickiness of trajectories in a perturbed Anosov system*, arXiv:nlin.CD/0511027 v1 14 Nov 2005.

In: Dynamical Systems
Editor: Jean Michel Tchuenche, pp. 51-81

ISBN: 978-1-62808-001-8

Chapter 3

KRYLOV AND SCHUR SUBSPACES IN CONSTRUCTION OF PREDICTION ERROR FILTER: COMPUTATIONAL COMPLEXITY AND PERFORMANCE COMPARISON

Hong Son Hoang and Rémy Baraille
SHOM/HOM, Toulouse, France

Abstract

The authors discuss the questions of complexity and performance of two subspace methods for filtering problem with very high dimensional system state, the Krylov subspace method and the dominant Schur subspace method. It will be shown that compared to the Schur subspace method, the Krylov subspace is computationally much less demanding and allows to construct the filter with nearly the same performance as that based on the Schur subspace. The advantages and disadvantages of these two subspaces in connection with the performance of the prediction error filter as well as its computational complexity will be studied in detail. These two subspace methods are also viewed from the point of view of the Ritz vector approach. As will be seen, for the very high dimensional systems, the Krylov subspace method will contribute to improvement of the filter performance by making possible utilization of the larger set of prediction error samples. Numerical results from the experiments with a simple dynamical model of moderate dimension as well as with data assimilation problems with very high dimensional Miami Isopycnic Coordinate Ocean Model (MICOM) will be also presented to show how work the algorithms in practice. Comparison of the performance of the prediction error filter (and its adaptive version) based on different sets of prediction error samples simulated by the two subspace algorithms are also exhibited.

1. Introduction

The goal of the reduced-order adaptive filter (ROAF), developed in [14], is to apply a reduction technique to seek an appropriate structure for the filter gain with pertinent parameters to be tuned in order to minimize the prediction error (PE) for the system output. It is found that the gain is composed of two operators, one is a projection from a reduced space onto a subspace of the full system state, another is the gain in the reduced space. The choice of these two operators is dictated by necessity to ensure the most important property of the designed filter, the filter stability. No real filter can sustain the constant growth of the estimation error that signifies existence of unstable eigenvalues of the filter fundamental matrix. Due to order reduction and random character of the estimated parameters, the following two important questions arise in what concerned with the stability of the filter: (i) to which extent the order reduction is allowable and how to construct a subspace of projection ? (ii) What parameters should be chosen as tuning parameters and how to constrain them to ensure the filter stability ?

In this paper we address the problem of construction of a filter using either Krylov Subspace (KrSS) or Schur Subspace (ScSS) to simulate the PE patterns which will serve as "data" to estimate the parameters of filter gain. The implementation and efficiency of this filter will be demonstrated from experiments on estimation and prediction of the state of (numerical) dynamical systems with moderate as well as with very high dimension. This concerns in particular a growing number of 3D hydrodynamics systems arising in operational oceanography to study the coastal management or climate change on marine ecosystems. The difficulties encountered here lay in the extreme large dimension of resulting system state (order of 10^{6-7}), not to say about the others like non-linearities and different types of uncertainty. The traditional optimal estimation methods, having to work with the matrices of 10^{12-14} elements, are simply inapplicable for solving such estimation problems. It is found that one way to ensure a filter stability is to construct the projection operator from dominant (unstable) eigenvectors or dominant Schur vectors (DScVs) of the system dynamics [15]. This task (finding DScVs) can be facilitated by applying the sampling procedure (SP) which generate, at low cost, the PE samples serving as "data" to estimate the filter gain [12]. For the question on a stable filter for strongly unstable spatially systems, see also [4], [10]. In [16] the question on a stability of the filter based on dominant singular vectors is studied. Generating the PE samples from a subspace of singular vectors is of less practical interest for very high dimensional systems since it requires a linear tangent model and its adjoint, the construction of which involves a very large amount of coding work. Moreover, the presence of discontinuities creates serious theoretical and numerical difficulties in adjoint coding.

In this context we would first mention an approach on Krylov subspace estimation [18] which proposes the algorithm for approximation to the estimation error variance using the conjugate search directions. Unlike the need to estimate the error covariance matrix (ECM) in the present paper, in [18] the ECM is supposed to be known a-priori and the task is to approximate the linear least-squares estimate by using a set of linearly independent functionals of the data that form basis for KrSSs. This method is efficient for the estimation problems with high dimensionality of the data vector and rapid decay of the eigenvalues of the ECM. The second class of filters, popular in the field of data assimilation in meteorology and oceanography, is the class of ensemble-based filters (EnBF) [6], [11] which is developed in the framework of the Kalman filter (KF). In this EnBF the prediction error ECM (denoted as ECM-PE) is estimated from an ensemble of model trajectories using the PE covariance represented by the ensemble spread [6], [11]. This ECM propagates in time on the basis of the Algebraic Riccati equation (ARE) which constitutes one of the key equations in the KF. In this regard, one of the main differences between the EnBF and the KF is seen in the way the ECM-PE is estimated : if in the KF the ECM-PE is computed from the priori known statistics of entering random processes, in the EnBF it is estimated from an ensemble of PE samples generated by model integrations from randomly perturbed system states. As a number of model integrations is constrained to be very limited in very high dimensional systems, the quality of the estimated ECM-PE is undoubtedly very poor, not to say on how one can ensure a stability of the EnBF. This class of EnBFs is a particular subclass of particle filters which is based on the Bayesian theorem for conditional density distribution and Monte-Carlo simulation [1]. There remains, however, a very long way to go and enormous efforts to do so that such particle filters become implementable for solving engineering filtering problems in very high dimensional systems since even for a moderate numerical model, a few hundred thousand simulations are usually required for well estimating conditional density distributions.

The paper is organized as follows. In section 2 the theory of a ROAF is outlined. The necessity to develop a ROAF arises from the need to overcome the difficulties in the design of a filter for state estimation problem in very high dimensional systems. As the reduction cannot be done without limitation, guaranteeing a stability of the filter becomes a most important task in the design of a ROAF. It is seen from there the role of KrSS and ScSS as projection subspace in ensuring the filter stability. Recall that the dominant invariant subspace (IvSS - the subspace of dominant eigenvectors) constitutes a basical projection subspace to ensure a stability of the filter. The reason we are interested in the KrSS (and ScSS) is that the computation of the IvSS is undesirable since the eigenvectors may be complex and their computation usually suffer from numerical instabilities. In Section 3 the question of a relationship between the ECM-PE and the projection operator in the filter gain is clarified. One sees that for high dimensional systems, a compromise between PE space and KrSS (or ScSS) is inevitable in the filter design: from the one hand, an optimal correction belongs to the space of the PE (with KrSS or ScSS as its subspaces) but, from the other hand, a stability of filter is guaranteed by the projection of the correction onto the IvSS. In this regard using the PE samples generated from KrSS and/or ScSS is a good compromise for the filter design. The Ritz eigenvectors and Riz eigenvalues which are the best approximations for the eigenvectors and eigenvalues from a some subspace are described in section 4. This general and nice approach is very useful for understanding the nature of the KrSS and ScSS. The SPs based on KrSS and ScSS are given here. Section 5 presents the numerical experiments on a simple dynamical model with moderate dimension. These experiments serve as good illustrations on how work these two SPs and what really results from application of these two algorithms. It is seen here that even for low dimensional stationary systems, with a quick convergence of the KVs, well chosen KVs as PE samples allow to design a high performance filter. The filtering problems for very high dimensional systems (oceanic numerical models) are presented in section 6. Conclusions are given in section 7.

2. Krylov Subspace and Schur Subspace

2.1. Motivation. Structure of Filter and Adaptation

To understand the significance of the KrSS and ScSS in solving the estimation problems for very high dimensional systems, let us look at the difficulties arising in the field of data assimilation in oceanography (and meteorology).

For simplicity, consider the standard linear filtering problem for a time-invariant system

$$x(k+1) = \Phi x(k) + w(k), k = 0, 1, 2, ... \tag{1}$$

$$z(k+1) = Hx(k+1) + \epsilon(k+1), k = 0, 1, 2, ... \tag{2}$$

here $x(k)$ is the n-dimensional system state at $k := t_k$ assimilation instant, Φ is the (nxn) fundamental matrix, $z(k)$ is the p-dimensional observation vector, H is the (pxn) observation matrix, w, ϵ are the model and observation noises. We assume $w(k)$, $\epsilon(k)$ are uncorrelated sequences of zero mean and time-invariant covariance Q and R respectively. For the class of filters of the structure

$$\hat{x}(k+1) = \hat{x}(k+1/k) + \delta\hat{x}(k+1), \delta\hat{x}(k+1) = K(k+1)\zeta(k+1), \hat{x}(k+1/k) = \Phi\hat{x}(k), \tag{3}$$

where $\zeta(k+1) = z(k+1) - H\hat{x}(k+1/k)$ is the innovation vector, $\hat{x}(k+1)$ is the filtered (or analysis) estimate, $\hat{x}(k+1/k)$ is the prediction for $x(k+1)$ at the assimilation instant $k+1$, the gain which will yield an unbiased minimum variance estimate for the system state is given by (see (2.4) in [12])

$$K(k+1) = E[e_p(k+1)\zeta^T(k+1)]E[\zeta(k+1)\zeta^T(k+1)]^{-1} \tag{4}$$

$E(.)$ is the mathematical expectation operator, $e_p(k+1) = \hat{x}(k+1/k) - x(k+1)$ is the PE for the system state $x(k+1)$, $\zeta^T(k+1)$ is the transpose of $\zeta(k+1)$.

For the system (1)(2) the gain $K(k+1)$ takes the form

$$K(k+1) = M(k+1)H^T[HM(k+1)H^T + R]^{-1} \tag{5}$$

The KF is given by (3)(5) with the ECM-PE $M(k)$ satisfying the ARE [17]. It is seen that for the state $x(k)$ of dimension of order $10^6 - 10^7$, it is impossible to apply the KF since in (5) $M(k)$ is composed from $10^{12} - 10^{14}$ equations.

A common approach to overcome such difficulties is to use an reduced-order filter (ROF). Most ROFs are designed on the basis of model order reduction (MOR) [3], i.e. obtaining first a model with smaller dimension while preserving certain properties of the original system. All the filter equations are next written out on the basis of a reduced order model (ROM). We want to stress that the reduction we will follow here (see for detail [12]) is different from the MOR approach: here the reduction is performed only on the subspace of the PE for the system state.

For the reviews on different approaches to approximating the gain in an KF and its modifications for nonlinear filtering problems in very high dimensional systems, see [8]. It is well known [17] that under the standard conditions, the KF is very reliable and efficient for state estimation in dynamical system but it is limited to a relatively restricted class of linear Gaussian problems. The major difficulties arising in the field of filtering of distributed dynamical systems, in particular in data assimilation in meteorology and oceanography, are related to non-linearity of the model equations which, together with a very high dimensionality of the system state and uncertainties of the model errors, make the problem of application of the KF to be extremely delicate [8].

For these reasons the traditional KF approach based on a MOR has been abandoned in [14] and a new Reduced-Order Adaptive Filter (ROAF) has been proposed. In this approach the filter gain structure has to be selected a-priori with a number of pertinent parameters θ to be adjusted to optimize the filter performance (i.e. to minimize the PE for the system output). This idea is expressed mathematically as follows: let a member set of filter gain

$$K = K(\theta), \theta \in \Theta \tag{6}$$

be given. Find an optimal gain $K^* = K(\theta^*)$ where θ^* is the solution of

$$J = J[\theta] = E[\Psi(\zeta(k))] \rightarrow min_\theta, \Psi(\zeta(k)) := ||\zeta(k)||^2 \tag{7}$$

Selecting improper structures for the filter gain may lead to instability of the filter and fast growth of estimation error : as the gain elements are functions of the random estimate $\theta(k)$, there is a high risk that the filter will become unstable if no special measures are imposed on the gain structure to ensure its stability during the optimization process. Mention that even the KF may suffer from the problem of divergence [7].

Before going to the question on the choice of the member set for the filter gain, let us return to the problem of the construction of an ROM. Let $x_e(k)$ be an estimate for the reduced state $x_r(k) = L_r x(k)$ of the dimension $n_e << n$. This estimate can be produced recursively in time by the ROF

$$x_e(k+1) = \Phi_e x_e(k) + K_e \zeta(k+1) \tag{8}$$

where $\Phi_e \in R^{n_e x n_e}$ is the fundamental matrix of the equation for the reduced state $x_r \in R^{n_e}$,

$$x_r(k+1) = \Phi_e x_r(k) + w_r(k), \tag{9}$$

$w_r(k)$ represents the model error in the reduced space.

By assuming $x = P_r x_r$, $P_r \in R^{n x n_e}$ be given, the estimate $\hat{x}$ for the full state x can be recovered by [3]

$$\hat{x}(k) = P_r x_e(k) \tag{10}$$

It is not hard to write out the KF from (9) and the observation equation

$$z(k+1) = H_e x_r(k) + v_r(k), H_e := HP_r \tag{11}$$

conditioned that the statistics of $w_r(k), v_r(k)$ are more or less well approximated. If the dimension of $x_r(k)$ is more or less moderate, it is possible to involve an ARE in the reduced space to estimate the ECM-PE in the reduced space. In practice, however, for the systems with dimension n of the order $10^6 - 10^7$, such approach should lead to large estimation error or even to filter divergence since a maximum allowable dimension n_e should be of order $O(100)$ which seems to be very small compared to the system dimension n.

2.2. On the Projection Operator P_r

Let us turn our attention on the choice of the operator P_r which has been assumed, up to now, to be given a-priori. We will show now that P_r plays the most important role in the filter design for very high dimensional systems. As P_r has the meaning of an operator which projects the estimate of the reduced state from the reduced space into the full space, let $L_r P_r = I_{n_e}$. It is therefore reasonable to take $P_r = L_r^+$ where L_r^+ is the pseudoinverse of L_r. Define $e_f(k) := \hat{x}(k) - x(k)$ as a filtered error (FE) for the full state at the time instant k where from (10) we have $x_e(k) = L_r\hat{x}(k) = P_r^+\hat{x}(k)$ and from (8) one obtains

$$\hat{x}(k+1) = \hat{x}(k+1/k) + K\zeta(t+1), K = P_r K_e, \quad (12)$$

$$\hat{x}(k+1/k) := \tilde{\Phi}\hat{x}(k), \zeta(t+1) := z(k+1) - H\hat{x}(k+1/k), \tilde{\Phi} := P_r \Phi_e P_r^+. \quad (13)$$

It is not hard to check that for $\tilde{\Phi} = P_r \Phi_e P_r^+$, the FE satisfies the equation

$$\begin{aligned} e_f(k+1) &= L(k)e_f(k) + \tilde{w}_f(k), L(k) := (I - KH)\Phi, \\ \tilde{w}_f(k) &= v(k+1) + [I - KH][\delta\Phi\hat{x}(k) + w(k)], \delta\Phi := \tilde{\Phi} - \Phi. \end{aligned} \quad (14)$$

Thus if we assume that $\tilde{w}_f(k)$ is a bounded sequence (in probabilistic sense), the FE will be bounded if the fundamental matrix $L(k)$ in the equation (13) is stable (in this paper we will understand the stability in the mean square sense, with the induced vector norm as $\sqrt{E||x||^2}$ and the Frobenius norm for a deterministic matrix). As $K = P_r K_e$, under most favorable conditions for the system (1)(2) (for example, the system is controllable and detectable), stability of the ROF (8) does not imply a stability of (13) since the stability of $L(k)$ depends on P_r. Thus the optimality (and stability) of the ROF (8) tells us very little about the stability of the full order filter (FOF) for $\hat{x}(k)$. By this reason the question on stability of the FOF is of primary importance compared with the optimality of the ROF. The optimality of the ROF has no significant impact on optimality of the FOF. An optimality of the FOF must be redefined as done in (7) [12].

The question on stability of the filter has been devoted in [13]. The FOF in [14] is not constructed on the basis of (8)(10) but by postulating directly the filter to be of the form

$$\hat{x}(k+1) = \Phi\hat{x}(k) + K\zeta(k+1), \; K = P_r K_e \quad (15)$$

It means that a filter like (8) for the reduced state x_r is excluded from the algorithm for generating $\hat{x}$. It is seen now that the equation for the FE is simplified as

$$\begin{aligned} e_f(k+1) = L(k)e_f(k) + w_f(k), L(k) := (I - KH)\Phi, \\ w_f(k) = v(k+1) + (I - KH)w(k) \end{aligned} \quad (16)$$

Compared (16) with (14) reveals that the forcing term in (14) includes in addition the error resulting from introducing a dynamical representation for the reduced state $x_r(k)$. In the filter (16) no direct estimation of the reduced state is performed. The reduction is introduced only through the projection of the innovation $\zeta(k)$ from the observation space onto the reduced space R^{n_e} and next from R^{n_e} into the full state R^n. We will refer to $span[P_r]$ (a linear space spanned by the columns of P_r) in the future as a *gain subspace*.

The study, conducted in this paper, is based on the main result obtained in [12], [14] which states that if the gain subspace $span(P_r)$ covers maximal invariant unstable subspace (including neutral eigenvectors) of Φ then there exists a gain $K = P_r K_e$ (with an appropriate parametrization, see (6)) which will ensure a stability of the filter (15)(6). As a consequence we will concentrate our attention on derivation of numerical algorithms for estimating $span(P_r)$ in order to construct an efficient filtering algorithm for estimating the state of very high dimensional systems.

3. Covariance of Prediction Error and PE Sampling Procedure

3.1. Covariance of Prediction Error

Return to the original equation for the filter gain (5). As seen from Eqs (3),(5), the correction $\delta\hat{x}(k+1)$ is computed by $\delta\hat{x}(k+1) = K(k+1)\zeta(k+1)$ hence the optimal filter gain projects the innovation onto the subspace spanned by the columns of the ECM for $e_p(k+1)$. As it is impossible to calculate this ECM for very high dimensional systems, one simple and efficient way to overcome this difficulty consists of approximating it through the error samples

$$M(k+1) = E[e_p(k+1)e_p^T(k+1)] \approx \frac{1}{N_s}\sum_{i=1}^{N_s} e_p^i(k+1)e_p^{i,T}(k+1) \quad (17)$$

where $e_p^i(k+1)$ is the *i-th* sample, e_p^T is the transpose of e_p. Thus when there is no feasibility to obtain the covariance $M(k)$ in a conventional way, a more simple way is to produce (or to simulate) the samples of the PE $e_p(k)$ and to use them to approximate the required statistics. That approach is widely known and is applied, for example, in the design of an EnBF [6][11]. However, as known from the Monte-Carlo and particle filtering approaches, even for a system with moderate dimension, a voluminous amount of samples should be required for well approximating the required statistics, not to say on the case of very high dimensional systems. As example, for meteorological and oceanic data assimilation problems, the samples $e_p^i(k+1)$ are usually simulated by runing the numerical model and taking the difference between the model solutions and the mean value (or climatology). These samples are then used to approximate $M(k) := M$ which, in many applications, is kept as time-invariant due to extremely high computational cost in simulating additional samples during assimilation. We want to stress that at the present, as the maximally allowable ensemble size N_s is of order $O(100)$, the gain subspace constructed from this ensemble is degraded seriously in such EnBF with high risk to be divergent. For the other possibilities to generate the samples during assimilation, see [6], [12]. We remark that in such filters the samples are first generated in a reduced space (using the ARE in the reduced space for evolving in time the ECM $M_e(k)$) and next the samples in the full space are obtained by integration of the numerical model. Compared to the time-invariant case, this approach requires in addition $O(100)$ times model integrations over the period of interest. To have the idea on how is the computational time required for such simulation, let the system state be of dimension 10^7 (as for the ocean model HYCOM (The Hybrid Coordinate Ocean Model) at SHOM (Toulouse, France)) for the Bay of Biscay configuration [2]. With a parallel code, 64 processors, during 5 days it is possible to simulate only about 100 error samples for 5 $days$ (ds) prediction period !!!

Thus, if for the systems with moderate dimensions there is no difficulty in simulating a large ensemble of PE samples, such effort makes no sense for very high dimensional systems. Exploring other approaches, more adapted to high dimensional systems, becomes an imperative task before going to construction of an efficient filtering algorithm. This concerns *first* the structure of the filter gain. As only a small amount of PE samples can be simulated for practical assimilation systems, the gain should be chosen a-priori with the structure which can span the unstable subspace of the system dynamics. With an appropriate parametrization, the unknown parameters in the filter gain will be estimated from a few simulated PE samples, which may be generated either before assimilation (off-line) or during assimilation process (on-line) [12]. *Secondly*, in order to limit a number of model integrations, the PE samples must be generated in an "optimal" way such that they should contain as maximally as possible the information on the PE. This second task can be solved by applying the PE sampling procedure proposed in [12].

3.2. PE Samples and PE Sampling Procedure

To better understand the role of PE samples in estimation of the filter gain, consider a simple linear filtering problem. For a today modern oceanic numerical model with a spatial grid (300 × 400 × 40) and 6 variables (temperature, salinity, two velocity components, layer thickness and sea surface anomaly), the state dimension is of order 10^7. The dynamical matrix has 10^{14} elements. Storing these real elements, in double precision, would require 8 × 100000 gigabytes. It is clear that it is impossible to store such matrices and to apply directly the well-known KF for estimating the oceanic circulation even under most favorable conditions. In similar situations an another approach to the filter design is inevitable.

As noted in the section 3.1, the ECM $M(k)$ can be approximated given an ensemble of PE samples $\{e^i(k)\}, k = 1, ..., N_s$. To see how one can generate the PE samples, suppose at the moment k we are given some estimate $\hat{x}(k)$ for the system state $x(k)$. If the true value of the system state is $x^*(k)$, the FE is $e_f(k) = \hat{x}(k) - x^*(k)$. At the next instant $k+1$, the PE is obtained by integration of the model,

$$e_p(k+1) = \hat{x}(k+1) - x^*(k+1) = \Phi\hat{x}(k) - \Phi x^*(k) = \Phi e_f(k) \tag{18}$$

(for simplicity, let us consider the case of perfect dynamical model). Thus if we put $e_f(k+1) := e_p(k+1)$ and repeat the above procedure, we obtain a sequence of vectors $\{e_p(k)\}, e_p(k) = \Phi e_p(k-1) = \Phi^k e_f(0)$ representing the PE for the considered dynamical system,

$$K_m(\Phi, e_f(0)) = span\{e_f(0), \Phi e_f(0), \Phi^2 e_f(0), ..., \Phi^{m-1} e_f(0)\} \tag{19}$$

The subspace $K_m(\Phi, e_f(0))$ is known as a KrSS of dimension m. Usually a normalization procedure is applied to $e_f(k)$, for example, $e_f(k)$ is normalized to have the norm 1 (in l_2 sense, for example). The KV $\Phi^l e_f(0)$ will approach the leading eigenvetor of Φ as l becomes large. For more detail on the PE sampling procedure, see [12].

Comment 3.1. For a system with moderate dimension, the samples $\Phi^k e_f(0), k = 1, 2, ..., m$ will become quickly parallel one to other since they converge to the first dominant eigenvector as m increases. If we want to take the KVs as approximations to the PE samples, the KVs should be selected before they become parallel. A simple example for such selection will be presented in section 5.1. Anyway, in low dimensional case, there is no constraint to well approach a dominant invariant subspace hence to ensure a stability of the filter.

Comment 3.2. For a very high dimensional system, the KVs from a KrSS of dimension $m = O(100)$ are far from being parallel and they can be considered as good candidates for approximating the PE samples from a dominant invariant subspace. Mention that involving some orthogonalization scheme in the Krylov procedure is necessary in the modern iterative methods for finding a few eigenvalues of large sparse matrices or solving large systems of linear equations (see the next section). As will be seen from numerical experiments in sections 5-6, the patterns taken from KrSS are sufficiently well representative for the PE and are efficient in estimating the filter gain.

Comment 3.3. For non-linear systems the "optimal" PE samples indicate the directions of most rapid growth of PE of a linear tangent operator.

4. Ritz Values and Ritz Vectors. Krylov Subspace and Schur Subspace

Theoretical results on the stability of a ROF in [12] require the computation of a dominant invariant subspace. Finding numerically such subspace for a very high dimensional system can be well performed based on the KrSS method and is summarized in this section.

4.1. Ritz Values and Ritz Vectors [9]

Basically eigenvalue approximations from a subspace are known as Ritz values. For $\Phi \in R^{n \times n}$, the set of all possible Ritz values is given by

$$W(\Phi) := [x^T \Phi x : x \in R^n, x^T x = 1]. \tag{20}$$

It is interesting to note that for the Hermitian part of $\Phi, H(\Phi) = (\Phi + \Phi^T)/2$, the relationship $W(H(\Phi)) = Re[W(\Phi)]$ holds, where Re denotes the real part. The Ritz value associated with v is the scalar θ such that $\Phi v - \theta v$ is orthogonal to the space spanned by v, i.e., $v^T(\Phi v - \theta v) = 0$. A Ritz value is optimal in the sense that it minimizes the norm of $\Phi v - \theta v$. The Ritz value represents the action of Φ restricted to the subspace spanned by v. For a Ritz value to be a good approximation to an eigenpair, the residual $\Phi v - \theta v$ should be small. For Ritz values generated from a subspace spanned by the columns of a matrix V, the residual is orthogonal not only to the Ritz vector, but also to any vector in the subspace. Hence $V^T(\Phi v - \theta v) = 0$. Since any $x \in span(V)$ can be written as Vy, $V^T(\Phi Vy - Vy\theta) = 0$, which is written as

$$Hy = V^T Vy\theta, H = V^T \Phi V. \tag{21}$$

Thus the problem of determining Ritz values from a subspace is equivalent to solving the generalized eigenvalue problem (21). If the columns of V are orthonormal, then $V^TV = I$, and the problem of determining Ritz values from a subspace reduces to an ordinary eigenvalue problem.

4.2. Why Krylov Subspace

Return to the problem of determining a few leading eigenvalues λ_i (and eigenvectors u_i) of the very large sparse real unsymmetric fundamental matrix Φ in (1). This problem is of considerable practical importance since it has a significant number of applications in scientific and engineering computing, including mathematical models in economics, Markov chain modelling of queueing networks, bifurcation problems ... In particular, as mentioned in section 2.1, leading eigenvectors play an important role in the construction of the operator P_r in (15) and to stabilize the ROFs. One sees that for an $(n \times m)$ matrix V, with $k \leq m << n$, $V^TV = I_m$ and $S = span\{v_1, ..., v_m\}$, the Rayleigh-Ritz procedure (21) consists in computing $H = V^T\Phi V$ and then solving the eigenvalue problem $Hy_i = \theta_i y_i$. The approximate eigenpairs $\tilde{\lambda}_i$, $\tilde{u_i}$ of the original eigenvalue problem are then $\tilde{\lambda}_i = \theta_i$, $\tilde{u_i} = Vy_i$ which are Ritz values and Ritz vectors, respectively. The Ritz values and vectors are the best possible approximations in that subspace. Matrix H is the projection of Φ onto S.

It is important to stress that the above process works better when the subspace S approximates a certain *invariant subspace* of Φ, for example, when S is equal to the KrSS associated with matrix Φ, i.e. $K_m(\Phi, x_0)$ (see (19)). That is why the KrSS plays a remarkable role in the algorithms for estimating a dominant invariant subspace. It is known that for the KrSS $S := K_m(\Phi, x_0)$ the method of Arnoldi allows to compute an orthogonal basis of V_m of S and at the same time computes the projected matrix H, all this in an efficient and numerically stable way. For the purpose of estimating the filter gain, however, we do not follow in this paper the Arnoldi approach since the eigenvectors may be complex. Moreover our interest is to generate the PE samples which are not the eigenvectors (see the next section). Computation of the KrSS Eq. (19) (or the ScSS) hence has the objective to exploit the KVs from $K_m(\Phi, x_0)$ as samples for the PE which will be used to approximate the filter gain.

Comment 4.1. As the sequence $\{\Phi^m e_f(0)\}$ in (19) tends to the dominant eigenvector, it is sometimes of interest to re-initialize the initial vector and to repeat the Krylov procedure with the objective to approach the another (for example, second) dominant eigenvector. Suppose x_1, λ_1 are eigenvector and eigenvalue obtained through (19) for sufficient large m. One way to approach the second dominant eigenvector and eigenvalue is put $\Phi := \Phi - x_1x_1^T\lambda_1$ and repeat the Krylov algorithm with a new $e_f(0)$ orthogonal to x_1. Here $x_1x_1^T\lambda_1$ represents Φ_e of the rank 1 (see Eq. (13)).

4.3. Orthogonal (Schur) Iteration Method

Historically the orthogonal iteration method aims to compute approximations to a block of eigenvalues at once. They can be advantageous when the dominant eigenvalue of Φ is a multiple eigenvalue or if the wanted eigenvalues are clustered. Choosing the blocksize to be L $(L > 1)$, the block Arnoldi method allows for finding the L eigenvalues of Φ of largest moduli.

As our objective is to generate the PE samples, the computation of eigenvalues has no importance for the construction of the projection subspace. In contrast, we are strongly interested in the algorithm in [12] based on Schur decomposition for generating PE samples.

The Schur decomposition reads as follows: If Φ is complex, then its Schur decomposition is $\Phi = XTX^*$, where X is unitary and T is upper triangular. If Φ is real (which is of our interest here) its Schur decomposition is

$$\Phi = XTX^T \tag{22}$$

where X is orthogonal and T is upper quasi-triangular (with (1×1) and (2×2) blocks t_{ii} on its diagonal). The columns of X are called the ScVs of Φ. The eigenvalues of Φ appear on the diagonal of T; complex conjugate eigenvalues of a real Φ correspond to 2-by-2 blocks on the diagonal of T.

On the basis of the Schur decomposition (22), the PE sampling algorithm is proposed in [12] which is iterative and uses the orthogonalization iteration procedure: Let L be an integer number satisfying $1 \leq L \leq n$. Given an $n \times L$ matrix X_0 with orthonormal columns, the method of orthogonal iteration generates the sequence of matrices X_k,

$$S_k = \Phi X_{k-1}, X_k G_k = S_k, k = 1, 2, ..., \tag{23}$$

where X_k is orthonormal. Thus, at k iteration, the columns of X_k are orthonormal vectors which are derived by : (i) integration of the model from each column of X_{k-1} to produce S_k; (ii) orthonormalization of L columns of S_k. One sees that the PE patterns (columns of S_k) belong to the space spanned by the vectors of X_k.

Suppose we have a Schur decomposition (22) and let

$$X = [X1, X2], X1 \text{ is of } (n \times L), \qquad T = \begin{pmatrix} T_{11} & T_{12} \\ 0 & T_{22} \end{pmatrix} \tag{24}$$

the eigenvalues $\lambda_i, i = 1, ..., n$ of Φ are ranged in a descreasing order of their absolute values, i.e. the L leading are located in T_{11}. It is well known that under mild conditions, the distance between $D_L(\Phi) := R[X1]$ and $R[X_k]$ is of order $O(|\frac{\lambda_{L+1}}{\lambda_L}|^k)$ where $R[X_k]$ denotes the linear space spanned by the columns of X_k where X_k is defined in the procedure (23) [9] (section 7.3). Thus the method allows us to generate PE patterns (the columns of X_k) developing in the direction of the subspace spanned by the columns of $X1$. As $X_k G_k = S_k$, G_k is non-singular, $span(S_k) \subset span(X_k)$ hence $span(S_k)$ develops in the directions of DScVs too. In the further the columns of S_k will be referred to as S-vectors (*SVs.*) It is easy to see that the algorithm (23) subject to $L = 1$ will generate the KVs.

Comment 4.2. For a real dynamics matrix Φ, the samples produced by the procedure (23) are always real. As the columns of $X1$ are the ScVs associated with L leading eigenvalues of Φ (they will be called the L DScVs) the columns of S_k are developed in the directions of most growing PE error of the system.

Comment 4.3. Generally speaking the ScVs are not eigenvectors. However by applying the algorithm (23) we generate the columns of X_k approaching the ScVs with dominant magnitudes. Let $\alpha_i, i = 1, ..., L$ be magnitudes obtained by applying G_k. One way to better approaching the "eigenvectors" if desired is to solve the system $\Phi x = \alpha_i x$ from which one obtains

$$x_i = (I - AA^+)s_i, A := \Phi - \alpha_i I. \tag{25}$$

5. Application of Krylov Subspace and Schur Subspace in Filtering

In the previous section we summarized the algorithms for constructing the KrSS and ScSS. These algorithms are very useful in solving many problems of linear algebra like computing approximate eingepairs of the given matrix Φ, solving the linear equation $\Phi x = b$ ... But the most important applications of these algorithms are found in the industrial field where the problems with very large-sized matrices are encountered frequently. For these matrices the classical method based on finding the roots of the characteristic polynomial of the matrix is infeasible.

5.1. PE Samples from Different PE Sampling Procedures

As an useful application of these two subspace methods, we will procede here to the problem of the design of a Prediction Error Filter (PEF). The PEF is originally proposed in [12] and its efficiency to different filtering problems has been demonstrated in [12]. The need to simulate the PE samples arise from the impossibility to apply traditional optimal filtering algorithms to the state estimation problem for very high dimensional systems.

The objective of this subsection is to demonstrate how work really the proposed PEF algorithms jointly with the KrSS and ScSS. Efficiency of the PEFs with the gains, computed on the basis of the different sets of PE samples, will be compared in the next subsection.

To examine in detail the different generated ensembles of PE samples, let us first consider a simple filtering problem (1)(2) with

$$\Phi = \begin{pmatrix} 0.4 & 0 & 0.2 \\ 0.3 & 0.8 & 0.3 \\ 0.3 & 0.2 & 0.5 \end{pmatrix}, H = \begin{pmatrix} 1 & 0 & 0 \end{pmatrix} \tag{26}$$

The model error $w(k)$ and the observation error $v(k)$ are supposed to be gaussian time uncorrelated sequences with zero mean and covariance I. The observations $z(k)$ are available at time instants $k = 1, 2, ..., 200$. The initial state $x(0)$ has the mean $\bar{x}(0) = (1, 1, 1)^T$ and covariance I.

Table 1 shows the eigenvalues and eigenvectors of the fundamental matrix Φ which all are real simple eigenvalues.

Let $\delta x(0)$ be a normalized (is of unit l_2 norm) initial perturbation. We will apply three following sampling procedures (SP) for generating the PE samples:

Table 1. Eigenvalues and eigenvectors of Φ

Eigenvalues	Eigenvectors
1	$(0.14744259, 0.88465136, 0.44232637)^T$
0.5	$(-0.53452235, 0.80178380, -0.26726133)^T$
0.2	$(-0.70710611, 0, 0.70710737)^T$

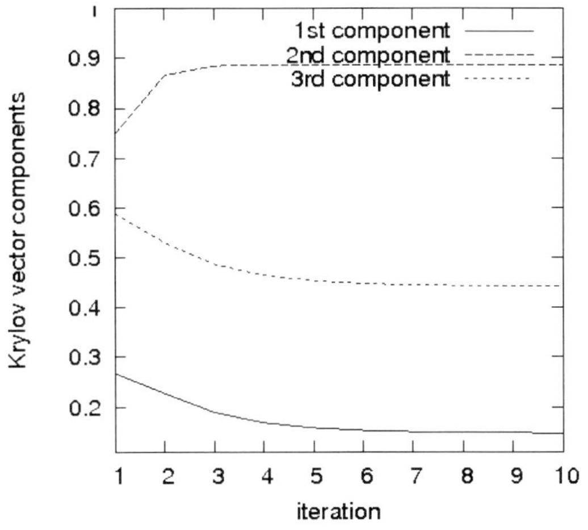

Figure 1. Krylov vectors produced by Algorithm SP1 during iterations. One sees a quick convergence of the KVs to the first dominant eigenvector of Φ.

(SP1) simulates the KVs from Krylov subspace: $K_m[\Phi, \delta x^n(1)]$ (19) subject to l_2 normalization : $\delta x(k+1) = \Phi \delta x^n(k)$, $\delta x^n(k) = \delta x(k)/||\delta x(k)||_2$.

(SP1.1) is similar to SP1 with exception that occasionally $\delta x^{n,o}(k)$ is orthogonalized with respect to the last $\delta x^{n,o}(k-1)$ and $\delta x(k+1) = \Phi \delta x^{n,o}(k)$.

(SP2) First a set of orthonormal vectors $\delta x^{o,n,i}(1), i = 1, ..., n_e, n_e > 1$ is chosen. At the following steps we compute $\delta x^i(k+1) = \Phi \delta x^{o,n,i}(k)$, $\delta x^{o,n,i}(k) = \delta x^i(k+1)/||\delta x^i(k+1)||_2$, $< \delta x^{o,n,i}(k), \delta x^{o,n,j}(k) >= \delta_{ij}$, δ_{ij} is a Kronecker symbol. Mention that for $n_e = 1$, SP2 reduces to SP1.

5.1.1. Algorithm SP1

The results obtained by Algorithm SP1 are shown in Figs 1-2. In Fig. 1 we display 3 components of $\delta x(k), k = 1, ..., 10$. It is seen that the vectors $\delta x(k)$ become quickly parallel as k increases and converge to the first eigenvector. For this raison in the assimilation experiment in the next section we will take only 3 vectors produced at the beginning of the Krylov subspace (namely at iterations $i = 2, 3, 4$) as PE samples to perform an ensemble of PE samples (denoted as EnKV, see Table 2). Fig. 2 shows the magnitude of KVs which tends to the first dominant eigenvalue of Φ.

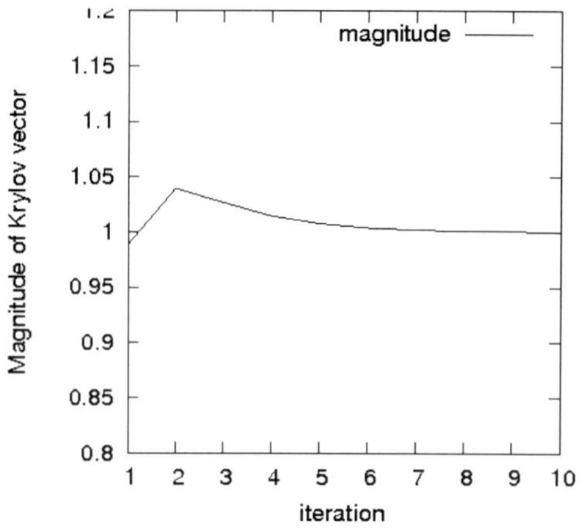

Figure 2. The magnitude of KVs resulting from application of Algorithm SP1. It converges to the first dominant eigenvalue of Φ.

Table 2. KVs resulting from SP1

No iteration	Krylov vector
2	$(0.15271021, 0.88536781, 0.44785392)^T$
3	$(0.15799367, 0.88589454, 0.45339835)^T$
4	$(0.16855025, 0.88630694, 0.46452612)^T$

One sees from Table 2 that if we pick samples as KVs after about 10 iterations, they all will be almost parallel and there is no interest to take more than one KV. For the assimilation experiments, only three vectors in Table 2 will participate in the estimation of the ECM dispite the fact that an ensemble of larger size can be collected. This is intentionally done to be in accordance with very high dimensional systems framework since for such systems one can only afford a very limited number of simulations (compared to the system dimension). Notice that perhaps it is more interesting to take first 2 KVs at the beginning and the third at the end of iteration since the latter is closer to the first eigenvector. Another strategy is to select other dominant directions by following Comments 4.1-4.2. Then the eigenvectors can be approached by using Eq. (25) as shown in Table 3. This is less interesting for very high dimensional systems since the produced KVs are rather independent than parallel and one cannot simulate perturbations until convergence (not to say on nonlinear systems, see section 5).

We remark that if the eigenvalues can be well approached by applying a restarted SP1 (Commnent 4.1), the produced KVs are in general far from the eigenvectors. That is why the use of (25) (which is simply a definition of the eigenvector given its eigenvalue) is necessary to seek the corresponding eigenvectors. However, looking at (25) one recognizes that finding a pseudo-solution of the system $Ax = s_i$, is a hard task for very high dimensional

Table 3. KVs and their magnitudes from restarted SP1 subject to (25)

Cycle	Magnitude	Krylov vector
2nd	0.5	$(-0.535, 0.802, -0.267)^T$
3rd	0.2	$(-0.707, -1.647E - 06, 0.707)^T$

Table 4. Eigen-values and eigen-vectors of Φ

Iteration	Magnitude	Krylov vector
1	1.1180339	$(1.0606601, 0.35355338)^T$
4	1.1180341	$(0.53758729, -0.98030615)^T$
8	1.1180339	$(-1.0916182, -0.24159807)^T$
10	1.1180340	$(-0.84824938, 0.72833574)^T$

systems. For the present example, existence of simple real eigenvalues makes the considered eigenvalue problem much easier. For a real Φ, having the complex eigenvalues, we are rather interested in derivation of a real Schur decomposition since computing the complex eigenvalues and eigenvectors represents a very hard and unstable task.

Let us pay the attention to the case of the existence of complex eigenvalues. Consider the dynamics matrix

$$\Phi = \begin{pmatrix} \alpha & \beta \\ -\beta & \alpha \end{pmatrix}, \tag{27}$$

One observes that Φ has two complex conjugate eigenvalues $\mu_{1,2} = \alpha i \beta$ and for $\alpha = 1, \beta = 0.5$ they have the magnitude 1.118. Table 4 shows that SP1 produces the KVs with the magnitude which converges quickly and its limit coincides with the exact magnitude of the two eigenvalues. However the KVs do not converge and Fig. 3 reveals that they are periodic.

We remark that if we were able to compute $v_1 := u+iv$ and $v_2 := u-iv$ as eigenvectors of Φ, the projection operator P_r can be constructed as $P_r = [v_1, v_2]$.

5.1.2. SP1.1

This procedure is similar to SP1 with exception that occasionally $\delta x^{n,o}(k)$ is orthogonalized with respect to the last $\delta x^{n,o}(k-1)$ and $\delta x(k+1) = \Phi \delta x^{n,o}(k)$. The PE samples produced by such procedure is shown in Fig. 4 for the dynamics (26) where the orthogonalization is performed after each 10 iterations. For the present model such orthogonalization procedure seems to be incapable of generating the more informative PE samples.

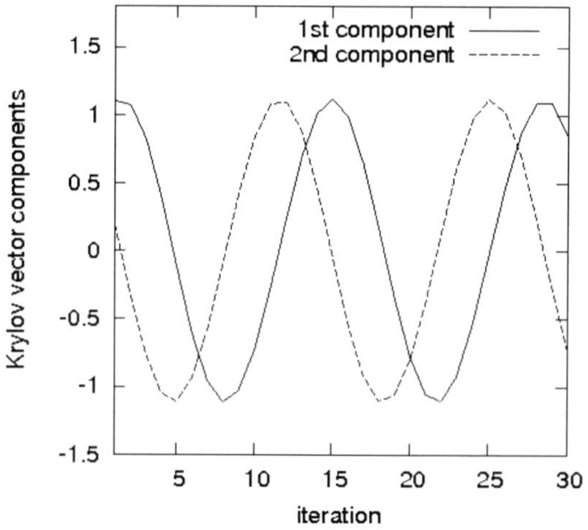

Figure 3. Krylov vectors produced by SP1 during iterations: they are periodic functions of iteration.

5.1.3. SP2

Return to the system (26). This procedure is an application of the algorithm (23) subject to $n_e = 3$. Fig. 5 shows that the produced values for the magnitude of three PE samples converge to the three eigenvalues of Φ.

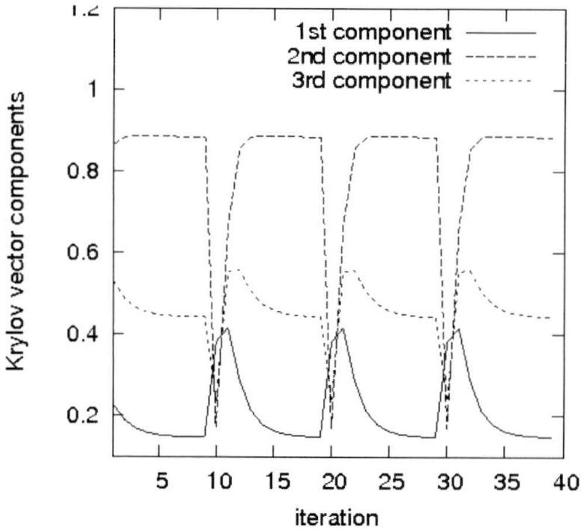

Figure 4. The KVs resulting from application of Algorithm SP1.1. Due to very quick convergence after each orthogonalization, this procedure is not capable of creating a new ensemble of more rich PE samples.

The columns of the matrix $S_{20} = [s1, s2, s3]$ obtained after 20 iterations are displayed in Table 5. The ensemble of these three vectors (denoted as EnSV) will be used as SVs to

Table 5. Column vectors (PE samples) in matrix S_{20}

Component	1st component	2nd component	3rd component	Magnitude
1st column	0.14744237	0.88465184	-0.41943580	1
2nd column	-0.39914078	-6.08849376E-02	-0.41943580	0.5
3rd column	-0.13764948	-0.18353246	9.17662978E-02	0.2

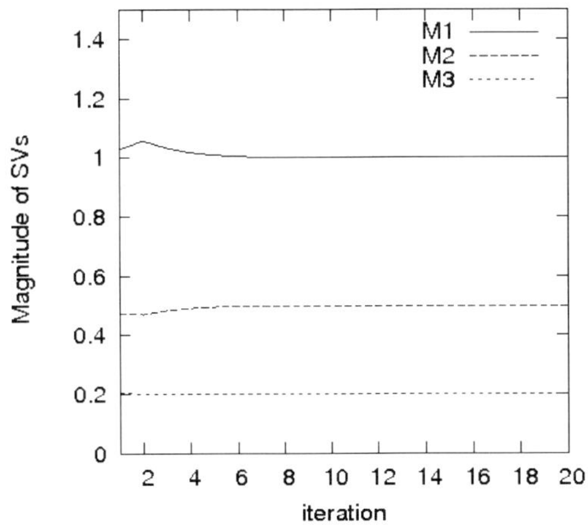

Figure 5. The magnitudes of PE samples produced by SP2 simulating ScVs for the dynamics (26).

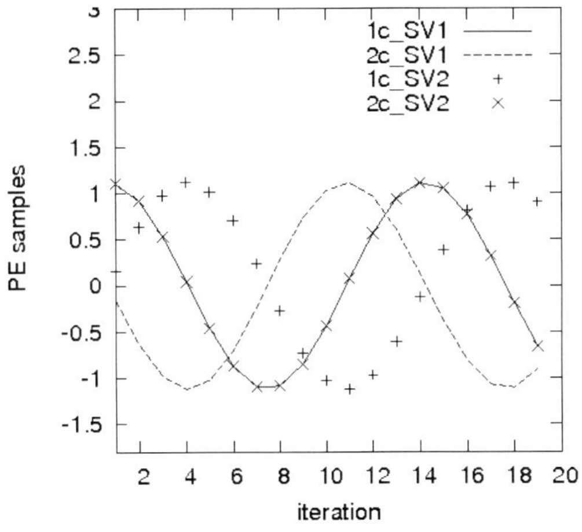

Figure 6. The PE samples produced by SP2 simulating two ScVs for the dynamics (27).

estimate the ECM in the assimilation experiments. We note that only the samples of the 1st column $s1$ converge to the 1st eigenvector. As to the magnitudes of the SVs, they all converge to the corresponding eigenvalues (see also Fig. 5).

We have also applied the Schur procedure SP2 to the dynamics (27) and the samples for two columns $s1, s2$ are displayed in Fig. 6. It is seen that these two vectors are orthogonal.

5.2. Assimilation Experiment for the Problem (26)

Let $U^{[i]} := [\delta x^1, ..., \delta x^i], i = 1, .., n_e$ where δx^j represents a PE sample. In the experiment to follow, the different PEFs (and their adaptive versions denoted as APEFs) will be used to estimate the system state. These filters are of the form (15) subject to corresponding gains $K^{[n_e]} = U^{[n_e]} K_e$. The gain in the reduced space is supposed to be time-invariant and is given by [14]

$$K_e = M_e H_e^T [H_e M_e H_e^T + R]^{-1}, H_e = HU^{[n_e]}. \tag{28}$$

It is seen that the gain in the full space is expressed as

$$K = U^{[n_e]} K_e = MH^T [HMH^T + R]^{-1}, M := U^{[n_e]} M_e U^{[n_e],T}. \tag{29}$$

To parameterize the gain with the objective to make an adaptation, let us decompose M as

$$M = P_r P_r^T \tag{30}$$

Using this decomposition the gain can be written as

$$K = P_r K_e, K_e = P_r^T H^T [HMH^T + R]^{-1}, \tag{31}$$

For adaptation purpose the gain is parametrized as [15]

$$K = P_r \Theta K_e, \Theta := diag[\theta(1), ..., \theta(n_e)] \tag{32}$$

where $\Theta = I$ corresponds to a non-adaptive PEF (denoted simply as PEF).

Two ensembles of PE samples (EnKV and EnSV) are used to perform $U^{[n_e]}$, one consists of the KVs displayed in Table 4 and other - SVs in Table 5.

5.2.1. PEF with KVs and SVs

In Table 6 we display RMSEs (Root Mean Square Errors) resulting from applying two PEFs, one is the PEF with the ECM estimated from the ensemble EnKV (denoted as PEF-KR) and other - from EnSV (PEF-SV). There are displayed also the errors produced by the KF which is the best filter in the mean squared sense, and the value called *persistence* calculated as the difference between two successive observations. This quantity (persistence) serves in many studies as the reference for qualifying the performance of the assimilation algorithms. The PEF-KRi (or PEF-SVi) signifies that only i first samples from EnKV (or EnSV) are used to perform $U^{[i]}$ (see Eqs (28), (29)). It is seen that all the filters produce the PE of much lower level than the persistence. The PEF-KR1 and PEF-SV1 (and PEF-KR2

Table 6. RMSE from different filters

Filters	Innov	Prediction Error (PE)	Filtered Error (FE)
Persistence	3.1773844	...	...
PEF-KR1	1.6827954	6.5358558	6.4286880
PEF-KR2	1.6540412	6.0428667	5.8824434
PEF-KR3	1.6301951	5.6236868	5.4171858
PEF-SV1	1.6841538	6.5569544	6.4518890
PEF-SV2	1.6377684	6.0636859	5.9361033
PEF-SV3	1.6354678	6.0160542	5.8840165
KF	1.5874269	4.9211411	4.6443896

and PEF-SV2) produce the estimates of nearly same quality. In contrast, the PEF-KR3 behaves better than PEF-SV3. There is a significant improvement of performance of PEF-KR2 (and of PEF-SV2) with respect to that of the PEF-V1 (PEF-SV1). With all the filters, the performance becomes better with larger ensemble size. The best performance, as expected, is produced by the KF.

5.2.2. APEF and KF

With the gain, estimated by a few number of PE samples, there is no surprise that the performance of the PEF degrades in comparison with that of the KF. The adaptive filter (AF) in [12] is just proposed with the objective to improve the performance of a non-adaptive filter in such situation. In fact the gain design outlined in subsection 5.2 has the objective to correctly parameterize the filter gain which subsequently is tuned to minimize the innovation $\zeta(k)$ (see (13)). The adaptation algorithm has the form

$$\theta(k+1) = \theta(k) - \gamma(k+1)\nabla_{\theta(k)}\Psi[\zeta(k+1)], k = 1, 2, ... \tag{33}$$

where $\gamma(k+1)$ is a scalar (or matrix) sequence guaranteeing a convergence of the algorithm of stochastic approximation, $\Psi[\zeta]$ is a sample of the cost function,

$$J[\zeta] = E[\Psi(\zeta)], \tag{34}$$

$E(.)$ is a mathematical expectation. In the experiments we put simply $\gamma(k) = \frac{1}{k}$ and $\theta(1) = (1, ..., 1)^T$. For more details, see [15].

The estimation results produced by the AFs are shown in Table 7. It is evident that adaptation allows to reduce significantly the estimation errors as seen from Table 8. The smaller the size of an ensemble of samples used in the PEF, the larger the error reduction is achieved by adaptation. To see how efficient is the adaptation mechanism, in Table 8 we display the percentage of error reduction produced by the AFs compared to their non-adaptive versions. Roughly speaking, the adaptation allows to reduce about 16-18 % of the estimation error for the system state. It is interesting to note that the APEF-KR3 is capable of producing the estimates with slightly lower error levels compared to what does the KF. To

Table 7. RMSE from different APEFs and KF

Filters	Innov	Prediction Error (PE)	Filtered Error (FE)
APEF-KR1	1.5866156	5.3564944	5.2924905
APEF-KR2	1.5759704	5.0116334	4.8803282
APEF-KR3	1.5686542	4.7401309	4.5585990
APEF-SV1	1.5871948	5.3739438	5.3123569
APEF-SV2	1.5740148	5.0961671	5.0146542
APEF-SV3	1.5733693	5.0596652	4.9755583
KF	1.5874269	4.9211411	4.6443896

Table 8. RMSE reduction

Filters	Innovation	Prediction Error (PE)	Filtered Error (FE)
APEF-KR1	5.7 %	18 %	18 %
APEF-KR2	4.7 %	17 %	17 %
APEF-KR3	3.8 %	16 %	16 %
APEF-SV1	5.8 %	18 %	18%
APEF-SV2	3.9 %	16 %	16 %
APEF-SV3	3.8 %	16 %	15 %

see what really happened in the APEF-KR3 and KF, Fig. 7 shows the curves representing the time evolution of the gain elements in these two filters. Here the curve AF-i (and KF-i) denotes the element $K(i, 1)$ of the gain in the APEF-KR3 (and KF). One sees that if the corrections for the 2nd and 3rd components of the system state are almost the same in these two filters, a larger correction has been made for the 1st component in the AF. That is due to the fact that only the 1st component of the system state is observed (see Eq. (26)) and the AF tries to minimize the PE for this component. The curve AF-1 grows quickly from the beginning up to about 120 iterations and seems to be stabilized thereafter. As to the KF, the solution of the ARE converges very quickly and as consequence the gain coefficients are stabilized after abour only 20 iterations. The value of $K(1, 1)$ in the AF is much greater than that in the KF which allows to reduce a more estimation error for the 1st component of the system state. Fig. 8 displays the RMSEs of the FE produced by the different employed filters. It is seen that the KF and APEF-KR3 (with 3 KVs) behave identically from the beginning up to $k = 100$ but after that the APEF-KR3 starts to perform better and better (compared to the KF).

6. Altimetric Data Assimilation in the Ocean Model MICOM

We present briefly here the results on application of the algorithms SP1 and SP2 for the design of a filter gain which will be employed next to estimate the oceanic circulation using

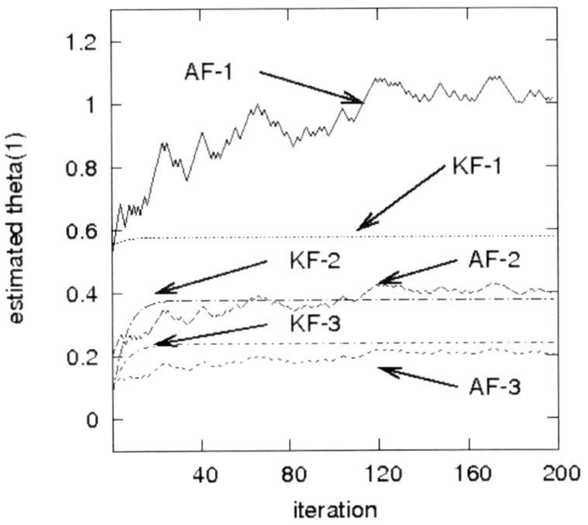

Figure 7. Evolution of the gain elements in the APEF-KR3 and KF: The curve AF-i (and KF-i) denotes the element $K(i, 1)$ of the gain K. If AF-i, KF-i, i = 2,3 are nearly the same, the element AF-1 is much greater than KF-1. This is related to the minimization of the innovation $\zeta(k)$ for the first component of the state vector.

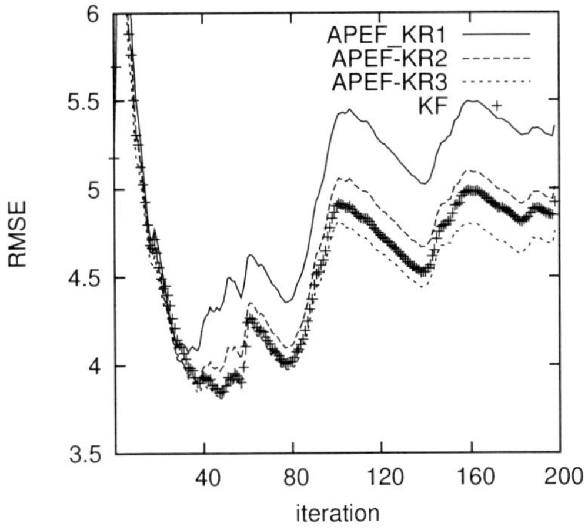

Figure 8. RMSEs of the FE produced by different filters.

the altimetric sea surface height (SSH) data.

6.1. MICOM Model and Observations

The Miami Isopycnal Coordinate Ocean Model (MICOM), used for the twin experiment in this paper, is identical to that described in [14]. The model configuration is a domain situated in the North Atlantic from 30^0 N to 60^0 N and 80^0 W to 44^0 W; for the exact model domain and some main features of the ocean current (mean, variability of the SSH, velocity

...) produced by the model, see [14]. The grid spacing is about 0.2^0 in longitude and in latitude, requiring $N_h = N_x \times N_y = 25200$ ($N_x = 140$, $N_y = 180$) horizontal grid points. The number of layers in the model is $N_L = 4$. It is configured in a flat bottom rectangular basin ($1860km \times 2380km \times 5km$) driven by a periodic wind forcing. The model relies on one prognostic equation for each component of the horizontal velocity field and one equation for mass conservation per layer. We note that the state of the model is $x := (h, u, v)$ where $h = h(i, j, lr)$ is the thickness of lr^{th} layer, $u = u(i, j, lr), v = v(i, j, lr)$ are two velocity components. The layer stratification is made in the isopycnal coordinates, i.e. the layer is characterized by a constant potential density of water.

The experiment to be presented here is a standard test of a so-called "twin-data" assimilation experiment in which the observations are generated by the model started from a "climatological" state. The observations are the SSH stored each 10 $days$ (ds). Mention that the period of 10 $days$ is equivalent to 800 model time steps. Totally there are 72 observations covering the period of 3 years.

6.2. Reduced-Order Filter and Gain Structures

The filter used for assimilating SSH observations is of the form

$$\hat{x}(k+1) = F[\hat{x}(k)] + K(k+1)P_{oi}\zeta(k+1), k = 0, 1, ... \tag{35}$$

where $\hat{x}(k+1)$ is the filtered estimate for $x(k+1)$ - the system state at $(k+1)$ assimilation instant, $F(.)$ represents the integration of the MICOM non-linear model over 10 ds, $K(k+1)$ is the filter gain, $\zeta(k+1)$ is the innovation vector. Generally speaking, as the observations are available not at all model grid points, P_{oi} is chosen as an interpolation operator which interpolates the SSH from observed points to all surface grid points. The gain K is symbolically given by $K = (K_h, K_u, K_v)^T$ with K_u, K_v representing the operators which produce the correction for the velocity (u, v) from the layer thickness correction $K_h P_{oi} \zeta$ using the geostrophy hypothesis. Thus the filter produces first the correction $\hat{\delta h}$ for the layer thickness h. The corrections $(\hat{\delta u}, \hat{\delta v})$ for two velocity components (u, v) are obtained by using the hypothesis on quasi-gepstrophy.

From (35) it is seen that strictly speaking the gain is composed from the product $K(t+1)P_{oi}$. For more detail, see Comment 6.1 below. As SSH observations are linear functions with respect to h and if we consider $\zeta' := P_{oi}\zeta$ instead of ζ, the innovation ζ' is then given in all horizontal grid points and the observation operator H now is of the form

$$H = [I_p, ..., I_p] \tag{36}$$

where I_p is the unit matrix of dimension pxp ($p = N_h$ is the number of all horizontal grid points). We have then the filter gain (for the h component) at k assimilation instant

$$K_h(k) = \Sigma_{12}\Sigma_2^{-1}, \Sigma_{12} := M(k)H^T, \Sigma_2 = HM(k)H^T + R(k) \tag{37}$$

with $M(k)$ - the ECM for $h(k)$, R is the covariance of the observation error (see also (15) for comparison). In the experiment, $R = 0$.

6.3. Structure of the ECM

As it is impossible to generate a sufficient large number of PE samples (compared to the size of $h = 25200$), it has no sense to apply the formula (17) to estimate the ECM $M(k)$. The following structure for the ECM will be used

$$M := [m_{kl} I_p]_{k,l=1}^{N_L} \tag{38}$$

Thus m_{kl} represents the PE covariance between k and l layers for the layer thickness variable. This structure has the advantage that the ECM M is non-singular if the matrix $[m_{kl}]_{k,l=1}^{N_L}$ is. The elements m_{kl} will be estimated using the PE patterns generated by the procedures SP1 and SP2. For the ensemble of L patterns $\delta h_p^k(i,j,lr), k = 1, ..., L$ from the set $S_k(L = 1), k = 1, ..., L$, the elements m_{kl} are estimated through

$$m_{lm'} = \frac{1}{L} \sum_{k=1}^{L} \mu_{lm'}^k, \tag{39}$$

$$\mu_{lm'}^k = \frac{1}{p} \sum_{i,j} \delta h_p^k(i,j,l) \delta h_p^k(i,j,m')$$

where i, j span all horizontal grid points whose number is equal to p.

The gain now is estimated as

$$K_h = [K_h(1), ..., K_h(N_l)]^T \otimes I_p,$$
$$K_h(l) = [\sum_{m'=1}^{N_L} m_{lm'}/s], s = \sum_{lm'=1}^{N_L} m_{lm'} \tag{40}$$

where $\otimes$ denotes the Kronecker product.

Comment 6.1. The gain structure described above can be obtained rigorously from (5) by assuming that the ECM M_h for the layer thickness h has a vertical and horizontal separable structure, $M_h = M^h M^v$, with M^v defined by (38) and $M^h = Cov[(i,j),(i',j')]$. Represent $M^h = DCD$ where D is diagonal whose diagonal elements are the standard deviations, and C is the corresponding correlation matrix. In the experiment to follow C is assumed to be homogenous and isotropic and has the form $C := e^{-\frac{d}{L}}$. Here $d = \sqrt{(i-i')^2 + (j-j')^2}$, L is the correlation length, (i,j) span all horizontal grid points. The operator P_{oi} then follows directly from this assumption. For simplicity, we let $L = 25$. Mention that better filter performance can be obtained by using the parameter L estimated from PE samples $\delta h_p^k(i,j,l)$.

6.4. Simulation of PE Samples

6.4.1. PE Samples

In Fig. 9 we display the magnitudes of the SVs (1st and 20nd). Mention that the MICOM is a non-linear ocean model and the KV and SV calculated here are understood as such obtained for a linear tangent model. One sees that the magnitude of the 20nd SV is much less than that of the 1st SV but most of the time it is greater than 1 (unstable). It is expected that for the model with dimension 25200 the number of unstable SVs would be very large. It is therefore unimaginable to calculate all the unstable and neutral SVs in order to project the correction onto the subspace spanned by all unstable and neutral SVs. That is why we have chosen the structre (38) for the ECM. The values of the 1st and 2nd SVs at the grid point (50,50,1) are displayed in Fig. 10 from which one observes also their non-stationary character. The values and magnitude of the KV are similar to those of the 1st SV and are not presented here. For estimating the elements of the ECM m_{kl} in (39) we have used the PE samples from the KV. From (40) the gain elements $K_h(k), k = 1, ..., 4$ are calculated and they are presented in Fig. 11.

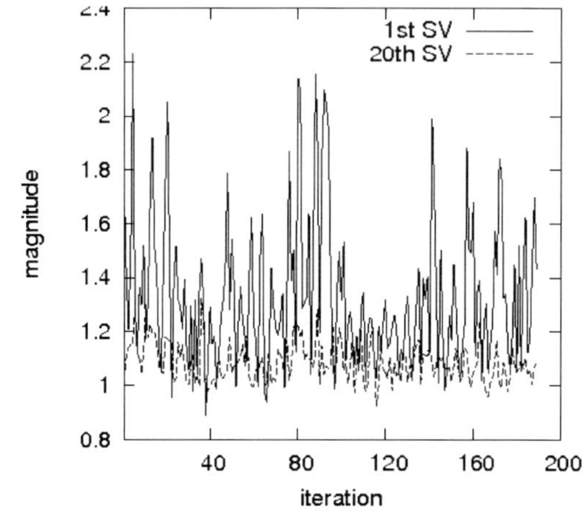

Figure 9. Magnitudes of the 1st and 20nd SV by SP2 during iterations. The magnitude of the 1st SV is much greater than that of the 20nd. They are non-stationary and unstable. That is due to non-linearity and unstability of the oceanic circulation. The same picture is observed for the magnitude of the KV produced by SP1.

6.4.2. Online SP2 Samples

Generally speaking, the PE at the time instant $k + 1$ depends on the system state $x(k)$ at the previous instant. In [12] in order to capture this dependence and to improve the filter performance, it is proposed to simulate the PE samples during assimilation period. By this way, the ECM $M(k)$ can be updated during assimilation. Suppose that $M(0)$ is the ECM obtained by (39) from the PE samples simulated off-line, i.e. by the model without assimilation. At the assimilation instant k, let $\hat{x}(k)$ be the estimate for the system state. Adding the perturbations $\delta\hat{x}^i(k), i = 1, ..., L$ to the system state and integrate the model from $\hat{x}(k)$ and $\hat{x}(k)+\delta\hat{x}^i(k)$ will produce an ensemble of L PE samples. These PE samples can be produced by either SP1 or SP2. Such on-line generated PE samples can be used to obtain the on-line estimated ECM $M(k)$. One way to update the ECM is to combine the

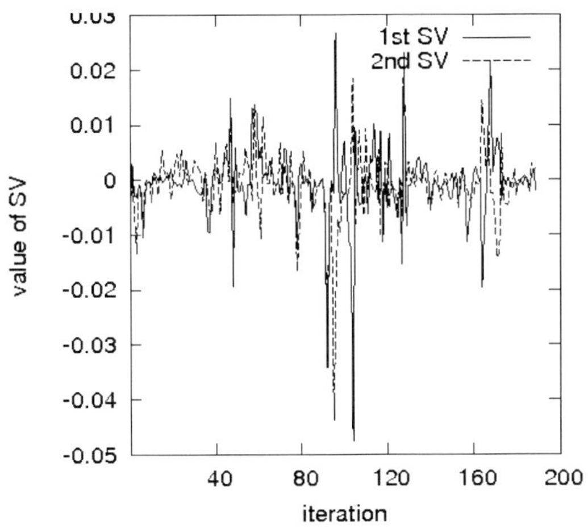

Figure 10. The value of the 1st SV and 2nd SV at the grid point (50,50,1). As for the magnitude there is no convergence and the SVs are non-stationary.

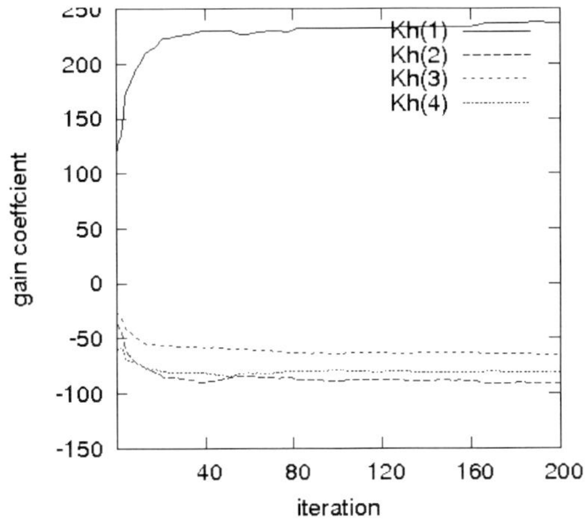

Figure 11. The gain coefficients $K_h(l), l = 1, ..., 4$ calculated from KVs simulated by SP1. They converge very quickly after about 80 iterations independently on a strong time-varying character of simulated KVs.

off-line $M(0)$ and on-line $M(k)$ in order to creat a new ECM $M_c(k)$,

$$M_c(k) = cM(0) + (1 - c)M(k), c \in [0 : 1] \tag{41}$$

The filter gain now is updated as done in (40) subject to $M_c(k)$. In (41) the parameter c reflects our confidence on the PE samples, developed by the model; the confidence on PE samples simulated from the current estimated system state is given by $(1 - c)$. The parameter c may be time-varying. For example, letting $c = c(k) = 1/k$ signifies that the information on $M(0)$ will be forgotten as more and more observations will be received and introduced in the filter.

Another way is proposed in [12] which has the objective to use the horizontal structure of the PE samples,

$$M(k) = U^{[L]}U^{[L],T}, U^{[L]} := [\delta h_p^1, ..., \delta h_p^L] \tag{42}$$

Combining (41) with (42) results in a new matrix $M_c(k)$ for which a simple formula like (40) is not valid for the computation of the gain. For a more economic way to compute the correction $K\zeta(k)$, see [12].

6.5. Adaptive Filter

There are different ways to parametrize the gain in order to make the adaptation as done in the previous section. For more details, see [14]-[15]. For a given ECM M, let us decompose $M = P_r P_r^T$ and then the parametrization can be done as follows

$$K_h = Pr\Theta K_e, K_e = He^T[H_e He^T + R]^{-1}, H_e := HP_r,$$
$$\Theta = diag[\theta(1), ..., \theta(N_l)]. \tag{43}$$

The optimal parameters $\theta(i)$ are tuned according to (33) to minimize the objective function (34).

6.6. Numerical Results

6.6.1. The PEF and Cooper-Haines Filter (CHF)

To see how works efficiently the PEF, we have applied, along with the PEF, the CHF [5] to solve the assimilation problem. That filter (CHF) is widely used in oceanic data assimilation and is reputed as an efficient and robust filter. It is constructed on the basis of the hypothesis on conservation of the linear potential vorticity. The gain in the CHF has the structure (40) and from the mentioned above hypothesis it implies

$$K_h = [(1-\alpha), 0, 0, \alpha)]^T \otimes I_p \tag{44}$$

For the present MICOM model, $\alpha = -184.964966$. Table 9 presents the estimation errors resulting from the CHF and PEF. Here $e_h(p)(m)$(or $e_h(f)(m)$ ...) signifies the predicted RMSE (denoted as RMS-PE) and filtered RMSE (RMS-FE) for the variable h with the unit in m. In the first series of experiments, the observations are available at all grid points on the surface. One sees that the PEF has reduced the SSH RMS-PE by about 60 % compared with that of the CHF. The reduction is of order 50 % (for both RMS-PE and RMS-FE) for the layer thickness. As to the velocity errors, the reduction is by order 40 %. To illustrate a more realistic situation when the along track SSH observations are available, Table 10 displays the errors produced by the CHF, PEF and PEF-on (which uses the on-line SP procedure subject to $L = 2$). The observations now are available only at the points

Table 9. Estimation errors

Variables	CHF	PEF	reduction
$ssh(p)(cm)$	6.24	2.71	57 %
$e_h(p)(m)$	8.61	4.12	52 %
$e_h(f)(m)$	7.74	3.74	51 %
$e_u(p)(cm/s)$	5.84	3.37	42 %
$e_u(f)(cm/s)$	3.46	2.03	41 %
$e_v(p)(cm/s)$	5.42	2.97	45 %
$e_v(f)(cm/s)$	3.47	2.11	39 %

Table 10. Estimation errors

Variables	CHF	PEF	APEF	PEF-on
$e_h(p)(m)$	10.24	6.86	6.65	7.076
$e_h(f)(m)$	9.47	7.13	6.96	6.86
$e_u(p)(cm/s)$	7.19	5.77	4.87	4.92
$e_u(f)(cm/s)$	5.99	4.77	4.28	4.35
$e_v(p)(cm/s)$	7.25	5.85	5.12	5.09
$e_v(f)(cm/s)$	6.34	5.20	4.6	4.62

$i_o = 1, 11, ..., 131; j_o = 1, 11, ..., 171$. One sees that with a less number of observations, the PEF still behaves much better than the CHF but with a less error reduction. The filter PEF-on in the last column of Table 10 denotes the PEF with the gain computed on the basis of (41)(42) subject to $\alpha = 0.3$. The columns of $U^{[L]}$, $L = 2$ consist of two SVs simulated at each assimilation instant k. It is seen that the filter PEF-on allows to improve considerably the performance of the PEF. The experiments with $L > 2$ do not allow to produce a more significant improvement on the filter performance.

6.6.2. PEF and APEF

In Table 10 the estimation errors produced by the APEF are also displayed. It is seen that the APEF is capable of reducing the PE and FE errors for all variables. The time averaged PE for the SSH and total velocity (u, v) of the CHF, PEF and APEF are plotted in Figs 12 and 13. One finds that if in the CHF the PE decreases from the beginning up to $k = 50$, this tendency is reversed after this period. In contrast, in the PEF and APEF, the PE decreases during all assimilation period. It is interesting to note that if the adaptation allows for decreasing more quickly the SSH PE in the APEF (compared to the PEF) after about 10 iterations, this effect happens later, only after 20 iterations, for the velocity error.

7. Summary and Conclusion

We have presented two subspace methods for generating the PE samples which serve as important sources of information for the construction of the gain subspace. The sampling

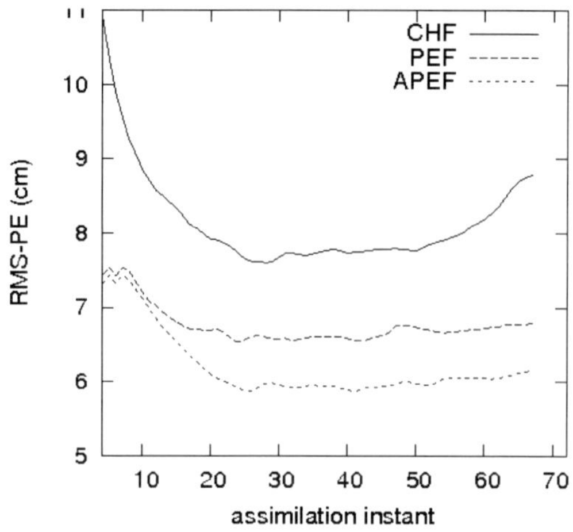

Figure 12. Time averaged RMSE of the innovation produced by the CHF, PEF and APEF. One sees (i) the PEF behaves much better than the CHF; (ii) the adaptation has a very positive impact on minimizing the PE of the innovation.

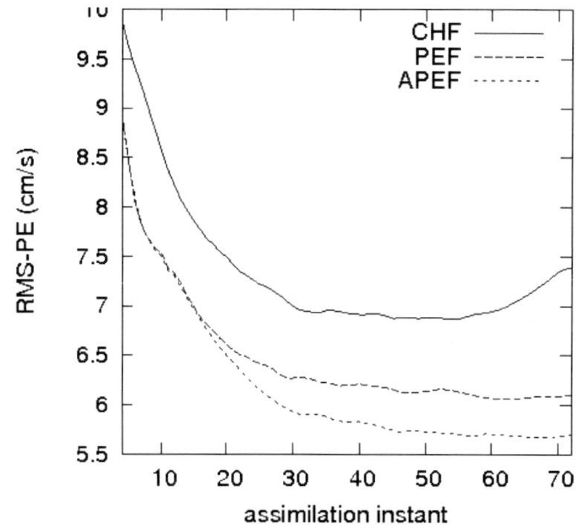

Figure 13. Time averaged RMS-PE for the total (u, v) velocity produced by the CHF, PEF and APEF. In the CHF, after the first period of error decreasing, the error restarts to increase after about 50 iterations.

procedure is introduced with the objective to estimate the ECM-PE that is impossible to compute by following the optimal filtering theory. For very high dimensional systems, such difficulty naturally occurs due to unrealistically high computational cost and huge memory requirement. It is undoubted that this sampling approach is in some sense closely related to the Monte-Carlo method and particle filtering. The difference with the latters lies in the way the PE samples are generated: the subspace methods produce samples not by adding random perturbations to the current system state as the others do, but consciously simulate errors to occur in the directions of the fastest growing PEs. This allows us to make enormous savings in computational cost and memory requirements, and to ensure a high performance of the filter.

As shown from the simulated experiments, the samples produced by the KrSS or ScSS are very informative in the sense that only a few PE samples are sometimes sufficient to produce good estimates for the filter gain. If for the system with moderate dimension, involving more PE samples in the construction of the gain is proved to be beneficial for improving the filter performance (section 5), no such benefice is observed for the high dimensional system in section 6. This happens, of course, under condition that the ECM is sufficiently well parameterized (as in the form (38), for example). In fact each element of the ECM (40), representing the covariance of the layer thickness PE between two layers, is estimated as a spatial average of such covariance computed over all horizontal grid points. As this number is very large (it is equal to 25200), such obtained estimate is very robust with respect to adding more PE samples.

From the presented experiments, with the performances of filters, constructed on the basis of the KrSS and ScSS, one can conclude (with prudence) that the KrSS is preferable to the ScSS due to its less computational cost and relatively high quality of produced PE samples. This conclusion is drawn solely from the implemented experiments and more research is needed to confirm these findings. In this context, the question on how one can select the more informative PE samples from the KrSS is of great importance. As an additional confirmation on the role of the KrSS in the design of the PEF, we have successfully implemented the PEF, APEF for the HYCOM model (see section 3.1), in the coastal Bay of Biscay configuration, to assimilate the real observations (SSH, sea surface temperature, in-situ data for temperature and salinity). The high-quality estimation results, produced by these filters, confirm once more the role of the PE samples generated by the KrSS.

Another important fact is also noteworthy: the implementation of the ROAF does not require the solution of a nonlinear matrix equation like an ARE, whose implementation is out of question for very high dimensional systems. In this context the role of adaptation is remarkable as a method to compensate for the loss of filter performance due to a lack of the ARE in the PEF: by an appropriate parametrrization of the gain, the adaptation is proved to be a very useful and efficient tool for improving the filter performance. This is well demonstrated from the experiments with both moderate and high dimensional systems in sections 5-6 : the AFs are capable of dramatically reducing the estimation errors produced by the non-adaptive filters and occasionally, can yield the performance comparable with that of the optimal KF.

References

[1] M. S. Arulampalam, S. Maskell, N. Gordon, and T. Clapp, A Tutorial on Particle Filters for Online Nonlinear/Non-Gaussian Bayesian Tracking, *IEEE Trans. Signal Proc.*, Vol. 50, No 2, (2002), pp. 174-188.

[2] M. Assenbaum, S. Casitas, S. Correard , P. Craneguy and S. Raynaud, Simulation Gascogne-Manche et Ouest-Portugal-Gibraltar, *SHOM Internal report*. Contract funded by French Ministry of Defense, No 04.87.011.00.470.29.55, 2008.

[3] Y. Baram and G. Kalit, Order Reduction in Linear State Estimation under Performance Constraints. *IEEE Trans. on Automatic Control*, AC-32, 11, 1987.

[4] S. Cohn and R. Todling, Approximate Data Assimilation Schemes for Stable and Unstable Dynamics, *J. Meteorol. Soc. Japan*, 74, No. 1, (1996), pp. 63-75.

[5] M. Cooper and K. Haines, Altimetric assimilation with water property conservation. *J. Geophys. Res.*, 101 (1996), pp. 1059-1077.

[6] G. Evensen, The ensemble Kalman filter: Theoretical formulation and practical implementation, *Ocean Dynamics*, 53 (2003), pp. 343-367.

[7] R. Fitzgerald, Divergence of the Kalman filter. *IEEE Trans. Autom. Contr.*, 16, (1971), pp. 736-747.

[8] M. Ghil and P. Manalotte-Rizzoli, Data assimilation in meteorology and oceanography. *Adv. Geophys.*, 33 (1991), pp. 141-266.

[9] G.H. Golub and C.F. Van Loan, *Matrix Computations*, 3rd edn, Johns Hopkins University Press, 1996.

[10] M.J. Grote and A.J. Majda, Stable time filtering of strongly unstable spatially extended systems. *PNAS*, Vol. 103, no 20, (2006), pp. 7548-7553.

[11] T.M. Hamill, 2006: Ensemble-based atmospheric data assimilation, in *Predictability of Weather and Climate*, Ed. T. Palmer, Cambridge Press, 2006, pp. 124-156.

[12] H.S. Hoang, R. Baraille, "Prediction error sampling procedure based on dominant Schur decomposition: Application to state estimation in high dimensional oceanic model". *J. Appl. Math. Comput.*, 218, (2011) 3689-3709.

[13] H.S. Hoang, R. Baraille and O. Talagrand, On an adaptive filter for altimetric data data assimilation and its application to a primitive equation model MICOM, *Tellus*, 57A (2005), no 2, 153-170.

[14] H.S. Hoang H.S., P. De Mey, O. Talagrand and R. Baraille, A new reduced-order adaptive filter for state estimation in high dimensional systems, *Automatica*, 33 (1997), pp. 1475-1498.

[15] H.S. Hoang, O. Talagrand and R. Baraille, On the design of a stable adaptive filter for high dimensional systems. *Automatica*, 37 (2001), pp. 341-359.

[16] H.S. Hoang, O. Talagrand and R. Baraille, On the stability of a reduced-order filter based on dominant singular value decomposition of the system dynamics. *Automatica*, Vol. 45, (2009), pp. 2400-2405.

[17] A.H. Jazwinski, *Stochastic Processes and Filtering Theory*, AP,NY, 1970.

[18] M.K. Schneider and A.S. Willsky, Krylov subspace estimation, *SIAM J. Sci. Comput.*, Vol. 22, No 5, (2001), pp. 1840-1864.

In: Dynamical Systems
Editor: Jean Michel Tchuenche, pp. 83-133

ISBN: 978-1-62808-001-8

Chapter 4

High-Fidelity Modeling and Vibration-based System Identification and Damage Inspection of Flexible Structures

P. Frank Pai
Mechanical and Aerospace Engineering Dept.,
University of Missouri, Columbia, Missouri, US

Abstract

Structural systems are often built with unwanted/wanted nonlinearities, damage introduces additional nonlinearities, and aging aggravates nonlinearities. Hence, high-fidelity nonlinear modeling is needed in development of techniques for system identification and damage inspection of structures. This chapter presents a geometrically exact modeling technique that can fully account for geometric nonlinearities of beams by using Jaumann strains, exact coordinate transformations, and orthogonal virtual rotations. The derivations are presented in detail, and fully nonlinear governing equations and boundary conditions are presented. For damage inspection and system identification of thin-walled structures, this chapter presents a dynamics-based methodology that combines a boundary-effect evaluation method (BEEM) for space-wavenumber analysis of measured operational deflection shapes (ODSs) and a conjugate-pair decomposition (CPD) method for time-frequency analysis of time traces of measured points. BEEM can locate and estimate structural damage by processing ODSs measured by a full-field measurement system (e.g., a scanning laser vibrometer or a camera-based motion measurement system). BEEM is a nondestructive spatial-domain method based on sliding-window processing of ODSs and it works without using any structural model or historical data for comparison. Similar to the short-time Fourier transform and wavelet transform, CPD uses adaptive windowed regular harmonics and function orthogonality to perform time-frequency analysis of time traces by extracting time-localized regular/distorted harmonics. Moreover, solutions from perturbation analysis of nonlinear dynamical systems provide excellent guidance and formulas for quantitative estimation of nonlinearity and system parameters. Both BEEM and CPD are local spectral analysis based on adaptive curve fitting. The first estimation of the wavenumber for BEEM and the frequency for CPD can be obtained using an enhanced five-point Teager-Kaiser algorithm based on finite difference. Numerical examples are used to demonstrate how to use high-fidelity structural models to help the development of techniques for system identification and damage inspection. Numerical simulations and experimental results show that the combination of BEEM and CPD for space-wavenumber and time-frequency analyses provides an accurate and efficient tool for damage inspection and system identification of thin-walled structures. Moreover, current status and remaining challenges in modeling and vibration-based system identification and damage inspection of flexible structures are also addressed.

Keywords: High-fidelity structural modeling, dynamics-based damage inspection and system identification, full-field measurement techniques, time-frequency analysis, space-wavenumber analysis

I. Introduction

Structural systems are often built with unwanted/wanted nonlinearities, damage introduces additional nonlinearities, and aging aggravates nonlinearities [1-14]. Large elastic deformations introduce geometric nonlinearity, deformation-dependent material properties result in material nonlinearity, and other nonlinearities can be caused by backlash, clearances between mounting brackets, geometric constraints on deformation, misalignment of substructures, dry friction, and many types of nonlinear hysteretic damping, including aerodynamic damping, damping of shock absorbers for vehicles, and material damping of shape memory alloys and other materials [1-6]. Notorious nonlinear vibrations include wing and panel flutters, tail buffeting, limit-cycle oscillation of wings with attached weapons, and amplitude-dependent natural frequencies of pylon-store-wing assemblies and tail-fin assemblies in aircraft engineering [15-18]; low-frequency large-amplitude vibrations of solar panels of satellites, membrane wrinkling of space structures, complex deployment dynamics of deployable/inflatable structures, multiple possible deployed geometries of large scientific balloons, and odd opening modes of parachutes for landing space vehicles in inner- and outer-space engineering [4,6]; large-amplitude vibrations of high-tower buildings under gusty wind, flutter of suspended bridges, and whirling of power and communication cables in civil engineering [1-3,8-11]; and dynamics of leaf springs and shock absorbers of cars, brake squeal of cars, dynamics of microbeams in MEMS devices, and chattering of machining processes in mechanical engineering [1,2,5,6,19,20]. Moreover, highly flexible deployable/inflatable structures have been used in many space structural systems, advanced vehicles, mechanical and civil structures, and medical devices and implants in order to reduce weight/space, increase speed/efficiency, and/or provide special mechanisms [4,6]. Such flexible structures are built with wanted geometric nonlinearities.

For a regular on-earth one-dimensional (1D) or 2D structure, its dynamic response is often pre-assumed to be *linear* if the vibration amplitude is less than its *thickness*. In perturbation analysis of geometrically nonlinear structures, governing equations are expanded into polynomials in terms of displacement variables and then defines the highest power of displacement variables as the order of nonlinearity [1-3,5]. This also implies the use of vibration amplitude to define the magnitude of nonlinearity. However, recent study [21] revealed that it is more appropriate to define linear response to a harmonic excitation by giving a limit on the response's *continuous frequency bandwidth* instead of the *vibration amplitude* because nonlinear response can be due to a large vibration amplitude, boundary constraints, loading conditions, and other factors. Different combinations of these factors may result in different nonlinear responses even if the displacement is smaller than the structure's thickness. For example, a hinged-hinged beam behaves more nonlinearly than a clamped-free beam under the same magnitude of vibration amplitude [21]. Moreover, a high-frequency mode starts nonlinear behavior at an amplitude smaller than that of a low-frequency mode because of high curvature. Furthermore, numerical and experimental results show that a simple 1D or 2D structure with simple boundary conditions (e.g., clamped-free conditions) can behave linearly even when its transverse displacement is more than 10 times its thickness [6]. Hence, geometrically exact high-fidelity modeling without Taylor expansion is really needed in order to use such structural theories for static/dynamic simulations and for development of advanced techniques for system identification, damage inspection, and health monitoring.

Recent great advancement in sensor technology enables full-field non-contact measurements of dynamical systems, and hence fast full-field vibration-based system identification and damage inspection of large mechanical systems becomes an exponentially growing area of research and development. However, success of this engineering practice highly depends on the use of high-fidelity system modeling and identification techniques. *High-fidelity modeling technique* is needed for simulation-based recognition of key system parameters and damage indicators in order to develop system identification techniques, and *high-fidelity system identification technique* is needed for accurate extraction of dynamic characteristics and damage indicators and accurate estimation of system parameters. However, challenging issues exist, as presented next.

1. Full-Field Noncontact Measurement

Recent developments in full-field non-contact measurement technologies stimulate further advancement of dynamics-based system identification and damage inspection of large structures because fast and accurate measurements of time-varying full-field operational deflection shapes (ODSs) are valuable for such engineering applications [6]. Two major recent systems for full-field measurements are scanning laser vibrometers and camera-based motion analysis systems. The measured data from such a system contain detailed space- and time-domain dynamic characteristics of the measured structure, but the amount of data is often huge. A damaged structure shows discontinuities in its ODSs and spatial derivatives of ODSs, and nonlinearities and intermittent transient events exist in the time traces of measured points. Because damage inspection of structures is a difficult reverse engineering task, how to efficiently and accurately extract concise, clear, and unique system dynamic characteristics and health information from such a large set of data is very challenging. Hence, advanced signal processing techniques play the key role in using such full-field measurement systems for vibration-based damage inspection and system identification of large structures.

Laser scanners and other high-resolution shape capture techniques do not operate effectively on fast moving objects because they are based on line-by-line or even just point-by-point scanning. A scanning laser vibrometer (SLV) is believed to be the most accurate device for measuring small structural vibrations in today's metrology market. For example, the Polytec PSV-200 single-head SLV shown in Figure 1 can provide non-contact (measuring velocities of a dynamical system by checking the frequency shift of a back-scattered laser beam), remote (up to $30m$ away), large area scanning (up to $40^\circ \times 40^\circ$), dense (up to 512×512 points), high frequency bandwidth ($0.2Hz-20MHz$), and accurate (a velocity resolution of $0.1\mu m/s$) measurements, but the SLV is not suitable for measurement of flexible multibody systems [6]. The reasons are 1) it cannot measure static deformations, 2) it can only measure velocities parallel to the zero-degree laser beam, 3) it can measure the operational deflection shape of a structure only if the structure is undergoing a periodic motion with a known period, 4) its measurement accuracy is significantly reduced by spectral noise caused by non-smooth surface, large dynamic deflection, and reflection of the laser beam, and 5) the retro-reflective paint or tapes that are usually used to enhance the back-scattered laser light and to reduce the spectral noise may significantly change the dynamic characteristics of a flexible structure (e.g., a membrane) under testing. If the structure's vibration is periodic with a period T and the recording at each location is trigger-controlled to begin at nT (n is an integer) after the beginning recording time of the previous measurement point, the velocity profile at time $t = t_k$ can be assumed to be the distribution of the measured velocities of all measurement points at t_k. In other words, the velocity profiles can be constructed only if the vibration is periodic and the period is known. Moreover, when the structure moves significantly, the laser beam does not shoot at the same point, especially when the structure undergoes large transient rotations. Furthermore, this effect may introduce artificial nonlinear phenomena to the obtained data. On the other hand, a videogrammetry system simultaneously measures the instant 3D coordinates of all markers, and hence can provide true time-varying operational deflection shapes.

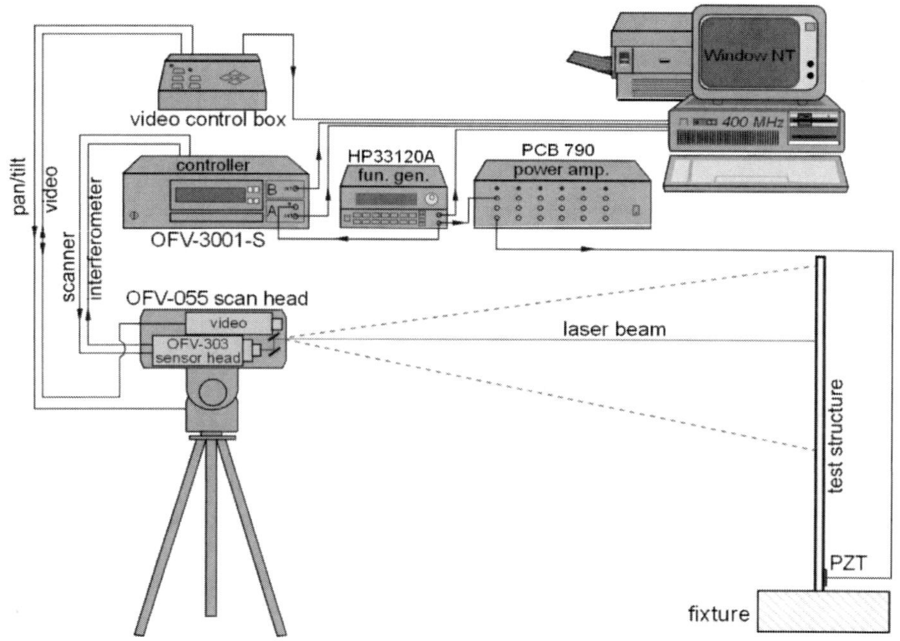

Figure 1. A Polytec PSV-200 scanning laser vibrometer.

Videogrammetry combines the functions of photogrammetry, computer vision, and digital image processing [22]. Photogrammetry is a technique for extracting 3D coordinates of object points from their two or more photographs taken from different angles and/or locations. A functional model of the imaging system in photogrammetry can be derived using condition equations that relate object points, image points, and imaging system parameters. The four condition equations are projectivity, collinearity, coplanarity, and scale restraint equations [22-25]. Several videogrammetry systems are commercially available today, but they are mainly designed for motion tracking of large multibody systems undergoing large motions. Applications of such systems have been used in production of movies, development of video games, modeling of complex human movement, locomotion analysis and lameness detection of animals, study of kinematics of sports, broadcast, stage production, and entertainment.

For example, Figure 2 shows an Eagle-500 motion analysis system set-up for measuring nonlinear vibrations of spinning flexible blades. The system uses 8 CMOS cameras to capture pictures of a structure when 8 visible red digital *light-emitting diode* (LED) strobes light up retro-reflective markers that are adhered on the structure. The cameras and strobes are synchronized to work at a speed between 0.1 to 2000 frames per second (FPS). For a frame rate between 0.1 and 500 FPS, a full resolution of 1280×1024 pixels is used. For a frame rate within 500~1000 (1000~2000) FPS, a resolution of 1280×512 (1280×256) pixels is used. Using triangulation techniques and the known focal lengths (after calibrations using an L-frame with 4 precisely located spherical markers and a T-wand with 3 markers) of the cameras and the known image coordinates of the bright points (caused by the retro-reflective markers) on the CMOS sensors inside the cameras, the Eagle-500 signal processing software computes and records the instant 3D object coordinates of the center of each retro-reflective marker that is seen by at least two cameras [26]. Hence, 3D time traces of all markers are available for performing dynamic animation using stick figures. Because the 3D coordinates of each marker are estimated by optimization using *bundle adjustment* [22-24] when two or more cameras see the marker, the measurement accuracy of the system is enough for dynamic testing of multi-body systems [25,26].

The often required measurement accuracy for most current applications of camera-based motion analysis systems is low, and such systems are often designed to track spherical markers of large sizes (e.g., 5 mm to 50 mm). *For vibration testing of structures, such large spherical markers are not appropriate because they are not directly adhered to structural surfaces and they cause serious complicated aerodynamic effects*. Furthermore, the measurement accuracy of such videogrammetry systems is not high enough for accurate vibration testing of structures because of several reasons. (1) The image resolution and frame rate of such cameras are low. (2) Each camera's parameters change with setup conditions, service time, and environments, and hence fast and accurate on-site calibration is needed but difficult. (3) Photogrammetry software developed for such systems are often for general purposes instead of structural testing. (4) Users without certain understanding of photogrammetry cannot appropriately setup such a system for high-accuracy measurements. Hence, to advance videogrammetry for vibration testing of structures and other deformable physical systems, special developments are needed in camera calibration, design of object markers, methodology for setting up a set of cameras, image processing, and identification and tracking of markers from 2D digital images [6,26].

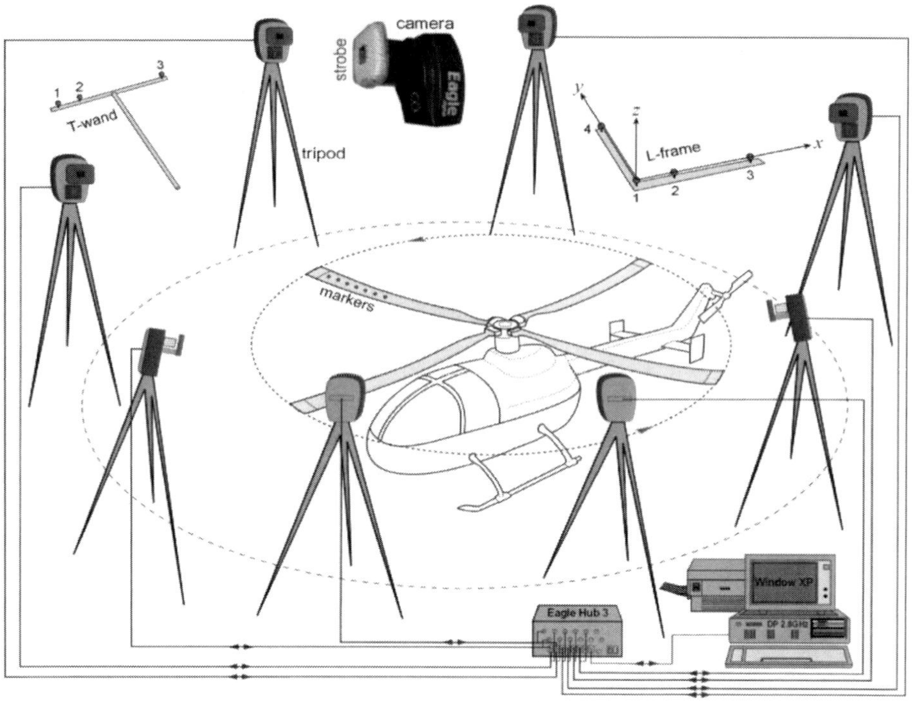

Figure 2. An EAGLE-500 eight-camera motion analysis system.

2. Geometrically Exact Modeling

Because a highly flexible structure (HFS) is designed to undergo large elastic deformations without plastic straining, it always springs back to its undeformed geometry. Moreover, because fully nonlinear total-Lagrangian strains allow structural modeling without using any floating coordinate systems for reference, a total-Lagrangian formulation with the use of fully nonlinear objective strain-displacement relations is most appropriate for modeling and analysis of HFSs [6]. However, the challenging tasks in deriving geometrically exact structural theories for HFSs include: 1) how to derive fully nonlinear strain-displacement relations that accurately account for large rotations and exclude rigid-body effects, 2) how to relate the chosen strains to their work-conjugate stresses using material constants that can be obtained from commonly performed experiments using engineering strain and stress measures, and 3) how to relate the governing equations derived using an energy approach (i.e., the extended Hamilton principle) to those derived using a vector approach (i.e., Newton's second law) [6].

Fully nonlinear total-Lagrangian strain-displacement relations are difficult to derive because large rotations are not vector quantities and are difficult to be related to displacements' spatial derivatives [27]. For exact description of large 3D rotations of a beam, one can use three Euler angles, two Euler angles, one principal rotation angle plus three direction cosines of the principal rotation axis, four Euler parameters, three Rodrigues parameters, and three modified Rodrigues parameters [27]. Unfortunately, such rotational variables are either sequence-dependent and/or spatially discontinuous because they are not mechanics-based variables. Hence, they are not appropriate for use as nodal degrees of freedom in total-Lagrangian finite-element modeling, and it is difficult to apply boundary conditions on such discontinuous and/or sequence-dependent rotational variables. This chapter will introduce mechanics-based variables for modeling of large rotations without singularity.

Green-Lagrange strains are often used in the literature and commercial finite-element codes to account for geometric nonlinearities in 1D and 2D structural theories [28-32]. It is mainly because their strain-displacement relations can be easily derived using the deformation gradient tensor without solving for the principal stretch ratios and their directions [6]. Unfortunately, because Green-Lagrange strains and their work-conjugate stresses (i.e., 2nd Piola-Kirchoff stresses) are energy (instead of geometric) measures, it is difficult to determine appropriate, constant material properties for use and to implement boundary conditions. Moreover, when Green-Lagrange strains are used, three Rodrigues parameters (or four Euler parameters) are needed for modeling large rotations. Unfortunately, because these math-based rotational variables cannot be related to spatial derivatives of displacement variables and have cusp problems (caused by discontinuous spatial derivatives), their relations to geometric boundary conditions are difficult to be derived and solved and they can result in artificial strain values. The most serious problem is that, when Green-Lagrange strains are used, the governing equations derived using an energy approach are different from those derived using a vector approach [6].

Here we demonstrate the derivation of a geometrically exact displacement-based total-Lagrangian beam theory that uses two mechanics-based rotational variables and three displacement variables without using any Euler angles. The beam theory fully accounts for geometric nonlinearities due to large rotations, large axial strain, initial curvatures, and transverse shear deformations by using Jaumann stress and strain measures, exact coordinate transformations, and the concept of orthogonal virtual rotations. The Jaumann strains are derived using a new concept of local displacements without polar decomposition analysis. A locally corotated reference frame is defined using the symmetry of Jaumann strains, and Jaumann strains are proved to be locally corotated, objective, geometric strains [6]. Because Jaumann strains are corotated engineering strains, Jaumann stresses are related to Jaumann strains by the material constants commonly obtained in experiments using small engineering stress and strain measures. Moreover, the governing equations derived using an energy approach are exactly the same as those derived using a vector approach. Furthermore, there is no need of coordinate transformation before updating strains, stresses, and displacements because it is a total-Lagrangian formulation with stresses and strains corrotated with the deformed structural geometry. Since highly flexible structures have thin thickness and hence transverse shear effects are usually small, we use an energy-consistent first-order shear-deformation theory to account for transverse shear strains in order to have efficient computation without significant loss of accuracy.

3. Damage Inspection and System Identification

For several decades structural engineers have been struggling with the development of dynamics-based methods for rapid damage inspection of large thin-walled structures [8-13]. Dynamics-based damage detection methods can be separated into *three groups* based on the complexity of sensor systems and signal processing methods being used. *Methods in the first group* require a full-field measurement method/tool (e.g., Moire interferometry, digital shearography, scanning laser vibrometers, and camera-based motion measurement systems), and they process the measured displacement, slope, or velocity field to compute curvatures and/or strains and then locate damage by examining abnormality or sudden change of these spatially distributed data [11,33,34]. *Methods in the second group* require simultaneous measurements of many points, and a well-calibrated structural model and a modal expansion/update method are used to locate damage [12,35]. *Methods in the third group* require simultaneous measurements of only few locations, process the measurement points' time traces to extract dynamic characteristics (i.e., natural frequencies, damping ratios, wave propagation characters, and nonlinear effects) to reveal the existence of damage, and use the traveling sequence of abnormality to locate damage [10,36-38]. The third approach with *time-of-flight analysis* is commonly used to locate damage in work, and it is more economical and practical than other methods for online monitoring of structural health. Except built-in nonlinearities, damage introduces additional nonlinearities, and aging aggravates nonlinearities of structures. Hence, dynamics of structural systems is often nonlinear [7-14]. For example, dry friction between fractured surfaces of a structure is a nonlinear effect, and the breathing of a crack causes nonlinear intermittent transient response. Also, demountable/retractable civil structures are built with nonlinearities caused by loose joints, and the usage/aging aggravates such nonlinearities. Because parametric models of such nonlinearities are often difficult to determine, non-parametric system identification is more feasible for such systems. Hence, the challenges for the third approach are how to extract dynamic characteristics from nonlinear noise-contaminated signals and how to correlate the responses measured at just few physical locations to derive damage indicators and locate damages.

Hence, we recommend the use of operational deflection shapes from the first approach for full-field damage inspection and pointwise time traces from the third approach for online health monitoring and system identification. For full-field damage inspection we introduce a boundary-effect evaluation method (BEEM). For system identification and online health monitoring we introduce a conjugate-pair decomposition (CPD) method. In this chapter we also show that time-frequency analysis is most appropriate for nonlinearity identification, and propose a signal processing and identification method that combines time-frequency analysis and perturbation solutions for parametric and non-parametric identification of nonlinear dynamical systems.

II. Geometrically Exact Modeling Technique

Here we show how to derive a geometrically exact beam theory that can account for arbitrarily large elastic deformations. One can follow the same approach to derive for highly flexible 2D structures [6].

An initially curved beam undergoing large deformation requires three coordinate systems to describe its movement, as shown in Figure 3a. The *abc* is a fixed rectangular coordinate system used for reference, the *xyz* is a fixed orthogonal curvilinear coordinate system used to describe the fixed undeformed beam geometry, and the $\xi\eta\zeta$ is a moving orthogonal curvilinear coordinate system fixed on the moving deformed beam. Let $\mathbf{i}_a$, $\mathbf{i}_b$, and $\mathbf{i}_c$ be the unit vectors of the *abc* system; $\mathbf{i}_x$, $\mathbf{i}_y$, and $\mathbf{i}_z$ be the unit vectors of the *xyz* system; and $\mathbf{i}_1$, $\mathbf{i}_2$, and $\mathbf{i}_3$ be the unit vectors of the $\xi\eta\zeta$ system. Moreover, *u*, *v*, and *w* represent the absolute displacements of the observed reference point *O* with respect to (w.r.t.) the *x, y*, and *z* axes, respectively, and *s* denotes the undeformed arc length along the reference line starting from the beam root. Figure 1b shows that the displacement vector $\mathbf{D}$ of an arbitrary point on the observed cross section consists of a rigid-body motion that moves a rectangle of side lengths *y* and *z* on the *yz* plane to that on the $\eta\zeta$ plane and a small local relative displacement vector $\mathbf{u}$ w.r.t. the $\eta\zeta$ plane. The $\mathbf{u}$ accounts for out-of-plane shear and torsional warpings and in-plane warping due to Poisson's effect and others. For the Euler-Bernoulli beam theory, out-of-plane warpings are neglected.

The two coordinate systems *abc* and *xyz* are related by a transformation matrix $[T^o]$ as

$$\{\mathbf{i}_{xyz}\} = [T^o]\{\mathbf{i}_{abc}\},\quad \{\mathbf{i}_{xyz}\} \equiv \begin{Bmatrix} \mathbf{i}_x \\ \mathbf{i}_y \\ \mathbf{i}_z \end{Bmatrix}, \{\mathbf{i}_{abc}\} \equiv \begin{Bmatrix} \mathbf{i}_a \\ \mathbf{i}_b \\ \mathbf{i}_c \end{Bmatrix} \tag{1}$$

One can use Eq. (1) and the orthonormality of $\mathbf{i}_x$, $\mathbf{i}_y$ and $\mathbf{i}_z$ to obtain

$$\{\mathbf{i}'_{xyz}\} = [k]\{\mathbf{i}_{xyz}\},\quad [k] \equiv \begin{bmatrix} \mathbf{i}'_x\cdot\mathbf{i}_x & \mathbf{i}'_x\cdot\mathbf{i}_y & \mathbf{i}'_x\cdot\mathbf{i}_z \\ \mathbf{i}'_y\cdot\mathbf{i}_x & \mathbf{i}'_y\cdot\mathbf{i}_y & \mathbf{i}'_y\cdot\mathbf{i}_z \\ \mathbf{i}'_z\cdot\mathbf{i}_x & \mathbf{i}'_z\cdot\mathbf{i}_y & \mathbf{i}'_z\cdot\mathbf{i}_z \end{bmatrix} \equiv \begin{bmatrix} 0 & k_3 & -k_2 \\ -k_3 & 0 & k_1 \\ k_2 & -k_1 & 0 \end{bmatrix} = [T^0]'[T^0]^T \tag{2}$$

where $(\ )' \equiv d(\ )/ds$, and $[T^0]^{-1} = [T^0]^T$ because $[T^o]$ is a unitary matrix. The k_1, k_2, and k_3 are initial curvatures w.r.t. axes *x*, *y*, and *z*, respectively. $[T^0]$ and k_i can be calculated from the known undeformed beam geometry defined w.r.t. the rectangular coordinate system *abc*.

The deformed reference axis ξ can be exactly and explicitly described by displacements *u*, *v* and *w*, as shown next. In Figure 3b, if $\overline{OP} = ds$, the corresponding deformed vector $\mathbf{\bar{O}\bar{P}}$ can be used to define $\mathbf{i}_1$ as

$$\mathbf{i}_1 \equiv T_{11}\mathbf{i}_x + T_{12}\mathbf{i}_y + T_{13}\mathbf{i}_z = \frac{\mathbf{\bar{O}\bar{P}}}{(1+e)ds} = \frac{\mathbf{i}_x + u'\mathbf{i}_x + v'\mathbf{i}_y + w'\mathbf{i}_z + u\mathbf{i}'_x + v\mathbf{i}'_y + w\mathbf{i}'_z}{1+e}$$
$$T_{11} = \frac{1+u'-vk_3+wk_2}{1+e},\ T_{12} = \frac{v'+uk_3-wk_1}{1+e},\ T_{13} = \frac{w'-uk_2+vk_1}{1+e} \tag{3a}$$

$$e = \sqrt{(1+u'-vk_3+wk_2)^2 + (v'+uk_3-wk_1)^2 + (w'-uk_2+vk_1)^2} - 1 \tag{3b}$$

$$\{\mathbf{i}_{123}\}=[T]\{\mathbf{i}_{xyz}\},\ [T]\equiv\begin{bmatrix}T_{11} & T_{12} & T_{13}\\ T_{21} & T_{22} & T_{23}\\ T_{31} & T_{32} & T_{33}\end{bmatrix},\{\mathbf{i}_{123}\}\equiv\begin{Bmatrix}\mathbf{i}_1\\ \mathbf{i}_2\\ \mathbf{i}_3\end{Bmatrix} \tag{4a}$$

where Eq. (2) is used, e is the axial strain on the ξ axis, and the expression of e is derived from the identity $\sqrt{T_{11}^2+T_{12}^2+T_{13}^2}=1$. The deformed system $\xi\eta\zeta$ can be related to the undeformed system xyz as

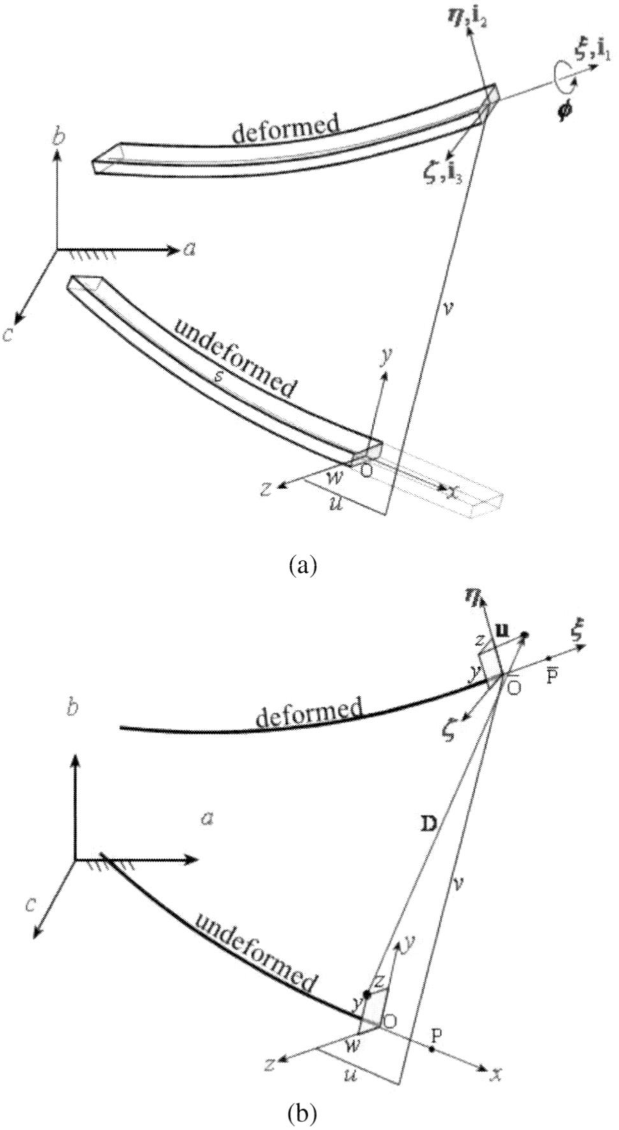

Figure 3. An elastic beam undergoing large rigid-elastic deformation: (a) three coordinate systems, and (b) displacement components.

$$T_{31} \equiv T_{12}T_{23} - T_{13}T_{22},\ T_{32} \equiv T_{13}T_{21} - T_{11}T_{23},\ T_{33} \equiv T_{11}T_{22} - T_{12}T_{21} \tag{4b}$$

Eq. (3a) shows that the direction cosines T_{1i} of the deformed reference axis ξ are fully described by u, v, w, u', v' and w'. Because $\mathbf{i}_3 = \mathbf{i}_1 \times \mathbf{i}_2$, when T_{1i} and $T_{2i} (i = 1, 2, 3)$ are known, T_{3i} can be obtained as shown in Eq. (4b). Since $|\mathbf{i}_2| = 1$ and $\mathbf{i}_1 \cdot \mathbf{i}_2 = 0$, we have

$$T_{22} = \pm\sqrt{1 - T_{21}^2 - T_{23}^2},\ T_{22} = \frac{-T_{11}T_{21} - T_{13}T_{23}}{T_{12}} \tag{4c}$$

The first one of Eq. (4c) determines the value of T_{22} and the second one determines the sign and the value of T_{22} for double checking. If $T_{12} = 0$, use the condition of smooth continuity of the T_{22} of two neighboring material points to determine the sign of T_{22}. Eq. (3b) enforces $T_{11}^2 + T_{12}^2 + T_{13}^2 = 1$, Eq. (4c) enforces $T_{21}^2 + T_{22}^2 + T_{23}^2 = 1$ and $\mathbf{i}_1 \cdot \mathbf{i}_2 = 0$, and Eq. (4b) enforces $\mathbf{i}_3 \cdot \mathbf{i}_1 = \mathbf{i}_3 \cdot \mathbf{i}_2 = 0$ and $T_{31}^2 + T_{32}^2 + T_{33}^2 = 1$. Hence, $[T]$ can be presented in terms of the eight variables u, v, w, u', v', w', T_{21} and T_{23}, and there are no singularity problems. Although the number of variables can be reduced to seven or even six, but each approach has its own singularity problems [27, 39].

For example, Figure 4 shows that, if the system *xyz* is rotated to the system $\xi\eta\zeta$ by three angles α_3, α_2, ϕ in sequence, we have

$$[T] = \begin{bmatrix} 1 & 0 & 0 \\ 0 & \cos\phi & \sin\phi \\ 0 & -\sin\phi & \cos\phi \end{bmatrix} \begin{bmatrix} T_{11} & T_{12} & T_{13} \\ -T_{12}/\sqrt{1-T_{13}^2} & T_{11}/\sqrt{1-T_{13}^2} & 0 \\ -T_{11}T_{13}/\sqrt{1-T_{13}^2} & -T_{12}T_{13}/\sqrt{1-T_{13}^2} & \sqrt{1-T_{13}^2} \end{bmatrix} \tag{5}$$

where α_3 rotates the axes x and y to $\bar{x}$ and $\bar{y}$, the α_2 ($|\alpha_2| \leq \pi/2$) rotates the axes $\bar{x}$ and z to ξ and $\bar{z}$, and the ϕ rotates the axes $\bar{y}$ and $\bar{z}$ to η and ζ. Eqs. (5) and (3a) show that $[T]$ can be presented in terms of the seven variables u, v, w, u', v', w' and ϕ. However, this beam theory is valid only if $|\alpha_2| \leq \pi/2$ because $\cos\alpha_2 = \sqrt{1-T_{13}^2}$ is used in Eq. (5). To extend the theory for $|\alpha_2| \leq \pi$ (i.e., the full range for any possible large rotations), one needs to consider the possibility of $\cos\alpha_2 = -\sqrt{1-T_{13}^2}$, which requires tracking of the actual, complicated rotation sequence of the three rotations [27]. Moreover, the $[T]$ matrix reveals that the singular points of this beam theory happen at $T_{13} = \pm 1$. Hence, this beam theory is only valid for modeling and analysis of rotor blades and other beam-like structures that mainly undergo large rigid-elastic deformations on the *xy* plane [40].

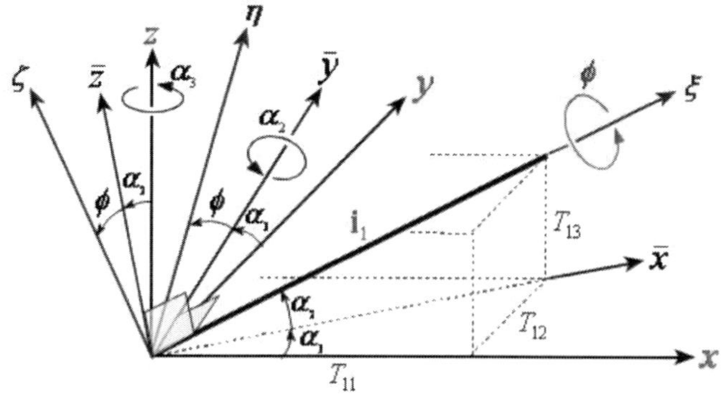

Figure 4. Three Euler angles and their kinematic relations.

After [*T*] is derived, one can use Eqs. (4a) and (2) and the identity $[T]^{-1}=[T]^T$ to obtain that

$$\{\mathbf{i}'_{123}\}=[K]\{\mathbf{i}_{123}\},\ [K]\equiv\begin{bmatrix}0 & \rho_3 & -\rho_2\\ -\rho_3 & 0 & \rho_1\\ \rho_2 & -\rho_1 & 0\end{bmatrix}=[T]'[T]^T+[T][k][T]^T \tag{6a}$$

where

$$\rho_1\equiv\mathbf{i}_2'\cdot\mathbf{i}_3=\sum_{i=1}^{3}\left(T'_{2i}T_{3i}+T_{1i}k_i\right)$$

$$\rho_2\equiv-\mathbf{i}_1'\cdot\mathbf{i}_3=\sum_{i=1}^{3}\left(-T'_{1i}T_{3i}+T_{2i}k_i\right)$$

$$=\frac{-1}{1+e}\left[T_{31}\left(u'-vk_3+wk_2\right)'+T_{32}\left(v'+uk_3-wk_1\right)'+T_{33}\left(w'-uk_2+vk_1\right)'\right]+\sum_{i=1}^{3}T_{2i}k_i$$

$$\rho_3\equiv\mathbf{i}_1'\cdot\mathbf{i}_2=\sum_{i=1}^{3}\left(T'_{1i}T_{2i}+T_{3i}k_i\right)$$

$$=\frac{1}{1+e}\left[T_{21}\left(u'-vk_3+wk_2\right)'+T_{22}\left(v'+uk_3-wk_1\right)'+T_{23}\left(w'-uk_2+vk_1\right)'\right]+\sum_{i=1}^{3}T_{3i}k_i \tag{6b}$$

Here, ρ_1 is the deformed twisting curvature w.r.t. the ξ axis, and ρ_2 and ρ_3 are the deformed bending curvatures with respect to the η and ζ axes, respectively. Eqs. (3a)-(6b) completely describe the $\xi\eta\zeta$ system w.r.t. the *xyz* system, the deformed beam geometry, and deformed curvatures.

After the undeformed and deformed beam geometries are fully described, a geometrically exact beam theory can be derived using the extended Hamilton principle, i.e.,

$$0=\int_0^t(\delta K_e-\delta\Pi+\delta W_{nc})dt \tag{7}$$

where t is the time, K_e the kinetic energy, Π the elastic energy, and W_{nc} the non-conservative work due to external loads. Virtual rotations $\delta\theta_i(i=1,2,3)$ w.r.t. the axes ξ,η and ζ are needed in order to derive δK_e, $\delta\Pi$ and δW_{nc}. Virtual rotations $\delta\theta_i$ result in variations of $\mathbf{i}_k$ as

$$\begin{Bmatrix}\delta\mathbf{i}_1\\ \delta\mathbf{i}_2\\ \delta\mathbf{i}_3\end{Bmatrix}=\begin{bmatrix}0 & \delta\theta_3 & -\delta\theta_2\\ -\delta\theta_3 & 0 & \delta\theta_1\\ \delta\theta_2 & -\delta\theta_1 & 0\end{bmatrix}\begin{Bmatrix}\mathbf{i}_1\\ \mathbf{i}_2\\ \mathbf{i}_3\end{Bmatrix} \tag{8}$$

Then, it follows from Eqs. (3b), (8), (4a) and (3a) that

$$\begin{aligned}
\delta e &= T_{11}\delta\left(1+u'-vk_3+wk_2\right)+T_{12}\delta\left(v'+uk_3-wk_1\right)+T_{13}\delta\left(w'-uk_2+vk_1\right)\\
\delta\theta_1 &= \delta\mathbf{i}_2\cdot\mathbf{i}_3=\sum_{i=1}^{3}T_{3i}\delta T_{2i}\\
\delta\theta_2 &= -\delta\mathbf{i}_1\cdot\mathbf{i}_3=-\sum_{i=1}^{3}T_{3i}\delta T_{1i}\\
&=\frac{T_{33}k_2-T_{32}k_3}{1+e}\delta u+\frac{T_{31}k_3-T_{33}k_1}{1+e}\delta v+\frac{T_{32}k_1-T_{31}k_2}{1+e}\delta w-\frac{T_{31}}{1+e}\delta u'-\frac{T_{32}}{1+e}\delta v'-\frac{T_{33}}{1+e}\delta w'\\
\delta\theta_3 &= \delta\mathbf{i}_1\cdot\mathbf{i}_2=\sum_{i=1}^{3}T_{2i}\delta T_{1i}\\
&=\frac{T_{22}k_3-T_{23}k_2}{1+e}\delta u+\frac{T_{23}k_1-T_{21}k_3}{1+e}\delta v+\frac{T_{21}k_2-T_{22}k_1}{1+e}\delta w+\frac{T_{21}}{1+e}\delta u'+\frac{T_{22}}{1+e}\delta v'+\frac{T_{23}}{1+e}\delta w'
\end{aligned} \tag{9a}$$

Eq. (9a) shows that $\delta e,\delta\theta_3$ and $\delta\theta_2$ are explicitly related to $\delta u',\delta v'$ and $\delta w'$ in the form

$$\begin{Bmatrix}\delta e\\ (1+e)\delta\theta_3\\ -(1+e)\delta\theta_2\end{Bmatrix}=[T]\begin{Bmatrix}\delta u'\\ \delta v'\\ \delta w'\end{Bmatrix}-[T][k]\begin{Bmatrix}\delta u\\ \delta v\\ \delta w\end{Bmatrix} \tag{9b}$$

One can use Eq. (6b) and the *Kirchhoff kinetic analogy* between curvatures ρ_i and angular velocities ω_i $(\equiv\dot{\theta}_i)$ to derive the angular velocity vector $\boldsymbol{\omega}$ of the $\xi\eta\zeta$ system as

$$\boldsymbol{\omega} \equiv \dot{\theta}_1\mathbf{i}_1 + \dot{\theta}_2\mathbf{i}_2 + \dot{\theta}_3\mathbf{i}_3,\ \dot{\boldsymbol{\omega}} = \ddot{\theta}_1\mathbf{i}_1 + \ddot{\theta}_2\mathbf{i}_2 + \ddot{\theta}_3\mathbf{i}_3$$
$$\dot{\theta}_1 \equiv \dot{\mathbf{i}}_2 \cdot \mathbf{i}_3 = \sum_{i=1}^{3} \dot{T}_{2i}T_{3i},\ \dot{\theta}_2 \equiv \dot{\mathbf{i}}_3 \cdot \mathbf{i}_1 = \sum_{i=1}^{3} \dot{T}_{3i}T_{1i},\ \dot{\theta}_3 \equiv \dot{\mathbf{i}}_1 \cdot \mathbf{i}_2 = \sum_{i=1}^{3} \dot{T}_{1i}T_{2i} \tag{10}$$

where $\dot{\theta}_i \equiv \partial\theta_i / \partial t$, and $\dot{\mathbf{i}}_k = \boldsymbol{\omega} \times \mathbf{i}_k$ and $\boldsymbol{\omega} \times \boldsymbol{\omega} = 0$ are used. Because the local displacement vector $\mathbf{u}$ in Figure 3b is due to in-plane and out-of-plane warpings, which are negligibly small for the kinetic energy of a flexible beam. Without $\mathbf{u}$, the displacement vector $\mathbf{D}$ in Figure 3b and its time derivatives and variation are given by

$$\begin{aligned}
\mathbf{D} &= u\mathbf{i}_x + v\mathbf{i}_y + w\mathbf{i}_z + y\mathbf{i}_2 + z\mathbf{i}_3 - y\mathbf{i}_y - z\mathbf{i}_z \\
\dot{\mathbf{D}} &= \dot{u}\mathbf{i}_x + \dot{v}\mathbf{i}_y + \dot{w}\mathbf{i}_z + \boldsymbol{\omega} \times (y\mathbf{i}_2 + z\mathbf{i}_3) \\
\ddot{\mathbf{D}} &= \ddot{u}\mathbf{i}_x + \ddot{v}\mathbf{i}_y + \ddot{w}\mathbf{i}_z + \dot{\boldsymbol{\omega}} \times (y\mathbf{i}_2 + z\mathbf{i}_3) + \boldsymbol{\omega} \times \left[\boldsymbol{\omega} \times (y\mathbf{i}_2 + z\mathbf{i}_3)\right] \\
\delta\mathbf{D} &= \delta u\mathbf{i}_x + \delta v\mathbf{i}_y + \delta w\mathbf{i}_z + y(\delta\theta_1\mathbf{i}_3 - \delta\theta_3\mathbf{i}_1) + z(\delta\theta_2\mathbf{i}_1 - \delta\theta_1\mathbf{i}_2)
\end{aligned} \tag{11a}$$

Hence, we have

$$\begin{aligned}
\int_0^t \delta K_e dt &= \int_0^t \int_0^L \int_A \delta(\rho\dot{\mathbf{D}} \bullet \dot{\mathbf{D}}/2) dA ds dt = -\int_0^t \int_0^L \int_A \rho\ddot{\mathbf{D}} \bullet \delta\mathbf{D} dA ds dt \\
&= -\int_0^t \int_0^L (m\ddot{u}\delta u + m\ddot{v}\delta v + m\ddot{w}\delta w + A_{\theta_1}\delta\theta_1 + A_{\theta_2}\delta\theta_2 + A_{\theta_3}\delta\theta_3) ds dt
\end{aligned} \tag{11b}$$

where A is the cross-sectional area, L is the beam length, and

$$\begin{Bmatrix} A_{\theta_1} \\ A_{\theta_2} \\ A_{\theta_3} \end{Bmatrix} \equiv \begin{Bmatrix} j_1\ddot{\theta}_1 - (j_2 - j_3)\dot{\theta}_2\dot{\theta}_3 \\ j_2\ddot{\theta}_2 - (j_3 - j_1)\dot{\theta}_3\dot{\theta}_1 \\ j_3\ddot{\theta}_3 - (j_1 - j_2)\dot{\theta}_1\dot{\theta}_2 \end{Bmatrix}, \begin{Bmatrix} j_1 \\ j_2 \\ j_3 \end{Bmatrix} \equiv \int_A \begin{Bmatrix} y^2 + z^2 \\ z^2 \\ y^2 \end{Bmatrix} \rho dA, m \equiv \int_A \rho dA \tag{11c}$$

Here the axes y and z are assumed to be the principal axes of the cross section and hence $\int_A \rho\{x, y, xy\} dA = \{0, 0, 0\}$ if ρ is constant over the cross section. The inertial terms A_{θ_i} in Eq. (11c) are the same as those of a rigid body, except that the rotary inertias here are for a unit axial length. Note that the rotary inertias j_k are often small, especially for flexible beams. If rotary inertias j_k are neglected, there are no nonlinear inertial terms because $A_{\theta_i} = 0$.

Eq. (11c) shows that the kinetic energy has a simple form when expressed in terms of angular velocities $\dot{\theta}_i$ w.r.t. the body-fixed orthogonal system $\xi\eta\zeta$. Moreover, for feedback control of a rigid-body system, it is more convenient to work with angular velocities defined w.r.t. the body-fixed axes because sensors measure angular motions and actuators apply torques w.r.t. the body-fixed axes. However, although the angular velocities are defined w.r.t.

three orthogonal axes and hence are convenient for use, they cannot be integrated to obtain three sequence-independent large rotation angles because large rotations are sequence-dependent. In other words, θ_i have no exact function forms and are called *quasi-coordinates* (i.e., not well defined coordinates). However, $\delta\theta_i$ and $\dot{\theta}_i$ have exact function forms, as shown in Eqs. (9a) and (10).

Jaumann strains B_{ij} are objective co-rotated engineering strains independent of rigid-body displacements, and they are most appropriate for modeling of highly flexible structures [6]. If the local relative small displacement vector $\mathbf{u}$ due to shear warpings is also included, fully nonlinear strain-displacement relations can be derived using the concept of local displacements relative to the deformed coordinate system $\xi\eta\zeta$ to be

$$\mathbf{u} = (g_5\gamma_5 + g_6\gamma_6)\mathbf{i}_1 \tag{12a}$$

$$\{\varepsilon\} = \begin{Bmatrix} e + z(\rho_2 - k_2) - y(\rho_3 - k_3) + g_5\gamma_5' + g_6\gamma_6' \\ -z(\rho_1 - k_1) + g_{6y}\gamma_6 \\ y(\rho_1 - k_1) + g_{5z}\gamma_5 \end{Bmatrix} = [S]\{\psi\}$$

$$\{\varepsilon\} \equiv \begin{Bmatrix} \varepsilon_{11} \\ \varepsilon_{12} \\ \varepsilon_{13} \end{Bmatrix},\ [S] \equiv \begin{bmatrix} 1 & 0 & 0 & 0 & z & -y & g_6 & g_5 \\ 0 & g_{6y} & 0 & -z & 0 & 0 & 0 & 0 \\ 0 & 0 & g_{5z} & y & 0 & 0 & 0 & 0 \end{bmatrix} \tag{12b}$$

$$\{\psi\} \equiv \{e,\ \gamma_6,\ \gamma_5,\ \rho_1\text{-}k_1,\ \rho_2\text{-}k_2,\ \rho_3\text{-}k_3,\ \gamma_6',\ \gamma_5'\}^T$$

where $g_5(y,z)$ and $g_6(y,z)$ are shear warping functions, $g_{5z} \equiv \partial g_5/\partial z$ and $g_{6y} \equiv \partial g_6/\partial y$. If the shear warping functions are assumed to be

$$g_5 = z,\ \ g_6 = y, \tag{12c}$$

then γ_5 and γ_6 represent the first-order shear rotation angles of the deformed cross-section w.r.t. the $\eta\zeta$ plane, as shown in Figure 5. For an isotropic beam having a rectangular cross-section with side lengths *b* and *h*, its shear warping functions can be derived to be [6]

$$g_5 = z - \frac{4}{3h^2}z^3,\ \ g_6 = y - \frac{4}{3b^2}y^3 \tag{12d}$$

where γ_5 and γ_6 are shear rotation angles of the deformed cross-section at the reference point (i.e., $\eta = \zeta = 0$). Minor terms due to nonlinear couplings between shear rotations and curvatures are neglected in Eq. (12b) because they are negligibly small terms for flexible beams [6]. If the minor inertia effects due to shear rotations are to be accounted for, one just needs to add the $\mathbf{u}$ in Eq. (12a) to the $\mathbf{D}$ in Eq. (11a). The Jaumann strains are related to their work-conjugate stresses, i.e., Jaumann stresses σ_{ij}, as

$$\{\sigma\}=[Q]\{\varepsilon\},\ \{\sigma\}\equiv\begin{Bmatrix}\sigma_{11}\\ \sigma_{12}\\ \sigma_{13}\end{Bmatrix},\ [Q]=\begin{bmatrix}E & 0 & 0\\ 0 & G & 0\\ 0 & 0 & G\end{bmatrix} \tag{13}$$

where E is Young's modulus and G is the shear modulus. An isotropic material is used here just for illustration. For anisotropic materials, the reduced material property matrix $[Q]$ is a full 3×3 matrix [6]. The $[Q]$ can be determined by experiments using small engineering strain and stress measures because Jaumann strains are co-rotated engineering strains [6]. On the other hand, if Green-Lagrange strains are used, the $[Q]$ needs to be determined by experiments using second Piola-Kirchhoff stresses and Green-Lagrange strains, which are nonlinear and computationally awkward and are not usually done in experiments.

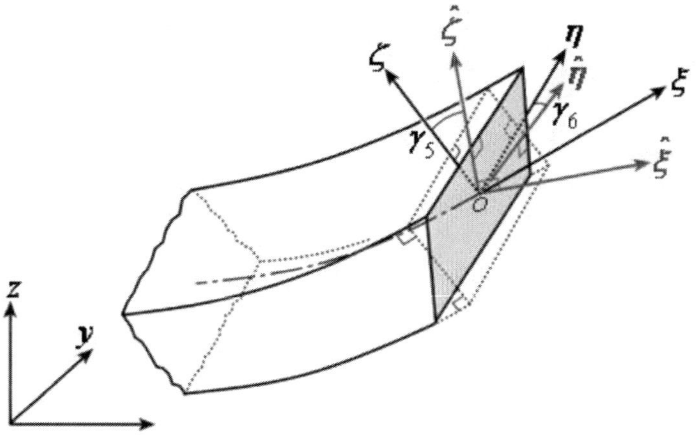

Figure 5. The cross section of a 3D first-order shear-deformable beam theory.

Using Eqs. (12b) and (13) we obtain

$$\begin{aligned}\delta\Pi &= \int_0^L \int_A \{\sigma\}^T\{\delta\varepsilon\}dAds\\ &= \int_0^L (F_1\delta e + M_1\delta\rho_1 + M_2\delta\rho_2 + M_3\delta\rho_3 + \lambda_2\delta\gamma_6 + \lambda_3\delta\gamma_5 - m_3\delta\gamma_6' + m_2\delta\gamma_5')ds\end{aligned} \tag{14a}$$

where the stress resultants are defined as

$$\begin{Bmatrix} F_1 \\ \lambda_2 \\ \lambda_3 \\ M_1 \\ M_2 \\ M_3 \\ -m_3 \\ m_2 \end{Bmatrix} \equiv \int_A \begin{Bmatrix} \sigma_{11} \\ \sigma_{12} g_{6y} \\ \sigma_{13} g_{5z} \\ \sigma_{13} y - \sigma_{12} z \\ \sigma_{11} z \\ -\sigma_{11} y \\ \sigma_{11} g_6 \\ \sigma_{11} g_5 \end{Bmatrix} dA = [D]\{\psi\},\ [D] \equiv \int_A [S]^T [Q][S] dA \tag{14b}$$

For an isotropic beam, if Eq. (12c) is used, we have

$$\begin{aligned} &\lambda_2 = \int_A \sigma_{12} dA = GAc_2\gamma_6,\ \ \lambda_3 = \int_A \sigma_{13} dA = GAc_3\gamma_5 \\ &M_1 = \int_A (\sigma_{13} y - \sigma_{12} z) dA = GI_1 c_1(\rho_1 - k_1),\ I_1 \equiv \int_A (y^2 + z^2) dA \end{aligned} \tag{14c}$$

where c_1 is a correction factor accounting for the decrease of torsional rigidity due to out-of-plane torsional warping and can be calculated using the theory of elasticity [41]. For example, $c_1 = 1$ for a circular cross section, and $c_1 < 1$ for non-circular cross sections. Moreover, c_2 and c_3 are *shear correction factors* accounting for nonuniform shear strains over the cross section [41].

It follows from Eqs. (6a) and (8) and $\mathbf{i}'_k \cdot \mathbf{i}_k = 0$ that $\delta\rho_i$ are related to $\delta\theta_i$ as [6]

$$\begin{Bmatrix} \delta\rho_1 \\ \delta\rho_2 \\ \delta\rho_3 \end{Bmatrix} = \begin{Bmatrix} (\delta\theta_1)' - \rho_3\delta\theta_2 + \rho_2\delta\theta_3 \\ (\delta\theta_2)' + \rho_3\delta\theta_1 - \rho_1\delta\theta_3 \\ (\delta\theta_3)' - \rho_2\delta\theta_1 + \rho_1\delta\theta_2 \end{Bmatrix} = \begin{Bmatrix} \delta\theta_1 \\ \delta\theta_2 \\ \delta\theta_3 \end{Bmatrix}' - [K] \begin{Bmatrix} \delta\theta_1 \\ \delta\theta_2 \\ \delta\theta_3 \end{Bmatrix} \tag{15}$$

If q_1, q_2 and q_3 are distributed forces along the *x*,*y* and *z* axes and q_4, q_5 and q_6 are distributed torsional and bending loads along the ξ, η and ζ axes, we have

$$\delta W_{nc} = \int_0^L [q_1\delta u + q_2\delta v + q_3\delta w + q_4\delta\theta_1 + q_5\delta\theta_2 + q_6\delta\theta_3] ds \tag{16}$$

Substituting Eqs. (11b), (14a), (15) and (16) into Eq. (7) and integrating by parts yields

$$0=\int_0^L\begin{bmatrix}F_1\delta e+\{m\ddot{u}-q_1,m\ddot{v}-q_2,m\ddot{w}-q_3\}\{\delta u,\delta v,\delta w\}^T\\-(M_1'+M_3\rho_2-M_2\rho_3-A_{\theta_1}+q_4)\delta\theta_1+(\lambda_2+m_3')\delta\gamma_6+(\lambda_3-m_2')\delta\gamma_5\\-(M_2'-M_3\rho_1+M_1\rho_3-A_{\theta_2}+q_5)\delta\theta_2+(-M_3'-M_2\rho_1+M_1\rho_2+A_{\theta_3}-q_6)\delta\theta_3\end{bmatrix}ds \tag{17a}$$
$$+[M_1\delta\theta_1+M_2\delta\theta_2+M_3\delta\theta_3-m_3\delta\gamma_6+m_2\delta\gamma_5]_0^L$$

The inertia-induced internal transverse shear forces F_2 and F_3 are given by [6]

$$F_2\equiv\frac{1}{1+e}\left(-M_3'-M_2\rho_1+M_1\rho_2+A_{\theta_3}-q_6\right),\ F_3\equiv\frac{1}{1+e}\left(M_2'-M_3\rho_1+M_1\rho_3-A_{\theta_2}+q_5\right) \tag{17b}$$

Replacing the $\delta e,\delta\theta_2$ and $\delta\theta_3$ in Eq. (17a) with Eq. (9b), using Eq. (17b), and taking integration by parts yields

$$0=\int_0^L\begin{bmatrix}-\left((\{F_1,F_2,F_3\}[T])'+\{F_1,F_2,F_3\}[T][k]\right)\{\delta u,\delta v,\delta w\}^T\\+\{m\ddot{u}-q_1,m\ddot{v}-q_2,m\ddot{w}-q_3\}\{\delta u,\delta v,\delta w\}^T\\-(M_1'+M_3\rho_2-M_2\rho_3-A_{\theta_1}+q_4)\delta\theta_1+(\lambda_2+m_3')\delta\gamma_6+(\lambda_3-m_2')\delta\gamma_5\end{bmatrix}ds \tag{17c}$$
$$+[M_1\delta\theta_1+M_2\delta\theta_2+M_3\delta\theta_3+\{F_1,F_2,F_3\}[T]\{\delta u,\delta v,\delta w\}^T-m_3\delta\gamma_6+m_2\delta\gamma_5]_0^L$$

Setting the coefficients of $\delta u,\delta v,\delta w,\delta\theta_1,\delta\gamma_6$ and $\delta\gamma_5$ to zero yields the following governing equations:

$$\frac{\partial}{\partial s}\left([T]^T\begin{Bmatrix}F_1\\F_2\\F_3\end{Bmatrix}\right)-[k][T]^T\begin{Bmatrix}F_1\\F_2\\F_3\end{Bmatrix}+\begin{Bmatrix}q_1\\q_2\\q_3\end{Bmatrix}=\begin{Bmatrix}m\ddot{u}\\m\ddot{v}\\m\ddot{w}\end{Bmatrix} \tag{18a}$$

$$M_1'+M_3\rho_2-M_2\rho_3+q_4=A_{\theta_1} \tag{18b}$$

$$m_3'=-\lambda_2 \tag{18c}$$

$$m_2'=\lambda_3 \tag{18d}$$

The corresponding boundary conditions are to specify:

$$
\begin{array}{lll}
\delta u = 0 & \text{or} & F_x (\equiv F_1 T_{11} + F_2 T_{21} + F_3 T_{31}) \text{ known} \\
\delta v = 0 & \text{or} & F_y (\equiv F_1 T_{12} + F_2 T_{22} + F_3 T_{32}) \text{ known} \\
\delta w = 0 & \text{or} & F_z (\equiv F_1 T_{13} + F_2 T_{23} + F_3 T_{33}) \text{ known} \\
\delta \theta_1 = 0 & \text{or} & M_1 \text{ known} \\
\delta \theta_2 = 0 & \text{or} & M_2 \text{ known} \\
\delta \theta_3 = 0 & \text{or} & M_3 \text{ known} \\
\delta \gamma_6 = 0 & \text{or} & m_3 \text{ known} \\
\delta \gamma_5 = 0 & \text{or} & m_2 \text{ known}
\end{array}
\tag{19}
$$

The governing equations (18a), (18b) and (17b) can also be derived using a vector approach based on Newton's second law and the free-body diagram of a differential beam element [6]. This shows that the energy formulation starting from the extended Hamilton principle (i.e., Eq. (7)) is fully correlated with the vector formulation, and governing equations obtained from these two different approaches are essentially the same. On the other hand, if Green-Lagrange strains and second Piola-Kirchhoff stresses are used in the extended Hamilton principle and the material stiffness matrix shown in Eq. (13) is used, one can never show that the so-obtained governing equations are the same as those from the vector formulation [6].

Eqs. (18a-d), (17b), (14b), (12b), (3b), (6b) and (9b) show that this geometrically exact beam theory is completely and explicitly described by $u, v, w, T_{21}, T_{23}, \gamma_5$ and γ_6. This is a 16th-order system because there are 8 boundary conditions need to be specified at each of the two ends, as shown by Eq. (19). If the influences of transverse shear deformations γ_5 and γ_6 are neglected, it reduces to a 12th-order system. For fully nonlinear finite-element modeling and analysis, u, v, w, u', v', w', T_{21}, T_{23}, γ_5 and γ_6 can be used as nodal degrees of freedom at each node [39]. *How simulation results from such a geometrically exact beam theory can be used to recognize key parameters and dynamic characteristics for system identification and damage detection will be demonstrated in Sec. V by numerical examples.*

III. Full-field Damage Inspection Technique

Small damage introduces local boundaries to a structure, and discontinuities of slopes, curvatures and other higher-order spatial derivatives exist at such boundaries. Extraction of such discontinuities can clearly reveal damage locations. Hence, detailed damage inspection of a stopped structure can be performed by processing the structure's operational deflection shapes (ODSs) measured by a full-field sensor system. A boundary-effect evaluation method (BEEM) is developed based on this concept for full-field damage inspection. To demonstrate it without complex derivations we will use operational deflection shapes (ODSs) of simple 1D structures in the following derivation. For 2D structures, the same process can be followed.

For a beam subjected to a harmonic excitation at a frequency Ω, its standing operational deflection shape $W(x)$ of a uniform segment without any loading points or damage has the following form [34]:

$$\begin{aligned} W(\bar{x}) &= a_1 \cos\beta x + b_1 \sin\beta x + a_2 \cosh\beta x + b_2 \sinh\beta x \\ &= C_1 \cos\beta\bar{x} - D_1 \sin\beta\bar{x} + C_2 \cosh\beta\bar{x} + D_2 \sinh\beta\bar{x} \\ &= C_1 \cos\beta\bar{x} - D_1 \sin\beta\bar{x} + C_2^* e^{\beta\bar{x}} + D_2^* e^{-\beta\bar{x}} \end{aligned} \tag{20}$$

$$C_1 \equiv \sqrt{a_1^2 + b_1^2}\cos(\beta x_m - \phi),\quad C_2 \equiv a_2 \cosh\beta x_m + b_2 \sinh\beta x_m,\quad \tan\phi \equiv b_1 / a_1,\quad \beta^4 \equiv m\Omega^2 / EI$$

$$D_1 \equiv \sqrt{a_1^2 + b_1^2}\sin(\beta x_m - \phi),\quad D_2 \equiv a_2 \sinh\beta x_m + b_2 \cosh\beta x_m,\quad C_2^* \equiv (C_2 + D_2)/2,\ D_2^* \equiv (C_2 - D_2)/2$$

where $\bar{x}$ $(\equiv x - x_m)$ is a shifted local coordinate, x_m is the location under observation, E is Young's modulus, I is the area moment of inertia, and m is the mass per unit length. Note that a_i and b_i are constants but C_i, D_i, C_2^* and D_2^* are functions of x_m. The ODS shown in Eq. (20) is actually caused by two traveling waves moving along opposite directions and two evanescent waves at the two ends of a beam because [42]

$$We^{-j\Omega t} = \frac{\sqrt{a_1^2 + b_1^2}}{2} e^{j(\beta x - \phi - \Omega t)} + \frac{\sqrt{a_1^2 + b_1^2}}{2} e^{-j(\beta x - \phi + \Omega t)} + \frac{a_2 + b_2}{2} e^{\beta x - j\Omega t} + \frac{a_2 - b_2}{2} e^{-\beta x - j\Omega t} \tag{21}$$

where $j \equiv \sqrt{-1}$. For a beam, its curvature is defined as $\kappa \equiv d\theta / dx$, where θ is the structure's rotation angle. Because the wavenumber β of an ODS is directly related to its wavelength λ as $\beta = 2\pi / \lambda$, we have $\beta = \int_0^\lambda \kappa dx \big/ \lambda$. Hence, β represents the average curvature over one wavelength, and κ is the pointwise wavenumber.

If the wavenumber is known, the coefficients C_i and D_i for the point at $\bar{x} = 0$ (i.e., $x = x_m$) can be extracted using the following sliding-window least-squares curve-fitting method. The *2n+1* data points around $x = x_m$ are used to minimize the following fitting error E_{rror}.

$$E_{rror} \equiv \sum_{i=-n}^{n} \alpha_i (W_i - Y_i)^2 \tag{22}$$

where W_i represents $W(\bar{x}_i)$, Y_i is the experimental data at $\bar{x}_i$, and α_i is a weighting factor chosen to be $\alpha_i = 1/(1 + |99i/n|)$. The coefficients C_i and D_i are determined from four algebraic equations to minimize E_{rror}. The wavenumber β in Eq. (20) can be estimated by picking up a representative wavelength λ of the measured ODS to obtain $\beta = 2\pi / \lambda$.

After C_i and D_i are obtained, it follows from Eq. (20) that, at $\bar{x} = 0$ (i.e., $x = x_m$),

$$W(0)=C_1+C_2^*+D_2^*=C_1+C_2 \qquad W'(0)=\beta(-D_1+C_2^*-D_2^*)=\beta(-D_1+D_2)$$
$$W''(0)=\beta^2(-C_1+C_2^*+D_2^*)=\beta^2(-C_1+C_2) \qquad W'''(0)=\beta^3(D_1+C_2^*-D_2^*)=\beta^3(D_1+D_2) \tag{23}$$

Eqs. (23) and (20) reveal that C_1 represents the central solution of W that exists over the whole spatial domain $0 \le x_m \le L$ of the beam, C_2 represents the boundary solution of W caused by boundary constraints, $D_1\beta$ represents the central solution of W' $(\equiv dW/dx)$, and $D_2\beta$ represents the boundary solution of W'. Moreover, C_1 and C_2 also separate W'' into central and boundary solutions, and D_1 and D_2 also separate W''' into central and boundary solutions. Furthermore, C_2^* represents the boundary solution (i.e., an evanescent wave) caused by the boundary at $x=L$, and D_2^* represents the boundary solution caused by the boundary at $x=0$ [34]. In other words, solving four algebraic equations for C_i and D_i results in answers for the four unknowns $W(0)$, $W'(0)$, $W''(0)$, and $W'''(0)$ with each one being decomposed into two or three parts (see Eq. (23)) and hence we have more than four damage indicators for locating and estimating damage. Note that $W(0)$, $W'(0)$, $W''(0)$, and $W'''(0)$ cannot be decomposed into two or three parts if a polynomial function or a spline curve is used to fit an ODS. Calculation of strains of a beam only requires $W''(0)$, but this signal decomposition method can also provide $W'''(0)$ and other higher-order derivatives. For a high-frequency ODS without damage, the boundary solutions should be zero at locations away from boundaries. Because damage creates new boundaries to a structure, if damage occurs at points away from boundaries, boundary solutions should exist around the damage and hence can reveal damage locations. Even if damage occurs at or close to regular boundaries, damage-induced boundary solutions are different from those due to regular boundary constraints and hence can also reveal damage locations [34].

The difference between the maximum elastic energy density Π_{den} and the maximum kinetic energy density T_{den} under steady-state harmonic vibration can be derived from Eqs. (20) and (23) as

$$T_{den}-\Pi_{den}=\frac{1}{2}m(W\Omega)^2-\frac{1}{2}EIW''^2=2m\Omega^2C_1C_2=2m\Omega^2(C_1C_2^*+C_1D_2^*) \tag{24}$$

Eq. (24) shows that, if there is no damage, $C_2=0$ and $T_{den}-\Pi_{den}=0$ at points away from boundaries. In other words, $T_{den}-\Pi_{den}\neq 0$ (or $C_1C_2\neq 0$) at a location away from boundaries indicates there is damage.

It follows from Eq. (20) that

$$A\equiv\sqrt{C_1^2+D_1^2}=\sqrt{a_1^2+b_1^2},\ \phi=\beta x_m-\tan^{-1}\frac{D_1}{C_1}=\tan^{-1}\frac{b_1}{a_1},\ C_2^*D_2^*=\frac{1}{4}\left(C_2^2-D_2^2\right)=\frac{1}{4}\left(a_2^2-b_2^2\right) \tag{25}$$

Eq. (25) reveals that A is a constant representing the amplitude of the two central solutions, and ϕ is a constant representing the phase delay of the central solution C_1 with respect to the corresponding cosine function. Moreover, $C_2^*D_2^*$ is a constant representing the multiplication of the right and left boundary solutions. If it is a high-frequency ODS, we have $C_2^*D_2^*=0$ because C_2^* and D_2^* are zero except around the right and left boundary points, respectively. Moreover, the *Sectional Standard Deviation* (SSD) and the overall *Standard Deviation* (SD) of the curve fitting can also indicate damage, and they are defined as

$$SSD \equiv \sqrt{\sum_{i=-n}^{n}[W(\bar{x}_i)-Y(\bar{x}_i)]^2/(2n+1)} \Big/ W_{\max}, \quad SD \equiv \sqrt{\sum_{m=1}^{M}[W(x_m)-Y(x_m)]^2/M} \Big/ W_{\max} \tag{26}$$

where M is the total number of points measured on the beam and $W_{\max}$ is the maximum of $|W|$. Because $C_2^*D_2^*, A, \beta, \phi$ and SSD are constant at any location of any low- or high-frequency ODS, they are five excellent damage indicators that can show damage locations at boundaries as well as any other locations and can be used for estimation of unknown boundary conditions [34].

Because ϕ is constant, it follows from Eq. (25) that

$$\beta = \frac{d}{dx}\left(\tan^{-1}\frac{D_1}{C_1}\right) \tag{27}$$

After C_1 and D_1 are obtained from the sliding-window curve fitting, the wavenumber of the point at $x=x_m$ can be re-estimated and updated using finite difference on Eq. (27), and this wavenumber can also be used as the initial guess of β for the next point. To reduce the influence of noise, one can average it over 4 or more data points. Even before C_1 and D_1 are obtained, β can be estimated by using the Teager-Kaiser algorithm (TKA) [43,44].

This model-independent method works well because localized boundary solutions can be separated from central solutions and because each coefficient has a physical meaning. Furthermore, BEEM provides 10 damage indicators for 1D structures, including $C_2, D_2, C_2^*, D_2^*, C_1C_2$, $C_2^*D_2^*, A, \beta, \phi$ and SSD. They can be used to check against each other to ensure the damage detection and diagnosis results. The wavenumber β is a new damage indicator that improves the sensitivity of our old BEEM [34], and it is similar to the use of the time-varying instantaneous frequency ω of a measurement point's time trace for damage detection, as shown later in Sec. IV.

For 2D structures (e.g., plates and shells), a functional form of ODSs similar to Eq. (20) exists [34], and BEEM can also work for detecting defects in 2D structures by using similar derivations. However, because BEEM requires the use of ODSs from full-field measurements but in-work moving structures can only be monitored using responses measured at just a few measurement points, this spatial-domain method is not appropriate for in-work damage detection and health monitoring of moving structures. Hence, a method for *time-frequency analysis* of time traces of spatially sparse measurement points is needed and introduced next.

IV. Online Health Monitoring and System Identification Techniques

Damage introduces intermittent response and nonlinearity to measured dynamic signals and causes distorted harmonic response of a structure under a harmonic excitation. Extraction of such intermittent response, nonlinearity and wave distortion can reveal the existence of damage and damage locations and can be used for online health monitoring and system identification. However, a general dynamic response of a discrete or continuous system contains multiple modal vibrations, and modal vibration of a nonlinear system is often amplitude- and frequency-modulated response [1-6]. Hence, how to separate these modes is one tough task, and how to track the time-varying amplitude and instantaneous frequency (*IF*) of each separate mode is another difficult job that requires advanced time-frequency analysis [45,46]. Next we present a conjugate-pair decomposition (CPD) method for time-frequency analysis and compare it with the short-time Fourier transform (STFT) and Hilbert-Huang transform (HHT).

1. Short-Time Fourier Transform

The short-time Fourier transform (STFT) processes a sub-record length T of a time signal sampled over $0 \le t \le \hat{T}\ (>>T)$, and the so-obtained frequency spectrum's peak frequency can be taken as the signal's instantaneous frequency (*IF*) [47]. Another approach for extracting *IF* is to use analytic functions [45]. However, time-frequency analysis from both approaches is difficult to be accurate, as explained next.

Any time function $u(t)$ defined over $0 \le t \le T$ can be extended by repeating itself to become a periodic function, and the extended function can be expressed in terms of regular harmonics with constant amplitudes A_i and frequencies ω_i by using the discrete Fourier transform (DFT) as [48]

$$u(t) = a_0 + 2\sum_{i=1}^{N/2}\left(a_i \cos\omega_i t + b_i \sin\omega_i t\right) = a_0 + \sum_{i=1}^{N/2} A_i \cos\theta_i$$
$$a_i \equiv \frac{1}{N}\sum_{k=1}^{N} u_k \cos\omega_i t_k,\ b_i \equiv \frac{1}{N}\sum_{k=1}^{N} u_k \sin\omega_i t_k,\ A_i \equiv 2\sqrt{a_i^2+b_i^2},\ \theta_i \equiv \omega_i t - \tan^{-1}\frac{b_i}{a_i} \tag{28}$$

where $\omega_i \equiv 2\pi i/T$, $u_k \equiv u(t_k)$, $t_k \equiv (k-1)\Delta t$, $k = 1,...,N$, N is the number of samples (assumed to be even here), Δt is the sampling interval, $1/\Delta t$ is the sampling frequency, $T(=N\Delta t)$ is the sub-record length and also the temporal resolution of STFT, $\Delta f(=1/T)$ is the frequency resolution, and $1/(2\Delta t)(=N/(2T))$ is the highest (or Nyquist) frequency available from the frequency spectrum $U(\omega_i)$ $(\equiv a_i - jb_i)$ where $j \equiv \sqrt{-1}$.

Because the signal is treated as a periodic function, $u_{N+1} = u_1$ is implicitly assumed in DFT. If sudden change (i.e., discontinuity) exists between u_N and u_1 due to sampling and/or transient effects, the ω_1 $(=2\pi/T)$ harmonic is not a real component. In order to account for this discontinuity, many high-frequency harmonics exist in Eq. (28) and result in the so-called *leakage* problem in the frequency domain and *Gibbs' phenomenon* at the two data ends in the time domain (the so-called *edge effect*) [48,49]. Because the lowest-frequency harmonic in Eq. (28) has a frequency $\omega_1 = 2\pi/T$, the T needs to cover at least one period of the signal's major harmonic in order to identify the signal's *IF* from the spectrum. According to Shannon's sampling theory, at least two data points (i.e., *N*=2) within $0 \le t \le T$ is needed for accurate frequency identification. Unfortunately, the existence of other high-frequency minor harmonics in the signal would require $N >> 2$ in order to account for the influence of high-frequency harmonics. Even if the *T* is exactly equal to one period of the major harmonic, the existence of other low- and/or high-frequency harmonics can cause discontinuity between u_N and u_1. Then, the *T* needs to cover several periods of the major harmonic and $N >> 2$ is needed in order to reduce the edge effect by averaging, and the so-obtained peak frequency from the spectrum is more an averaged value over the T, instead of the actual *IF*.

Because *T* is the temporal resolution and $\Delta f = 1/T$ is the frequency resolution, compromise between the temporal and spectral resolutions is needed in order to make $T\Delta f = 1$. The frequency resolution can be increased by zero padding to artificially increase *T*. The temporal resolution can be increased by using a smaller *T* (but larger than the period of the major component). Unfortunately, the edge effect due to discontinuity becomes more serious when *T* is smaller. Hence, time history tapering by using window functions is often performed before STFT, but signal tapering reduces the accuracy of the extracted amplitude value. Other methods for improving frequency spectra from STFT include data overlapping and spectral averaging. Unfortunately, because these techniques are not well documented, their improvements on short-time frequency spectra are very limited and results are often user-dependent [45-49].

Another approach for extracting *IF* is to use analytic functions, as shown below. The leakage problem and edge effect can be eliminated by extending the two data ends with smooth curves to make u, $\dot{u}$, and $\ddot{u}$ continuous between u_N and u_1. (Of course, this process can only be done case by case because it is difficult to program such a signal processing process.) Hence, Eq. (28) indicates that it is theoretically possible to exactly present an arbitrary signal in terms of regular harmonics of constant frequencies and amplitudes by using Fourier transform with appropriate sampling and data extension. After Eq. (28) is obtained, because A_i and $\dot{\theta}_i$ are constant, the exact conjugate part of $u(t)$, $v(t)$, can be obtained using Hilbert Transform (HT) as [49]

$$v(t) \equiv HT\left(a_0 + \sum_{i=1}^{N/2} A_i \cos\theta_i\right) = \sum_{i=1}^{N/2} A_i \sin\theta_i \tag{29}$$

Moreover, it follows from Eqs. (28) and (29) that

$$\dot{u}(t) = -\sum_{i=1}^{N/2} \omega_i A_i \sin\theta_i, \quad \dot{v}(t) = \sum_{i=1}^{N/2} \omega_i A_i \cos\theta_i \tag{30}$$

Then, one may calculate the instantaneous frequency ω using the phase angle θ on the u-v plane and finite difference as

$$\omega = \frac{d\theta}{dt} = \frac{d\left(\tan^{-1} v/u\right)}{dt} = \frac{u\dot{v} - v\dot{u}}{u^2 + v^2} \approx \frac{u\Delta v/\Delta t - v\Delta u/\Delta t}{u^2 + v^2} \tag{31}$$

Because Eq. (31) can give the *IF* for each data point by using only two data points, the temporal resolution is essentially Δt, instead of *T*. Hence, one can use a much larger *T* to reduce the edge effect without changing the temporal resolution. However, this approach is not suitable for online frequency tracking because Eq. (28) needs to process a long data length *T* in order to reduce the edge effect in $v(t)$ and to have an accurate estimation of *IF* using Eq. (31). More seriously, Eq. (31) gives inaccurate or even erroneous *IF* when the signal is not dominated by just one major harmonic and/or any moving average exists in the signal [46,50]. On the other hand, Hilbert-Huang transform (HHT) uses the empirical mode decomposition (EMD) for signal decomposition and then uses Hilbert transform (HT) and Eq. (31) for accurate time-frequency analysis of each decomposed component [45,51-54].

2. Hilbert-Huang Transform

Fourier and wavelet transforms use pre-determined harmonics and wavelets, respectively, as basis functions and the conformality between the basis functions and the signal under analysis to extract components. On the other hand, because HHT uses EMD to extract components without using any pre-determined basis functions, it is essentially different from Fourier and wavelet transforms [45]. EMD is a data-driven signal decomposition technique that sequentially extracts zero-mean regular/distorted harmonics from a signal, starting from high- to low-frequency components, and it is a dyadic filter equivalent to an adaptive wavelet [54,55].

The first step of HHT is to use EMD to sequentially decompose a signal $u(t)$ into n intrinsic mode functions (IMFs) $c_i(t)$ and a residual $r_n(t)$ as [45]

$$u(t) = \sum_{i=1}^{n} c_i(t) + r_n(t), \tag{32}$$

where c_1 has the shortest characteristic time scale and is the first extracted IMF. The characteristic time scale of c_1 is defined by the time lapse between the extrema of u. Once the extrema are identified, compute the upper envelope by connecting all the local maxima using a natural cubic spline, compute the lower envelope by connecting all the local minima using another natural cubic spline, subtract the mean of the upper and lower envelopes, m_{11}, from the signal, and then treat the residuary signal as a new signal. Repeat these steps for K times until the left signal has a pair of symmetric envelopes (i.e., $m_{1K} \approx 0$), and then define c_1 as

$$c_1 \equiv u - m_{11} \cdots - m_{1K} . \tag{33}$$

This sifting process eliminates low-frequency riding waves, makes the wave profile symmetric, and separates the highest-frequency IMF from the current residuary signal. During the sifting process for each IMF, a deviation D_v is computed from two consecutive sifting results as

$$D_v \equiv \sqrt{\sum_{i=1}^{N} [c_{1k}(t_i) - c_{1k-1}(t_i)]^2 \Big/ \sum_{i=1}^{N} c_{1k-1}^2(t_i)} , \tag{34}$$

where, for example, $c_{1k} \equiv u - m_{11} \ldots - m_{1k}$, $t_i = (i-1)\Delta t$, and $T(= N\Delta t)$ is the sampled period. A systematic method to end the iteration is to limit D_v to be a small number (e.g., $D_v = 0.001$) and/or to limit the maximum number of iterations. However, the dyadic property of EMD is valid only if the iteration number is fixed at about 10 because the dyadic property would break down if the iteration number is too high or too low [45].

After c_1 is obtained, define the residual r_1 $(\equiv u - c_1)$, treat r_1 as a new data set, and repeat the steps shown in Eq. (33) to obtain other c_i $(i = 2, \ldots, n)$ as

$$c_i = r_{i-1} - m_{i1} \cdots - m_{iK} , \; r_{i-1} \equiv u(t) - c_1 \cdots - c_{i-1} . \tag{35}$$

The whole sifting process can be stopped when the residual r_n $(= r_{n-1} - c_n)$ becomes a monotonic function from which no more IMFs can be extracted. For data with a trend, the last IMF r_n should be the trend and has at most one local extremum.

The second step of HHT is to perform Hilbert transform and compute the time-varying frequency ω_i and amplitude A_i of each c_i. After all $c_i(t)$ are extracted, one can perform Hilbert transform (HT) to obtain $d_i(t)$ $(= HT(c_i))$ from each $c_i(t)$ and then define $z_i(t)$ and use Eq. (32) to obtain

$$z_i(t) \equiv c_i(t) + jd_i(t) = A_i e^{j\theta_i},\ A_i \equiv \sqrt{c_i^2 + d_i^2},\ \theta_i \equiv \tan^{-1} d_i / c_i,\ \omega_i \equiv \dot{\theta}_i$$
$$u(t) = r_n + \mathrm{Re}\left(\sum_{i=1}^{n} [c_i(t) + jd_i(t)]\right) = r_n + \mathrm{Re}\left(\sum_{i=1}^{n} A_i(t) e^{j\theta_i(t)}\right). \tag{36}$$

Comparing the time-varying $A_i(t)$ in Eq. (36) with the constant $A_i\left(\equiv 2\sqrt{a_i^2 + b_i^2}\right)$ in Eq. (28) reveals that HHT is a time-varying signal decomposition method and it allows the use of distorted harmonics to decompose a complex signal into just a few IMFs. Because the moving average of each $c_i(t)$ is removed by the EMD process, the accuracy of instantaneous frequency ω_i from Eq. (36) is far more accurate than that from Eq. (31).

Unfortunately, because of the use of Hilbert transform (which is based on Fourier transform) for $d_i(t)$, the accuracy of time-frequency analysis of HHT seriously suffers from the edge effect, especially when the data length is short. Hence, a long data length is often required for HHT to process the whole data length to obtain accurate instantaneous frequencies for points away from two data ends. Next we introduce a conjugate-pair decomposition (CPD) method that can provide fast, localized time-frequency analysis without the edge effect because it processes only a few time-domain data points without using Hilbert transform.

3. Conjugate-Pair Decomposition

As shown later in Sec. V, the ten damage indicators of BEEM shown in Sec. III can clearly indicate damage locations because discontinuous spatial derivatives exist at damaged locations. Similarly, when a measurement location is subjected to impacts and/or discontinuous loading caused by intermittent contact of damaged surfaces and/or erratic external loads, its velocity and acceleration can be discontinuous. In view of Eq. (20) and in order to improve our old conjugate-pair decomposition (CPD) method for frequency tracking and structural damage detection [50], we assume a time signal $u(t)$ have the following form for sliding-window curve fitting:

$$u(\bar{t}) = C_1 \cos\omega\bar{t} - D_1 \sin\omega\bar{t} + C_2^* e^{\omega\bar{t}} + D_2^* e^{-\omega\bar{t}} + C_0 \tag{37a}$$

where ω is assumed to be known from estimation or from the previous time instant, $\bar{t}\ (\equiv t - t_m)$ is a shifted time, and t_m is the time instant under analysis. Because transient dynamic response often contains a moving average, the C_0 is a constant to capture the moving average. Moreover, the C_2^* and D_2^* intend to capture the left-side $(\bar{t} < 0)$ and right-side $(\bar{t} > 0)$ local transient effects caused by discontinuity at $t = t_m$. The sliding-window least-squares curve-fitting process shown in Sec. III can be used to determine the 5 coefficients in Eq. (37a) using two or more data points (i.e., $n \geq 2$ in Eq. (22)) from each side of $t = t_m$. Because the ω is assumed to be known in Eq. (37a), the algebraic equations for least-squares fitting are linear ones, instead of nonlinear ones when ω is unknown. One can use the Teager-Kaiser algorithm (TKA) [43,44] to provide the initial guess of ω or use the frequency spectrum of a small section of the beginning time trace.

Because the moving average of a signal may change with time, we also consider the use of a quadratic polynomial to better capture the moving average and assume that

$$u(\bar{t}) = C_1 \cos \omega \bar{t} - D_1 \sin \omega \bar{t} + C_0 + C_2 \bar{t} + C_3 \bar{t}^2 \tag{37b}$$

However, because CPD depends on local least-squares curve-fitting using just a few data points, the orthogonality between the functions used in Eq. (37b) (or Eq. (37a)) cannot be fully enforced. When more unknowns are used in Eq. (37b) (or Eq. (37a)), the estimations become less accurate. Since the oscillation frequency and amplitude are the two most important parameters of a dynamical signal, it is important to have accurate extraction of C_1 and D_1. Hence, we recommend the use of Eq. (37b) without C_2 and C_3 for CPD analysis.

Similar to Eq. (27) for the estimation of wavenumber, the instantaneous frequency ω can be estimated using

$$\omega = \frac{d}{dt}\left(\tan^{-1} \frac{D_1}{C_1} \right) \tag{38}$$

After C_1 and D_1 are obtained from sliding-window fitting, the instantaneous frequency ω of the current point can be updated using finite difference on Eq. (38). This value can also be used as the initial guess of ω for the next point. To reduce the influence of noise, one can average it over 4 or more data points. Because the initial guess of ω is updated using Eq. (38), this method is able to adapt and capture the actual variation of frequency [50]. Moreover, the accuracy of ω for each time instant can be improved by running several iterations of the fitting process and updating the value of ω for a time instant. For online health monitoring, CPD can track the instantaneous frequency of an arbitrary signal by processing three or more most recent data to estimate its instantaneous frequency and amplitude.

If the signal contains p (>1) harmonic components, one can assume that

$$u(\bar{t}) = \sum_{i=1}^{p} \left(C_i \cos \omega_i \bar{t} - D_i \sin \omega_i \bar{t} \right) + C_0 \tag{39}$$

Then similar derivations can be obtained by following the same procedure. However, more components in a signal would require more data points and hence a larger window length for least-squares curve fitting. Then the obtained frequency is more an average over the window length, instead of an instantaneous frequency. On the other hand, one can use EMD to decompose the signal into p IMFs, and then use CPD to perform accurate time-frequency analysis of each IMF. Since EMD often introduces edge effects into each IMF, one can assume the form in Eq. (37a) for each IMF and then the boundary solutions C_2^* and D_2^* can be captured and separated from the central solutions C_1 and D_1. Note that an IMF from EMD analysis should have no moving average (i.e., $C_0 = 0$).

4. Guidance from Perturbation Solutions

One major challenge in dynamics-based system identification is how to account for the influences of nonlinearities [3,8-10,14]. Here we show how to combine the use of time-frequency analysis and perturbation solutions for accurate system identification. We consider the following Duffing oscillator subjected to a harmonic excitation at a frequency Ω close to its linear natural frequency ω and the corresponding steady-state second-order perturbation solution [1]:

$$\ddot{u} + 2\varsigma\omega\dot{u} + \omega^2 u + \alpha u^3 = F\cos(\Omega t) \tag{40a}$$

$$u(t) = a_1\cos(\Omega t - \phi) + a_3\cos(3\Omega t - 3\phi) = A\cos(\hat{\Omega}t - \phi) \tag{40b}$$

$$A(t) \square a_1 + a_3\cos(2\Omega t - 2\phi), \qquad a_3 = \frac{\alpha a_1^3}{32\Omega^2} \square a_1 \tag{40c,d}$$

$$\hat{\Omega}(t) \square \Omega + \frac{2\Omega a_3}{a_1}\cos(2\Omega t - 2\phi) \tag{40e}$$

$$\left(\frac{3\alpha}{8\omega}\right)^2 a_1^6 - \frac{3\alpha\sigma}{4\omega}a_1^4 + \left(\sigma^2 + \varsigma^2\omega^2\right)a_1^2 - \left(\frac{F}{2\omega}\right)^2 = 0 \tag{40f}$$

$$\varsigma = \left(\frac{3\alpha a_1^2}{8\omega^2} - \frac{\sigma}{\omega}\right)\tan\phi, \qquad \hat{\omega} = \omega + \frac{3\alpha}{8\omega}a_1^2 \tag{40g,h}$$

where ϕ is the phase difference between the forcing function and the response, and ς, α, F and ϕ are constants. The amplitude a_1 is a nonlinear function of F, ς, ω and α, as shown in Eq. (40f) from first-order perturbation analysis [1]. Eq. (40b) shows that $u(t)$ consists of two synchronous harmonics (i.e., reaching maxima at the same time), but, because $a_1 \square a_3$, $u(t)$ would appear as one distorted harmonic with an amplitude A and a frequency $\hat{\Omega}$ varying at a frequency 2Ω. This amplitude- and frequency-modulation phenomenon can be used to determine the order (cubic or other) of nonlinearity. Moreover, if

$\alpha > 0$, $a_3/a_1 > 0$, $\hat{\Omega}$ and A are at their maxima when $u(t)$ is at its maxima or minima (i.e., $\dot{u} = 0$), and $\hat{\Omega}$ and A are at their minima when $\dot{u}(t)$ is close to its maxima or minima (i.e., $u = 0$). If $\alpha < 0$, $a_3/a_1 < 0$, $\hat{\Omega}$ and A are at their minima when $u(t)$ is at its maxima or minima (i.e., $\dot{u} = 0$), and $\hat{\Omega}$ and A are at their maxima when $\dot{u}(t)$ is close to its maxima or minima (i.e., $u = 0$). This phenomenon can be used to determine the type (hardening or softening) of nonlinearity, and the magnitude of nonlinearity (i.e., α) can be estimated using Eq. (40d) and the time-domain data $a_1(t)$ and $a_3(t)$ as $\alpha = 32\Omega^2 a_3/a_1^3$. Moreover, one can use Eq. (40g) to estimate the damping ratio ς, use Eq. (40h) to estimate the free undamped amplitude-dependent natural frequency $\hat{\omega}$, and use Eq. (40f) with a known $\sigma(\equiv \Omega - \omega)$ to obtain F and then estimate the mass as $m = F_0/F$, where F_0 is the known excitation amplitude.

The major point here is that, with time-varying frequency and amplitude from time-frequency analysis and the guidance of perturbation solutions, the order, type and magnitude of nonlinearity and all system parameters ω, ς, m and $\hat{\omega}$ can be estimated from just one steady-state response to a harmonic excitation at $\Omega \approx \omega$. Moreover, numerical and experimental results show that all these characteristics under a harmonic excitation also exist in free transient vibrations [56]. For systems with different orders of nonlinearity, similar perturbation solutions can be derived for identification of different nonlinearities [1,5]. For a system with quadratic nonlinearity, for example, its steady-state response can be shown to be a distorted harmonic with its amplitude and frequency varying at a frequency Ω.

V. Numerical and Experimental Results

Here we use examples to demonstrate: 1) how to use numerical simulations from high-fidelity structural models to recognize key dynamic characteristics for system identification and damage inspection, 2) some experimental verifications, 3) how to use BEEM for full-field structural damage inspection, and 4) how to use CPD and HHT for identification and monitoring of nonlinearity and system parameters.

1. High-Fidelity Structural Modeling for Development of Identification/Inspection Techniques

We consider a $1000 \times 25.4 \times 6.4 mm^3 (= L \times b \times h$ along the x, y, and z axes) 2024-T851 aluminum beam hinged at $x = 0.2L \,\&\, 0.8L$ and subjected to two harmonic base excitations $w(0,t) = w(L,t) = 3.0\sin(\Omega t)\ mm$ along the z direction. The material properties are Young's modulus $E = 72.4\ GPa$, Poisson's ratio $\nu = 0.33$, and mass density $\rho = 2780\ kg/m^3$. The first four linear natural frequencies $\bar{\omega}_i$ are computed to be 32.91 Hz, 72.15 Hz, 110.77 Hz and 221.10 Hz. Transient analysis by direct numerical integration using the Newmark-beta method is performed using an in-house fully nonlinear finite-element code GESA (Geometrically Exact Structural Analysis) [6]. GESA has been implemented with geometrically exact 1D (as shown in Sec. II for highly flexible beams) and 2D structural theories for modeling and analysis of highly flexible structures undergoing large elastic deformations, and its accuracy has been verified through many numerical and experimental studies [6]. The beam here is modeled using twenty fully nonlinear beam elements (*eBeam29n* in [6]), and the damping matrix is computed using a modal damping ratio of 0.03 for each mode. If the excitation frequency is $\Omega = 32.91\,Hz$ ($=\bar{\omega}_1$), Figure 6 shows the steady-state response $w(0.5L,t)$ at Node 11 and its time-varying frequency and amplitude from HHT analysis. The only intrinsic mode function (IMF) extracted by empirical mode decomposition (EMD) is the signal itself. Because Figs. 6a,c,d show that ω_1 and A_1 modulate exactly at 2Ω and ω_1 and A_1 are at their maxima when $w(0.5L,t)$ is at its maxima or minima, the beam has a hardening cubic nonlinearity (explained right after Eqs. (40a)-(40h)). Figure 6d shows that the continuous frequency band of $w(0.5L,t)$ is 30.3~35.6 Hz but that of $w(0.1L,t)$ is 32.5~33.3 Hz. Figures 6c,d also show that $w(0.1L,t)$ and $w(0.35L,t)$ have about the same amplitude but very different frequency bandwidths. On the other hand, $w(0.35L,t)$ and $w(0.5L,t)$ have about the same frequency bandwidth but very different amplitudes. The reason is that, because the beam is hinged at $x = 0.2L \,\&\, 0.8L$, $w(0.35L,t)$ and $w(0.5L,t)$ are under the stretching effect caused by large deformation but $w(0.1L,t)$ is not. Hence, the frequency variation of a signal is valuable for detection of nonlinearity even under a small-amplitude vibration like this case. For regular on-earth structures, a common understanding is that the amplitude of a linear vibration needs to be less the structure's thickness. Figure 6d shows that, although the vibration amplitude of $w(0.5L,t)$ is less than *h/2*, the frequency variation is large (>15%). *Hence, it is better to define linear vibration by giving a limit on the continuous frequency bandwidth, instead of the vibration amplitude.* When CPD with *p=2* in Eq. (39) is used to decompose the $w(0.5L,t)$ shown in Figure 6a into two components, we obtain $a_1 = 2.153\,mm$ and $a_2\left(\equiv\sqrt{C_2^2 + D_2^2}\right) = 0.0878\,mm$. Using Eq. (40d) we obtain $\alpha = 32\Omega^2 a_2 / a_1^3$ $= 0.2815\Omega^2 = 12037.3$ (with $w(x,t)$ in mm). One can use this nonlinear coefficient in the study of idealized single-mode nonlinear vibration.

Next we consider this beam clamped at both ends and subjected to two equal harmonic forces at $x = L/20 \,\&\, 19L/20$. The first four linear natural frequencies $\bar{\omega}_i$ are 33.6 Hz, 92.5 Hz, 181.4 Hz, and 300.0 Hz. The two forces are $100\sin\Omega t$ N with $\Omega = \bar{\omega}_1 = 33.6$ Hz. Figure 7 shows the steady-state response $w(0.5L,t)$ at Node 11 and its time-varying amplitude and frequency from HHT analysis. Because Figures 7a,c,d show that ω_1 and A_1 modulate exactly at 2Ω and ω_1 and A_1 are at their maxima when $w(0.5L,t)$ is at its maxima or minima, the beam has a hardening cubic nonlinearity. Figure

7d shows that the continuous frequency band of $w(0.5L,t)$ is just 1.0 Hz (33.1~34.1 Hz) but that of $w(L/20,t)$ (where the force is applied) is 1.8 Hz (32.7~34.5 Hz). On the other hand, if the two nodes at $x = L/20 \& 19L/20$ are forced to undergo a harmonic motion $w(L/20,t) = w(19L/20,t) = 0.2\sin\Omega t$ mm, the steady-state response $w(0.5L,t)$ is about the same as that shown in Figure 7a but its continuous frequency band is 1.2 Hz (33.0~34.2 Hz). These two cases indicate that a mechanical system's response to a force can be different from that to a base excitation. For dynamic testing, most mechanical shakers intend to provide defined forces, but the dynamical shaker-structure interaction often makes it impossible. This problem needs to be considered in performing delicate nonlinear system identification.

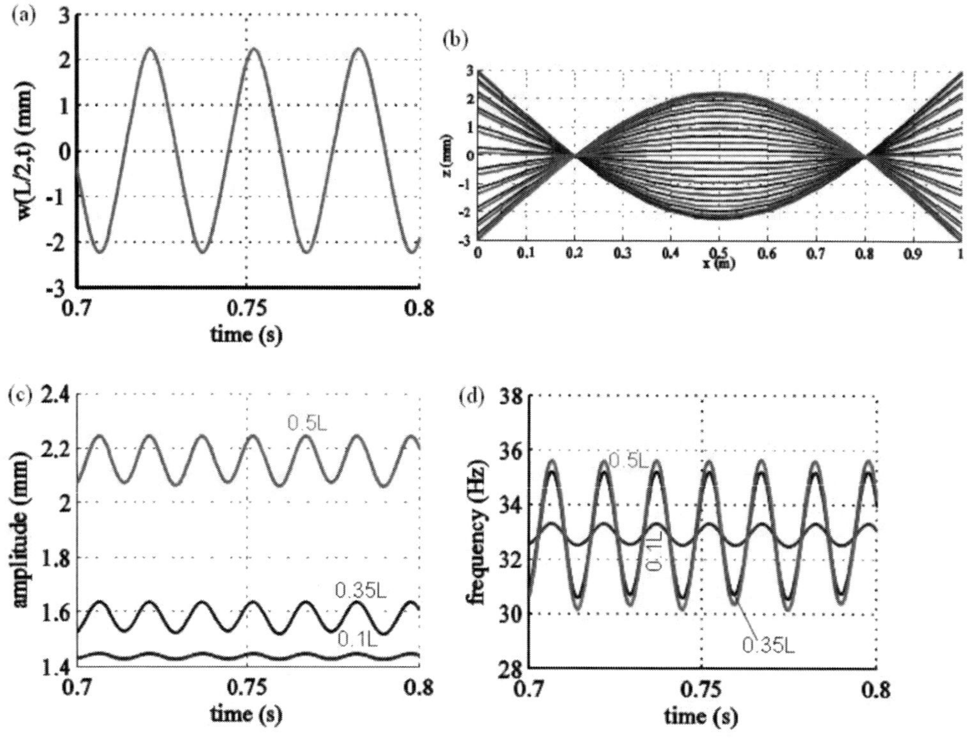

Figure 6. HHT analysis of response of a hinged-hinged beam: (a) $w(0.5\,L,t)$, (b) 25 ODSs during one period, and (*c,d*) $A_1(t)$ and $w_1(t)$ of $w(0.1\,L,t)$, $w(0.35L,t)$ and $w(0.5L,t)$.

Note that the continuous frequency band in Figure 7d is much smaller than that in Figure 6d but the amplitude in Figure 7a (larger than the beam thickness) is larger than that in Figure 6a. This shows again that it is more appropriate to define linearity by giving a limit on the continuous frequency bandwidth, instead of on the vibration amplitude. Considering Figures 7d and 6d, one may think that nonlinearity identification on a clamped-clamped beam is going to be difficult because the frequency variation is small. However, experimental results presented later show that the proposed time-frequency analysis is accurate for delicate identification of nonlinearities even for clamped-clamped beams undergoing low-frequency vibration.

Next we consider the spinup of a clamped-free highly flexible rotor blade subject to a linearly increasing angular speed $\Omega = 2\pi t$ rad/s starting from its static equilibrium position, as shown in Figure 8a with $r = 0$. The titanium alloy blade has $L \times b \times h = 479.0 \times 50.8 \times 0.45 mm^3$, $E = 127\ GPa$, $\nu = 0.36$ and $\rho = 4430\ kg/m^3$. Twenty fully nonlinear beam elements (*eBeam29n* in [6]) are used. The static vertical displacement of the rotor tip due to gravity is computed to be $w = -12.6\ cm$, and the experimental value was $w = -12.7\ cm$[6]. For transient analysis using direct numerical integration, the time step Δt is initially set at 0.002s, and it is automatically adjusted by comparing the previous step's number of iterations for convergence with a reference number of iterations (30 is used in this work). The rotor tip's vertical displacement in Figure 8b shows that the spinning-induced centrifugal force raises up the rotor when the spinning speed Ω increases.

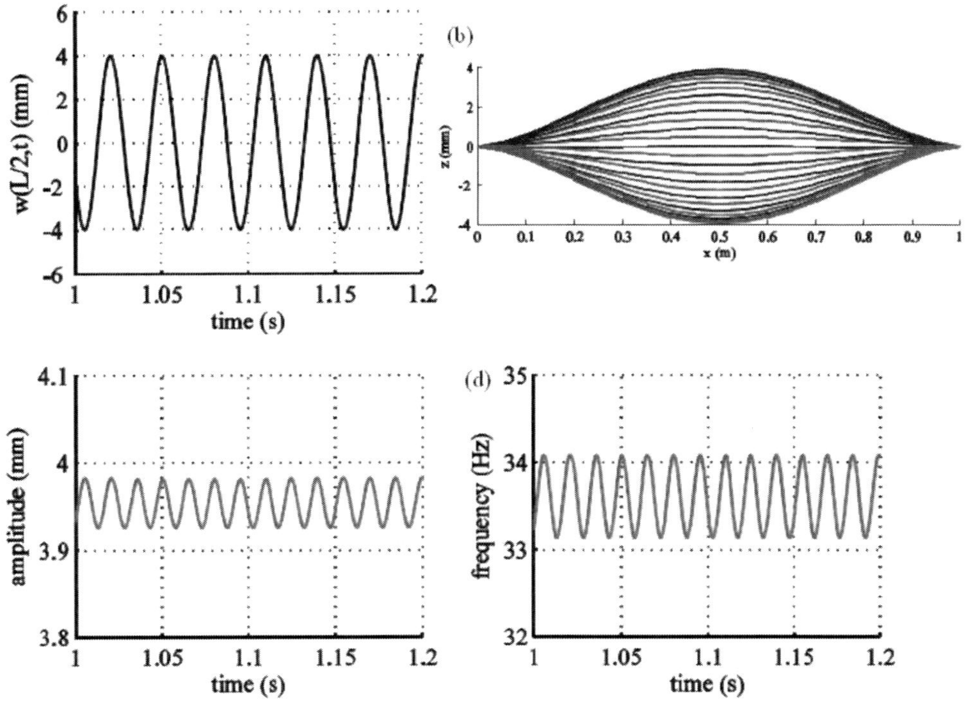

Figure 7. HHT analysis of response of a clamped-clamped beam: (a) $w(0.5L,t)$, (b) 25 ODSs during one period, and (c,d) $A_1(t)$ and $w_1(t)$ of $w(0.5\ L,t)$, $w(0.35L,t)$ and $w(0.5L,t)$ from HTT.

Figures 8c and 8d show the time-varying frequencies from HHT and CPD analyses on the first two IMFs c_1 and c_2 extracted by EMD from the signal shown in Figure 8b. The first IMF is caused by the second and other higher-order transverse modal vibrations, and the second IMF is caused by the first transverse modal vibration. The amplitude of c_1 is less than

0.3 mm and the amplitude of c_2 is about *5 mm.* Because there is no damping added in the numerical simulation, propagating high-frequency waves are small but never disappear. Hence, the time-varying frequencies of IMFs (especially c_1) are not expected to be smooth. However, Figures 8c and 8d show that the average frequency of c_2 increases from about 1.5 Hz to 4.5 Hz and that of c_1 increases from about 10 Hz to 15 Hz. These frequency ranges of c_1 and c_2 agree well with those numerically exact answers (the dotted lines) obtained from multiple shooting analysis with the assumption of steady-state vibrations at different constant spinning speeds [6].

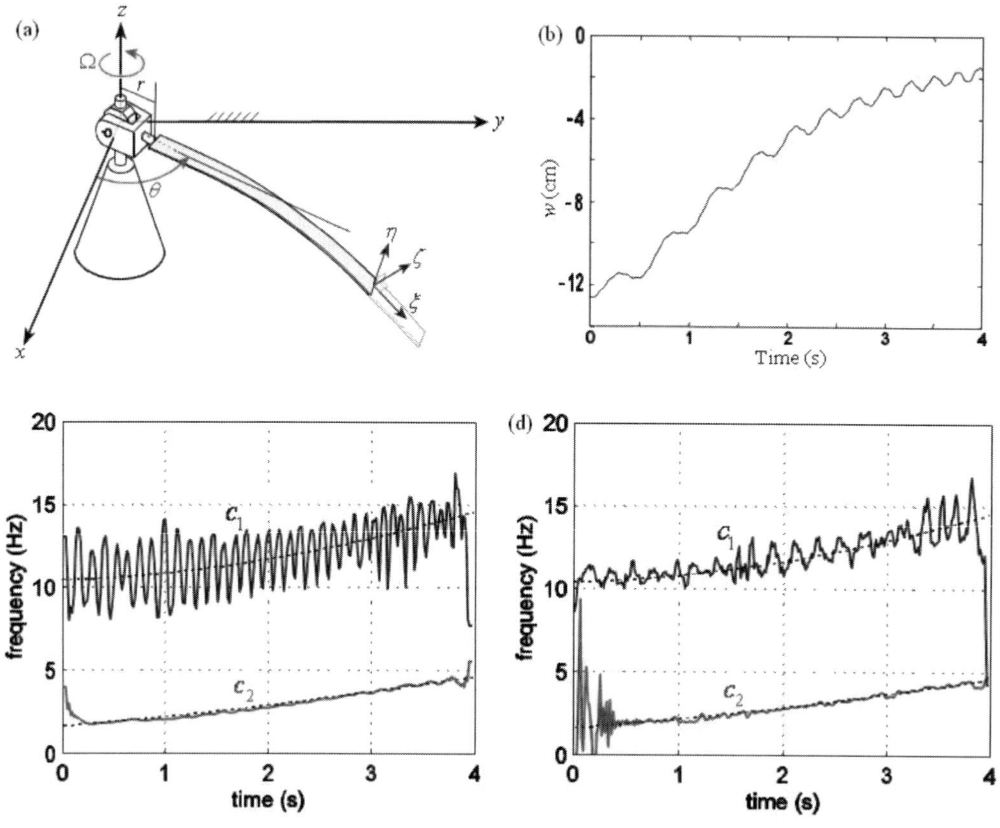

Figure 8. Spinup of a helicopter blade with initial static deformation and gravity under a constant acceleration: (a) coordinate systems, (b) the time-varying vertical displacement of the tip end, (c) time-varying frequencies from HHT analysis, and (d) time-varying frequencies from CPD analysis.

The numerical results from the three examples shown in Figures 6-8 demonstrate that the proposed geometrically exact beam theory can provide high-fidelity noise-free nonlinear responses for delicate identification of nonlinearity and dynamic characteristics. Such high-fidelity structural models are needed for numerical experiments toward development and validation of nonlinear system identification methods.

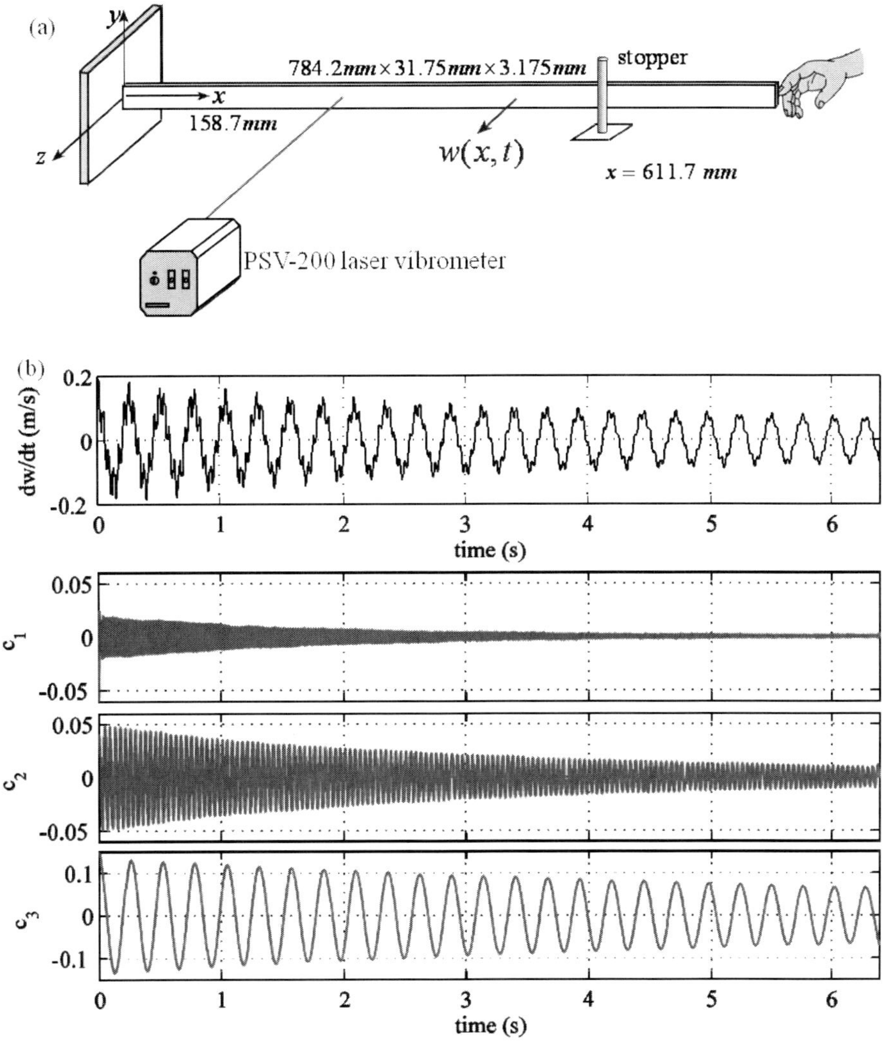

Figure 9. Transient vibration of a clamped-free beam: (a) setup (no stopper), (b) $w(0.1587\text{m},t)$, and (c) IMFs from EMD.

2. Some Experimental Verifications

For experimental verification, we consider a horizontal clamped-free $784.2\times31.75\times3.175\ mm^3$ steel beam subject to an initial static transverse tip displacement about *11.5 cm*, as shown in Figure 9a (without the stopper). Figure 9b shows the transverse vibration velocity of a point at $x = 158.7\ mm$ measured using a Polytec PSV-200 scanning laser vibrometer with a sampling frequency of 2560Hz (i.e., $\Delta t = 1/2560$ s). The first four natural frequencies $\overline{\omega}_i$ are identified from the signal's power spectral density (PSD, Figure 10c) to be 3.82 Hz, 24.1 Hz, 67.5 Hz, and 132.0 Hz, and the mass density is measured to be

$\rho = 6951\, kg / m^3$. Hence, Young's modulus E is estimated by matching the first four experimental natural frequencies with the theoretical linear frequencies (3.85, 24.12, 67.53, 132.34 Hz) from the Euler-Bernoulli beam theory to be $E = 148\, GPa$.

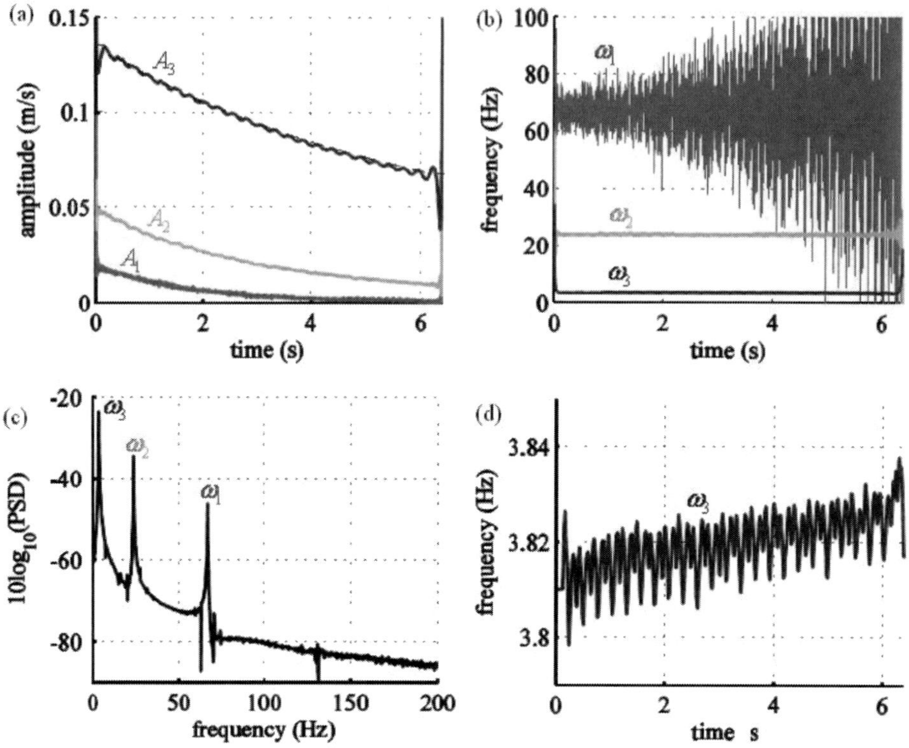

Figure 10. Time-frequency analysis of w(0.1587m,*t*): (a,b) amplitudes and frequencies of IMFs from HHT, (c) PSD of *w*(0.1587m,*t*), and (d) frequency of c_3 (*t*) ferom CPD.

Figure 9c shows the three IMFs sequentially extracted using EMD and the following signal conditioning method. We add $0.6\sin(\bar{\omega}_3 t)$ to the signal and then extract the third and other high-frequency modes and noise into c_1 and then obtain the second and first modal vibrations as c_2 and c_3, respectively. Figures 10a,b show the time-varying amplitudes and frequencies of IMFs from HHT analysis. Because the first IMF c_1 (mainly the third modal vibration) contains other high-frequency modes and noise of the original signal, its calculated frequency ω_1 $(\approx \bar{\omega}_3)$ shows significant variation in Figure 10b. Moreover, Figure 10b shows that, when an IMF's amplitude is small, its calculated frequency is inaccurate because a small c_i in Eq. (36) makes $\tan^{-1}(d_i / c_i)$ difficult to be accurate. For HHT, the inaccurate amplitudes and frequencies at the two data ends in Figures 10a,b are caused by the discontinuity-induced Gibbs' effect. The spectrum of c_1 (not included here) clearly reveals that the 3rd and other high-frequency modes are all extracted into c_1. Figures 10a,d show the time-varying amplitude (broken black line) and frequency of the first-mode vibration obtained by using CPD to process the c_3 in Figure 9c after down-sampling using $\Delta\tilde{t} = 40\Delta t$ and five data points ($4\Delta\tilde{t}$ cover about one period of c_3). Figure 10d clearly shows that the frequency increases about 0.02Hz when the amplitude decreases during 6.4 s, and the frequency varies at a frequency of $2\omega_3 (\approx 2\bar{\omega}_1)$. These two phenomena reveal that the first-mode vibration c_3 contains a softening cubic nonlinearity, which agrees with previous numerical results from multiple shooting analysis [6]. Moreover, Figures 9c and 10d show that the frequency is at its maximum when c_3 is at its maximum or minimum (i.e., locations of zero displacement), and the frequency is at its minimum when c_3 is at its zeros (i.e., locations of maximum or minimum displacement). This also indicates the existence of softening nonlinearity. This example demonstrates how delicate nonlinearity identification can be performed by using the proposed time-frequency analysis that combines HHT and CPD analyses.

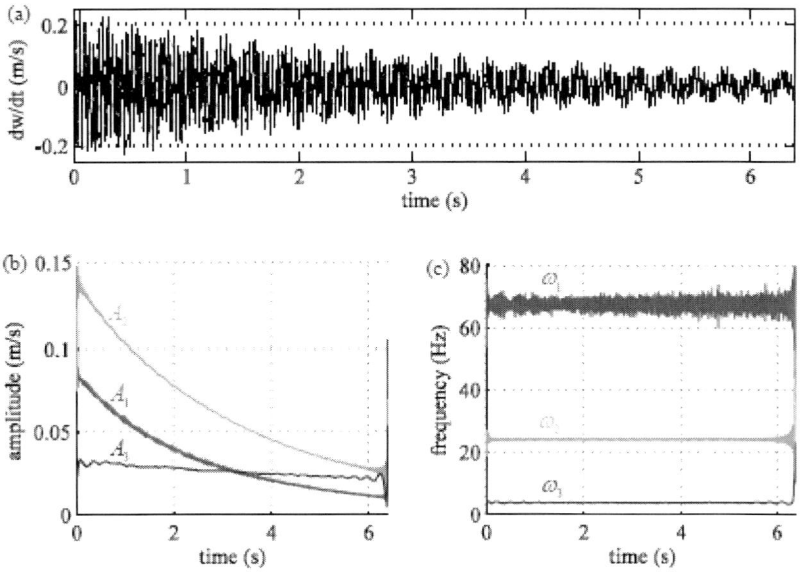

Figure 11. Transient vibration of the clamped-free beam in Figure 9a with stopper: (a) *w* (0.1587 m,*t*), and (b,c) amplitudes and frequencies of IMFs from HHT.

Next we consider the beam shown in Figure 9a with the stopper and an initial static tip displacement about 3.0 *cm*. The stopper is suddenly removed right after the end load is released. The stopper is set at *x/L=0.78* (i.e., the nodal location of the second mode shape) in order to have the second mode participate more in the vibration. Figure 11a shows the transverse vibration velocity measured at $x = 158.7\ mm$. Three IMFs are extracted by EMD with the use of a signal conditioning method, which adds $0.6\sin(\bar{\omega}_3 t)$ to the signal before extracting the first IMF. Figures 11b,c show the time-varying amplitudes and frequencies of the IMFs. The frequency of c_1 in Figure 11c shows less variation than that in Figure 10b just because its amplitude (Figure 11b) is larger than that in Figure 10a. However, the frequency of c_1 in Figure 11c still varies within a large frequency band because c_1 contains the third mode, other high-frequency modes, and measurement noise. Again, the HHT's edge effects shown in Figures 11b,c can be significantly reduced by using CPD to process the IMFs.

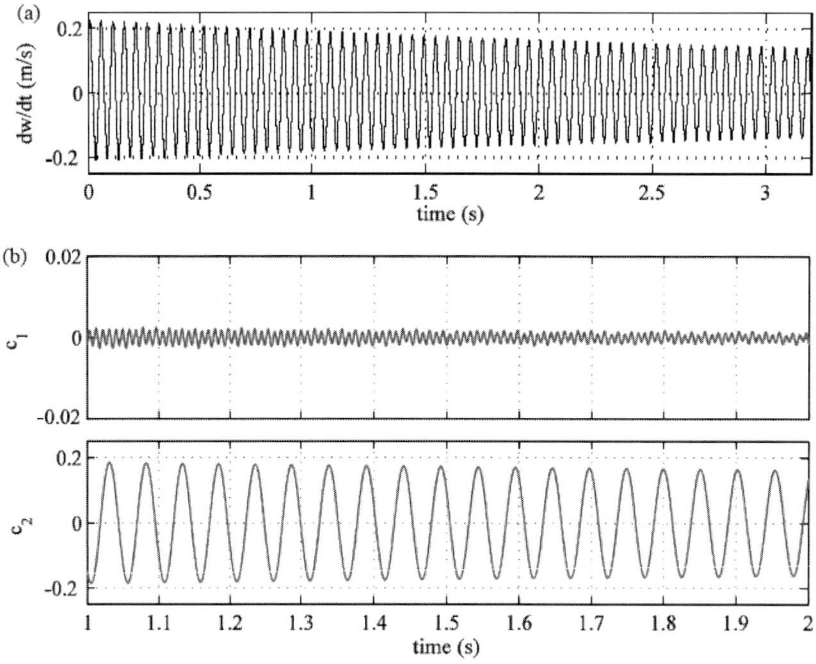

Figure 12. Transient vibration of a clamped-clamped beam: (a) w $(0.5L, t)$, and (b) IMFs from EMD.

Next we consider the beam shown in Figure 9a with clamped-clamped boundary conditions and a length L=859.2 mm. The first four natural frequencies $\bar{\omega}_i$ are identified from the signal's power spectral density (PSD) to be 19.5 Hz, 54.5 Hz, 107.7 Hz, and 179.9 Hz, and the theoretical linear frequencies from the Euler-Bernoulli beam theory are 20.4 Hz, 56.2 Hz, 110.2 Hz, and 182.3 Hz. The mid-span point of the beam is given an initial static displacement about 2 *mm*, and a sampling frequency of 5120 Hz is used. Figures 12a,b show the transverse vibration velocity measured at $x = L/2$ and the 2 IMFs extracted from the signal using EMD with the signal conditioning method by adding $0.6\sin(\bar{\omega}_3 t)$ to the signal before extracting the first IMF. Because it is a symmetric loading, the second mode is not excited and the c_1 and c_2 are the third and first modal vibrations, respectively. Figures 12b and 13a,b show that the amplitude and frequency of c_2 (i.e., the first mode) modulate at $2\bar{\omega}_1$

and they are at their maximum when c_2 is at its zero (i.e., locations of maximum or minimum displacement) and at their minimum when c_2 is at its maximum or minimum (i.e., locations of zero displacement). These clearly reveal the existence of a hardening cubic nonlinearity. Because the frequency from HHT analysis suffers from the edge effect, it cannot show a clear trend of frequency decrease when the amplitude decays. However, Figure 13c,d show the time-varying amplitude and frequency of the first-mode obtained by using CPD to process the c_2 in Figure 12b after down-sampling using $\Delta\tilde{t} = 16\Delta t$ and five data points to cover about one period of c_2. Figure 13d clearly shows that the frequency decreases about 0.1Hz when the amplitude decreases during 3.2 s, and the frequency varies at a frequency $2\omega_2 (\approx 2\bar{\omega}_1)$. These two phenomena again reveal that the first-mode vibration contains a hardening cubic nonlinearity. The maximum vibration amplitude can be calculated from Figure 13c,d to be as small as 1.8mm (< h=3.175mm), but the proposed method can still identify the hardening cubic nonlinearity.

Because the proposed system identification method is based on time-frequency analysis of pointwise time responses, it works for any signals measured on 1D or 2D structures or 3D solids. Of course, more experimental verifications on different mechanical systems with different nonlinearities are needed in order to validate this technology.

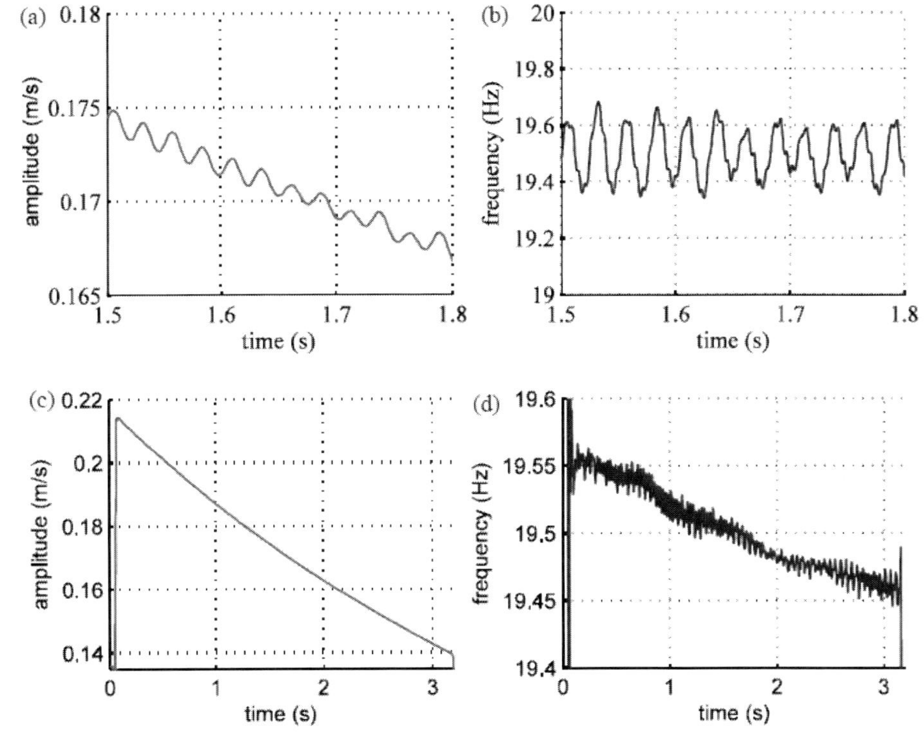

Figure 13. Time-frequency analysis of $c_2(t)$ in Figure 12B: (a,b) amplitude and frequency from HHT, and (c,d) amplitude and frequency from CPD.

3. BEEM for Full-Field Structural Damage Inspection

To show BEEM for full-field damage inspection of structures we consider a $558.8\times 25.4\times 6.4mm^3$ $(=L\times b\times h)$ 2024-T4 aluminum beam with a through-the-width crack having a crack width $c=1.0mm$ and a crack depth $e=1.28mm$ $(e/h=0.2)$ at $x=L/2=279.4mm$. The material properties are Young's modulus $E=73.1GPa$, Poisson's ratio $\nu=0.33$, and mass density $\rho=2780\ kg/m^3$. The total number of measurement points is chosen to be 500 (= M in Eq. (26)). Moreover, n = 4 (see Eq. (22)) and $\Delta x=L/(M-1)$ (i.e., the space between two adjacent measurement points) are used, and hence the sliding-window length used in Eq. (22) is $2n\Delta x$. Figure 14 shows the 10 damage indicators obtained using the BEEM shown in Sec. III to process the 7th mode shape under clamped-clamped boundary conditions and obtained using the Euler-Bernoulli beam theory and the spectral element method [6]. Figures 14e-g show that, except around the damage location, $C_2^*D_2^*=0$, A and β are constant, and ϕ is sectionally constant with a sudden change from -45° to -45.9° at the damage location. The results also show that the central solutions C_1 and D_1 are smooth harmonic functions, the boundary solutions C_2 and $C_1C_2\left(=(T_{den}-\Pi_{den})/(2m\Omega^2),\ \text{see Eq. (24)}\right)$ are continuous but have a small cusp at the damage location. Note that the right-end boundary solution C_2^* and the left-end boundary solution D_2^* are well separated. The thin black lines in Figures 14c–h are obtained for comparison by using $\Delta x=4L/499$ (i.e., quadrupled the sliding-window length). Note that the use of a longer sliding-window length reduces the peaks and variations (especially $C_2^*D_2^*$) of 9 indicators but increases the peak of SSD. We note that all C_i, D_i, C_2^* and D_2^* are continuous because the four functions used in the data fitting (see Eq. (20)) are continuous. However, because W' and W''' are discontinuous [34], the use of continuous functions to fit such a discontinuous function results in Gibbs' phenomenon. The sudden and significant sign changes of D_2, C_2^* and D_2^*, sudden change of ϕ, and localized peaks of $C_2, C_1C_2, A, C_2^*D_2^*, \beta,$ and SSD at the damage location reveal that these ten solutions are all sensitive damage indicators and are mainly caused by Gibbs' phenomenon. Gibbs' phenomenon also makes C_1 and D_1 non-smooth at the damage location, and the deviation of C_1 from the ODS also reveals the damage location. Hence, Gibbs' phenomenon enables this new boundary-effect evaluation method (BEEM) for accurate identification of small damage and defects.

If the estimated β used in Eq. (20) is too large (small), the ϕ in Figure 14g would gradually increase (decrease) and oscillate, and the β obtained from Eq. (27) and shown in Figure 14f would oscillate. These two phenomena can be used to determine whether the used β value is accurate. Moreover, if the TKA is used to process the so-obtained C_1, the obtained β is accurate and does not oscillate. For example, the exact value from the spectral element analysis is $\beta = 42.1375$. If an inaccurate estimation $\beta = 40$ is used in Eq. (20), the obtained ϕ continuously changes from -46.5° to -112.8°, the β from Eq. (27) varies between 40 and 44.4 with an average of 42.2, the β from the original four-point TKA varies between 42.13 and 42.16, and the β from an improved five-point TKA (see [46]) varies between 42.135 and 42.143. These results indicate that, even if an inaccurate β is used in Eq. (20), the so-obtained C_1 is still very accurate. This new approach of using TKA to estimate β dramatically improves the accuracy of the original BEEM [34]. With the exact value of $\beta = 42.1375$, the seventh natural frequency can be calculated as $\omega_7 = \beta^2 \sqrt{EI/m} = 2677.2Hz$. If no damage, $\beta = 42.1653$ and $\omega_7 = 2680.7$ Hz. Because the crack only reduces ω_7 by 0.13% and β by 0.066%, it is very difficult to use a single value of β to determine the existence of damage, but the spatial distribution of β in Figure 14f clearly indicates the existence and location of damage.

Because a clamped-free boundary condition is easier than a clamped-clamped boundary condition in setting up experiments without inappropriate pre-straining, we set up a clamped-free $608.0 \times 18.9 \times 4.8 mm^3$ aluminum beam to be excited at 1556.25 Hz (around the 8th natural frequency) by an integrated $50.8 \times 25.4 \times 0.38\ mm^3$ QP10N PZT patch at $x = 0$ (see Fig. 1), as shown in Figure 15. Three cracks were cut on the beam at $x/L = 0.244, 0.489, 0.747$ with a crack width $c = 1\ mm$ and a crack depth ratio $e/h = 0.13, 0.28, 0.16$. Figure 16 shows the 10 damage indicators obtained using $\Delta x = 3L/500$, and the thin black lines are obtained for comparison using $\Delta x = 6L/500$. Note that significant sign changes of D_2, C_2^* and D_2^*, the sudden drop of ϕ, and the peaks of C_2, C_1C_2, A, $C_2^*D_2^*$, β and SSD clearly indicate the three damage locations. The deviations of C_1 and D_1 from the ODS also reveal the damage locations. The deviations and peaks around $x=0$ in Figures 16c-h between the two sets of damage indicators obtained from using two different sliding-window lengths are due to the actuation through a PZT patch at the fixed end.

The operational deflection shape in Figure 16a is actually spatial distribution of velocities (in *m/s*) directly measured by the scanning laser vibrometer. The corresponding vibration amplitude of Figure 16a has a value of $81.8\ \mu m$. When the vibration amplitude is small, the experimental spectral noise becomes relatively significant. Experimental results confirm that, when the input voltage amplitude to the PZT actuator increases, the curves in Figure 16 become more clear.

The PSV-200 scanning laser vibrometer used in the experiments is a sensitive velocity sensor system with a velocity resolution as high as $0.1\ \mu$m/s and a high frequency bandwidth (0.2 Hz to 20 MHz). Hence, it can accurately record high-frequency but small-amplitude structural vibrations like the structural responses excited by a small PZT patch in our experiments. For high-frequency short-duration signals, one needs to make sure the chosen sampling frequency is high enough. For structures of complex geometries, the proposed method can be used to perform localized area-by-area inspection because the method does not require any boundary conditions to be known. The earlier version of the BEEM had been used to inspect complex structures like rectangular and circular isotropic plates with multiple damages, composite laminates with internal delaminated areas, and actual aircraft wing panels with inside ribs and multiple damages [34].

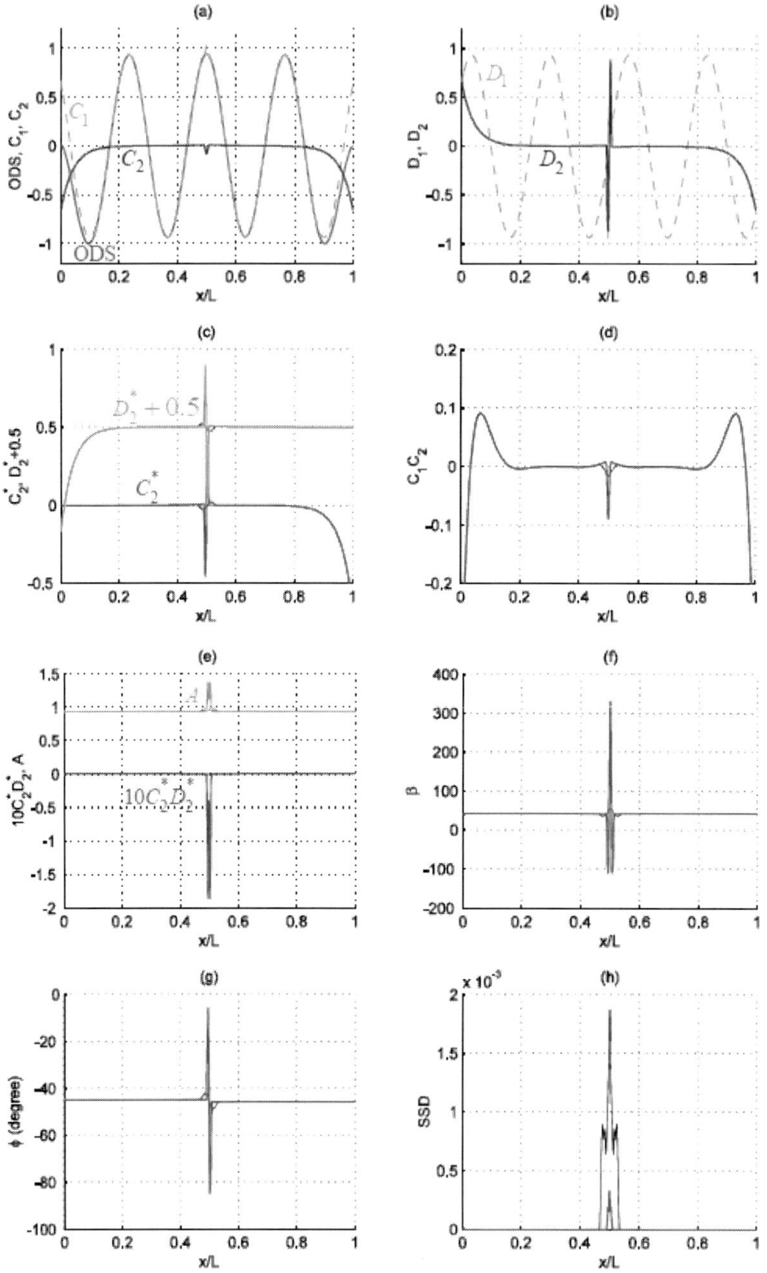

Figure 14. Damage indicators from BEEM analysis of the 7th mode shape of a clamped-clamped beam: (a) ODS, C_1 and C_2, (b) D_1 and D_2, (c) C_2^* and D_2^*, (d) C_1 C_2, (e) C_2^* D_2^* and A, (f) β, (g) ϕ, and (h) SSD.

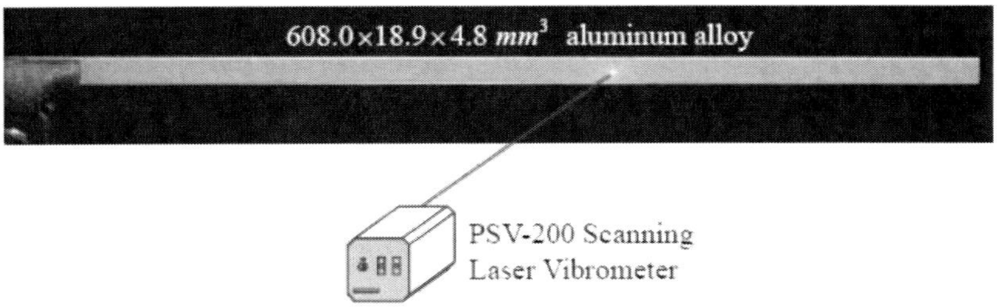

Figure 15. Experimental setup for measuring operational deflection shapes.

4. System Identification and Damage Monitoring

To show HHT and CPD for capturing damage-induced transient events we consider the following nonlinear non-stationary system:

$$
\begin{gathered}
m\left(\ddot{u}+2\varsigma\omega\dot{u}+r\omega^2(u+0.05u^3)\right)=F_0,\ u(0)=2,\dot{u}(0)=0 \\
r=\begin{cases}
0.98-0.02\tanh\dfrac{t-5}{t_r} & \text{for } 0\le t\le 7.5 \\
0.94-0.02\tanh\dfrac{t-10}{t_r} & \text{for } 7.5< t\le 12.5 \\
0.90-0.02\tanh\dfrac{t-15}{t_r} & \text{for } 12.5< t\le 17.5 \\
0.86-0.02\tanh\dfrac{t-20}{t_r} & \text{for } 17.5< t
\end{cases} \\
\omega=2\pi,\varsigma=0.01,m=2,t_r=0.05
\end{gathered}
\tag{41}
$$

where $\tanh[(t-t_a)/t_r]$ is a function that changes from *-1* to *+1* around $t=t_a$ with a transition period determined by t_r. In addition, there is an impact force $F_0=5\ N$ applied between $25\le t\le 25.05\ s$. $\Delta t=0.1$ s is used and a vector of normally distributed zero-mean noise with a small standard deviation of 0.003 is added to the signal obtained from direct numerical integration using the Runge-Kutta method. The PSD in Figure 17b shows a wide frequency bandwidth, which is caused by the multiple transient events. Figure 17c shows that the frequency tracked by CPD clearly shows the impact instant at *t=25s* and the 4 time instants when the stiffness suddenly decreases by 4% and the natural frequency drops by 2% (because the vibration frequency is proportional to $\sqrt{r}$). Figure 17c also shows that the free vibration frequency decreases when the amplitude decreases, indicating a hardening–type nonlinearity. Figure 17d shows that the frequency extracted by HHT does not clearly show the 5 transient events (especially the impact event) because the low sampling rate (10 Hz), noise, and the edge effect introduced by Hilbert transform (HT) decrease the accuracy. On the other hand, a low sampling rate is favorable to CPD, CPD averages out noise because it is based on curve fitting, and the edge effect does not exist in CPD because HT is not used. This example shows that CPD is better than HHT for capturing transient events. The accuracy of Figure 17d from HHT can be improved by using a smaller Δt , but, for this case here, the results can never be as good as those from CPD because it is a transient response.

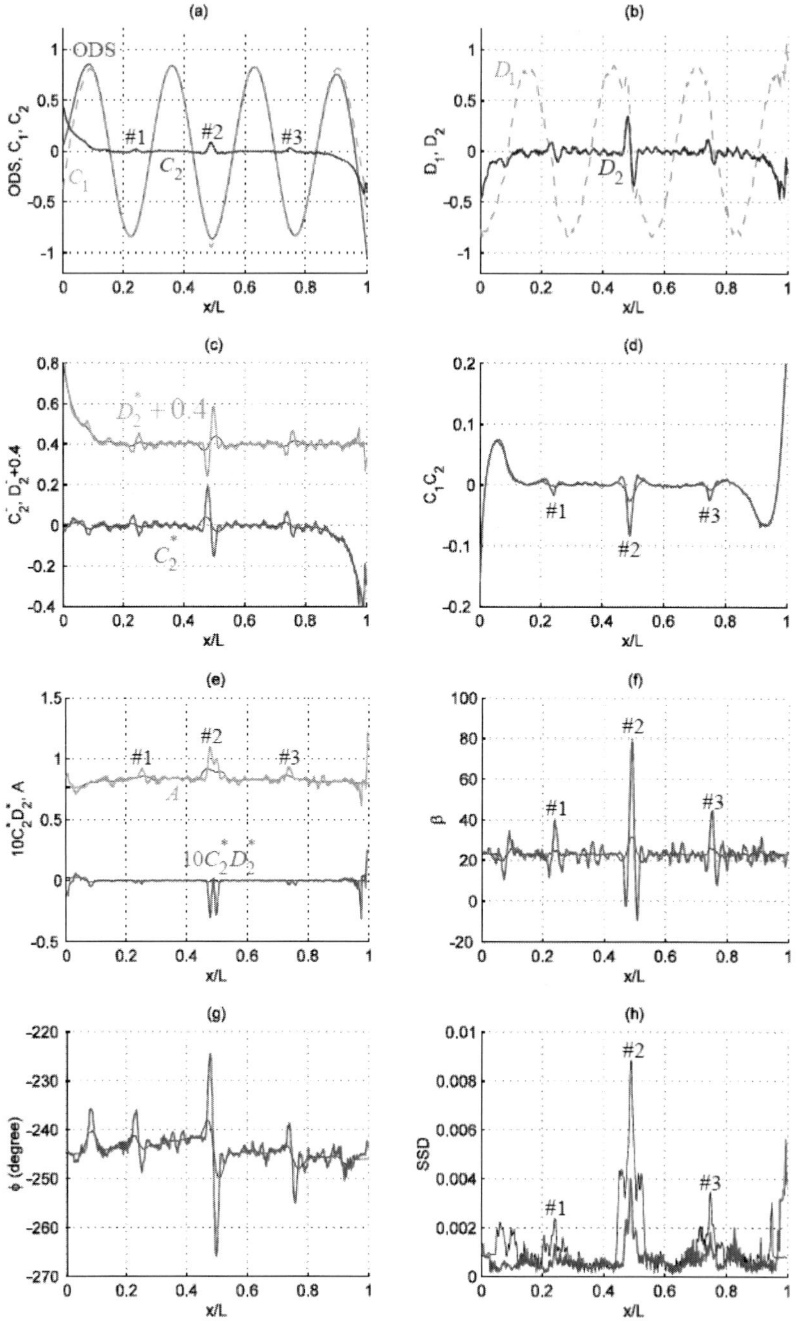

Figure 16. Damage indicators from BEEM analysis of an experimental 1556Hz ODS with 3 cracks: (a) ODS, C_1 and C_2, (b) D_1 and D_2, (c) C_2^* and D_2^*, (d) C_1 C_2, (e) C_2^* D_2^* and A, (f) β, (g) ϕ, and (h) SSD.

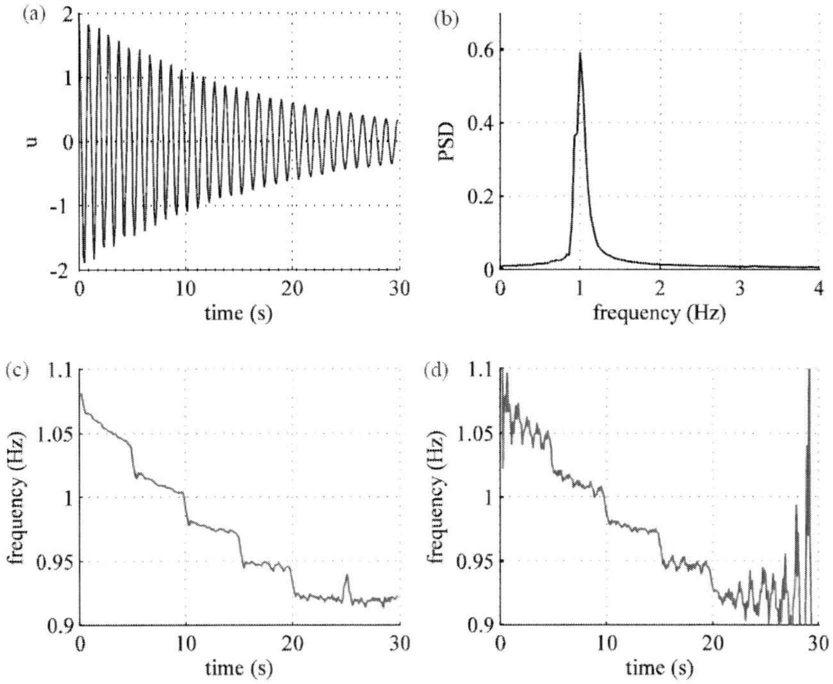

Figure 17. Time-varying frequency of the oscillator shown in Eq. (41): (a) $u(t)$, (b) PSD, c) from 5-point CPD, and (d) from HHT.

To show the use of time-frequency analysis for identification of nonlinearity and system parameters, we consider the following nonlinear oscillator:

$$m\left(\ddot{u}+2\varsigma\omega\dot{u}+\omega^2 u+\alpha u^3\right)=F_0\cos\Omega t \tag{42a}$$

$$m=2,\ \omega=2\pi,\ \varsigma=0.01,\ \alpha=10,\ F_0(=Fm)=50,\ \Omega=\omega \tag{42b}$$

Direct numerical integration is performed with $\Delta t=1/30$ s for 150 s to damp out any transient effect. Figure 18b shows that the steady-state response consists of several harmonics equally separated by 2Ω (= 2 Hz), indicating a frequency modulated at 2Ω. Figures 18c,d show that the time-varying frequency and amplitude from HHT analysis modulate at 2Ω, indicating a cubic nonlinearity (see Eqs. (40c) and (40e)). Moreover, because the frequency and amplitude are at their maxima when u is at its local maxima or minima and are around their minima when $u=0$, we know that $\alpha>0$. If only one pair of conjugate functions are used in Eq. (39), Figures 18c,d show that the time-varying amplitude and frequency from CPD do modulate at 2Ω and the amplitude is at its maxima when u is at its local extrema, indicating a hardening cubic nonlinearity. However, the time-varying frequency does not clearly indicate a hardening nonlinearity and the variations of amplitude and frequency are not the same as those from HHT analysis because they are different by definition (see Eqs. (36) and (38)). Unfortunately, HHT is not able to decompose the time-varying amplitude shown in Figure 18c into a_1 and a_3 that are needed for using perturbation solutions to estimate system parameters. On the other hand, CPD is able to. The amplitudes and frequencies from CPD using two pairs of conjugate functions with $\omega_1 = \Omega$ and $\omega_2 = 3\Omega$ in Eq. (39) are constant and $a_1 = \sqrt{C_1^2 + D_1^2} = 1.4834$ and $a_2 = \sqrt{C_2^2 + D_2^2} = 0.0286$. Because there are five unknowns in Eq. (39), at least 5 data points are needed for frequency extraction and 7 points are used here. One can use Eq. (40d) to estimate $\alpha = 32\Omega^2 a_2 / a_1^3 = 11.07$. Furthermore, we obtain $\phi = 0.0440$ from Figure 18a and Eqs. (40a) and (40b) and $\varsigma = [3\alpha a_1^2/(8\omega) - \sigma]\, tan\phi / \omega = 0.0102$ from Eq. (40g) with $\sigma(= \Omega - \omega) = 0$. Then Eq. (40f) with $\sigma = 0$ is used to obtain $F = 27.127\ (= F_0 / m)$ and hence $m = F_0 / F = 1.843$ is estimated. The key point here is that all α, ς and m can be estimated with an error less than 10% from just one steady-state response to a harmonic excitation at $\Omega \approx \omega$. It follows from Eq. (40h) that the undamped natural frequency under a vibration amplitude $a_1 = 1.4834$ is $\hat{\omega} = 1.21\omega$, which indicates that the magnitude of nonlinearity is strong. Because the perturbation solutions are derived based on the assumption of weak nonlinearity, the estimation accuracy increases when the nonlinearity coefficient α and the excitation force F_0 decrease. Also, the accuracy of estimations can be further improved when more sets of responses to different F_0 and/or Ω are available.

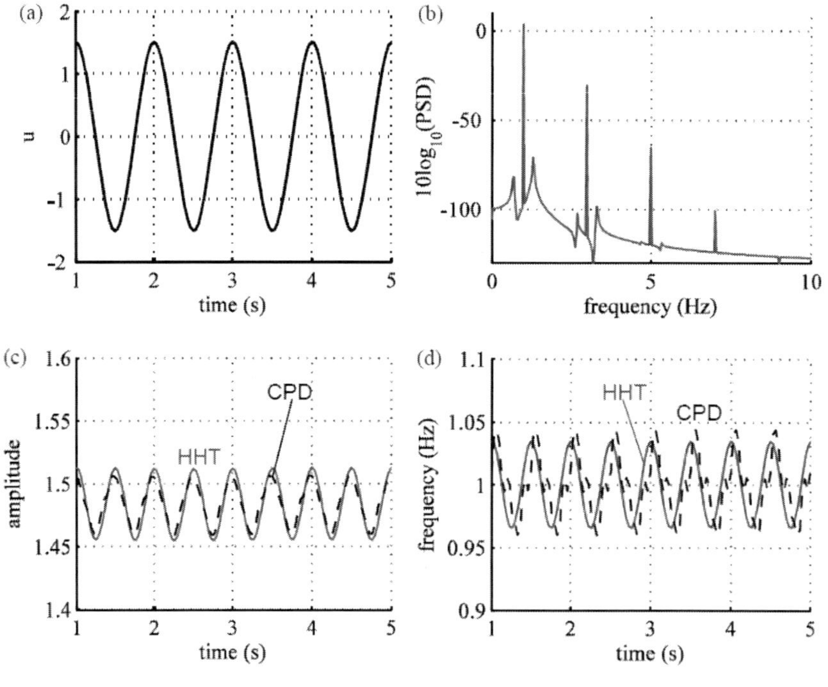

Figure 18. Time-frequency analysis of the steady-state response of Eqs. (42a) and(42b): (a) $u(t)$, (b) PSD, (c) amplitude, and (d) frequency.

To demonstrate nonparametric identification, we consider Eq. (42a) with

$$\omega = 2\pi,\ \varsigma = 0.01,\ \alpha = 10,\ F = 0,\ u(0) = 0,\ \dot{u}(0) = 30. \tag{43}$$

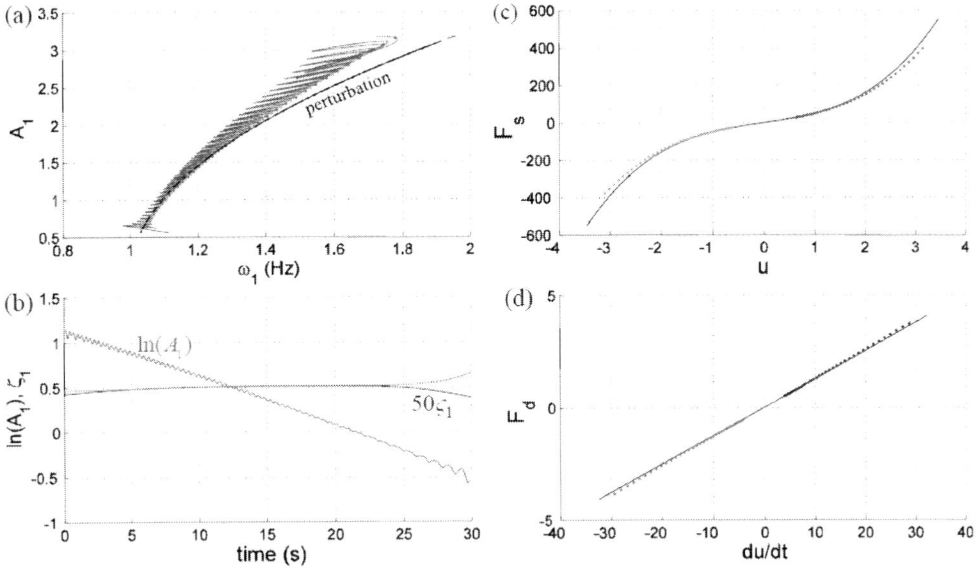

Figure 19. Nonlinearity identification of Eqs. (42a) and (43): (a) $w_1 - A_1$, (b) ln (A_1) and (c) u-F_s curve, and (d) u-F_d curve. The black line in (a) represents the perturbation soluton. The black solid lines in (c) and (d) represent the exact spring and damping forces, respectively.

The only IMF, $c_1 (= u)$, is the free damped distorted harmonic. Figure 19a shows its ω_1-A_1 curve from HHT. When A_1 is large, the backbone curve from the perturbation solution Eq. (40h) for a weakly nonlinear oscillator deviates from the actual backbone curve (the green dots) because this is a strong nonlinear system. The errors around the two ends of the time-varying damping ratio ς_1 (≈ 0.01) in Figure 19b are caused by the edge effect. The nonlinear specific spring force (i.e., force per unit mass) F_s in Figure 19c is calculated as $F_s = \omega^2 A$, where the A and ω are the instantaneous amplitude and frequency when u is at its extrema (i.e., peaks or valleys). The specific damping force (i.e., force per unit mass) F_d in Figure 19d is calculated as $F_d = 2\varsigma\omega B$, where the ς, ω and B are the instantaneous damping ratio, frequency, and amplitude when $\dot{u}$ is at its extrema. Figure 19d shows that the estimated F_d agrees well with the exact one $(F_d = 2\varsigma\omega\dot{u})$. Figure 19c shows that the estimated F_s shows a minor deviation from the exact one $(F_s = \omega^2 u + \alpha u^3)$ because the amplitude and frequency of c_1 modulate at $2\omega_1$. Further decomposition of u using the proposed CPD with $p=2$ and $\omega_2 = 3\omega_1$ in Eq. (39) yields an F_s coinciding with the exact curve in Figure 19c. The key point here is that time-frequency analysis of just one transient response can provide the order (cubic), type (hardening), α and ω (from Eq. (40h) and Figures 19a&19c), and ς (from Figures 19b&19d) with decent accuracy.

Conclusion

Because nonlinearity identification plays the key role in system identification and damage inspection of structures, high-fidelity nonlinear structural models are needed for providing accurate noise-free nonlinear responses for numerical experiments toward development of dynamics-based identification/inspection techniques. A geometrically exact beam theory is derived to show how to accurately account for geometric nonlinearities in structural modeling, and the nonlinear beam theory is used to illustrate time-varying dynamic characteristics of nonlinear systems. One derived new concept is that linear vibration of a structure is better to be defined by giving a limit on its continuous frequency bandwidth of response to a single-frequency excitation, not on its vibration amplitude. However, what should be the bandwidth limit is a topic that requires further study.

A dynamics-based methodology for identification/inspection of thin-walled structures is proposed. It combines a boundary-effect evaluation method (BEEM) for full-field damage inspection by processing operational deflection shapes (ODSs) measured by a scanning laser vibrometer (or other full-field measurement systems), and a conjugate-pair decomposition (CPD) method for system identification and damage inspection/monitoring by performing time-frequency analysis of pointwise time traces of measured points. Both BEEM and CPD are nondestructive, use no structural models or historical data for comparison, and work on any structural materials.

Damage introduces new boundaries to a structure, and influences of such boundaries on steady-state high-frequency ODSs are spatially localized effects. The BEEM takes advantage of these characteristics in performing area-by-area extraction of damage-induced boundary effects from ODSs to reveal damage locations. The BEEM decomposes an ODS into central and boundary solutions by using a sliding-window least-squares data-fitting technique. Numerical and experimental results show that the derived ten damage indicators for beams are sensitive because of Gibbs' phenomenon.

Responses of damaged structures often contain nonlinearities and transient events. The CPD method analyzes time-domain data and uses perturbation solutions for estimation of parameters. CPD uses one or more pairs of windowed adaptive harmonics and function orthogonality to extract time-varying frequency and amplitude of a dynamical signal. Unlike HHT, CPD processes only time-domain data and is free from the edge effect caused by Gibbs' phenomenon due to Hilbert transform. Moreover, CPD is good at capturing transient events. Results show that a nonlinear single-degree-of-freedom system can be completely identified by time-frequency analysis of its steady-state response to a single-frequency excitation, and complete nonparametric identification can be done by time-frequency analysis of its free damped vibration. This is impossible for any other methods in the literature.

For signals from high-frequency sampling and without significant discontinuity between the beginning and ending points, HHT can provide more accurate time-varying frequencies and amplitudes. Hence, how to combine HHT and CPD for better time-frequency analysis is a topic worth study. The proposed system identification methodology using time-frequency analysis and perturbation solutions need to be further studied and experimentally verified for dynamical systems with other types and orders of nonlinearities.

Acknowledgment

This work has been supported by the National Science Foundation through Grants CMMI-1039433 and IIP-1026883 and the NASA Headquarters through Grant NNX09AV08A. The support is gratefully acknowledged.

References

ABAQUS-6.9 *Theory Manual,* Dassault Systèmes Simulia Corp., Providence, RI, 2009.

Adams, D.E., *Health Monitoring of Structural Materials and Components: Methods with Applications,* Wiley-Interscience, New York, 2007.

Antman, S.S., *Nonlinear Problems of Elasticity*, Springer, 1995.

Balageas, D., Fritzen, C.P., and Güemes, A., *Structural Health Monitoring*, International Society for Technology in Education (ISTE), London, UK, 2006.

Bendat, J. and Piersol, A., *Random Data: Analysis and Measurement Procedures*, 3rd Ed., John Wiley & Sons, New York, 2000.

Bismarck-Nasr, M.N., "Finite Elements in Aeroelasticity of Plates and Shells," *Applied Mechanics Reviews* (49), s17-s24, 1996.

Brigham, E.O., *The Fast Fourier Transform*, Prentice-Hall, Englewood Cliffs, New Jersey, 1974.

Chang, F.K. (editor), *Structural Health Monitoring: Advances and Challenges in Design and Implementation*, Destech Publishing, *Proceedings of the 5th International Workshop on Structural Health Monitoring*, Stanford, CA, September 14-17, 2005.

Collacott, R.A., *Structural Integrity Monitoring*, Chapman and Hall, New York, 1985.

Doebling, S.W., Farrar, C.R., Prime, M.B., and Shevitz, D.W., *Damage Identification and Health Monitoring of Structural and Mechanical Systems from Changes in Their Vibration Characteristics: A Literature Review,* Report No. LA-13070-MS, Los Alamos National Laboratory, 1996.

Dowell, E.H., "Panel Flutter: A Review of the Aeroelastic Stability of Plates and Shells," *AIAA Journal* (8), 385-399, 1970.

Dowell, E.H. and Ilgamov, M., *Studies in Nonlinear Aeroelasticity*, Spring-Verlag, New York, 1988.

Farrar, C.R., Worden, K., Todd, M.D., Park, G., Nichols, J., Adams, D.E., Bement, M.T., and Farinholt, K.M., *Nonlinear System Identification for Damage Detection*, Report No. LA-14353, Los Alamos National Laboratory, 2007.

Flandrin, P., Rilling, G., and Goncalves, P., "Empirical Mode Decomposition as a Filter Bank," *IEEE Signal Processing Letters* 11(2), 112–114, 2004.

Giurgiutiu, V., Zagrai, A., and Bao, J., "Piezoelectric Wafer Embedded Active Sensors for Aging Aircraft Structural Health Monitoring," *Structural Health Monitoring* 1, 41-61, 2002.

Graff, K.F., *Wave Motion in Elastic Solids*, Dover, New York, 1975.

Greaves, J., *EVaRT 4.6 User's Manual*, Motion Analysis Corporation, Santa Rosa, CA (http://www.motionanalysis.com).

Gruen, A., "Fundamentals of Videogrammetry — A Review," *Human Movement Science* 16 (2-3), 155-187, 1997.

Hahn, S.L., *Hilbert Transforms in Signal Processing,* Artech House, Boston, 1996.

Hodges, D.H., *Nonlinear Composite Beam Theory,* AIAA, Reston, Virginia, 2006.

Huang, N.E. and Attoh-Okine, N.O. (editors), *The Hilbert-Huang Transform in Engineering*, CRC Press, Boca Raton, FL, 2005.

Huang, N.E., Shen, Z., and Long, S.R., "A New View of Nonlinear Water Waves: The Hilbert Spectrum," *Annual Review of Fluid Mechanics* 31, 417-457, 1999.

Huang, N.E., Shen, Z., Long, S.R., Wu, M.C., Shih, H.H., Zheng, Q., Yen, N.C., Tung, C.C., and Liu, H.H., "The Empirical Mode Decomposition and the Hilbert Spectrum for Nonlinear and Non-Stationary Time Series Analysis," *Proc. R. Soc. Lond.* A 454, 903-995, 1998.

Huang, N.E., Wu, M.C., Long, S.R., Shen, S.S.P., Qu, W., Gloersen, P., and Fan, K.L., "A Confidence Limit for the Empirical Mode Decomposition and Hilbert Spectral Analysis," *Proc. R. Soc. Lond.* A 459, 2317-2345, 2003.

Ihn, J.B. and Chang, F.K., "Detection and Monitoring of Hidden Fatigue Crack Growth Using a Built-In Piezoelectric Sensor/Actuator Network: Part I and Part II," *Journal of Smart Materials and Structures* 13, 609-630, 2004.

Jenkins, C.H.M., *Gossamer Spacecraft: Membrane and Inflatable Structures Technology for Space Applications*, AIAA, Reston, Virginia, 2001.

Lee, S.H., *MSC/NASTRAN Handbook for Nonlinear Analysis,* Version 67, The MacNeal-Schwendler Corporation, 1992.

Maragos, P., Kaiser, J.F., and Quatieri, T.F., "Energy Separation in Signal Modulations with Application to Speech Analysis," *IEEE Transactions on Signal Processing* 41, 3024–3051, 1993.

Mei, C., Abdel-Motagaly, K. and Chen, R., "Review of Nonlinear Panel Flutter at Supersonic and Hypersonic Speeds," *Applied Mechanics Reviews* (52), 321-332, 2000.

Mikhail, E.M., Bethel, J.S., and McGlone, J.C., *Introduction to Modern Photogrammetry*, John Wiley & Sons, New York, 2001.

Moon, F.C., *Chaotic Vibrations, An Introduction for Applied Scientists and Engineers*, Wiley-Interscience, New York, 1987.

Nayfeh, A.H. and Mook, D.T., *Nonlinear Oscillations,* Wiley-Interscience, New York, 1979.

Nayfeh, A.H. and Pai, P.F., *Linear and Nonlinear Structural Mechanics*, Wiley-Interscience, New York, 2004.

Pai, P.F., "Circular Instantaneous Frequency," *Advances in Adaptive Data Analysis* 2(1), 39-64, 2010.

Pai, P.F., "Geometrically Exact Beam Theory without Euler Angles," *Int. J. of Solids and Structures* 48, 3075-3090, 2011.

Pai, P.F., *Highly Flexible Structures: Modeling, Computation and Experimentation*, AIAA, Reston, Virginia, 2007.

Pai, P.F., "Long Transient Phenomenon in Nonlinear Structural Vibration," *J. of Sound and Vibration* 321, 305-318, 2009.

Pai, P.F., "Three Kinematic Representations for Modeling of Highly Flexible Beams and Their Applications," *Int. J. of Solids and Structures* 48, 2764-2777, 2011.

Pai, P.F., "Three-Point Frequency Tracking Method," *Structural Health Monitoring* 8(6), 425-442, 2009.

Pai, P.F. and Huang, L., "Dynamics- and Laser-Based Boundary Effect Evaluation Method for Damage Inspection of One- and Two-Dimensional Structures," *Journal of Nondestructive Evaluation* 25(2), 83-105, 2006.

Pai, P.F., Ramanathan, S., Hu, J., Chernova, D.K., Qian, X., and Wu, G., "Camera-based Noncontact Metrology for Static/Dynamic Testing of Flexible Multibody Systems," *Measurement Science and Technology* 21(8), 085302 (14pp), 2010.

Pai, P.F., Sundaresan, M.J., and Nguyen, B.A., "Identification of Nonlinearities for Damage Inspection of Thin-Walled Structures," SPIE 19th International Symposium on Smart Structures and Materials & Nondestructive Evaluation and Health Monitoring, San Diego, California, March 11-15, 2012.

Photomodeler User's Manual Version 6.3, Eos Systems, Vancouver, Canada, 2009. (http://www.photomodeler.com).

Ratay, R.T. (editor), *Structural Condition Assessment*, Wiley-Interscience, New York, 2005.

Sohn, H., Farrar, C.R., Hemez, F.M., Shunk, D.D., Stinemates, D.W., and Nadler, B.R., *A Review of Structural Health Monitoring Literature*: 1996-2001, Report No. LA-13976-MS, Los Alamos National Laboratory, 2003.

Steinchen, W. and Yang, L., *Digital Shearography*, SPIE Press, Washington, DC, 2003.

Stepan, G., Szalai, R., Mann, B.P., Bayly, P.V., Insperger, T., Gradisek, J., and Govekar, E., "Nonlinear Dynamics of High-Speed Milling ---Analyses, Numerics, and Experiments," *Journal of Vibration and Acoustics* (127), 197-203, 2005.

Teager, *H*.M., "Some Observations *on O*ral Air Flow During Phonation," *IEEE Transactions on Acoustics, Speech and Signal Processing* 28, 599-601, 1980.

Timoshenko, S.P. and Goodier, J.N., *Theory of Elasticity*, third ed., McGraw-Hill, New York, 1970.

Villaggio, P., *Mathematical Models for Elastic Structures,* Cambridge University Press, 1997.

Wang, C.S. and Chang, F.K., "Built-in Diagnostics for Impact Damage Identification of Composite Structures," Proceedings of the 2nd International Workshop on Structural Health Monitoring, San Francisco, CA, 1999.

Worden, K. and Tomlinson, G.R., *Nonlinearity in Structural Dynamics: Detection, Identification and Modelling,* Institute of Physics Publishing, Bristol and Philadelphia, 2001.

Wu, Z. and Huang, N.E., "A Study of the Characteristics of White Noise Using the Empirical Mode Decomposition Method," *Proc. R. Soc. Lond.* A 460, 1597-1611, 2004.

Wu, G., He, X., and Pai, P.F., "Geometrically Exact 3D Beam Element for Arbitrary Large Rigid-Elastic Deformation Analysis of Aerospace Structures," *Finite Elements in Analysis & Design* 47, 402-412, 2011.

Zhao, M.X. and Balachandran, B., "Dynamics and Stability of Milling Process," *Int. J. of Solids and Structures* (38), 2233-2248, 2001.

Zimmermann, D.C. and Kaouk, M., "Structural Damage Detection Using a Minimum Rank Update Theory," *Journal of Vibration and Acoustics* 116(2), 222-231, 1994.

In: Dynamical Systems
Editor: Jean Michel Tchuenche, pp. 135-162

ISBN: 978-1-62808-001-8

Chapter 5

COMPLEXITY AND CHAOS THEORY IN SOCIAL SCIENCES: SCIENTIFIC PHILOSOPHY OR QUANTITATIVE SCIENCE?

Rosalia Condorelli
Department of Political and Social Sciences, Catania University, Catania, Italy

"Scientific genius is the capacity to be surprised."
Henry Poincarè

Abstract

This chapter deals with the possibility of applying *Complexity and Chaos Theory* to Social Sciences. There are many theorists in the field of the Physical Sciences who share Ruelle's thesis that it is more controversial to use nonlinear dynamical system analysis in the field of the Social Sciences. This is not a theoretical difficulty. No doubt even in non-physical systems we observe complex time evolutions and chaotic processes dynamics. No doubt even in social systems we observe dynamics so unstable and discontinuous that self-organizational processes no longer work ending up in *chaos*. According to Ruelle, the real problem arises when one shifts from the level of "scientific philosophy" to the domain of "quantitative science". Even though the time evolution of social systems may be as discontinuous and chaotic as physical ones, they cannot be analyzed in the same manner. The reasons for this have been duly presented, discussed and reviewed. Ruelle's main objection, the intrinsic impossibility of the Social Sciences to achieve basic equations or models of deterministic evolution "*really convincing*", cannot be accepted in its general absoluteness. Proof of this is the fact that in the *soft* Sciences we often find the successful application of *the logistic map*. Outside the physical sciences, this well-known non-linear model is the rule and not the exception (May, 1976). Contrary to Ruelle's assertion, the application of complexity and chaos theory in the Social Sciences is a real possibility. As a result, this chapter discusses the implications of Complexity and Chaos Theory for social science research.

Keywords: Complexity and Chaos Theory; nonlinear dynamical social systems; logistic map

Introduction

Complexity and Chaos Theory is the "new frontier" of modern science. We are now facing the eclipse of the classic Newtonian-Laplacian conception expressing linear, deterministic, relationships among real phenomena, and the emergence of a new science. This revolves around the acknowledgement of non-linearity and thus instability, uncertainty and unpredictability of real phenomena, in a word around the acknowledgement of their complexity. This means that real phenomena may be characterized by non-proportional or nonlinear relationships of cause and effect and may, therefore, be particularly sensitive to even the slightest changes in initial conditions (c.d. butterfly effect)[1]: in non-linear systems any "small" change in the cause variable can produce a "big", non-proportional, change in the effect variable, just as a "big" change in the cause variable can generate a "small" change in the effect variable. Therefore, unlike linear systems, nonlinear systems can generate very complex behavioral dynamics over time (stable, periodic and chaotic temporal behavior), occurring intermittently throughout the life of the system. During some time periods, behavior may reveal linear continuity. However, during other time periods, relationships between system's components may change, resulting in dramatic structural and behavioral changes (or bifurcations). Nonlinear systems, although governed by deterministic laws, can be so sensitive to even the slightest changes from environmental circumstances that they appear unstable and discontinuous. Positive feedback amplifies change, breaking up existing structures and behaviours, creating unexpected outcomes in the generation of new structures and behaviours. Thereby, new equilibrium structures, new stages of organization, and new orders, may spontaneously emerge (self-organization)[2] or even states of chaos, an erratic and

---

[1] Sensitive dependence on initial conditions, that is the circumstance by which even the slightest change in initial conditions lead to a significant change in a system's evolution, constitutes the primary characteristic which distinguishes nonlinear systems from linear ones. The latter are structurally stable; they do not change their behavior under small perturbations (slight changes in initial conditions or parameters) but respond proportionally, so much so that the effects of perturbations are damped and remain contained over time. Consequently, "linear systems exhibit smooth, regular, and well-behaved motion" (Elliot and Kiel, 1997a: 5) and are resilient to any attempts of destabilization. Naturally, this is a guarantee of system predictability. In other words, from diverse and possible starting conditions, the trajectories - or the total amount of states assumed by the system – which depart from each one of these conditions oscillate closely near each other. In fact, whatever the point of departure is, there will be rather similar previsions. We might say, for example, with a high degree of probability that, regardless of initial conditions, the state of the system in the temporal instant t_{100} has a value of x. However, in non-linear systems this is not so, especially when they are in a chaotic regime. These are structurally unstable systems and can change their behavior significantly with even the slightest perturbation, whose effect is not damped but may be amplified exponentially. This means that the trajectories coming from variations in starting conditions, first oscillate closely, but after a few iterations, diverge insignificantly. Thus, the uncertain knowledge of initial conditions and sensitivity to change in these conditions makes the system unpredictable in the long run. In effect, according to the value assumed by the system in the temporal instant t_0, predictions will be different every time. In other words, making predictions becomes intrinsically nonsensical.

The '*butterfly metaphor*' was coined by Lorenz (1963) precisely to explicate sensitive dependence on intitial conditions and the unpredictability of non-linear chaos systems. What realistic changes can the flapping of a butterfly's wings cause in atmospheric conditions? Yet the flapping of a butterfly's wings in Texas may generate a tornado in China. Poincaré, who wrote significant pages on dependence on initial conditions in *Science and Method* (1908), and later Lorenz suggested that weather systems are typical examples of unpredictable deterministic systems in the long term.

[2] "Self-organization is the spontaneous emergence of new structures and new forms of behavior in open systems far from equilibrium, characterized by internal feedback-loops and described mathematically by nonlinear equations" (Capra, 1996:85). It is important to note that this emergence of new structures is *spontaneous* or externally unforced from some hierachical controller.

seemingly random behavior, devoid of pattern (nonperiodic behavior) and of apparent order, in which uncertainty dominates and predictability breaks down (deterministic chaos).

Therefore, complex systems are far away from-equilibrium systems or dissipative systems (Prigogine and Stengers, 1984). Constituent parts of the system exchange information among themselves and with their environment and to adjust their behavior as a function of information they process. So, they dynamically transform themselves from one moment to the next, forced to abandon their previous reference state by internal instability and external perturbation. They then fluctuate between two or more new points of possible equilibrium, *irreversibly* evolve to a new configuration, reorganize themselves around a new reference point and maintain it until a new perturbation once again pushes them into new evolutionary trajectories. Or else, they flow into chaos and can remain chaotic, and eventually destroy themselves.

Emergence, novelty, surprise are the words we use to describe the new patterns of relationship, the new high-level properties of the system generated by collective dynamics of its components or by *nonlinear interactions* among its constituent parts. In this sense, "the behavior of the overall system cannot be obtained by summing the behavior of the constituent parts" (Holland, 1998: 122). In other words, the global characteristics or properties of an emergent system are properties of the "whole" that are not possessed by any of the individual parts making up the "whole". Therefore, they are not reducible to the characteristics of the lower-level constituents or subsystems and cannot be directly deducible by observing the properties of the system lower-level components. As McDaniel and Driebe emphasize, the consequent unpredictability of an emergent system is *fundamental*(2005:6).

Social systems are dissipative systems as well as physical systems.They can exhibit discontinuous changes over time, evolving toward greater and greater complexity. To apply the conception of dissipative systems to the study of society means regarding social systems as "inherently historical entities whose evolution is driven as much by internal instability as by external perturbation" (Harvey and Reed, 1997: 306). And the same conditions of sensitive dependence on initial conditions observed for natural systems are the foundation for their historicity.

Crutchfield is right when he says that this new approach, centered on non-linearity, has caused a "revolution" (1991: 22), involving numerous branches of science: Meteorology, Physics, Chemistry, Biology, Geology, Engineering, Medicine, as well as Economics, Sociology, and Political Science. The 'new' science is intrinsically interdisciplinary. It can describe fluid turbulence, earthquakes, epidemics as well as Stock Market crashes, political and social competition processes, population and crime dynamics, the spread of innovations, urban growth, or in general, the dynamics of new forms of social aggregation. In brief, no social scientist will deny that social phenomena are not linear just as the planets' orbits are not linear (Brown, 1995).

Naturally, this does not mean that social phenomena are constantly chaotic but they do have a potential for positive feedback and chaotic interludes. Brown illustrates these course changes from a linear standpoint in the following metaphor:

> "Social phenomena are like serpents. Not much is linear about them, particularly with regard to their movements through time. They follow not a direct path, but wind among the vagaries of the development of the broader human culture. This broader culture also is not fixed. It is like sand. The marks that are made on it by social

phenomena quickly blend into the historical evolution of this culture, mixing with the previous marks of all other phenomena, until it is difficult to say that any one mark is distinct from all of the others. We live in a human universe of movement and change. That which seems fixed may be only the ghost of a contrived static perspective that ignore the continuing evolution of human interdependencies. [..] most of our experiences are those of change within a background of further change; *our lives developing like the movements of serpents, serpents in the sand"* (Brown, 1995:10, italics not in original).

Of course, we do concede that basic discontinuity of natural and social processes is hardly a new discovery. Over two thousand years ago, Greek philosophers theorized about it. Even classic science was aware of atmospheric "disorder", in fluid turbulence, in running faucets, in falling leaves, in stones rolling down a cliff, in oscillations of magnetic waves, in plant and animal population dynamics, in heart beats, in the diffusion of nervous impulses and so on. These irregularities, however, were considered brain teasers or at worst, monstrosities (Gleick, 1987). Nowadays, thanks to ever more refined mathematics (Topological studies, Ergodocity theories, Differential Non-linear Equation Analysis) and computational instruments (computers able to run complex nonlinear programs), we are able to do what Poincaré was not able to do, that is, to theorize while controlling the terms of a problem. To sum up, the real *novelty* today is that we are able to formulate assertions on discontinuity and unpredictability of certain phenomena while ensuring rigorous empirical control over these assertions.

Nowadays, we are witnessing an explosive development in nonlinear system researching a wide range of disciplines. The Social Sciences too have studied discontinuity in their fields of research in a seemingly constructive and satisfying manner. This paper would like to clarify some of the perplexities as to the possibility *of applying* the techniques of *Complexity and Chaos Theory* (or *Nonlinear Dynamical System Theory)* to the field of the Social Sciences. These perplexities do not intend to assert theoretical difficulties. They refer to the convictions formerly expressed by David Ruelle in the 90s. According to Ruelle, one of the true pundits of the field, for various reasons, the impact of chaos in the Social Sciences concerns the level of 'scientific philosophy' and certainly not of 'quantitative science' (Ruelle, 1991: 79). This assertion not only applies to the Social Sciences but also to Biology, Ecology, and to all the so-called "soft" sciences.

Since however Social Sciences have indeed used this theory somewhat successfully, it seems opportune to clear up the contents of this strange question. Strange because, on one hand the use of the theory has had comforting results in the social research and continues to do so, yet Ruelle's position is generally accepted and widely shared by numerous scientists, expressing basic physical science orientation.

A Tantalizing Situation

From Ruelle's point of view, there are at least three main reasons that render the use of *Complexity and Chaos Theory* problematic in social science. However, it is extremely important to note that *in principle* Ruelle has no objection to the use of this theory in social science. According to the author, there is no doubt that discontinuity may be inherent both in physical as well as in social phenomena. There is no doubt that even in dissimilar systems from the physical systems we can observe such complex time evolutions that "it is *reasonable* to think that we have sensitive dependence on initial conditions" (Ruelle, 1990:244).

In terms of "scientific philosophy" nothing prevents phenomena like "the fate of empires", the "motion of billiards", the "motion of planets" or even the evolution of weather, from being grouped under the same heading, where *chaos* resolves, in the reconciliation of determinism and unpredictability that characterizes it, the vexed question of the rapport between determinism, chance, and human free will. Ruelle explicitly refers to Hadamard (1898) and above all to Duhem (1906) and Poincarè (1908), the first to theorize the peculiar sensitivity of dynamic systems to even small changes in initial conditions and, starting from their intrinsic unpredictability, to highlight their importance from a philosophic perspective (Ruelle, 1991: 45 e ss.).

According to Ruelle, however, problems arise when we turn from "scientific philosophy" to quantitative science. The Social Sciences in this case find themselves in a "tantalizing situation" (1991:85). If on one hand their evolution in time may be similar to the evolution of discontinuous physical systems, particularly chaotic, without possibility of control or predictability, on the other hand they cannot be analyzed in the same way. As mentioned above, there are three basic reasons for this.

The first reason refers to the main condition for a rigorous quantitative study of complexity and chaos in dynamic systems. Such studies require first of all a "quantitative knowledge of the system's dynamics". In other words, the underlying equations of motion that govern the system's temporal evolution need to be determined. Actually, in Ruelle's point of view, this knowledge can only be obtained in Astronomy, Hydro-dynamics and Economics but not in the Social Sciences, Biology or Ecology. In the latter cases, although it is possible to observe various complicated evolutions over time, "we don't have really convincing models of deterministic time evolution" (Ruelle, 1990:244), or equations of motion governing the behavior of the nonlinear systems, able to reproduce and reconstruct satisfyingly observed data series and explain its evolution.

A case in point is Economics:

> "There is little doubt that economics and finance give us examples of chaos and unpredictable behavior (in a technical sense). But it is difficult to say more, because we do not have here the kind of carefully controlled systems with which physicists like to experiment. Outside events which the economists call shocks, cannot be neglected. Earnest efforts have been made to analyze financial data (which are known with much better precision than economic data) in the hope of isolating a moderately complicated dynamical system. Such hopes, in my opinion, have failed" (1991:85).

Precision by which one may control the impact of external perturbations is significant in determining a possible equation of motion. When this control is successfully enabled as in hydro-dynamics or in electro-mechanics, the presence of chaos can be rigorously demonstrated. Otherwise, for some less well-controlled systems, one may confidently infer that chaos and unpredictability are present but "a precise demonstration is practically impossible" (1990:243). This is why social science, biology and ecology systems are still far from the realm of quantitative science.

In the light of these considerations, the second reason is of decisive importance for Ruelle.

In fact, if equations governing behavior of nonlinear systems are unknown, the only way to obtain information on the dynamic evolution is to extract it directly from the examined time series by the estimation of specific coefficients (Lyapunov exponent and correlation integral), able to ascertain eventual chaotic behavior. Yet the Social Sciences could not even rely on this method since they, *as a rule,* have time series whose features do not seem to allow a reliable estimate of these coefficients: *low precision* (and therefore *presence of noise*) and *limited size* of data (*small data sets).* It is hard to obtain long and uncorrupted, uncontaminated by random perturbations, time series. Usually they involve short time series, a short sequence of data, too short for accuracy and reliability of the estimates (from 200 to 800-1000 observations at best). In other words, since in such time series the dynamic information is limited by sample size and masked by noise, trying to use Grassberger and Procaccia's algorithm for calculating correlation integral and determining correlation dimension brought up *inconclusive* results according to Ruelle, unable to reliably distinguish between chaotic deterministic processes and stochastic processes. So, it would be unrealistic to use these methods for discriminating purposes, as was seen for example in market production studies (1990: 247).

We will return to this question in more detail as it constitutes one of the most controversial issues in chaos theory literature.

Finally, there is a third problem that according to Ruelle renders the use of *Complexity and Chaos Theory* problematic in social science. This third element would nevertheless have a negative influence even if we had long time series and basic equations of motion were known. Ruelle believes that social processes are characterized by a peculiar property: *persistence* in their dynamics. In other terms, he believes that these systems "learn" and "slowly change over time" (1991: 79). A periodic system, for example, would tend to keep up its state over long periods of time before a possible change.

These three elements together leave the Social Sciences – including biology and ecology - "back" that is at the level of scientific philosophy. While not excluding further progress in these fields, essentially they would still be, according to Ruelle, in the same condition as meteorology was in Poincaré's times.

In general Ruelle's objections are shared by many. For example, Kellert sees a certain gap between physical systems and social systems. Social systems, as well as those in epidemiology and ecology, are much less easily controllable and controlled. Therefore, the analysis of non-linear dynamics is a *more controversial* issue. In fact, "regarding the need for high quality time-series data, social systems are notoriously *uncooperative*…it is difficult to obtain even quantitative economic data that match the quality of data available to the physical sciences" (Kellert, 1995: 37).

This fact makes it hard to reliably reconstruct an attractor. As we know, this reconstruction is essential for calculating spatial correlation coefficients and Lyapunov exponents. Consequently, to the difficult identification of underlying data generating mechanism (chaotic or random), is added the difficulty of identifying and quantifying the most significant variables of the system, so as to reach a simple mathematical model. For Kellert, Social Sciences, therefore, are unable to apply the mathematical techniques of nonlinear dynamics and chaos theory and to discover chaos within them. Social systems are "a genuine challenge – though perhaps not an impossible one –to such attempts" (1995: 36).

For this reason, the use of Complexity and Chaos Theory in Social Science remains essentially in the metaphorical realm. Naturally, this is a limit which we must be constantly aware of.

To sum up, the unknowability of equations that model social processes and the limited number of data that the Social Sciences have to work with, are viewed as significant obstacles in using the techniques of nonlinear dynamics in their field. This is the common opinion of many authors such as Medio (1993), Lorenz (1993), Ramsey, Sayers e Rothman (1992), and in general is widely found in the hard science fields.

At this point, because of the relevance of the question, it is opportune to enter into more details and to expound the fundamentals of what has been called 'the curse of dimensionality" (Cheng e Tong, 1994).

The Curse of Dimensionality

A chaotic *dynamical system* gives rise to a *disorderly* distribution of constituent elements although the underlying mechanism governing its behavior is intrinsically deterministic. As observed by Schaffer, chaos makes us to rethink the very idea of randomness (1987: 233).Chaotic behavior is governed by a set of laws and yet it does not repeat itself over time and is thus a periodic, apparently irregular behavior, devoid of pattern in its longitudinal evolution, sensitively dependent on initial conditions. According to Johnson and Habyarimana (1997), of the different characteristics of chaotic processes that have been identified, the random-like nature of the processes is "perhaps the most universal one", and "this is true for both law order and high order chaotic processes" (1997: 290). Being so, at first sight it appears indistinguishable from a random series, produced by a entirely stochastic process. Thus the basic problem is to determine whether the observed dynamic is simply a stochastic or chaotic process.

Whereas conventional spectral analysis fails to distinguish chaos from random signal since both the signals have continuous broadband power spectra (Eckmann and Ruelle, 1985), Grassberger and Procaccia's algorithm tries to solve this problem (Grassberger and Procaccia, 1983a). It tries to distinguishbetween chaotic and stochastic dynamics by providing a reliable estimate of system's *dimension,* that is the number of variables that intervene to configure a system and therefore to produce the observed data series. In this respect, the authors' idea is rather simple: if data are seen as *points* in a multidimensional space, the system's chaotic nature can be ascertained when two conditions are complied with.

First, dynamical system must to give rise to an *attractor*[3]. Every point trajectory which can be described from close initial conditions ends up being 'attracted' toward the same

[3] Attractors are geometric structures described by the system in its evolution. These geometric shapes attract all the orbits starting from different initial conditions, hence the name attractors. These shapes can be discovered by using state *space or phase space mapping.* A phase space represents a graphic backdrop for presenting the motion of time –based data (Kiel and Elliott, 1997:27)We construct a XY graph and place on the horizontal axis or abscissa the observed values x_t and on the vertical axis or ordinate the same values measured with time delays x_{t-n} (1, 2,…3, and so on time lags). The most suitable time lag for reconstructing the attractor is found by experiments. Usually, however, it is set at equal 1. The values x_t and x_{t-n} constitute the coordinates of a point on a multidimensional space, and the set of points identified in this way describe an orbit which in the long run traces the typical attractor of the observed system. In fact, according to Takens theorem (1980) the geometric shape thereby constructed is topologically equivalent to the original attractor, that is the one that would be reconstructed if we knew exactly the number and the values of variables explaining the observed

geometric figure. This geometric object is defined attractor, namely "a set of points toward which complicated time paths starting in its neighborhood are attracted" (Baumol and Benhabib, 1989: 91). Obviously the set of points on the attractor have to be 'neighbors' or *correlated spatially*. Generally speaking, according to the non-linear mechanism, attractors assume different shapes and patterns. They represent the structures that describe and dominate a nonlinear system during its evolution (Kiel and Elliott, 1997: 28), storing information about a system's behavior over time. When nonlinear behavioral regime is stable, it converges to a steady state and do not leave this point. So attractor is a mathematical equilibrium or fixed point. When nonlinear behavior that can occur over time is periodic behavior, namely a cyclical behavior oscillating around a set of defined mathematical points in a periodic manner and creating a circular pattern, attractors are *limit cycles*. When a nonlinear dynamical system exhibits a chaotic, apparently random behavior, attractors take on surprising and strange shapes, whence comes the common definition of *strange attractors*. Attractors provide "a qualitative assessment of dynamic systems in motion" (Mosekilde, Aracil, and Allen, 1988: 21). In fact, there are theorists such as Kiel and Elliott who firmly believe that to understand nonlinear systems you merely have to look "at the pictures they generate" (1996: 26).

According to complexity theorists, strange attractors are the most evident proof that there is an *underlying order* within chaotic behavior. It is good evidence of the deterministic nature of chaos (*order in chaos*). Indeed, random processes do not give rise to any attractor.

Secondly, the number of variables which determine the process must be *finite*. Or, in more technical terms, the *dimension* of the system (and therefore the *dimension* of the attractor) must be *finite* besides being fractal. From the nonlinear dynamical point of view, chaotic processes would have a finite dimensional attractor (and therefore, a low number of degrees of freedom) whereas a truly *random process* distinguishes itself, primarily, by its infinitely large dimensionality (high number of degrees of freedom) (Grassberger and Procaccia, 1983b; Theiler, 1990).So given the observed data, a value finite and noninteger of the correlation dimensions implies that the dynamical process is governed by an underlying nonlinear deterministic mechanism.Usually a *correlation dimension* above 5 implies the presence of a random process (Sprott and Rowlands, 1995).

Grassberger–Procaccia algorithm (GPA) is a simple enough technique to provide us with knowledge of the maximum number of variables needed to model and explain its dynamical behavior and, consequently, to differentiate chaotic dynamics from stochastic dynamics. However, the estimation of the dimension requires *large data sets* with high precision.

Nevertheless, Grassberger and Procaccia's algorithm itself says nothing formally *theoretical* about the length of time series needed for reliable estimates of *correlation dimension*. More specifically, the two authors perform experiments and, after empirical trials and after compiling their results, reach certain conclusions: to determine the dimension of a

dynamical process (or if we knew the *governing equation of motion* of the nonlinear system). In fact, it maintains the properties of the original attractor, including the dimension. This topological equivalence is called by a complicated term *diffeomorphism*.

Every system has its own attractor, with its typical shape. For example, if a system is stable (i.e forced pendulum), the attractor is at a fixed point. If the system display periodic behavior the attractor is a limit cycle (i.e free pendulum). When the system is chaotic, the attractor assumes the strangest shapes, thus the attractor is called *strange*. In the case of logistic equations, the system's attractor is reconstructed considering a bi-dimensional space x_t / x_{t-1}. Setting the x_t values on the abscissa and x_{t-1} values on the ordinate, the points that are identified by the coordinates x_t and x_{t-1} form a hill-shaped attractor.

two-dimensional attractor like Hénon's, for instance, a series of 15000 data suffice; the same number would suffice for a one-dimensional attractor, like the hill-shaped attractor generated by the logistic map operating in a chaotic regime; instead 25000 data would be needed for a tri-dimensional attractor (Lorenz, Rossler…), and so on.

Smith (1988) set about this very ambitious task: he tried to elaborate a *rule* to fix the length of a time series from which the coefficient can be reliably estimated. In light of the fact that the larger the system's dimension is, the greater the number of data needed, Smith's conclusion was that this mathematical law can be identified in an exponential order of 42^D. This in layman's terms means that we need exceptionally long time series, able to reach a billion of data. For example, a dimension of D=6 would require for its reliable estimation "the analysis of a minimum of 5489031774 (uncorrelated) data points" (Smith, 1988: 287). However, we need at least 1764 data points even to detect a bi-dimensional attractor (D=2).

A second decisive contribution comes from Tsonis's (1992) systematic work. Here the constant is set at equal 10 and the required data must reach $10^{2+0.4D}$. It is easy to do the math and conclude that even in this case, although not billions, we need thousands of data points. If we want to reliably estimate a 6 dimension we need about 25200 data points, and a maximum of about 1600 for a 3 dimension. For a dimension of 2 the length of time series is however very close to the level of 700 data.

According to Smith, an empirical element can justify such high set limits. From calculations that we are unable to provide here, it seems that short time series introduce significant calculation distortions. The heaviest distortions produce substantially lower estimates than the actual *D*. Naturally, this distortion effect is more pronounced for higher dimension systems which might lead to the unpleasant conclusion they not able to distinguish random from chaotic processes (*ibidem*), mistaking a deterministic series for a random series. Only if the data sample approaches infinity does everything 'work' exactly as described by Grassberger and Procaccia, allowing an effective distinction between the two distributions (Smith, 1988:286).

However, there is another serious factor creating bias. Ramsey and Yuan (1989, 1992) ascertained that the estimates calculated on a series of 200, 400, 1000 and even 2000 data points, generated by iterating Hénon's two dimensional model, were significantly higher than the effective 1.21, risking therefore the opposite of what Smith had reported: mistaking a deterministic series for a random one. Dimension calculations were reliable starting only from 5000 data points. In conclusion, according to the authors, 30000 data points is the ideal number of observations needed to obtain 'reasonable estimates' (Ramsey and Yuan, 1989: 296). Once again the reliability of the estimate strongly depends on the availability of *long* and high quality, noise-free data (*large data set*).

This difficulty is crucial for not only the dimension estimates by GPA but also other quantities, such as the Lyapunov exponent, which is based on the time-delay reconstruction. Even in this case, in fact, estimate procedures require the reconstruction of the attractor or of the point trajectories that define the topological shape of the system. Given time series data from a deterministic dynamical system, it is often of interest to determine whether the underlying dynamics are regular or chaotic. For example, heart rate data measured in electrocardiograms are believed to be chaotic for the case of healthy patients but show regularity in the case of congestive heart failure. Lyapunov exponent is the usual test for distinguishing regular from chaotic dynamics in deterministic dynamical systems. The calculation of the maximal Lyapunov exponent allows one to detect whether a deterministic dynamical system is chaotic or non-chaotic.As we know, in fact, Lyapunov's exponent provides a measure of the system's sensitivity even at the slightest change in initial conditions, determining *quantitatively* if chaos exists in a deterministic system (Brown, 1995: 22). In this respect, it expresses the rate with which two trajectories departing from nearby initial conditions (*nearest neighbors)* diverge exponentially over time. Or, in the other word, it "measures the rate of divergence of nearby points during the evolution of a dynamical system" (McBarnett, 1996:187).

If the series is chaotic, the exponent must be positive. A positive value, in fact, indicates that a dynamical system is particularly sensitive to a very small change in initial conditions, so much so that in the presence of such a change the respective trajectories diverge exponentially quickly after only a few iterations. On the contrary, a negative value indicates that nearby initial conditions convergence exponentially fast and therefore system periodicity. A 0 value indicates stability: the trajectories departing from nearby initial conditions remain separate at the same identical distance throughout the system's entire time evolution. In short, a positive maximal Lyapunov exponent indicates divergence and chaos (if $\lambda > 0$ nearby trajectories separate exponentially), while negative exponents indicate convergence or stability (if $\lambda \leq 0$ nearby trajectories stay close to each other) (standard references include Wolf, 1985, 1986; Eckmann *et al*., 1986; Parker *et al*., 1989; Arbabanel *et al*., 1993; Schreiber, 1999; Kantz and Schreiber, 2004).

At this point, given a positive exponent, its consistency orients the researcher toward making a judgment on the possible *predictability* of the series. It can be interpreted as a *predictability loss*, where the loss of predictive power is measured in *bits per iteration*. When, for example, the exponent has a maximum 1 value, this means that one information bit is lost in each iteration. In this case, "if one measures with eight-bit accuracy, all predictive power is lost after 8 iterations" (McBarnett, 1996: 192) (after 8 years for yearly observations, after 8 months for monthly observations ... etc). If there is an 6 digit number, all information would be lost after 6 iterations, and so on according to the precision whereby the phenomena under study was being measured.

The positive Lyapunov exponent confirms that observed dynamics are governed by a nonlinear deterministic process since chaotic dynamic cannot be produced by a linear process. However, as well as the GP-algorithm, the computation of the concerned Wolf-algorithm requires a large number of data points uncontaminated by a random noise. Being able to reliably distinguish a chaotic system from a system with a regular behavior requires the use of thousands upon thousands of data points especially in the presence of continuous data (Wolf, 1985, 1986; Theiler, 1990).

To overcome this obstacle, other experts have tried to make some adjustments and modification to Wolf's original algorithm (Wright, 1984; Sano and Sawada, 1985; Sato, Sano and Sawada, 1987; Briggs, 1990; Nychka *et.al*., 1992). The result being that the issue was shifted on the choice of algorithms and on calculation biases that some algorithms have introduced (see Briggs, 1990; Hunt *et al*., 2003).

To sum up, from this perspective, there seems to be very good reasons for *not* using chaos theory techniques when dealing with small data sets. Evidently, this is a question of prime importance for social researchers as a series of 200 data points is the most frequent time series of the most frequent type of social research. However, the data requirement question, here referred to in its essentiality, seems to be under various aspects far from being a closed issue.

The Data Requirement Question: A Question Far from Being a Closed Issue

There are significant arguments against the Ruelle's objections. The first deals with the above mentioned question of the number of data points required for reliable estimation of the quantitative measures characterizing chaotic behavior. In fact, contrary to what certain theorists try to show, no argument is so controversial than the length of the time series needed for an accurate estimate of correlation dimension. Some authors have presented a clear-cut affirmation, stating that there is no *general answer* to this question, able to determine *univocally* and *definitively* the *"right"* length of a time series in order to ascertain the presence of deterministic chaos as its underlying generating mechanism (Albano *et al.*, 1987; Ramsey *et al.*, 1992). On one hand, it has been stated that reliable estimates need unlimited noiseless data - thousands of data points from a minimum of 10000 to a maximum of 15000, 25000, better yet 40000, up to 50000 data points (Smith, 1988; Ramsey and Yuan, 1989; Tsonis, 1992). On the other hand, it has also been stated that we can successfully work with *small data sets*. In fact, despite the unquestionable advantage of working with large data sets, a lot of modifications on the GPA have been made to obtain a valid estimation from small size data. The length of the series decreases from thousands to hundreds of data points, with sequences of 800, 500 observations, and even fewer. Especially in case of time series generated by difference equations (i.e Hénon map, Logistic map) even 200 observable data points can ensure reliable results (Abraham*et al.*, 1986; Albano*et al.*, 1987; Swinney, 1987).

Abraham and his colleagues have made some decisive remarks in this regard. Without any circumlocutions, in reference to correlation dimension, Lyapunov exponent and entropy measure, they state:

> Most papers describing these methods have suggested that relatively large data sets are required for the calculations. In many situations, large data sets cannot be obtained and in others the computations using large data sets are prohibitively long. We present evidence that some calculations can be successfully completed for small data sets(Abraham *e al.*, 1986: 217).

Again,

> "It is not our intention to argue that using small and noisy data sets is at all optimum or desirable in general. Rather, we wish to point out that in those situations where one is limited to small data sets either by the nature of the phenomenon being studied or by other technical constraints, it is still *possible* to calculate such quantities as dimensions. As always, small and noisy data sets impose severe limitations on the accuracy and precision of quantities calculated from them. However, as the examples below show, it remains possible to achieve enough accuracy and precision for the calculated dimensions to be used to distinguish between deterministic chaos and noise, as well as to distinguish between different kinds of chaotic behavior" (1986: 218).

Resulting in the following conclusion

> "[....] the results presented here demonstrate that it is possible reliably to perform calculations with 500 data points or fewer" (*ibidem*).

So, they demonstrate that it possible to calculate the correlation dimension from *short noisy time series*.In effect, these are the results. The first reported example refers to the correlation dimension calculation on 500 data points obtained from a Hénon map's simulation. The result was 1.28 with a 6% difference in respect to the 1.21 value obtained with 15000 data points. The same was performed using a series of 500 data points generated from a logistic equation: the correlation dimension had a 0.92 value with a 1% difference in respect to the 0.93 value obtained with 5000 data points. A same dimension of 1.2 was found both with 300 and 700 data points generated from a sinusoidal model and with a random error component added to each observation. And finally, results of the correlation dimension calculations with 6000 experimental data points measuring the output intensity of CO_2laser were reported. Even in this case, the results appeared "substantially unaltered" when the experimental study passed from 6000 to 500 data points (1986: 221).

Albano *et al.* (1987) echo Abraham's words.

Given the absence of a general answer to the data requirement question, the experimental approach as "the best approach" (1987: 211) to the problem makes it possible for them to set a new record. Even the number of 500 data points was successively reduced – in respect to time series generated by Hénon map 200, 100 and even to 50 observations seemed to suffice for reliable estimates of correlation dimension. According to the authors, this occurs because the correlation dimension "is a property of an integral, and integrals are numerically robust objects" (*ibidem*). Therefore, Albano has no doubts and his conclusion is lapidary. Far from suggesting that 50 data points are the ideal number, the results obtained, however, strongly question the previous claim that several millions data points are required to calculate the correlation dimension of the Hénon attractor and therefore, in more general terms, to resolve the dimension of any other attractor generated by a two-dimensional nonlinear dynamical system. This claim is in error by a factor of 10^3 or of 10^4(*ibidem*). And this would be true even if noisy experimental data were used.

Moreover, experimental studies evidenced that it is still possible to resolve reliably the dimension of the attractor for systems of differential equations or for continuous data even if we do not have thousands and thousands of data points. Jensen *et al,* for example, have successfully examined the Rayleigh-Benard system with only 2500 observations (1986). Kurth and Herzel, in their study on solar radio pulsations, estimated spatial correlation on 640 data points generated by Lorenz equation, and found a high degree of agreement between the literature values calculated with 25000 data points and their values. For Lorenz attractor the literature value of the dimension is 2.06 compared to a 2.09 estimate (1986). Swinney (1987) showed that the dimension of Couette-Taylor flow can be reliably estimated reducing the number of data from 20000 to merely 1000 data points.

On the basis of his experience with physical and chemical systems, Swinney too (1987) proposed an approximate empirical guideline for the number of data points required to estimate correlation dimensions of an attractor. As *a rule of thumb*, the author found that 10^D points can give reasonable estimates of attractor dimension. This confirms Albano's results for a low dimensional attractor.

Nevertheless, certain caution is necessary. According to Arbabanel (1996), it is, in fact, a good rule to look suspiciously at results obtained when the dimension being estimated is the greatest possible dimension that would be estimated with available data. For instance, for Eckmann and Ruelle (1992) the number of non-correlated observations needed to determine a given dimension must be greater than$10^{D/2}$. In this case, we need many more than 1000 data points to estimate the dimension of a six dimension attractor. Which means that if we were to estimate a 6 dimension with 1000 data points there would be good reasons to suspect that the analyzed system is a system whose dimensionality is higher than the estimated dimensionality or instead is essentially a random system? Only if the estimated dimension were much lower than the maximum estimate would there be good reasons to conclude that the observed time series data express a chaotic dynamic. In this respect, if 200 data points were available we would be able to give credit only to estimates of the dimension of the attractor D equal to 1 (2 dimension at the most).

This is an area of intense research activity. Further improved methods have been proposed for accurately estimating the correlation dimension using short time series of low precision.Usually the ability of the proposed methods to distinguish between chaotic and stochastic dynamics is still demonstrated through a number of simulated series generated by well-known models (Lorenz, Rössler, and Mackey–Glass equations, Hénon map and Ecological map). For example, Judd's algorithm uses 300 data points to reliably estimate a dimension equal to 8 (Judd, 1992, 1994) in low-dimensional models. The procedure proposed by Marzocchi *et al.* (1996) assess the practical efficiency of a modified Sugihara and May method (1990), allowing an effective discrimination between chaos and stochasticity in time series generated by the most common discrete parameter processes (logistic, Hénon, biological, Tent, trigonometric, and Ikeda maps)with only 50 data points and with a noise level ~ 10% of the signal amplitude.A simple, numerically robust and yet computationally less intensive new method has been proposed by Bhattacharya and Kanjilal (1999). The robustness of the proposed method (based on Singular Value Decomposition-SVD) has been demonstrated using 2000 data points from Lorenz and Rossler equations and against large amounts of additive correlated noise.

Several methods have been also developed for calculating the largest Lyapunov exponent from short time series in the presence of additive noise corruption. Amog them, Zeng *et al.*(1991)proposed an algorithm able to obtain a reasonable Lyapunov-exponent spectrum from 5000 data points from Lorenz equation, using the method to study the predictability of the atmosphere from observational meteorological data. Rosenstein' algorithm (1993) is a fast, easy to implement, method, robust to variations in number of data points, reconstruction delay, and noise level, and reliable for small data sets from a mimimum of 100 to a maximum of 1000 data points from all above mentioned models. Likewise 1000 data points is the data size sufficient to calculate reliably the diffusional measure test proposed by Johnson and Habyarimana (1998). Recently, Gottwald and Melbourne (2005) have introduced a binary test (the 0-1 test) for distinguishing regular from chaotic dynamics in deterministic dynamic systems which presents certain advantages over the traditional Lyapunov test for chaos. In particular, it works well for time series of 1000 data points' contaminated by measurement error. The sample entropy algorithm too is a robust index for detecting deterministic nonlinearity in relatively short time series corrupted by observational noise (Ramdani *et al.*, 2009).

Many recent studies (i.e Kitoh *et al.*, 2000; Yang and Qiong Wu, 2011; Yang *et a*l., 2012; Krishnamurthy, 2012) prove that the dimensional analysis and the reliable detection and quantitative description of nonlinear inherent determinism in time series of limited precision and relatively limited length are still of "unique value" and will be fruitful tools for the future study in the realm of the nonlinear analysis (Kito *et al.*, 2000: 182).

In conclusion, it seems that this brief *excursus* has brought up some noteworthy elements for social science research. Whatever else, one fact is evident: when a dynamic system is *low dimensional,* in particular one or two dimensions, the data requirement is not as high as maintained by many authors on the subject. For low dimensional systems we can, in fact, discriminate between chaotic and random systems, regular or chaotic process, by calculating reliable estimates of the correlation dimension and Lyapunov exponent even with a minimum of 200 data points.

This finding is also shown by McBarnett in his studies on the evolution of public opinion during the 1984 presidential primary election campaigns. For a simulated time series generated by low-dimensional nonlinear dynamic systems (i.e. Hénon map), he shows that estimates of the correlation dimension and Lyapunov exponent were robust for the various length of time series data, i.e. from 200 to 1000 data points and in presence of large amount of random noise (McBarnett,1997).A positive Lyapunov for Mondale's and Jackson's chances of winning time series, based on opinion data and modeled by one-dimensional system, confirmed that the evolution of public opinion during the 1984 presidential election was sensitively dependent on initial condition and unpredictable. In other words, it indicated an underlying chaotic dynamic process. In general, McBarnett was able to suggest that shift in public opinion can occur in dramatic fashion and that, therefore, it is possible to observe pattern of aperiodicity preventing reliable forecast of electoral results. Schaffer (1988) also obtained robust estimates with *small and noisy data sets* generated from a logistic equation.

As we shall see, this conclusion is especially relevant if it is connected to the second fundamental argument in opposition to the Ruelle's objections.

A Social Science Model: The Logistic Model

As we said, according to Ruelle Physical Science is the only one which allows us to know the basic equation of motion. In effect, we have many examples in this field. A few well-known examples are Lorenz model for chaotic behavior of weather conditions, Rossler equation for chemical reactions, and Couette-Taylor model for thermodynamic flows. However, if we take a look at the *soft* sciences, those that our author excludes from complex and chaos theory application, we discover that the theory has indeed been used and applied successfully. The success derives from the widespread and profitable use of the *logistic map*

Over thirty years ago, May, one of the major authorities in the study of the structure and dynamic behavior of the logistic map, stated that outside the physical sciences, this non-linear model *is the rule and not the exception,* foreseeing its systematic and fruitful use in the social science field (May, 1976:467).

In effect, this is what actually happened. In fact, its use has become highly pervasive. Every time a social phenomenon has been studied from a possible nonlinear perspective, the logistic equation appeared to be the most suitable for modeling its behavior and for reconstructing its underlying governing mechanism. This is precisely what happened in the field of Economics for the volatility market, in price analysis, in consumer consumption (Gaertner, 1987), in overlapping generations, in the relationship between company profits and advertising campaigns (Baumol and Quandt, 1985) and between production/productivity and investments (Baumol and Benhabib, 1989), in financial market trends and in complex competitive economy behavior (Guesnerie and Woodford, 1992), in Walrasian processes

(Bala and Majumdar, 1992; Hahn, 1992); in studies involving political processes, from political competition phenomena (McBarnett, 1996) to the arms race (Saperstein, 1984, 1997a, 1997b; Grossmann and Mayer-Kress, 1989; Campbell and Mayer-Kress (1997); and in Sociology, from drug use (Priesmeyer, 1995) to the spread of new technologies and informations (West, 1997), crime and infant mortality (Priesmeyer, 1995).

It has, among other things, been the base model for a whole series of equations representing the most complex derivations. As Dixon (1994) opportunely pointed out, we might even call it a real logistic equations family. Some of the most well-known derivations have allowed complex social processes to be modeled. Specifically, some logistic structure variations of the Lotka-Volterra predator-prey model have been employed successfully to model social processes such as competitive innovation diffusion (Sonis, 1984), urban-regional growth/decline dynamics (in describing the dynamic interaction between populations – predator- and incomes – prey, i.e Dendrinos and Mullally, 1981; Dendrinos, 1992), economic competition among firms (Bischi and Tramontana, 2009), political competition, for so-called electoral *volatility* (Brown, 1991), dynamics of revolutionary processes (in the analysis of the relationship among relative deprivation, revolution, State coercion, and foreign intervention, Tsebelis and Sprague, 1989, 2010), and social control processes (Huckfeldt, 1989, 1990; Condorelli, 2012). Huckfeldt was the first to simulate a Nonlinear Interdependence Competitive Model governing the dynamic interaction between noncompliance and coercion. This model was a "system of nonlinear difference equations, which specified predatory dynamics in the Lotka-Volterra tradition" (Huckfeldt, 1990:421).With some due adjustments, it was applied to the study of Noncompliance and Coercion relationships in Italy from 1985-2003 (Condorelli, 2012).

This being the situation, the discovery of this *regularity* is of relevant consequence for the social sciences. We can say that the logistic map *is* the structure of a non-linearly oriented social phenomenon. The most visible consequence of this situation refers to the *logic* underlying the social phenomena's evolution mechanism which is modeled by this equation. We can consider social phenomena as not being susceptible to unlimited growth but rather *bounded* in their development that is, limited by forces whose action reduces their growth rate thereby preventing an exponential increase. In fact, as is known, logistics is the result of an opportune adjustment by Verhulst of the Malthus' law for population growth (May, 1976; Braun, 1993; Kostelich and Armbruster, 1996). It reasonably expresses a limited growth process, within the framework of a limited resource environment. In other terms, it excludes an *ad infinitum* growth rate, implying a limit value (or *carrying capacity*) beyond which the system no longer grows. The behavior is expressed by the following equation known as *logistic map (or discrete logistic equation):*

$$x_{t+1} = k\, x_t\, (1 - x_t\,).$$

The equation indicates that x_{t+1} is a non-linear function of x_t, or rather $x_{t+1} = f(x_t)$. The formula tells us that the consistency of the x variable in t_{+1} time, for example, of a biological population, or an organizational population (social groups like burocratic, political, industrial apparatus and so on) depends on the consistency of the variable over time t, according to a *k* growth parameter. This parameter is not constant as it would be if the growth were linear, because, as said previously, the environment cannot sustain unlimited growth. Therefore, from this perspective, there is a maximum number of individuals that a certain population can gain in consequence of the limited environmental resources available. Thus, the more the population reaches its carrying capacity (maximum level of sustainability), the more the environment will discourage further growth. The non-linear 1- x_t component expresses the limit, the *restraint that* the environment poses on the x variable increase. The reasoning behind this is rather simple. In 1-x_t, 1 stands for 100% environment carrying capacity, its theoretical limit of sustainability. Thus, if x_t were 1% there would be 99% resources left to sustain greater population increase (1-0.01), and 99% multiplied by *k* (and then by x_t) can hardly lower the growth rate (if *k* were 2, the growth rate would be 99% of 2, or 1.98 x_t). On the contrary, if the population is 80% of the maximum value sustainable by the environment, there will be few resources to sustain further growth or rather 20% (or 1-0.80). The growth rate would be reduced by environment pressure (a 20% of 2 reductions or 0.4 x_t).Further expansion would be thusly impeded. The population falls; hence, with low values, growth still continues but at an ever decreasing rate, until xt stabilizes at a fixed value (if $k = 2$, x_t stabilizes at 0.50).

As observed by Marion, the whole question makes perfect logical sense, as well as naturally, mathematical sense (1999: 201).

Experts were fascinated by the surprisingly complex behavior by that which is the most simple of discrete non-linear one-dimensional systems. A *bizarre* behavior, said May (*ibidem*). In fact, the equation can describe a behavior that culminates at an equilibrium point, the so called *steady state,* or it can describe a periodic behavior with ever increasing cycle lengths, or it can have what seem to be totally random fluctuations. It is, however, basically important to note that such dynamic behavior depends exclusively on the *k* parameter value.

More specifically, when the value of *k* is between 0 and 3, the system stabilizes at *fixed point or point of equilibrium.* As *k* increases, system behavior becomes more and more complex. The stationary state loses stability as soon as *k* is more than 3: the system begins to fluctuate, becomes unstable but is able to evolve, reorganizing itself around a new configuration. So at 3.2 *bifurcates*, splitting up, dividing into two new fixed points. The system alternates between these two fixed points in a stable 2 period cycle. Obviously, given the two values which repeat sequentially, the model tells us that the *x* variable possesses a biennial, bi-semestral, bi-monthly, bi-weekly ... cyclical variation, according to time unit used.

Once again, cycle stability remains until the parameter passes a certain critical threshold: beyond this point, the system becomes unstable and the process 'bifurcates' creating a *new* cycle. It, initially stable, has a *new- doubled- period.* In fact, another interesting *bifurcation* is created around a *k* value of 3.5. The cycle time doubles to 4, so that now there are 4 values that alternate. In other words, 4 years, 4 months, or 4 days must pass in order to return to the initial value and to repeat the sequence. The period doubles for *k*=3.56, passing to a recursive sequence composed by eight values, and goes on and on this way till *k* keeps on increasing, creating bifurcation upon bifurcation, up to 16, 32, 64, 128, etc. cycle times. Naturally, the various cycles to which the process tends are the system's attractors.

In technical terms, what we are viewing is a real *period-doubling cascade*, or rather an infinite cycle sequence of period2^n.And it is exactly "this truly fantastic bifurcation phenomenon", as Smale and Williams defined it, which is widely acknowledged as the *road to chaos* (Feigenbaum, 1978). In fact, when *k* is in the range of 3.8 to 4, the system's behavior enters into a regime that Li and Yorke (1975) first named 'chaotic'. Values succeed one another in an erratic, a periodic manner, without ever repeating themselves, and therefore, appear to have no rules. *But randomness is merely apparent*, since the nature of mechanism or equation governing system behavior is still deterministic.

The dynamic behavior deriving from the logistic equation depending on the *k* value finds a good graphic portrayal in the well- known *bifurcation diagram*. In figure 1, below, we see clearly the system's bifurcation points or the values of unstable equilibrium – placed on the ordinate - that *x* exhibits as a function of *k* value- - placed on the abscissa.

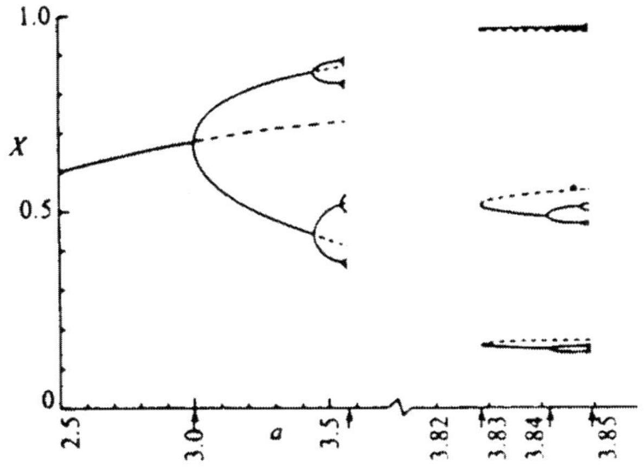

Figure 1. Stilization of bifurcation diagram (Source: Gleick, 1987)

We might say that *k* represents a sort of 'regulating device', when we turn it on we get a constant rise in dynamic behavior complexity: stationary -> periodic -> chaotic, with period doubling cascade as the mechanism generating chaos (Stewart, 1989: 177).

Therefore, it can be concluded that every phenomena whose process is modeled by the logistic map expresses an underlying deterministic nature, which is responsible for every bounded evolution change including any eventual erratic behavior. In other words, the phenomena hide an underlying order which has its own formal and geometric 'physiognomy'. As a rule, in the case of chaotic behavior, the underlying order takes on a bell-shaped attractor.

At this point, within such a delineated framework, we shall again reconsider the question regarding the relationship between the length of the time series and the accuracy of the estimated correlation dimensions. To this purpose, we tried to verify the hypothesis - H_0 – which establishes the need to use along time series (2000 time series data, for example) *versus* the alternative hypothesis – H_1– which implies the possibility of using *small data sets*. Starting from certain prefixed initial conditions and with a *k* value equal to 4, we iterated the logistic map in order to obtain by simulation time series from a minimum of 100 to a maximum of 2000 data points. Naturally, according to the H_0 hypothesis, 2000 data points should produce a dimension estimate close to the true dimension of the system, i.e. 1. Thus, if the estimate of the correlation dimension calculated on 200 observations were sensitively far from the one calculated on 2000 data points, the H_0hypothesis would be accepted. Table 1 shows the dimension of correlation D calculated using Grassberger and Procaccia's procedure. Each time series was constructed using an initial $x_0$value of0.9582139. Therefore, the generating equation is the following:

$$x_t = 0.9582139*4*(1- 0.9582139)$$

Table 1.

N	*D*
100	0.98
150	1.12
200	0.96
300	1.04
400	1.04
500	1.04
600	1.05
700	1.04
800	1.04
900	1.04
1000	1.04
1200	1.01
1400	1.01
1600	1.01
2000	0.98

Table 2.

N	*D*
150	0.80
200	0.93
300	0.98
400	0.98
500	0.97
600	0.98
700	0.98
800	0.98
900	0.98
1000	0.98
1200	0.98
1400	0.98
1600	0.98
2000	0.986

The results of the simulation, naturally fractals, clearly revealed the method to be *robust*. The estimate calculated on 200 data points differs from the one calculated on 2000 data points by only 3%. In any case the *slope* did not increase with the *embedding dimension* increase, activating the expected saturation process in the case of a chaotic system. Calculations were repeated starting with an $x_0$value of 0.160160.

These results show a general uniformity. The difference between 200 data point estimates and those calculated on 2000 data points is quite low, 5%.

To further ensure the procedure, for each time series we reconstructed the attractor in the *phase space*. We have previously mentioned that the attractor for a logistic system with chaotic behavior is bell shaped. In our case the outcome was more than favorable: the attractor constructed in this way presented this geometric shape for each one of the time series under examination.

Finally, we applied one of many control tests - the *surrogate data method* proposed by Scheinkman and LeBaron (1989). In this test, original data were mixed up just as we do with a pack of playing cards. This *shuffling* was used as an added test to distinguish deterministic non-linear processes from stochastic ones: after having shuffled the data around, the underlying determinism disappears so that if the dimension of correlation estimated on the shuffled data were similar to the one estimated on the original data, system's deterministic nature would be seriously in doubt. In our case, however, nothing of the kind happened in any of our samples: in the respective 'randomized' time series data, the correlation dimension resulted much higher than that calculated on the original data set (continuing to increase as the *embedding dimension* increased). Nor were we able to reconstruct any attractors.

In conclusion, as Abano *et al.* emphasized, when the underlying equation of motion that governs the behavior of a nonlinear system is the logistic map it is likely to infer system's dimension reliably even with small data sets of 100- 200 data points.

We must finally ask whether these results can change with the addition of an error component to a time series. The problem has already been studied by Schaffer (1988).The author showed beyond a shadow of a doubt that not even the addition of the random error or noisy component is able to "obscure" the estimate reliability. As we said, scientific literature reports similar results for Lyapunov exponent as well (Wolf, 1985; Rosenstein, 1993).

Complexity and Chaos Theory: Implications for Social Science Research

The success of the logistic map demonstrates how we can study social phenomena discontinuity scientifically, avoiding a purely metaphorical use of the complexity and chaos approach. The logistic equation allows us to pass from theory to practice and to revise traditional approach marked by a linear deterministic view implying predictability and control of social phenomena and therefore unsatisfactory analytical outcomes. In fact, the widespread use of the Non-linear Complexity Approach makes us question exactly the degree of our predictability and control capacity of natural and social systems.The heuristic importance of a nonlinear analytical approach now seems evident. In light of this, there are two main implications regarding complexity and chaos theory for social science research.

As already noted, the new science loosens the ties between determinism and predictability supported by Laplace's theory. The Newtonian-Laplacian weltanschauung portrayed the world as a machine with perfect, predictable clock-like precision. From this deterministic conception comes the predictability of phenomena over time. We all remember Laplace's famous saying "give me the state of the universe at any given time and I will predict the future". Now, however, determinism and predictability have to be kept separated due to the discovery that even if behavior is unexpected, discontinuous, unpredictable, there are nevertheless deterministic laws involved in its unfolding. From this perspective, *emergence* and *causality* are reconciled. This extraordinary discovery, that deterministic laws can produce behavior only apparently random, that disorder can spring from order in a view that does not separate them irreparably, makes the old issue of incompatibility between randomness and laws' determinism obsolete. In 1944, for example, in his famous letter to Max Born, Einstein wrote that not even the great success of quantum theory could convince him that randomness was at the base of everything. Einstein was convinced: God did not play dice with the world, rather everything obeys a law[4].Along with deterministic chaos comes a new landscape for mathematics, where we no longer ask whether God is playing dice with the world but rather what the rules of his game are (Stewart, 1989: 8).

The *first* implication of this new acquisition for scientific research and for social science research in particular is clear. Complex and chaotic systems can be *understood*, even if not *predictable.*). In other words, *social emergence* can be *explained.* On one hand, research has to reassess its ambitions regarding predictability, yet it can and must focus on discovering mechanisms and laws which structure and generate discontinuity (*basic equation governing system's behavior).* While, as Brown suggests, linear determinism has dominated our thinking process to an extent that it has hindered a more realistic representation of the human condition, to understand the nonlinearity, complexity, and interdependencies of our individual and collective life, without a doubt, constitutes the occasion to be….. *no longer passive riders on an evolutionary voyage that touches every level of our existence*(1995:146)…., the unavoidable condition *to understand how and why we live the way we do,..* and so *to gain true freedom to determinate our fate…* (*cit.*:144). *Where is,* asks the author, *self-determination among the blind*? (*ibidem).* Statements of this type are literally teeming in his work *Serpents in the Sand.* Knowing the laws or the basic equations of motion allow a better understanding of social systems. It also ensures another decisive possibility which introduces us the *second* implication.

Complexity and Chaos Theory introduces a *different way* of thinking about predictability. We could, in fact, come up against the objection that is, in the end, a waste of time insisting on the *Complexity and Chaos Theory,* studying its analytical tools only to conclude that the system is ….unpredictable. However, is it really possible to consider non-predictability and hence un-controllability of social phenomena devoid of any heuristic value? From a sociological point of view we can give a firm negative answer. It is significant to know whether a phenomena presents chaotic behavior and therefore cannot be controlled, or to know that it is on the *edge of chaos*, i.e. in a transition zone, approaching chaos, thus on the verge of not being controllable.

Let us note that certain modern perspectives seem to revolve around this category. For instance, during the last ten years, we have seen the trend of defining some phenomena like wars in terms of *loss of control* more than in the usual terms of the past like 'the will to go to war' or 'conspiracy' and so on. Saperstein's view is opportune in this regard.

> "I suggest that war be viewed as a breakdown in predictability: a situation in which small perturbations of initial conditions such as malfunctions of early-warning radar systems or irrational acts of individuals disobeying order lead to large unforeseen changes in the solution to the dynamical equations of model. There is no way then to predict the effect of the actions of any participant – analyst, planner, statesman or general – with any certainty. Certainty war, between roughly equivalent parties, has often been characterized by uncertainty in outcome; nations which start wars in great confidence very often end defeated. Typically, in the period immediately before the outbreak of a

[4] A. Einstein's letter to Max Born, 7 September 1944, – Einstein, A. (1969). *Albert Einstein, Hedwig und Max Born: Briefwechsel 1916–1955.* Munich: Nymphenburger Verlagshandlung.

war the nations attempt desperately to claw back as they slide over the precipice, although they by this time already have lost control of their own destinies: control and predictability are lost"(Saperstein, 1984: 303).

Historians, on their part, seem to support this perspective. For example, the problem of ascertaining what the causes of the outbreak of the Great War were is still an open issue. Recently, documents have been exhibited on the basis which scientists are led to believe that Hitler was rather far from wanting to commit to a world-wide conflict so soon, and without any delay.This perspective has ethical and political implications as well. For example, a politician proposes a new bill concerning a social phenomenon which was ascertained to have *chaotic behavior*. The politician would be obliged to admit to an *adequately* informed public opinion that the new bill might have consequences that no one is able to predict.

To sum up, non-predictability and uncontrollability cannot be considered of little interest. Consequentially knowing the *threshold* where phenomena turns into chaos is precious information, obviously in attempting to prevent and therefore to avert reaching an uncontrollable situation. As Elliott and Kiel ascertained, only "by better understanding the confluence of chance and determinism in social systems evolution we may better learn when and how to direct policy responses" (Elliott and Kiel, 1997: 68).

So, even if we are unable to predict future outcomes in a phenomenon's dynamic evolution (of a biological population, of a social or electoral competition, of industrial production, of spread of disease or of an technological invention), non-linear dynamics study through building nonlinear models allow us to 'predict' what conditions (*parameter values*) give way to certain types of dynamic behavior. We can know at *what values* a company goes bankrupt, a party loses the elections, a political system become instable, a war breaks out, crime waves uncontrolled and social control system disorganizes. In the same way we can know at *what values* a biological population (as well as a disease) dies out or evolves in a periodic or chaotic behavior. This type of knowledge is much more important for intervention policies (social or otherwise) rather than an exact knowledge of the system's future evolution outcomes.

For further clarification of this point, let us apply this argument to a social phenomenon like *noncompliance-coercion relationships* already quoted above, where the relationship is modeled according to a logistic structure. The study of this model tells us that at k values under 3.8-4 crime variability remains in a situation of possible controllability. In this situation any intervention aimed at influencing the system's current trend (in this specific case, assure a decrease in crime) can produce expected outcomes. Obviously, we need to avoid the situation from falling into a *chaotic range* and it is always the model which indicates that this can happen if the k value reaches 3.8-4.*This value marks an alarming loss in predictability and therefore in control.* It would be impossible, in fact, to predict what would happen in this case. Note, we are not saying that we would not know what to do in such a situation – or what line of intervention to take. We are saying something much different, namely that *social phenomenon flowing into chaos produces absolutely unpredictable outcomes for any intervention adopted.* It has been shown that if the crime rate reached this value, flowing into chaos, it would pull law enforcement agencies into chaos as well. In conclusion, the galloping, uncontrollable rise in crime would disorganize any form of intervention able to adapt the police force and their work to such an uncontrollable increase in crime. For instance, a law enforcement agency may increase the number of police officers but we have no idea what the outcome will be. Thus, measures taken to reduce crime might have a completely opposite outcome; they may produce even a rise in crime rates or an actual fall in these rates. The problem is that we are unable to predict what outcome will emerge. Priesmayer provides us with an exemplifying case, regarding cocaine use in the USA from 1985 to 1990. In his investigation, the logistic equation fits cocaine use with a value k of 3.6, dangerously close to the critical threshold of 3.8, a value inducing the author to the following disarming conclusion:

> Put simply, actions which decrease current use may contribute to higher future use or they may not; actions which contribute to higher current use may contribute to lowering use in the future or they may not. [...] Does it suggest that [...]attempts to lower cocaine use by aggressive intervention are far less certain? If cocaine use is not controllable in this way, what then is to be used to control cocaine use? (Priesmayer, 1990: 333).

In brief, no one is able to define the terms of a scenario. Or rather, any type of scenario is equally possible. Possible scenarios might mirror films such as *Escape from New York* or, on the contrary, film series where enforcement agencies become powerful managers of criminal enterprises. In an extreme case (not so extreme if we judge by current situations), crime agencies and police could form alliances creating strange symbiotic phenomena. This idea is suggested by the situation in Russia where, from my point of view, the actual situation appears dangerously on the edge of chaos or pre-chaotic. Within a few years, Russia's crime rate increased by 40%, causing a paradoxical situation. The rise in the crime rate led to a rise in police enforcement yet the reduction of crime was not forthcoming but there was more and better orchestrated collusion and corruption. In this scenario, it is not rhetorical to ask where crime control efforts will end up and what line of action to take against noncompliance.

To avoid similar dangers, I repeat that it is absolutely important *to know* and therefore to be able to *predict* the threshold of chaotic behavior for a social science researcher. Consequently, the study of models of nonlinear dynamic systems suitable for attaining this knowledge is also of prime importance.

Referring back again to Saperstein, for example, by using a simple bilateral logistic model, the author estimated that it would be sufficient to have a parameter over to 1 in one of the equations model for a *transition into chaos* or in other words for the *outbreak of war*. It is evident that this knowledge can give a significant contribution to any peace oriented policy, by controlling that the arms build-up stays below the critical threshold level. Saperstein did not only limit himself to a 'theoretical' study, but provided empirical examples. He analyzed the arms race of different couples of European nations during 1936-37, and between the USA and the URSS in the 70s and 80s. It was ascertained that the relationship between Nazist Germany and the Soviet Union were within "a chaotic region" (a=1.34, b=0.0657): in fact, this case history has shown us the intrinsic uncertainty connected to a state of war (1984: 305). In USA and URSS relations, the chaotic region was instead only touched.

Gottman *et al.'s* study on marital relations is also relevant. It presents these relationship as a discontinuous process (both stable and prone to catastrophe), and analyzes this discontinuity applying a nonlinear model using difference equations. The model indicates the *threshold* beyond which a marriage is expected to break up (Gottman *et al.*, 2005).

To conclude, from a heuristic point of view there is more than one valid reason to support the use of non-linear instruments in social system analysis.

Conclusion

We have reason to believe that Ruelle's main objection, that is, the intrinsic impossibility for the Social Sciences to attain *really convincing of deterministic time evolution* (1990: 244), cannot be accepted in its general absoluteness. Granted that we are dealing with systems that cannot be placed under control in the same terms as those of the physical sciences, but, as argued above, this condition does not seem to constitute, in absolute terms, an *impediment* for the theory's use in the social realm. On the contrary, this application already appears to be a *real* possibility.

Naturally, the greater the system's dimension increase, the more difficulties are encountered in attempts at modeling. However, some clarifications are inevitable. In the first place, we are dealing with attempts inserted into a theoretical framework which selects and therefore narrows the field of possible choices so that only certain variables and not others - only certain relations and not others can be investigated in the reconstruction of the system under study. In short, the construction of the model never takes place in a sort of disorienting theoretical void but always refers to a relationship of values which orients the researcher to explore certain paths rather than others. Therefore, it is not so obvious in what sense Ruelle and Kellert point out that the basic problem of the social science is to know many and which variables constitute the system.

In the second place, concerning possible estimate coefficient reliability, the data requirement for difference equation is far shorter than for systems of differential equations. Moreover, the problem of a small noisy data sets resolvable with the help of more efficient algorithms.

To close the discussion, some comments have to be directed toward the third argument which according to Ruelle impedes the use of the chaos theory for social science studies, that is, *persistence*. In our author's argumentation, persistence seems to be displayed as an intrinsic quality, distinctive of social phenomena, whereas, the physical sciences are characterized par excellence by a continuous and pervasive variability.

From my point of view, it is possible to advance the following counter-objection: the presumed persistence or *non*-persistence of a given process is a question of empirical verification, so that an aprioristic judgment would reveal itself to bean affirmation so binding as it is disputable.

Rather, the real problematic aspect of this issue was pointed out by Ruelle, Kellert and other social scientists like Huckfield, Grebogi and so on. It consists of a possible *simply metaphorical* use of the Complexity and Chaos Theory in the Social Sciences. It is likened to an anti-scientific orientation simply finalized toward easy instrumentalization. This danger is both serious and concrete. However, it is far from being considered *intrinsic* to the use of the Complexity and Chaos Theory. Avoiding the above attitude depends uniquely on the intellectual honesty of social researchers: it is up to them to choose whether to do social science or to use the theory as a pretext for ideological ends, which has nothing to do with seeking neither knowledge nor its progress.

References

Abraham, N.B., Albano, A.M., Das, B., De Guzman, G., Yong S., Gioggia, R.S, Puccioni, G.P., & Tredicce, J.R.(1986). Calculating the Dimension of Attractors from Small Data Sets. *Phisics Letters A,* 114, pp. 217-224.

Albano, A.M., Mees, A.I., De Guzman, G.C., & Rapp, P.E. (1987). Data Requirements for Reliable Estimation of Correlation Dimension. In H. Degn, A.V. Holden, & L. Olsen (Eds.), *Chaos in Biological Systems* (pp. 207-220). New York: Plenum.

Abarbanel, H.D.I., Brown, R., Sidorovich, J.J., & Tsimring, L.S. (1993). The analysis of observed chaotic data in physical systems, *Reviews of Modern Physics*, 65, pp. 1331–1392.

Arbabanel H. D.I. (1996).*Analysis of Observed Chaotic Data.* New York: Springer.

Bala, V., & Majumdar, M. (1992). Chaotic Tatonnement. *Economic Theory*, 2, pp. 437-445.

Baumol, W.J., & Benhabib, J. (1989). Chaos: Significance, Mechanism, and Economic Applications. *Journal of Economics Perspectives*, 3, pp. 77-105.

Baumol, W.J., & Quandt, R.E.(1985). Chaos Models and theirs Implications for Forecasting. *Eastern Economic Journal*, 11, pp. 3-15.

Bhattacharya, J., & Kanjilal, P.P. (1999) On the detection of determinism in a time series. *Physica D,* 132, pp. 100–110.

Bischi, G.I., &Tramontana, F. (2009). Three-dimensional discrete-time Lotka–Volterra models with an application to industrial clusters. *Communications in Nonlinear Science and Numerical Simulation.* doi:10.1016/j.cnsns. 2009.10.021.

Braun, M. (1993). *Differential Equations and Their Applications. An Introduction to Applied Mathematics.* New York: Springer.

Briant, P., Brown, R., & Arbabanel, H.D.I. (1990). Lyapunov Exponents from Observed Time Series. *Physical Review Letters*, 65, pp. 1523-1526.

Briggs, K. (1990). An improved method for estimating Lyapunov exponents of chaotic time series. *Physical Review A*, 151, pp. 27-32.

Brown, C. (1995). *Chaos and Catastrophe Theories.* Thousand Oaks, CA: Sage Publications.

Brown, C. (1991). *Ballots of Tumult. A Portrait of Volatility in American Voting.* Ann Arbor: The University of Michigan Press.

Brown, C. (1995). *Serpents in the Sand. Essays on the Nonlinear Nature of Politics and Human Destiny*. Ann Arbor: The University of Michigan Press.

Capra, F. (1996). *The Web of Life*. New York: Doubleday-Anchor book.

Campbell, D.K., & Mayer-Kress, G. (1997). Chaos and politics: Applications of nonlinear dynamics to socio-political issues. In: C. Grebogi, & J.A. Yorke (Eds.), *The impact of chaos on science and society* (pp.18-62). NewYork: United Nation University Press.

Cheng, B., & Thong, H. (1994). Orthogonal Projection, Embedding Dimension and Sample Size in Chaotic Time Series from a Statistical Perspective. *Philosophical Transaction of Royal Society of London*, 348, pp. 325-341.

Condorelli, R. (2007). *Complessità e Controllo sociale*. Catania-Roma: Bonanno.

Condorelli, R. (2012). Applied nonlinear dynamical systems in social science. A nonlinear model for social control system: an application to Italian coercion system. *Quality & Quantity*, doi: 10.1007/s11135-012-9666-7 (in press).

Crutchfield, J., Farmer, J.D., Packard, N.H., & Shaw, R.S. (1986). Chaos. *Scientific American,* 255, pp. 46-57.

Dendrinos, D., & Mullally, H. (1985). *Urban Evolution: Studies in the Mathematical Ecology of Cities.* New York: Oxford University Press.

Dendrinos, D. (1992). *The Dynamics of Cities.* London: Routledge.

Dixon, R. (1994). The logistic family of discrete dynamic models. In J. Credy, & V. Martin, *Chaos and Non-linear Models in Economics. Theory and Application*s (pp. 44-69). UK: Edward Elgar.

Duhem, P. (1906). *La théorie physique. Son objet et sa structure.* Paris: Chevalier et Rivière.

Eckmann J.P.,& Ruelle D. (1985). Ergodic theory of chaos and strange attractors. *Reviews of Modern Physics*, 57, pp. 617-656.

Eckmann, J.P., Kamphorst, S.O., Ruelle, D., & Ciliberto, S. (1986). Lyapunov exponents from time series. *Physical Review A*, 34, pp. 4971–4979.

Eckmann, J.P.,& Ruelle, D. (1992). Fundamental limitations for estimating dimensions and Lyapunov exponents in dynamical systems. *Physica D*, 56, pp. 185-187.

Elliott, E., & Kiel, D. L. (1997). Nonlinear Dynamics, Complexity, and Public Policy. Use, Misuse, and Applicability. In: R.A. Eve, S. Horsfall, & M.E. Lee (Eds.), *Chaos, Complexity, and Sociology. Myths, Models and Theories* (pp.64-78). Thousands Oaks, CA: Sage Publications Inc.

Feigenbaum, M. (1978). Quantitative Universality for a Class of Non-linear Transformations. *Journal of Statistical Physics*, 19, pp. 25-52.

Gaertner, W. (1987). Periodic and Aperiodic Consumer Behavior. *Applied Mathematics and Computation*, 22, pp. 233-254.

Gleick, J. (1987). Chaos: making a new science. New York: Viking Penguin Inc.

Gottman, J.M., Murray, J.D., Swanson, C.C., Tyson, R., & Swanson, K.R. (2005). *The Mathematics of Marriage. Dynamic Nonlinear Models.* Massachusetts: MIT Press.

Gottwald, G.A., &Melbourne, I. (2005). Testing for chaos in deterministic systems with noise. *Physica D,* 212, pp.100–110.

Grassberger, P., & Procaccia, I. (1983a). Measuring the Strangeness of Strange Attractors. *Physica D*, 9, pp. 189-208.

Grassberger, P., & Procaccia, I. (1983b). Caracterization of Strange Attractors. *Physical Review Letters*, 50, pp. 346-349.

Grossmann, S., & Mayer-Kress, G. (1989). Chaos in the international arms race. *Nature*, 337, pp. 701-704.

Guesnerie, R., & Woodford, M. (1992). Endogenous Fluctuations. In J.J. Laffont (Ed.), *Advances in Economic Theory: Proceedings of the Sixth World Congress of the Econometric Society*: Vol. 2 (pp. 289-412). Cambridge: Cambridge University Press.

Hadamard, J. (1898). Les surfaces à courbures opposeès et leurs lignes géodésiques. Journal de Mathématiques Pures et Appliquées, 4, pp. 27-73.

Hahn, F.R. (1992). Equilibrium Dynamics and Chaos. *Jahrbucher fur Nationalokonomie and Statistik*, 210, pp. 279-285.

Harvey, D.L., & Reed, M. (1997). Social Sciences as the Study of Complex Systems. In D.L. Kiel, & E. Elliott (Eds.), *Chaos Theory in the Social Science: Foundations and Applications* (pp. 295-323). Ann Arbor: The University of Michigan Press.

Holland, J. H. (1998.) *Emergence: From Chaos to Order.* Reading, MA: Addison-Wesley.

Huckfeldt, R. (1989). Noncompliance and the Limits of Coercion: The Problematic Enforcement of Unpopular Laws. *Mathematical and Computer Modelling,* 12, pp. 533-546.

Huckfeldt, R (1990). Structure, Indeterminacy and Chaos. A Case for Sociological Law. *Journal of Theoretical Politics,* 2, pp.413-433.

Hunt, H.W., Antle, J.M., &Paustian, K. (2003). False determinations of chaos in short noisy time series. *Physica D*, 180, pp.115-127.

Johnson, M., & Habyarimana, J. (1998). An operational test for distinguishing between complicated and chaotic behavior in deterministic systems. *Physica D,* 116, pp. 289-300.

Judd, K. (1992). An improved estimator of dimension and some comments on providing confidence intervals. *Physica D*, 56, pp. 216-228.

Judd, K. (1994). Estimating dimension from small samples. *Physica D,* 71, pp. 421-429.

Kantz, H., & Schreiber, T. (2004). *Nonlinear Time Series Analysis*, 2nd edn. Cambridge: Cambridge University Press.

Kellert, S. (1995). When Is the Economy Not Like the Weather? The Problem of Extending Chaos Theory to the Social Sciences. In A. Albert (Ed.), *Chaos and Society* (pp. 35-47). Amsterdam: IOS Press.

Kiel, D. L., & Elliott, E. (1997). Exploring Nonlinear Dynamics with a Spreadsheet: A Graphical View of Chaos for Beginners. In: L.D. Kiel, & E. Elliott (Eds.), *Chaos Theory in the Social Science: Foundations and Applications* (pp.19-29). Ann Arbor: University of Michigan Press.

Kitoh, S., Kimura, M., Mori, T., Takezawa, K., & Endo, S. (2000). A fundamental bias in calculating dimensions from finite data sets. *Physica D*, 141, pp. 171-182.

Kostelich, E.J., & Armbruster, D. (1996). *Introductory Differential Equations. From Linearity to Chaos.* New York: Addison-Wesley.

Krishnamurthy, K. (2012). A Method for Estimating the Lyapunov Exponents of Chaotic Time Series Corrupted by Random Noise Using Extended Kalman Filter. *Mathematical Modelling and Scientific Computation*, 283, pp.237-244.

Kurth, J., & Herzel, H.(1987). An attractor in a Solar Time Series. *Physica D*, 25, pp. 165-172.

Lorenz, H.W. (1993). *Nonlinear Dynamical Economics and Chaotic Motion.* Berlino: Springer.

Marion, R. (1999). *The Edge of Organization. Chaos and Complexity Theories of Formal Social Systems.* Thousand Oaks, CA: Sage Publications Inc.

Marzocchi, W., Mulargia, F., & Gonzato, G. (1996). Detecting low-dimensional chaos in time series of finite length generated from discrete parameter processes. *Physica D*, 90, pp. 31-39.

May, R.M. (1976). Simple mathematical models with very complicated dynamics. *Nature*, 261, pp. 459-467.

May, R.M., Leonard, W.J. (1975). Nonlinear Aspects of Competition between Three Species. *SIAM Journal of Applied Mathematics,* 29, pp. 243-253.

McBarnett, M. (1997). Complexity in the Evolution of Public Opinion. In: D.L. Kiel, & E. Elliott (Eds.), *Chaos Theory in the Social Science: Foundations and Applications* (pp. 165-196). Ann Arbor: The University of Michigan Press.

McDaniel, R.R, & Driebe, D.J. (2005). Uncertainty and Surprise. An Introduction. In R.R. McDaniel, &D.J. Driebe (Eds), *Uncertainty and Surprise in Complex Systems: Questions on Working with the Unexpected* (pp.3-12). New York: Springer.

Medio, A. (1993). *Chaotic Dynamics*. UK.: Cambridge University Press.

Mosekilde, E., Aracil, J., & Allen, P.M. (1988). Instabilities and Chaos in Nonlinear Dynamic Systems. *System Dynamics Review*, 4, pp. 14-55.

Nychka, D., Ellner, S., McCaffrey, D.,& Gallant, A.R. (1992). Finding chaos in noisy systems. *Journal of Royal Statistical Society B*, 54, pp. 399–426.

Parker, T.S., & Chua, L.O. (1989). *Practical Numerical Algorithms for Chaotic Systems*. New York: Springer.

Poincarè, H. (1908). *Science et méthode*. Paris: Ernest Flammarion.

Priesmeyer, R.H. (1995). Logistic Regression: A Method for Describing, Interpreting, and Forecasting Social Phenomenon with Nonlinear Equations. In A. Albert (Ed.), *Chaos and Society* (pp. 3329-338). Amsterdam: IOS Press.

Prigogine, I.,&Stegers, I. (1984). *Order out of chaos: Man's new dialogue with nature*. Flamingo.

Ramdani, S., Bouchara, F., & Lagarde, J. (2009). Influence of noise on the sample entropy algorithm. *Chaos*,doi.org/10.1063/1.3081406.

Ramsey, J.B., Sayers, C.L., & Rothman, P. (1992). The Statistical Properties of Dimension Calculations Using Small Data Sets: Some Economic Applications. In J. Benhabib (Ed.), *Cycles and Chaos in Economic Equilibrium* (pp. 394-428). New Jersey: Princeton University Press.

Ramsey, J.B., & Yuan, H. (1989). Bias and Error bars in Dimension Calculations and theirs Evaluation in some Simple Models. *Physics Letters A*, 134, pp. 287-297.

Rosenstein, M. T., Collins, J.J., & De Luca, C.J. (1993). A practical method for calculating largest Lyapunov exponents from small data sets. *Physica D*, 65, pp. 117-134.

Ruelle, D. (1990). Deterministic chaos: the science and the fiction. *Proceedings of Royal Society of London A*, 427, pp. 241-248.

Ruelle, D. (1991). *Chance and Chaos*. New Jersey: Princeton University Press.

Sano, M., & Sawada, Y. (1985). Measurement of the Lyapunov spectrum from a chaotic time series. *Physical Review Letters*, 55, pp. 1082-1085.

Saperstein, A.M.(1984). Chaos – a model for the outbreak of war. *Nature*, 309, pp. 303-305.

Saperstein A.M. (1997a). The prediction of Unpredictability: Applications of the New Paradigm of Chaos in Dynamical Systems to the Old Problem of the Stability of a System of Hostile Nations. In D.L. Kiel, & E. Elliott (Eds.), *Chaos Theory in the Social Science: Foundations and Applications* (pp. 139-163). Ann Arbor: The University of Michigan Press.

Saperstein, A.M.(1997b). The Origin of Order and Disorder in Physical and Social Deterministic Systems. In: R.A. Eve, S. Horsfall, & M.E. Lee (Eds.), *Chaos, Complexity, and Sociology. Myths, Models and Theories* (pp.102-122). Thousands Oaks, CA: Sage Publications Inc.

Sato, S., Sano, M., & Sawada, Y. (1987). Practical methods of measuring the generalized dimension and the largest Lyapunov exponent in high dimensional chaotic systems. *Progress of Theoretical Physics*, 77, pp.1-5.

Schaffer, W.M. (1987). Chaos in Ecology and Epidemiology. In H. Degn, A.V. Holden, & L. Olsen (Eds.), *Chaos in Biological Systems* (pp. 233-248). New York: Plenum.

Schaffer, W. M. (1988). *Dynamical Software*, Vol.2, Tuscon, AZ: Dynamical Systems Inc.

Scheinkman, J.A., & LeBaron, B. (1992). Nonlinear Dynamics and Stock Return. In J. Benhabib (Ed.), *Cycles and Chaos in Economic Equilibrium* (pp. 446-473). New Jersey: Princeton University Press.

Schreiber, T. (1999). Interdisciplinary application of nonlinear time series methods. *Physics Report*, 308, pp.1–64.

Smith, L. A. (1988). Intrinsec Limits on Dimension Calculations. *Physics Letters A*, 133, pp. 283-288.

Sonis, M. (1984). Dynamic Choice of Alternatives, Innovation Diffusion and Ecological Dynamics of the Lotka-Volterra Model. *London Paper in Regional Science,* 14, pp. 29-43.

Sprott, J.C, & Rowlands, G. (1995). *Chaos Data Analyzer*. New York: American Institute of Physics.

Stewart, I. (1989). *Does God play dice? The mathematics of chaos*. Cambridge: Blackwell.

Sugihara, G., & May, R.M.(1990). Nonlinear forecasting as a way of distinguishing chaos from measurement error in time series. *Nature,* 344, pp. 734 – 741.

Swinney, H.L. (1987), personal communication in Albano, A.M., Mees, A.I., De Guzman, G.C., & Rapp, P.E. (1987). Data Requirements for Reliable Estimation of Correlation Dimension. In H. Degn, A.V. Holden, & L. Olsen (Eds.), *Chaos in Biological Sistems* (pp. 207-220). New York: Plenum.

Takens, F. (1980). Detecting Strange Attractors in Turbulence. In A. Dold, & B.Eckmann (Eds.), *Lecture Notes in Mathematical Systems 898: Dynamical Systems and Turbulence* (pp. 366-381). New York: Springer.

Theiler, J. (1990). Estimating fractals dimension. *Journal of Optical Society of America A*, 7, pp. 1055-1073.

Tsebelis, G., & Sprague, J. (1989). Coercion and Revolution: Variations on a Predator-Prey Model. *Mathematical and Computer Modelling*, 12, pp. 547-559.

Tsebelis, G., & Sprague, J. (2010). Coercion and Revolution: Variations on a Predator-Prey model. Mathematical and Computer Disorder? *Criminology and Public Policy,* 9, pp.139-172.

Tsonis, A.A. (1992). *Chaos: From Theory to Applications*. New York: Plenum Press.

West, B.J. (1997). Chaos and Related Things: A Tutorial. *The Journal of Mind and Behavior*, 18, pp. 103-126.

Wolf, A., Swift, J.B., Swinney, H.L., & Vastano, J.A. (1985). Determining Lyapunov exponents from a time series. *Physica D,* 16, pp. 285-317.

Wolf, A. (1986). Quantifying Chaos with Lyapunov Exponents. In: A. Holden (Ed.), *Chaos* (pp. 273-290). Princeton, NJ: Princeton Univeristy Press.

Wright, J. (1984). Method for calculating a Lyapunov exponent. *Physical Review A*, 29, pp. 2924-2927.

Yang, C., &Qiong Wu, C. (2011). A robust method on estimation of Lyapunov exponents from a noisy time series. *Nonlinear Dynamics*, 64, pp. 279-292.

Yang, C., Qiong Wu, C., &Zhang, P. (2012). Estimation of Lyapunov exponents from a time series for *n*-dimensional state space using nonlinear mapping. *Nonlinear Dynamics*, 69, pp. 1493-1507.

Zeng, X., Eykholt, R., & Pielke, R.A. (1991). Estimating the Lyapunov exponent spectrum from short time series of low precision. *Physical Review Letters*, 66, pp. 3229-3232.

In: Dynamical Systems
Editor: Jean Michel Tchuenche, pp. 163-182
ISBN: 978-1-62808-001-8

Chapter 6

LOCAL STABILITY ANALYSIS OF AN AGE-STRUCTURED HEPATITIS B TRANSMISSION MODEL

***Ousmane M. Tessa*[1,*], *Bisso Saley*[1], *Herieth Rwezaura*[2] *and Jean M. Tchuenche*[2]**
[1]Department of Mathematics and Computer Science,
Abdou Moumouni University, Niamey, Niger
[2]Department of Mathematics, University of Dar es Salaam,
Dar es Salaam, Tanzania

Abstract

This chapter deals with the analysis of the local stability of a modified age-structured deterministic model for the transmission dynamics of Hepatitis B virus first proposed by Zhao *et al* [29]. The model exhibits two equilibria, namely: the disease-free equilibrium which is locally asymptotically stable if a certain disease threshold parameter $\mathcal{R}_0$ which depends solely on the parameters associated with the model is less than unity; a public health implication of this result is that Hepatitis B could be eliminated in the population - an endemic equilibrium which exists for $\mathcal{R}_0 > 1$ and is locally asymptotically stable under certain condition and consequently, the disease may persist.

Keywords: Local stability, age-structured model, Hepatitis B, perturbation, basic reproduction number

MSC 2000: 45D05, 92B05, 92D30, 92C60

1. Introduction

Hepatitis B is found throughout the world, but its prevalence varies greatly; it is especially high in Asia, sub-Saharan Africa, the south Pacific, as well as within specific populations

*E-mail addresses: musatesa@refer.ne, ousmane@musatesa.net (Corresponding Author)

in South America, the Middle-East, and the Arctic. It was first discovered in 1963 by Dr. Baruch Blumberg and colleagues, who identified a protein (the "Australia antigen") that reacted to antibodies from patients with hemophilia and leukemia [3]. The association of this protein with infectious hepatitis was discovered 3 years later by several investigators, and the virus was specifically seen by electron microscopy in 1970.

Hepatitis B virus (HBV) causes a serious health problem [29]. The World Health Organization (WHO) estimates that 2 billion people are infected worldwide, with approximately 5% of the world's population (or 350 million people) being carriers of chronic hepatitis B [27]. Every year, 1 to 2 million people die due to an infection by this virus. Therapy is administered to suppress viral replication and prevent progression of liver disease in chronic hepatitis B [3]. 15% of individuals carry the virus while 70 to 90% have come into contact with the virus in sub-Saharan Africa. Extensive studies in The Gambia showed that sibling-to-sibling transmission was of major importance rather than perinatal or vertical transmission ([15], [25]). In addition, it was found that infection was uncommon in infants under the age of 6 months and, in Africa Hepatitis B is predominantly acquired in early childhood, rather than during parturition or in adult life. However, the epidemiological situation of Hepatitis B is quite disturbing considering that approximately 90% of babies who contract the virus will become chronic carriers, and of those, 25% will die of liver cancer or cirrhosis [4].

Since 1982, a vaccine against hepatitis B has been available [3]) and is 95% effective in preventing infection and its chronic consequences. WHO recommends that all infants receive the hepatitis B vaccine, administered in three or four separate doses, as part of existing routine immunization schedules in 179 countries (by July 2011), compared with 31 countries in 1992 ([28]). The cost-effectiveness of adding universal or selective vaccination strategies in these countries depends on the ability to sufficiently identify, reach and fully vaccinate persons in various risk groups, each one being justified on the basis of its economic evaluation [2]. These factors are a barrier in developing countries as a large number of individuals may not afford prophylaxis, therapy or vaccination.

Mathematical modeling is one of the best approaches for gaining insight into some key parameters that drive the disease dynamics. It also provides a framework within which the maximum information may be drawn from all available knowledge about the clinical and population epidemiology of the disease [5]. Our model is a modified/extended version of the one first proposed in [29]. The aim of this study is to carry out a detailed qualitative analysis of the age-structure model (first order partial differential equations) as well as its age-independent version which was not considered in [29]. It is our view that this study provides an in-depth analysis of the qualitative dynamics of HBV. Additionally, there are some important differences between our model and that presented in [6]. For instance in [6], the authors consider the gender whilst we do not. Their recovery rates are constants contrary to ours which are age dependent.

The paper is organized as follows: Section 2 introduces an age-structure compartmental model to study the dynamics of HBV in the presence of a vaccine. The reproductive numbers are computed in Section 3, and the qualitative behavior of the disease-free steady state is also studied. Section 4 concludes the chapter.

2. The Model

The population is homogeneously mixing and a variable portion is experiencing exponential dynamics. Following the classical assumption [8], we formulate an age-structure model based on the characteristics of HBV. Our compartmental model is expressed by a set of first-order partial differential equations. A population of size $N(a,t)$ is stratified into five compartments, namely: (1) susceptibles $S(a,t)$, (2) latent $L(a,t)$, (3) temporary HBV carrier $T(a,t)$, (4) chronic HBV carrier $C(a,t)$ and (5) immune individuals $R(a,t)$, where a represents the age and t is the length of follow-up.

After acquiring an acute (for which treatment is supportive) HBV infection through an adequate and effective contact with a temporary or a chronic HBV carrier, a susceptible individual moves to the latent class during an average period of 45 days; after which, he or she becomes a temporary HBV carrier for 90 days on average. The infected host clears HBV, recovers and becomes lifelong immune, otherwise progresses to become a chronic carrier. Only temporary and chronic carriers of HBV are capable of spreading the disease to those in the susceptible class. A small proportion of chronic carrier may clear HBV and join the immune class ([1], [29]).

Vaccination is an effective way to control the transmission of HBV. So, a proportion of susceptibles enter directly into the immune class after being successfully immunized with hepatitis B vaccine (see model flowchart). All neonates are assumed to be susceptible. We exclude vertical incidence in our model, which implies that newborns are not infected by their mothers. Sex differences are ignored.

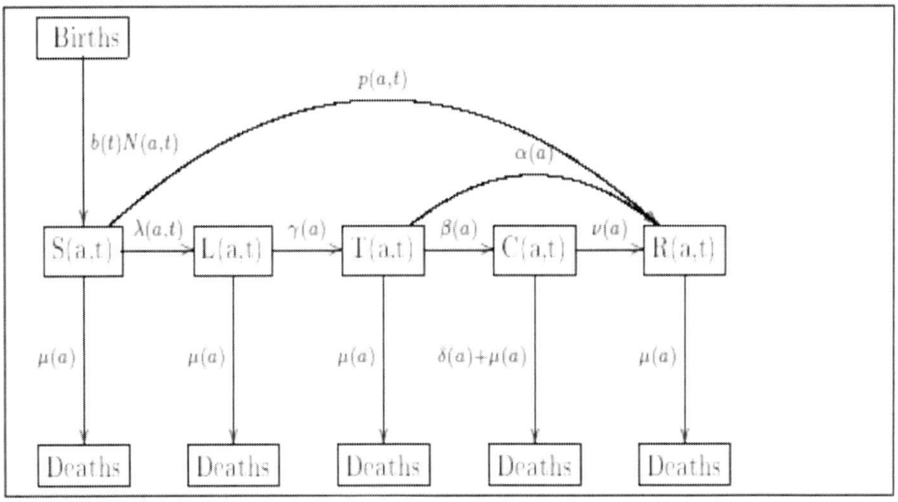

Figure 1. Age-structure compartmental model of HBV's transmission.

Parameters description

- $b(t) = b$ is the constant birth rate;
- $\lambda(a,t)$ is the force of infection

- $\gamma(a) = \gamma$ is the constant rate of transition from latent period to temporary HBV carrier;
- $\alpha(a) = \alpha$ is the constant rate of transition from temporary carriers to immune class;
- $\beta(a)$ is the rate of transition from temporary HBV carrier to chronic carrier;
- $\nu(a)$ is the rate of HBV clearance in chronic carriers class;
- $\delta(a)$ is the additional mortality rate related to HBV;
- $\mu(a)$ is the age-specific natural mortality rate;
- $p(a,t)$ the successful vaccinated proportion against HBV.

Under the above assumptions and parameters, the spread of the disease can be described by the following system of partial differential equations:

$$\begin{aligned}
\frac{\partial S(a,t)}{\partial a} + \frac{\partial S(a,t)}{\partial t} &= b(t)N(a,t) - [\lambda(a,t) + p(a,t) + \mu(a)]\, S(a,t) \\
\frac{\partial L(a,t)}{\partial a} + \frac{\partial L(a,t)}{\partial t} &= \lambda(a,t)S(a,t) - [\gamma(a) + \mu(a)]\, L(a,t) \\
\frac{\partial T(a,t)}{\partial a} + \frac{\partial T(a,t)}{\partial t} &= \gamma(a)L(a,t) - [\beta(a) + \alpha + \mu(a)]\, T(a,t) \\
\frac{\partial C(a,t)}{\partial a} + \frac{\partial C(a,t)}{\partial t} &= \beta(a)T(a,t) - [\nu(a) + \mu(a) + \delta(a)]\, C(a,t) \\
\frac{\partial R(a,t)}{\partial a} + \frac{\partial R(a,t)}{\partial t} &= p(a,t)S(a,t) + \alpha T(a,t) + \nu(a)C(a,t) - \mu(a)R(a,t)
\end{aligned} \tag{1}$$

with initial and boundary conditions

$$\begin{aligned}
& S(0,t) = \int_0^{a_+} b\,[S(a,t) + L(a,t) + T(a,t) + C(a,t) + R(a,t)]\, da \\
& L(0,t) = T(0,t) = C(0,t) = R(0,t) = 0, \\
& S(a,0) = S_0(a); L(a,0) = L_0(a); T(a,0) = T_0(a); C(a,0) = C_0(a); R(a,0) = R_0(a); \\
& N(a,0) = N_0(a)
\end{aligned} \tag{2}$$

By adopting the separable intercohort constitutive form for the force of infection, we can express $\lambda(a,t)$ by

$$\lambda(a,t) = m(a)\widehat{\lambda}(t) \text{ where } \widehat{\lambda}(t) = \int_0^{a_+} \theta(a) \frac{L(a,t) + T(a,t) + C(a,t)}{N(a,t)} da.$$

a_+ is the maximum age of individual ($[0, a_+]$ is the life span), such that any of the variables evaluated at this value is zero, that is for instance, $S(a_+, t) = 0$. The functions m and θ are respectively the age-specific contagion and age-specific infectiousness.

Summing the equations in (1) leads to the following relations for the population dynamics affected by the disease

$$\begin{aligned} \frac{\partial N(a,t)}{\partial a} + \frac{\partial N(a,t)}{\partial t} &= [b - \mu(a)]\, N(a,t) - \delta(a)C(a,t) \\ N(0,t) &= \int_0^{a_+} bN(a,t)da \end{aligned} \tag{3}$$

Consider the following assumptions:

(i) $a_+ > 0$ and all the initial and boundary conditions are non-negative, $(a,t) \in [0, a_+] \times \mathbb{R}_+$

(ii) μ is a non-negative and non-decreasing smooth function such that

$$\int_0^{a_+} \mu(a)da = \infty, \mu = 0 \text{ on } \mathbb{R} \setminus (0, a_+) \text{ and } \forall\, \alpha \in [0, a_+],\ \tilde{\mu}(\alpha) = \int_0^{\alpha} \mu(v)dv.$$

2.1. Well-Posedness

Denote by $\Gamma = \left\{(a,t) \in (0, a_+) \times (0, \tilde{T})/a - t = a_0 - t_0\right\} = \left\{(a_0 + p, t_0 + p)/p \in (0, a_+ - a_0)\right\}$ with $a_0, t_0 \in \{0\} \times (0, a_+) \cup (0, \tilde{T}) \times \{0\}$ any characteristic line; then the following result which we state without proof holds:

Lemma 1. *Assume that (i) and (ii) above hold, then for any $\tilde{T} > 0$, system (1) has a unique solution $U(t,a) = (S(a,t), L(a,t), T(a,t), C(a,t), R(a,t))$ on $(0, \tilde{T})$ such that*

$$(S(a,t), L(a,t), T(a,t), C(a,t), R(a,t)) \in \mathcal{C}[0, \tilde{T}]$$

for any characteristic line Γ of the equation $a - t = a_0 - t_0$ and $\int_0^{a_+} U(a,t)da \in L^2(0, \tilde{T})$

2.2. Qualitative Property

Let $n(t) = \int_0^{a_+} N(a,t)da$ be the total population at time t. From Lemma 2.2 of Fragnelli *et al.*, [7], we state without proof the following result:

Lemma 2. *Assume (i) and (ii) hold, then the solution of (1) satisfies $N(a,t) \geq 0$ on $(0, \tilde{T})$, $\forall t$.*

Moreover, if there exists $\delta \in [0,1]$ such that $n_0 \leq \{1 - \delta - N(0,t)\} \int_0^{a_+} e^{-\tilde{\mu}(s)} ds$ and $n(t) \leq (1 - \delta)N(a,t)$.

The last inequality means that the total population at time t when $\delta = 0$ is always larger than the total population whenever a disease outbreak is present.

Furthermore, the inequality $n(0) \leq (1-\delta)N_0(a) > 0$ simply means that the population remain non-negative.

The proof of Lemma (2.) is not obvious and we do not attempt it here. This result although harder to prove is somehow obvious from the practical point of view because whenever there is a disease outbreak, the total population is expected to decline.

More details on qualitative properties of such systems can be found in Fragnelli *et al.* [7]. In relation to the literature, various authors have worked on age-structured models, formulated for different purposes. For instance, Sowunmi [18] modeled heterosexual population with age-structure and gestation period and his results on local exponential stability of steady state as well as boundedness can be found in [19]. Tchuenche [20] analyzed a genetically transmitted disease with age-strucuture and genotypic birth functions by writing his differential equations in a vector form mathematical tractability and convenience. The use of vector notation proves to be convenient for analytical purposes, and it is essentially an idealization that should come logically as an extension of the basic algebraic theory [20]. In order to prove the existence of mild solutions, the system can be transformed into an abstract Cauchy problem [12, 22]]. Disease persistence or endemicity has been considered by Thieme [23], while Mugisha and Luboobi [16] analyzed the effect of vertical transmission in the transmission of an age-structured model.

Our analysis differs from previous ones in various ways. First of all, our model equations are different; they are a modified and improved version of the model first proposed by Zhao et al., [29], but was not qualitatively analyzed. Secondly, we derived the explicit expression of the threshold parameter at the endemic equilibrium.

2.3. Asymptotic Behaviour

From the system of equations (3), describing the dynamics of the total population, we get similar one to the well-known McKendrick-Von Foerster type model (See, Tchuenche [21]), and we state that

$$\frac{\partial N(a,t)}{\partial a} + \frac{\partial N(a,t)}{\partial t} \leq [b - \mu(a)]\, N(a,t)$$

Thereafter, using the method of characteristics or the Laplace transform technique, its solution is given by:

$$N(a,t) = N_0(a-t)H(t-a)e^{-\{\tilde{\mu}(a)-\tilde{\mu}(a-t)\}} + N(0,t-a)H(a-t)e^{-\tilde{\mu}(a)}$$

where $H(\cdot)$ is the Heaviside function (in the direction of continuity of each branch of the solution $t < a$ and $t > a$, respectively.

Following Tchuenche [20], we can show after some tedious calculations (using appropriate norm and applying the classical Gronwall's Lemma) that

$$\|N(t,\cdot)\| \leq \bar{N}_0 e^{\{1-\tilde{\mu}(a)\}}$$

where

$$\bar{N}_0 = e^{-\tilde{\mu}(a)} \int_0^{a+} N_0(s)e^{-\tilde{\mu}(s)}ds$$

is a positive constant depending on the parameters of the equation.

Since our model is well-posed and admits solutions, for convenience, we normalize the system by introducing the following fractions :

$$s(t,a) = \frac{S(a,t)}{N(a,t}, \; l(t,a) = \frac{L(a,t)}{N(a,t}, \; x(t,a) = \frac{T(a,t)}{N(a,t},$$
$$c(t,a) = \frac{C(a,t)}{N(a,t}, \text{ and } r(t,a) = \frac{R(a,t)}{N(a,t}.$$

Subsequently, we obtain a simplified system in terms of proportion by making the following hypothesis:

(i) Suppose that the additional mortality rate related to HBV is equal to zero, i.e. $\delta(a) = 0$ and

(ii) As latent individuals remain in this class during an average period of 45 days, we can merge this compartment to that of temporary immune without loss of generality.

$$\begin{aligned}
\frac{\partial s(a,t)}{\partial a} + \frac{\partial s(a,t)}{\partial t} &= b - \left[b + m(a)\widehat{\lambda}(t) + p(a,t)\right] s(a,t) \\
\frac{\partial x(a,t)}{\partial a} + \frac{\partial x(a,t)}{\partial t} &= m(a)\widehat{\lambda}(t)s(a,t) - [\beta(a) + \alpha + b]\, x(a,t) \\
\frac{\partial c(a,t)}{\partial a} + \frac{\partial c(a,t)}{\partial t} &= \beta(a)x(a,t) - [\nu(a) + b]\, c(a,t) \\
\frac{\partial r(a,t)}{\partial a} + \frac{\partial r(a,t)}{\partial t} &= p(a,t)s(a,t) + \alpha x(a,t) + \nu(a)c(a,t) - br(a,t)
\end{aligned} \tag{4}$$

with initial and boundary conditions

$$\begin{aligned}
&s(0,t) = \frac{S(0,t)}{N(0,t)} = 1; \quad x(0,t) = c(0,t) = r(0,t) = 0 \\
&s(a,0) = s_0(a); \quad x(a,0) = x_0(a); \quad c(a,0) = c_0(a); \quad r(a,0) = r_0(a)
\end{aligned} \tag{5}$$

Therefore, we have

$$s(a,t) + x(a,t) + c(a,t) + r(a,t) = 1 \text{ and } \widehat{\lambda}(t) = \int_0^{a_+} \theta(a)\,[x(a,t) + c(a,t)]\, da \tag{6}$$

2.4. Time-Independent System

At steady state system (4) reduces to

$$\begin{aligned}
\frac{ds(a)}{da} &= b - \left[b + m(a)\widehat{\lambda} + p(a)\right] s(a) \\
\frac{dx(a)}{da} &= m(a)\widehat{\lambda}s(a - [\beta(a) + \alpha + b]\, x(a) \\
\frac{dc(a)}{da} &= \beta(a)x(a) - [\nu(a) + b]\, c(a) \\
\frac{dr(a)}{da} &= p(a)s(a) + \alpha x(a) + \nu(a)c(a) - br(a)
\end{aligned} \tag{7}$$

and from (5), its initial and boundary conditions are

$$s(0) = 1 \text{ and } x(0) = c(0) = r(0) = 0 \tag{8}$$

Using the time independent process, according to equation (6) we have

$$s(a) + x(a) + c(a) + r(a) = 1 \text{ and } \widehat{\lambda} = \int_0^{a_+} \theta(a)\,[x(a) + c(a)]\, da \tag{9}$$

System (7) always has a trivial disease-free equilibrium given by

$$s^0(a) = 1,\ x^0(a) = c^0(a) = r^0(a) = 0. \tag{10}$$

3. Stability of the Disease-Free State

To study the local stability of the disease-free equilibrium, we have to linearize system (7) about (10).

Let $\overline{s}(a,t)$, $\overline{x}(a,t)$, $\overline{c}(a,t)$ and $\overline{r}(a,t)$ be the perturbations of $s(a,t), x(a,t), c(a,t), r(a,t)$ respectively, i.e.

$$\begin{aligned}
s(a,t) &= s^0(a) + \overline{s}(a,t), \quad x(a,t) = x^0(a) + \overline{x}(a,t) \\
c(a,t) &= c^0(a) + \overline{c}(a,t), \quad r(a,t) = r^0(a) + \overline{r}(a,t)
\end{aligned}$$

Then, the perturbations satisfy the following system of equations:

$$\begin{aligned}
\frac{\partial \overline{s}(a,t)}{\partial a} + \frac{\partial \overline{s}(a,t)}{\partial t} &= b - \left[b + m(a)\widehat{\lambda}(t) + p(a,t)\right] \overline{s}(a,t) \\
\frac{\partial \overline{x}(a,t)}{\partial a} + \frac{\partial \overline{x}(a,t)}{\partial t} &= m(a)\widehat{\lambda}(t)\overline{s}(a,t) - [\beta(a) + \alpha + b]\, \overline{x}(a,t) \\
\frac{\partial \overline{c}(a,t)}{\partial a} + \frac{\partial \overline{c}(a,t)}{\partial t} &= \beta(a)\overline{x}(a,t) - [\nu(a) + b]\overline{c}(a,t) \\
\frac{\partial \overline{r}(a,t)}{\partial a} + \frac{\partial \overline{r}(a,t)}{\partial t} &= p(a,t)\overline{s}(a,t) + \alpha\overline{x}(a,t) + \nu(a)\overline{c}(a,t) - b\overline{r}(a,t)
\end{aligned} \tag{11}$$

Without any ambiguity, let the exponent of growth be ω, then consider the exponential solutions of system (11) of the form :

$$\overline{s}(a,t)=\overline{s}(a)e^{-\omega t};\quad \overline{x}(a,t)=\overline{x}(a)e^{\omega t};\quad \overline{c}(a,t)=\overline{c}(a)e^{\omega t} \text{ and } \overline{r}(a,t)=\overline{r}(a)e^{\omega t}.$$

The functions $\overline{s}(a),\overline{x}(a),\overline{c}(a),\overline{r}(a)$ and ω satisfy the following ODE system :

$$\frac{d\overline{s}(a)}{da} = be^{-\omega t}-(\omega+b+m(a)\widehat{\lambda}+p(a))\overline{s}(a) \tag{12}$$

$$\frac{d\overline{x}(a)}{da} = m(a)\widehat{\lambda}\overline{s}(a)-(\omega+\beta(a)+\alpha+b))\overline{x}(a) \tag{13}$$

$$\frac{d\overline{c}(a)}{da} = -(\omega+v(a)+b]\overline{c}(a)+\beta(a)\overline{x}(a) \tag{14}$$

$$\frac{d\overline{r}(a)}{da} = -(\omega+b)\overline{r}(a)+v(a)\overline{c}(a)+\alpha\overline{x}(a)+p(a)\overline{s}(a) \tag{15}$$

$$\widehat{\lambda} = \int_0^{a+}\theta(a)\left[\overline{x}(a)+\overline{c}(a)\right]da \tag{16}$$

$$\overline{s}(0) = \overline{x}(0)=\overline{c}(0)=\overline{r}(0)=0.$$

Solving equations (13) and (14), we obtain

$$\overline{x}(a)=\widehat{\lambda}\left(\int_0^a m(\eta)\overline{s}(\eta)e^{(\omega+\alpha+b)(\eta-\tau)}e^{\int_\tau^\eta\beta(\zeta)d\zeta}d\eta\right)\times e^{-(\omega+\alpha+b)(a-\tau)}e^{-\int_\tau^a\beta(\zeta)d\zeta} \tag{17}$$

$$\overline{c}(a)=\left(\int_0^a\beta(\mu)\overline{x}(\mu)e^{\int_\tau^\mu(\omega+b+\nu(\xi))d\xi}d\mu\right)\times\ e^{-(\omega+b)(a-\tau)}e^{-\int_\tau^a\nu(\xi)d\xi} \tag{18}$$

Substituting the relation (17) into (18), we obtain :

$$\begin{aligned}\overline{c}(a) &= \widehat{\lambda}\int_0^a\left[\beta(\mu)\int_0^\mu m(\eta)\overline{s}(\eta)e^{\int_\tau^\eta(\omega+\alpha+b+\beta(\zeta))d\zeta}d\eta e^{-\alpha(\mu-\tau)}e^{-\int_\tau^\mu(\beta(\xi)-\nu(\xi))d\xi}\right]d\mu \\ &\quad \times e^{-(\omega+b)(a-\tau)}e^{-\int_\tau^a\nu(\xi)d\xi}\end{aligned} \tag{19}$$

Substituting equations (17) and (19) into (16), it follows that

$$\begin{aligned}\widehat{\lambda} &= \int_0^{a+}\theta(a)\left[\overline{x}(a)+\overline{c}(a)\right]da \\ &= \widehat{\lambda}\int_0^{a+}\theta(a)e^{-(\omega+b)(a-\tau)}\left[\int_0^a m(\eta)\overline{s}(\eta)\left(e^{-\alpha(a-\tau)}e^{-\int_\tau^a\beta(\xi)d\xi}\right.\right. \\ &\quad \left.\left.+\int_0^\eta\beta(\mu)e^{-\alpha(\mu-\tau)}e^{-\int_\tau^\mu(\beta(\xi)-\nu(\xi))d\xi}d\mu\times e^{-\int_\tau^a\nu(\xi)d\xi}\right)d\eta\right]da\end{aligned} \tag{20}$$

and by dividing both sides of (20) by $\widehat{\lambda}$, we obtain the characteristic equation

$$1 = \int_0^{a+} \theta(a) e^{-(\omega+b)(a-\tau)} \left[\int_0^a m(\eta)\overline{s}(\eta) \times \left(e^{-\alpha(a-\tau)} e^{-\int_\tau^a \beta(\xi)d\xi} + \int_0^\eta \beta(\mu) e^{-\alpha(\mu-\tau)} e^{-\int_\tau^\mu (\beta(\xi)-\nu(\xi))d\xi} d\mu \times e^{-\int_\tau^a \nu(\xi)d\xi} \right) d\eta \right] da \tag{21}$$

where

$$\overline{s}(a) = be^{-\omega t} \times \int_0^a \left(e^{(\omega+b)(\mu-\tau)} e^{\int_\tau^\mu (m(\xi)\widehat{\lambda}+p(\xi))d\xi} \right) d\mu \times e^{-(\omega+b)(a-\tau)} \times e^{-\int_\tau^a (m(\xi)\widehat{\lambda}+p(\xi))d\xi}$$

Denote the expression in the right hand side of (21) by $F(\omega)$ and set the basic reproductive number as $\mathcal{R}_0 = F(0)$. Thus,

$$\mathcal{R}_0 = \int_0^{a+} \theta(a) e^{-b(a-\tau)} \left[\int_0^a m(\eta)\overline{s}_0(\eta) \times \left(e^{-\alpha(a-\tau)} e^{-\int_\tau^a \beta(\xi)d\xi} + \int_0^\eta \beta(\mu) e^{-\alpha(\mu-\tau)} e^{-\int_\tau^\mu (\beta(\xi)-\nu(\xi))d\xi} d\mu \times e^{-\int_\tau^a \nu(\xi)d\xi} \right) d\eta \right] da \tag{22}$$

where

$$\overline{s}_0(\eta) = b \times \int_0^\eta \left(e^{b(\mu-\tau)} e^{\int_\tau^\mu (m(\xi)\widehat{\lambda}+p(\xi))d\xi} \right) d\mu \times e^{-b(\eta-\tau)} \times e^{-\int_\tau^\eta (m(\xi)\widehat{\lambda}+p(\xi))d\xi}$$

It is worth noting that vaccination leads to a decrease of $\mathcal{R}(\omega)$, the vaccination-induced reproductive number. That is $\mathcal{R}(\omega) \leq \mathcal{R}_0$, for any ω.

If we denote ω_0 as the unique negative real solution of equation (21), ω_0 is the dominant real parts of the roots of $F(\omega)$ as found in ([12]). Therefore, we have the following local stability result with $\mathcal{R}_0$ given by (22) above. as quoted by Li *et al.*, [12]:

Theorem 3. *The disease-free equilibrium of system (4) is locally asymptotically stable (LAS) if* $\mathcal{R}_0 < 1$ *and unstable if* $\mathcal{R}_0 > 1$.

We claim the following global stability result of the disease-free equilibrium.

Theorem 4. *If* $\mathcal{R}_0 < 1$, *then the disease-free equilibrium is globally asymptotically stable (GAS).*

The proof is based on Fatou's Lemma and is similar to that of Theorem 3.2 by Li and Fang [11]

4. Existence and Stability of the Endemic States

The common method used to find an endemic steady state for age-structured models consists of obtaining expressions for time independent solutions $(s^*(a), x^*(a), c^*(a), r^*(a))$ of equation (4) that satisfies

$$\frac{ds^*(a)}{da} = b - [b + m(a)\lambda^* + p(a)]\, s^*(a) \tag{23}$$

$$\frac{dx^*(a)}{da} = m(a)\lambda^* s^*(a) - [\beta(a) + \alpha + b]\, x^*(a)) \tag{24}$$

$$\frac{dc^*(a)}{da} = \beta(a)x^*(a) - [\nu(a) + b]\, c^*(a) \tag{25}$$

$$\frac{dr^*(a)}{da} = p(a)s^*(a) + \alpha x^*(a) + \nu(a)c^*(a) - br^*(a) \tag{26}$$

with initial and boundary conditions

$$s^*(0) = 1, x^*(0) = c^*(0) = r^*(0) = 0,$$

$$s^*(a) + x^*(a) + c^*(a) + r^*(a) = 1, \tag{27}$$

$$\lambda^* = \int_0^{a+} \theta(a)\, [x^*(a) + c^*(a)]\, da. \tag{28}$$

Solving equation (23) and (24), we obtain

$$s^*(a) = e^{-\lambda^* \int_0^a m(\sigma)d\sigma} \times e^{-\int_0^a p(\sigma)d\sigma} \times e^{-ba} \left(1 + b\int_0^a e^{\int_0^\eta (b+m(\xi)\lambda^* + p(\xi))d\xi} d\eta\right) \tag{29}$$

$$x^*(a) = \lambda^* \left(\int_0^a m(\eta)s^*(\eta)e^{\int_\tau^\eta (\beta(\sigma)+\alpha+b)d\sigma} d\eta\right) \times e^{-(b+\alpha)(a-\tau)} e^{-\int_\tau^a \beta(\sigma)d\sigma} \tag{30}$$

$$c^*(a) = \int_0^a \beta(\tau)x^*(\tau)e^{-\int_\tau^a (\nu(\sigma)+\mu(\sigma))d\sigma} d\tau \tag{31}$$

Substituting relation (30) into (31) and changing the order of integration, we obtain

$$\begin{aligned} c^*(a) = {} & \lambda^* \int_0^a \left[\beta(\mu)\left(\int_0^\mu m(\eta)s^*(\eta)e^{\int_\tau^\eta (\beta(\sigma)+\alpha+b))d\sigma} d\eta\right) \times e^{-\alpha(\mu-\tau)} \right. \\ & \left. \times e^{-\int_\tau^\mu (\beta(\sigma)-\nu(\sigma))d\sigma}\right] d\mu \times e^{-b(a-\tau)} \times e^{-\int_\tau^a \nu(\sigma)d\sigma} \end{aligned} \tag{32}$$

Then, the relation (28) can be rewritten as

$$\begin{aligned}\lambda^* &= \int_0^{a_+} \theta(a)\left[x^*(a)+c^*(a)\right] da \\ &= \lambda^* \int_0^{a_+} \Bigg[\theta(a)e^{-b(a-\tau)} \times \int_0^a m(\eta)s^*(\eta)e^{\int_\tau^\eta(\beta(\sigma)+\alpha+b))d\sigma} \\ &\times \left(e^{-\alpha(a-\tau)} \times e^{-\int_\tau^a \beta(\sigma)d\sigma}\right. \\ &+ \left.\left(\int_0^\eta \beta(\mu)e^{-\alpha(\mu-\tau)}e^{-\int_\tau^\mu(\beta(\sigma)-\nu(\sigma))d\sigma}d\mu\right) e^{-\int_\tau^a \nu(\sigma)d\sigma}\right) d\eta\Bigg] da\end{aligned} \tag{33}$$

Dividing both sides by λ^*, we obtain the following characteristic equation :

$$\begin{aligned}1 &= \int_0^{a_+} \Bigg[\theta(a)e^{-b(a-\tau)} \times \int_0^a m(\eta)s^*(\eta)e^{\int_\tau^\eta(\beta(\sigma)+\alpha+b))d\sigma} \times \\ &\left(e^{-\alpha(a-\tau)} \times e^{-\int_\tau^a \beta(\sigma)d\sigma} + \left(\int_0^\eta \beta(\mu)e^{-\alpha(\mu-\tau)}e^{-\int_\tau^\mu(\beta(\sigma)-\nu(\sigma))d\sigma}d\mu\right)\right. \\ &\left. e^{-\int_\tau^a \nu(\sigma)d\sigma}\right) d\eta\Bigg] da\end{aligned} \tag{34}$$

Substituting equation (29) into relation (34), we obtain

$$\begin{aligned}1 = &\int_0^{a_+} \Bigg[\theta(a)e^{-b(a-\tau)} \int_0^a m(\eta)e^{-\lambda^*\int_0^\eta m(\sigma)d\sigma}e^{-\int_0^\eta(b+p(\sigma))d\sigma}e^{\int_\tau^\eta(\beta(\sigma)+\alpha+b))d\sigma} \\ &\times \left(1+b\int_0^\eta e^{\int_0^\xi(b+m(\sigma)\lambda^*+p(\sigma))d\sigma}d\xi\right) \\ &\times \left(e^{-\alpha(a-\tau)}e^{-\int_\tau^a \beta(\sigma)d\sigma} + \left(\int_0^\eta \beta(\mu)e^{-\alpha(\mu-\tau)}e^{-\int_\tau^\mu(\beta(\sigma)-\nu(\sigma))d\sigma}d\mu\right)\right. \\ &\left.\times e^{-\int_\tau^a \nu(\sigma)d\sigma}\right) d\eta\Bigg] da\end{aligned} \tag{35}$$

The expression on the right hand-side of (35), $G(\lambda^*)$, is a continuous and monotone decreasing function of λ^*. The model system (4) has an endemic equilibrium given by equations (29, 30, 32) if equation (35) has a positive solution λ^*. Setting $G(0) = R_0$ since whenever $\mathcal{R}_0 > 1$, we get $G(0) > 1$.

From equations (28),(30),(32) and the relation $x^*(a) + c^*(a) < 1$, it follows that

$$\begin{aligned}\lambda^* \int_0^a m(\tau)e^{-\lambda^*\int_0^\tau m(\sigma)d\sigma}e^{-\int_0^\tau p(\sigma))d\sigma}&\Big((1+\theta(a))e^{-\gamma(a-\tau)} \\ &+\gamma\int_\tau^a e^{-\gamma(s-\tau)}e^{-\int_s^a(\beta(\sigma)+\alpha+\mu(\sigma))d\sigma}ds\Big)\, d\tau < 1\end{aligned} \tag{36}$$

Then, for any $\lambda^* > 0$, we have the following inequality

$$\begin{aligned}\lambda^* G(\lambda^*) &= \int_0^{a+} \left[\theta(a)\lambda^* \int_0^a m(\tau)e^{-\lambda^* \int_0^\tau m(\sigma)d\sigma} e^{-\int_0^\tau p(\sigma))d\sigma} \times \right. \\ &\quad \times \left((1+\theta(a))e^{-\gamma(a-\tau)}\right. \\ &\quad \left.\left. + \gamma \int_\tau^a e^{-\gamma(s-\tau)} e^{-\int_s^a (\beta(\sigma)+\alpha+\mu(\sigma))d\sigma} ds\right) d\tau \right] da \qquad (37) \\ &< \int_0^{a+} \theta(a)da\end{aligned}$$

By setting $\theta^+ = \displaystyle\int_0^{a+} \theta(a)da$, we have $G(\theta^+) < 1$ as $G(0) > 1$ and $G(\lambda^*)$ is a continuous and monotone decreasing function of λ^*. Therefore, the equation $G(\lambda^*) = 1$ has a unique positive solution $\overline{\lambda}^* \in (0, \theta^+)$.

Furthermore, since $\overline{\lambda}^* \leq \theta^+$, the model system of equations in (4) has a unique endemic equilibrium which is given by the unique solution of equation (35) corresponding to $\overline{\lambda}^*$.

Let $\widehat{s}(a,t)$, $\widehat{x}(a,t)$, $\widehat{c}(a,t)$, $\widehat{r}(a,t)$, $\widehat{p}(a,t)$ and $\widehat{\lambda}(t)$ be the perturbations of $s(a,t)$, $x(a,t)$, $c(a,t)$, $r(a,t)$, $p(a,t)$ and $\lambda(t)$, respectively. That is;

$$\begin{array}{ll} s(a,t) = \widehat{s}(a,t) + s^*(a) & x(a,t) = \widehat{x}(a,t) + x^*(a) \\ c(a,t) = \widehat{c}(a,t) + c^*(a) & r(a,t) = \widehat{r}(a,t) + r^*(a) \\ p(a,t) = \widehat{p}(a,t) + p^*(a) & \lambda(t) = \widetilde{\lambda}(t) + \lambda^* \end{array}$$

The perturbations satisfy the following system of equations:

$$\left\{\begin{array}{l} \dfrac{\partial \widehat{s}(a,t)}{\partial t} + \dfrac{\partial \widehat{s}(a,t)}{\partial a} = -\left[b + m(a)\lambda + p(a,t)\right]\widehat{s}(a,t) - \left[m(a)\widetilde{\lambda}(t) + \widehat{p}(a,t)\right] s^*(a) \\ \frac{\partial \widehat{x}(a,t)}{\partial t} + \frac{\partial \widehat{x}(a,t)}{\partial a} = m(a)\lambda\widehat{s}(a,t) + m(a)\widetilde{\lambda}(t)s^*(a) - (\beta(a) + \alpha + b)\,\widehat{x}(a,t) \\ \frac{\partial \widehat{c}(a,t)}{\partial t} + \frac{\partial \widehat{c}(a,t)}{\partial a} = \beta(a)\widehat{x}(a,t) - (\nu(a) + b)\,\widehat{c}(a,t) \\ \frac{\partial \widehat{r}(a,t)}{\partial t} + \frac{\partial \widehat{r}(a,t)}{\partial a} = p(a,t)\widehat{s}(a,t) + \widehat{p}(a,t)s^*(a) + \alpha\widehat{x}(a,t) + \nu(a)\widehat{c}(a,t) - b\widehat{r}(a,t) \\ \widetilde{\lambda}(t) = \displaystyle\int_0^{a+} \theta(a)\left[\widehat{x}(a,t) + \widehat{c}(a,t)\right] da \\ \lambda^* = \displaystyle\int_0^{a+} \theta(a)\left[x^*(a) + c^*(a)\right] da. \end{array}\right. \qquad (38)$$

Considering the following exponential solutions for the perturbed system (38)

$$\begin{array}{ll} \widehat{s}(a,t) = \widetilde{s}(a)e^{\lambda t} & \widehat{x}(a,t) = \widetilde{x}(a)e^{\lambda t} \\ \widehat{c}(a,t) = \widetilde{c}(a)e^{\lambda t} & \widehat{r}(a,t) = \widetilde{r}(a)e^{\lambda t} \\ \widetilde{\lambda}(t) = \widetilde{\lambda}e^{\lambda t} & \widehat{p}(a,t) = \widetilde{p}(a)e^{\lambda t} \end{array}$$

we obtain

$$\begin{cases} \dfrac{d\widetilde{s}(a)}{da} = -\lambda\widetilde{s}(a) - [b + m(a)\lambda + p(a)]\,\widetilde{s}(a) - (m(a)\,\widetilde{\lambda} + \,\widetilde{p}(a))s^*(a) \\ \\ \dfrac{d\widetilde{x}(a)}{da} = -\lambda\widetilde{x}(a) - (\beta(a) + \alpha + b))\,\widetilde{x}(a) + m(a)\lambda\widetilde{s}(a) + m(a)\widetilde{\lambda}s^*(a) \\ \dfrac{d\widetilde{c}(a)}{da} = -\lambda\widetilde{c}(a) - (\nu(a) + b)\,\widetilde{c}(a) + \beta(a)\widetilde{x}(a) \\ \dfrac{d\widetilde{r}(a)}{da} = -\lambda\widetilde{r}(a) - b\widetilde{r}(a) + p(a,t)\widetilde{s}(a) + \,\widetilde{p}(a)s^*(a) + \alpha\widetilde{x}(a) + \nu(a)\widetilde{c}(a) \\ \widetilde{\lambda} = \displaystyle\int_0^{a_+} \theta(a)\,[\widetilde{x}(a) + \widetilde{c}(a)]\,da \\ \lambda^* = \displaystyle\int_0^{a_+} \theta(a)\,[x^*(a) + c^*(a)]\,da \\ \widetilde{s}(0) = \widetilde{l}(0) = \widetilde{x}(0) = \widetilde{c}(0) = \widetilde{r}(0) = 0 \end{cases} \tag{39}$$

Assuming that $\widetilde{\lambda} \neq 0$ and setting

$$\overline{s}(a) = \frac{\widetilde{s}(a)}{\widetilde{\lambda}}, \overline{x}(a) = \frac{\widetilde{x}(a)}{\widetilde{\lambda}}, \overline{c}(a) = \frac{\widetilde{c}(a)}{\widetilde{\lambda}}, \overline{r}(a) = \frac{\widetilde{r}(a)}{\widetilde{\lambda}}, \overline{p}(a) = \frac{\widetilde{p}(a)}{\widetilde{\lambda}}$$

then, system (39) now reads :

$$\frac{d\overline{s}(a)}{da} = -[\lambda + b + m(a)\lambda + p(a)]\,\overline{s}(a) - [m(a) + \,\overline{p}(a)]\,s^*(a) \tag{40}$$

$$\frac{d\overline{x}(a)}{da} = -[\lambda + \beta(a) + \alpha + b)]\,\overline{x}(a) + m(a)\lambda\overline{s}(a) + m(a)s^*(a) \tag{41}$$

$$\frac{d\overline{c}(a)}{da} = -[\lambda + \nu(a) + b)]\,\overline{c}(a) + \beta(a)\overline{x}(a) \tag{42}$$

$$\frac{d\overline{r}(a)}{da} = p(a)\overline{s}(a) + \alpha\overline{x}(a) + \nu(a)\overline{c}(a)) - [\lambda + b]\,\overline{r}(a) + \overline{p}(a)s^*(a) \tag{43}$$

$$1 = \int_0^{a_+} \theta(a)\,[\overline{x}(a) + \overline{c}(a)]\,da \tag{44}$$

$$\lambda^* = \int_0^{a_+} \theta(a)\,[x^*(a) + c^*(a)]\,da \tag{45}$$

$$\overline{s}(0) = \overline{x}(0) = \overline{c}(0) = \overline{r}(0) = 0$$

Solving the following equations (40, 41, 42, 43), we get

$$\overline{s}(a) = -\left(\int_0^a (m(\sigma)+\overline{p}(\sigma))s^*(\sigma)e^{\int_\tau^\sigma(\lambda+b+m(\xi)\lambda+p(\xi))d\xi}d\sigma\right) \times e^{-(\lambda+b)(a-\tau)} \times e^{-\int_\tau^a(m(\xi)\lambda+p(\xi))d\xi} \tag{46}$$

$$\overline{x}(a) = \left(\int_0^a (\lambda m(\sigma)\overline{s}(\sigma)+m(\sigma)s^*(\sigma))\, e^{\int_\tau^\sigma(\lambda+b+\alpha+\beta(\xi))d\xi}d\sigma\right) \times e^{-\int_\tau^a(\lambda+b+\alpha+\beta(\xi))d\xi} \tag{47}$$

$$\overline{c}(a) = \left(\int_0^a \beta(\sigma)\overline{x}(\sigma)e^{\int_\tau^\sigma(\lambda+b+\nu(\xi))d\xi}d\sigma\right) \times e^{-\int_\tau^a(\nu(\xi)+b+\lambda)d\xi} \tag{48}$$

We can rewrite $\overline{x}(a)$ as

$$\overline{x}(a) = \left[\lambda\int_0^a \left(m(\sigma)\overline{s}(\sigma)e^{\int_\tau^\sigma(\lambda+b+\alpha+\beta(\xi))d\xi}\right)d\sigma + \int_0^a \left(m(\sigma)s^*(\sigma)e^{\int_\tau^\sigma(\lambda+b+\alpha+\beta(\xi))d\xi}\right)d\sigma\right] \times e^{-\int_\tau^a(\lambda+b+\alpha+\beta(\xi))d\xi} \tag{49}$$

Henceforth, by substituting (46) into (49) and changing the order of integration we obtain :

$$\overline{x}(a) = -\lambda\int_0^a m(\sigma)\left(\int_0^\sigma (m(\eta)+\overline{p}(\eta))s^*(\eta)e^{\int_\tau^\eta(\lambda+b+m(\xi)\lambda+p(\xi))d\xi}d\eta \times e^{-\int_\tau^\sigma(m(\xi)\lambda+p(\xi)-\beta(\xi)-\alpha)d\xi}\right)d\sigma \times e^{-\int_\tau^a(\lambda+b+\alpha+\beta(\xi))d\xi} + \int_0^a m(\sigma)s^*(\sigma)e^{\int_\tau^\sigma(\lambda+b+\alpha+\beta(\xi))d\xi}d\sigma \times e^{-\int_\tau^a(\lambda+b+\alpha+\beta(\xi))d\xi} \tag{50}$$

Substituting (50) into (48) and changing the order of integration we obtain:

$$\overline{c}(a) = -\lambda\int_0^a\left\{\beta(\sigma)\int_0^\sigma\left[m(\eta)\left(\int_0^\eta(m(\xi)+\overline{p}(\xi))s^*(\xi)e^{\int_\tau^\xi(\lambda+b+m(\rho)\lambda+p(\rho))d\rho}d\xi\right) \times e^{-\int_\tau^\eta(m(\xi)\lambda+p(\xi)-\beta(\xi)-\alpha)d\xi}d\eta\right] \times e^{-\int_\tau^\sigma(\beta(\rho)+\alpha-\nu(\rho))d\rho}\right\}d\sigma \times e^{-\int_\tau^a(\nu(\sigma)+b+\lambda)d\sigma} + \int_0^a\beta(\sigma)\left(\int_0^\sigma m(\eta)s^*(\eta)e^{\int_\tau^\eta(\lambda+b+\alpha+\beta(\xi))d\xi}d\eta \times e^{-\int_\tau^\sigma(\beta(\rho)+\alpha-\nu(\rho))d\rho}\right)d\sigma \times e^{-\int_\tau^a(\lambda+b+\nu(\sigma))d\sigma} \tag{51}$$

Summing these last expressions of $\overline{x}(a)$ and $\overline{c}(a)$, we obtain

$$\overline{x}(a)+\overline{c}(a)=\left\{\begin{array}{l} -\lambda\int_0^a m(\sigma)\left(\int_0^\sigma (m(\eta)+\overline{p}(\eta))s^*(\eta)\times e^{\int_\tau^\eta(\lambda+b+m(\xi)\lambda+p(\xi))d\xi}d\eta\right.\\ \left.\times e^{-\int_\tau^\sigma(m(\xi)\lambda+p(\xi)-\beta(\xi)-\alpha)d\xi}d\sigma\right)\times e^{-\int_\tau^a(\beta(\xi)+\alpha+b+\lambda)d\xi}\\ -\lambda\int_0^a\left[\beta(\sigma)\int_0^\sigma m(\eta)\int_0^\eta[((m(\xi)+\overline{p}(\xi))s^*(\xi)\right.\\ \left.\times e^{\int_\tau^\xi(\lambda+b+m(\rho)\lambda+p(\rho))d\rho}\right)d\xi\\ \left.\times e^{-\int_\tau^\eta(m(\xi)\lambda+p(\xi)-\beta(\xi)-\alpha)d\xi}\right]d\eta\times e^{-\int_\tau^\sigma(\beta(\rho)+\alpha-\nu(\rho))d\rho}\Big]d\sigma\\ \times e^{-\int_\tau^a(\nu(\xi)+b+\lambda)d\xi} \end{array}\right\}$$

$$+\left\{\begin{array}{l} \int_0^a m(\sigma)s^*(\sigma)e^{\int_\tau^\sigma(\lambda+b+\alpha+\beta(\xi))d\xi}d\sigma e^{-\int_\tau^a(\beta(\xi)+\alpha+b+\lambda)d\xi}+e^{-\int_\tau^a(\nu(\xi)+b+\lambda)d\xi}\times\\ \int_0^a\beta(\sigma)\left[\int_0^\sigma m(\eta)s^*(\eta)e^{\int_\tau^\eta(\lambda+b+\alpha+\beta(\xi))d\xi}d\eta\times e^{-\int_\tau^\sigma(\beta(\rho)+\alpha-\nu(\rho))d\rho}\right]d\sigma \end{array}\right\} \tag{52}$$

By collecting the terms with factor λ, we can rewrite $\overline{x}(a)+\overline{c}(a)$

$$=-\lambda e^{-(b+\lambda)(a-\tau)}\left\{\begin{array}{l} \int_0^a\left(\int_0^\sigma(m(\eta)+\overline{p}(\eta))s^*(\eta)e^{\int_\tau^\eta(\lambda+b+m(\xi)\lambda+p(\xi))d\xi}d\eta\right)m(\sigma)\\ e^{-\int_\tau^\sigma(m(\xi)\lambda+p(\xi)-\beta(\xi)-\alpha)d\xi}d\sigma\times e^{-\int_\tau^a(\beta(\xi)+\alpha)d\xi}\\ +\int_0^a\left[\int_0^\sigma\left[\int_0^\eta\left((m(\xi)+\overline{p}(\xi))s^*(\xi)e^{\int_\tau^\xi(\lambda+b+m(\rho)\lambda+p(\rho))d\rho}d\xi\right)m(\eta)\right.\right.\\ \left.\left.\times e^{-\int_\tau^\eta(m(\xi)\lambda+p(\xi)-\beta(\xi)-\alpha)d\xi}\right]d\eta\beta(\sigma)\times e^{-\int_\tau^\sigma(\beta(\rho)+\alpha-\nu(\rho))d\rho}\right]d\sigma\\ \times e^{-\int_\tau^a\nu(\xi)d\xi} \end{array}\right\} \tag{53}$$

$$+e^{-(b+\lambda)(a-\tau)}\left\{\begin{array}{l} \int_0^a\left(m(\sigma)s^*(\sigma)e^{\int_\tau^\sigma(\lambda+b+\alpha+\beta(\xi))d\xi}\right)d\sigma\times e^{-\int_\tau^a(\beta(\xi)+\alpha)d\xi}+\\ \int_0^a\left[\int_0^\sigma\left(m(\eta)s^*(\eta)e^{\int_\tau^\eta(\lambda+b+\alpha+\beta(\xi))d\xi}d\eta\right)\beta(\sigma)\right.\\ \left.\times e^{-\int_\tau^\sigma(\beta(\rho)+\alpha-\nu(\rho))d\rho}\right]d\sigma\times e^{-\int_\tau^a\nu(\xi d\xi} \end{array}\right\}$$

$$\overline{x}(a) + \overline{c}(a) = e^{-(b+\lambda)(a-\tau)} \left[-\lambda U(a,\lambda) + V(a,\lambda)\right] \tag{54}$$

$$\text{where } U(a,\lambda) = \left[\begin{array}{l} \int_0^a \left(\int_0^\sigma (m(\eta) + \overline{p}(\eta))s^*(\eta) \times e^{\int_\tau^\eta (\lambda+b+\alpha+\beta(\xi))d\xi} d\eta\right) \\ \times m(\sigma) e^{-\int_\tau^\sigma (m(\xi)\lambda + p(\xi) - \beta(\xi) - \alpha)\, d\xi} d\sigma \times e^{-\int_\tau^a (\beta(\xi)+\alpha)d\xi} + \\ \int_0^a \left[\int_0^\sigma \left[\left(\int_0^\eta (m(\xi) + \overline{p}(\xi))s^*(\xi) e^{\int_\tau^\xi (\lambda+b+m(\rho)\lambda+p(\rho))d\rho} d\xi\right)\right.\right. \\ \left.\left. \times m(\eta) e^{-\int_\tau^\eta (m(\xi)\lambda+p(\xi)-\beta(\xi)-\alpha)\,d\xi}\right] d\eta\right] \beta(\sigma) e^{-\int_\tau^\sigma (\beta(\rho)+\alpha-\nu(\rho))d\rho} d\sigma e^{-\int_\tau^a \nu(\xi)d\xi} \end{array}\right]$$

$$\text{and } V(a,\lambda) = \int_0^a \left[\int_0^\sigma m(\eta)s^*(\eta) e^{\int_\tau^\eta (\lambda+b+\alpha+\beta(\xi))d\xi} d\eta\right] \beta(\sigma) \times e^{-\int_\tau^\sigma (\beta(\rho)+\alpha-\nu(\rho))d\rho} d\sigma \times e^{-\int_\tau^a \nu(\xi)d\xi}$$

$$+ \int_0^a m(\sigma)s^*(\sigma) e^{\int_\tau^\sigma (\lambda+b+\alpha+\beta(\xi))d\xi} d\sigma \times e^{-\int_\tau^a (\beta(\xi)+\alpha)d\xi}$$

Substituting the value of $\overline{x}(a) + \overline{c}(a)$ from (54) into (44), we get

$$1 = \int_0^{a+} \theta(a)\left[\overline{x}(a) + \overline{c}(a)\right] da = \int_0^{a+} \theta(a) e^{-(b+\lambda)(a-\tau)} \left[-\lambda U(a,\lambda) + V(a,\lambda)\right] da \tag{55}$$

$$Q(\lambda) = -\lambda \int_0^{a+} \theta(a) e^{-(b+\lambda)(a-\tau)} U(a,\lambda) da + \int_0^{a+} \theta(a) e^{-(b+\lambda)(a-\tau)} V(a,\lambda) da \tag{56}$$

By integrating the first expression of (56), we can say that this part is a decreasing function of λ due the presence of the factor $\lambda e^{-(b+\lambda)}$.

For the second part, we need to consider the patterns of the epidemiology of Hepatitis B. For instance $p(a)$ increases, we have a steady falling of β and $\lambda \longrightarrow 0$, as a matter of fact coherent with the impact of the vaccination. If we consider the situation $\lambda \longrightarrow \infty$, we see that α decreases as well, and the exponents of the exponential in the integrand go towards zero.

Bearing in mind the above computations, we can state the following result, which is similar to that of Li and Fang [11].

Lemma 5. *For $0 \leq \tau \leq a \leq a_+$, $Q(\lambda)$ is a decreasing function of λ, which tends to zero as $\lambda \longrightarrow \infty$ and $Q(0) < 1$.*

The proof is immediate and follows the same method in [11].

Theorem 6. *If $\mathcal{R}_0 > 1$ and Lemma 3 above is satisfied, then, the EE is LAS and the disease may persist.*

This gives us an indication of why the disease is still present in the population despite all international eradication efforts. Li it et al., [12] discussed the potential benefits of optimal vaccination and its role in curtailing an epidemic in an age-structured model. Consequently, we cite their result here (Theorem 4.1, p. 35 [12]). For more details and proof, see [12].

Theorem 7. *There are two possible optimal vaccination strategies*

(i) one-age strategy : vaccinate the susceptible population at exactly age a, say.

(ii) two-age strategy : vaccinate part of the susceptible population at age a_1, and the remaining susceptible at a later age a_2.

Conclusion

We investigated the local stability of equilibria of an age-structured model with continuous vaccination program (where a fraction of susceptible individuals is vaccinated per unit time) for the transmission dynamics of Hepatitis B in a community. The model which incorporates some essential features of Hepatitis B dynamics enables us to explicitly compute the disease threshold parameter commonly referred to as the reproduction number (using some robust techniques of dynamical systems) for which the DFE is locally asymptotically stable if this threshold quantity is less than unity and unstable otherwise. Whenever this threshold number is greater than unity, the disease may persist. Our results generalize those presented in ([12]), where vital dynamics is not considered.

The proposed model may be refined in various ways: by explicitly adding - a treated class - a vaccinated group (instead of the current continuous vaccination strategy adopted in this study). Also, both cohort and continuous vaccination can be considered. Furthermore, since the incubation period of HBV ranges from 45 to 160 days (mean = 100 days), a time delay could be appropriately incorporated into the model equations. The analysis (mathematical tractability) of the resulting model with age structure may become a daunting task. However, by discarding the age-structure, one can investigate the impact of some intervention strategies by analyzing a deterministic model with vaccination and treated classes. Another way to extend the model is to relax some of the age-dependent coefficients by making them constants.

References

[1] R.M. Anderson, R.M. May Age-related changes in the rate of disease transmission: implications for the design of vaccination programs. *J Hygiene* 1985;94:365-436.

[2] P. Beutels Economic evaluations of hepatitis B immunization : a global review of recent studies (1994-2000), *Health Economics* 2001, Vol. 10: 751-774.

[3] B. S. Blumberg, H. J. Alter and S. Visnich. A "new" antigen in leukemia sera. *JAMA*. 1965 Vol. 191, 541-546

[4] CDC. A Comprehensive Immunization Strategy to Eliminate Transmission of Hepatitis B Virus Infection in the United States, *MMWR* 2005, 54(RR16);1-23, `http:\\www.cdc.gov\ncidod\diseases\hepatitis\b\acip.htm` (Accessed 23/8/07).

[5] S. Deuffic,1 L. Buffat, T. Poynard and A-J. Valleron. Modeling the Hepatitis C Virus Epidemic in France. *Hepatology* 1999 Vol. 29, 1595-1601.

[6] W. J. Edmunds, G. F. Medley and D. J. Nokes The transmission dynamics and control of hepatitis b virus in the Gambia. *Stat. Med.* 1996 Vol. 15, 2215-2233.

[7] G. Fragnelli, P. Martinez and J. Vancostenoble Qualitative properties of a population dynamics system describing pregnancy, *Math. Mod. Meth. Appl. Sci.* (2005)Vol. 15, No. 4, 507-554.

[8] H. W. Hethcote The mathematics of infectious diseases. *SIAM REVIEW* 2000 Vol. 42, No. 4, pp. 599-653

[9] J. Guckenheimer Dynamical systems theory for ecologists : A brief introduction. In, *Ecological Time Series*, J. Steele and T. Powell (eds.), pp. 55-69, Raven Press, N.Y., 1994.

[10] Incidence of acute hepatitis B-United States, 1990-2002. *MMWR Morb Mortal Wkly. Rep.* 2004 Vol. 52, 1252-1254.

[11] X-Z Li and B. Fang Stability of an age-structured SEIR epidemic model with infectivity in latent period, *Appl. Appl. Math. in press.*

[12] X-Z Li, G, Gupur and G-t Zhu Existence and uniqueness of endemic states for the age-strucutred MSEIR epidemic model, *Acta Math. Appl. Sinica*, English Series, Vol. 18, No. 3, 441-454.

[13] H. S. Margolis, M. J. Alter and S. C. Hadler Hepatitis B : evolving epidemiology and implications for control. *Semin. Liver Dis.* 1991 Vol 11, 84-92.

[14] E. E. Mast, M. J. Alter and H. S. Margolis. Strategies to prevent and control hepatitis B and C virus infections : a global perspective. *Vaccine.* 1999 Vol. 17, 1730-1733.

[15] M. E. Mendy Changes in viral load and HBsAg and HBeAg status with age in HBV chronic carriers in The Gambia. *Virology Journal* (2008), 5:49

[16] J.Y.T Mugisha and L.S. Luboobi Modelling the effect of vertical transmission in the dynamics of HIV/AIDS in an age-structured population, *South Pac. J. Nat. Sci.* (2003) Vol. 21, 82-90.

[17] Recommendation of the Immunization Practices Advisory Committee (ACIP). Recommendations for Protection Against Viral Hepatitis. *MMWR Morb Mortal Wkly Rep.* 1985 Vol. 34, 313-335.

[18] C.O.A. Sowunmi A model of heterosexual population dynamics with age structure and gestation period, *J. Math. Anal. Appl.* (1993) Vol. 2, 390-411.

[19] C.O.A. Sowunmi Stability of steady state and boundedness of a 2-sex population model, *Nonlinear Anal.* (Erratum) (2002) Vol. 15, 903-920.

[20] J. M. Tchuenche Theoretical Population Dynamics Model of a Genetically Transmitted Disease : Sickle-Cell Anaemia, *Bull. Math. Biol.* (2007) Vol. 69, No. 2, 699-730.

[21] J. M. Tchuenche A Note on the MacKendrick Von Foerster Type Model in a Population with an Additional Structure. *Int. J. Math. Educ. Sci. Technol.* (2003) Vol. 34, No. 3, 463-470.

[22] J.M. Tchuenche Patient-dependent effects in disease control : A mathematical model, *ANZIAM J.* (2007) Vol. 48, 583-596.

[23] H.R. Thieme Disease extinction and disease persistence in age-structured epidemic models, *Nonlinear Anal.* (2001) Vol. 41, 6181-6194.

[24] P. Van den Driessche and J. Watmough Reproduction numbers and sub-threshold endemic equilibria for compartmental models of disease transmission. *Math. Biosci.* 2002 vol. 180, 29–48.

[25] H. C. Whittle, H. Inskip, A.K. Bradley, K. McLaughlan, F. Shenton, W. Lamb, J. Eccles, B.A. Baker, A.J. Hall The pattern of childhood hepatitis B infection in two Gambian villages. *J Infect Dis* 1990,161:1112-1115.

[26] World Health Organization *The Yaoundé Declaration on the elimination of Hepatitis B Infection*. Geneva, Switzerland : World Health Organization; 1992.

[27] World Health Organization. *Geographic Prevalence of HbsAg*, `http:\\www.who.int\vaccines-surveillance\graphics\htmls\hepbprev.htm` (Accessed 23/8/07).

[28] World Health Organization. *Hepatitis B, Fact sheet N204*, `http:\\www.who.int\mediacentre\factsheets\fs204\en\` (Accessed 08/8/12).

[29] S. Zhao, Z. Xua and Y. Lub A Mathematical model of hepatitis B virus transmission and its application for vaccination strategy in China. *Int. J. Epidemiol.* 2000 Vol. 29, 744-752.

In: Dynamical Systems
Editor: Jean Michel Tchuenche, pp. 183-200
ISBN: 978-1-62808-001-8

Chapter 7

PARAMETRIC RESONANCE AND HOMOCLINIC CHAOS IN A BULLARD-TYPE DYNAMO

***O. Foupouapouognigni**, *M. Siewe Siewe*† *and C. Tchawoua*‡**
Université de Yaoundé I, Faculté des sciences, Département de Physique,
Laboratoire de Mécanique, Yaoundé, Cameroun

Abstract

The authors show that magnetic flux created by radial current in a Bullard disk dynamo modeled by bistable oscillator under parametric excitation can be used to investigate generic properties such as instability or bifurcation of more realistic mean field dynamos. Using the theory of dynamical systems, a differential equation is constructed to study the nonlinear vibration of the magnetic flux. Based on the harmonic balance method, the first approximate solution and corresponding solution of the primary parametric resonance are obtained, followed by the stability analysis of the steady state. Magnetic fluxes created by radial current driven by parametric excitation with constant frequency can display stationary, saddle-node bifurcations, and hysteresis. The Melnikov method is used to detect the parametric threshold of instability of the dynamo, i.e., the value from which chaotic reversals of the generated current are observed. Finally this system is integrated numerically via the fourth-order Runge-Kutta to show fractal dimension spectra and basin boundaries associated with fast dynamo magnetic field. These results have implication for the expected behavior of stellar dynamos.

1. Introduction

The remarkable advances in nonlinear dynamics over the past two decades have revealed various systematic patterns of behavior that recur in many different nonlinear problems. Many textbooks on instabilities and bifurcation phenomena have been published since the early 1980s when chaos and more generally nonlinear physics became fashionable. The

*E-mail address: oumarfoupouagnigni@yahoo.fr
†E-mail address: martinsiewesiewe@yahoo.fr
‡E-mail address: ctchawa@yahoo.fr

question of how a bifurcation is modified when the control parameter fluctuates is an important question regarding the problem of the self-generation of magnetic field by means of a conducting fluid, known as the dynamo instability [1].

Magnetic fields are ubiquitous in the Cosmos. They permeate galaxies, cause stellar eruptions and shield planets from cosmic radiation. These fields, however, are not due to permanent magnets but are sustained dynamically through processes driven by the motions of electrically conducting fluids within the astrophysical objects. These processes are sometimes called fluid dynamos or, more precisely, homogeneous or magnetohydrodynamic dynamos. Their study is an important topic in geo- and astrophysics and gives rise to a host of interesting mathematical and computational questions.

Parametric resonance has been proposed in the somewhat different context of spiral galaxies as a promoter of bisymmetric magnetic field structure. Because of this, the possibility that fluctuations excite the magnetic field by the parametric resonance mechanism cannot be excluded [2]. A possibility of the parametric resonance between the oscillatory solution of the dynamo equations and oscillating parameters of the dynamo system may appears. The idea of the parametric resonance in dynamo system has been formulated by [3]. The parametric resonance problem in the dynamo system is however rather different from the analogous problem in the oscillatory system. The dynamo system is governed by different types of equations than the oscillatory system. It mean that the excitation conditions of parametric resonance for the dynamo system can differ from the excitation conditions for the oscillatory system [4].

In recent years, academic as well as industrial interests on the subject of chaos has been growing as more and more systems from various disciplines exhibit the well known signature of chaos. From the Lorentz system [5] to applications in Lasers [6], disk dynamo [7], the important of the understanding of the theory of chaos is well established and reinforced as the understanding grows deeper.

In this paper, the discussion will focus on two specific issues: first, the connection between parametric resonance and local bifurcation. The connection between chaotic behavior and global (homoclinic) bifurcations are also considered. We aim to study how a bifurcation is modified when the control parameter fluctuates. This is an important question regarding the problem of self-generation of a magnetic field by means of a conducting fluid, known as the dynamo instability. After introducing the disk-dynamo model, we derive the analogy with a parametric oscillator (section II) and in section III, both the effect of the parametric excitation on the steady-state oscillations in an magnetic flux created by radial current, the frontier between turbulent or chaotic and periodic oscillations are analytically calculated. We also address the effect of the parametric excitation on the route to chaos cause by homoclinic bifurcation by means of analytical method based on the Melnikov theory and numerical simulations (section IV). The concluding remark is given in section V.

2. The Model and Governing Equations

The Bullard (or homopolar) dynamo [8] is the first example of a magnetic instability triggered by a conductor in motion. The experimental device is depicted on figure 1: a conducting disk rotates around its axis at the angular velocity Ω and a small magnetic field is applied in the vertical direction inducing a current in the disk from the axis towards the edge of the conducting disk. Then this current flow in a conducting loop which, if orientated in an appropriate way, induced a magnetic field orientated in the same direction as the initial one. Thus an infinitesimal magnetic field can be amplified by this mechanism, leading to a dynamo. Here we study a modified version of the original Bullard dynamo where an azimuthal current at the edge of the disk and mechanical friction are permitted. The azimuthal current has been introduced by Moffatt [9] who showed that the hypothesis of a purely radial current would violate the conservation of magnetic flux in the case of a perfectly conducting disk. According to Hide [10], neglecting the mechanical friction is unwarranted for it would render the disk-dynamo structurally unstable. Three equations are enough to describe this system [12, 13]:

$$\begin{cases} \dot{X}(\tau') = r(Y - X) \\ \dot{Y}(\tau') = XZ - (1+m)Y + mX \\ \dot{Z}(\tau') = g_0(1 - (1+m)XY + mX^2) - fZ \end{cases} \tag{1}$$

where, X and Y are respectively the magnetic fluxes created by the radial current and azimuthal current, Z is the angular velocity of the disc. g_0 proportional is the externally applied torque, f proportional is the mechanical friction, r and m are the positive constant depending on the electrical properties of the system. The two first equations represent the effect of the rotating disk on the magnetic field and the third equation is the evolution of the angular velocity, subject to Joule heating and externally applied torque (the term proportional to g_0) and a mechanical friction (the term proportional to f).

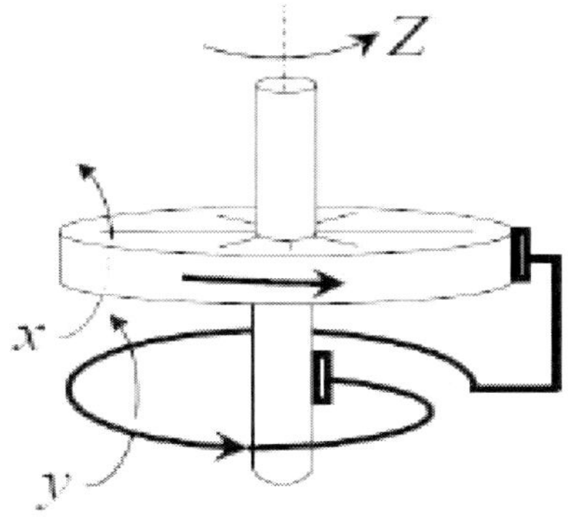

Figure 1. Bullard simple homopolar disc.

Consider the case in which the forces of friction are very high compared to the inertias of the disc i.e., ($\dot{Z} = 0$), the third equation of the system Eq.(1) is written

$$Z = Z_0(1 - (1+m)XY + mX^2) \tag{2}$$

where Z_0 is the positive constants. By combining the two first equation of system Eq.(19), we obtain

$$\ddot{X} + (Z_0(m+1)X^2 + (m+1+r))\dot{X} - (r(Z_0 - 1))X + rZ_0X^3 = 0 \tag{3}$$

let

$$\beta = \frac{Z_0(m+1)}{m+1+r}, x = \sqrt{\beta}X, \quad \text{and} \quad \tau = (m+1+r)^2\tau'$$

we obtain

$$\ddot{x} + c_1(1+x^2)\dot{x} - \xi x + c_2 x^3 = 0. \tag{4}$$

where,

$$c_1 = \frac{1}{m+1+r}, c_2 = \frac{rZ_0}{(m+1+r)^4\beta}, \xi = \frac{r(Z_0-1)}{\beta(m+1+r)^4}.$$

The dynamics of the magnetic flux has breadth the disc is described by Eq. (4). Where, x is the magnetic flux created through the disc, $\ddot{x}$ and $\dot{x}$ are respectively the derivative first and second order, we assume that the coefficients in these Eq. (4) has positive, small values.

3. Parametric Excitation

We consider a parameterized Duffing oscillator with nonlinear damping which govern the dynamic of the magnetic flux through its disc.

$$\ddot{x} + c_1(1+x^2)\dot{x} - \xi x + c_2 x^3 = 0 \tag{5}$$

Here we are interesting in the generation of the magnetic flux through the disk by the parametric resonance mechanism characterized by the unsteadiness of the disk rotation, it agrees to take ξ under the form

$$\xi = \xi_0 + \mu\cos(2\omega\tau) \tag{6}$$

where ξ_0 related to the fundamental frequency of the system and μ is the amplitude of the periodic perturbation (e.g., pendulum with vertical oscillations of their pivot, fast spinning disks with time-varying loads [11]). A parametric resonance (also called swing excitation) is observed when the frequency of the perturbing velocity pattern is twice the natural frequency of the dynamo. Likewise, parametric excitations have been found in a Bullard-type disk dynamo model with periodic modulation of the disk rotation. Therefore, inserting Eq.(6) into Eq. (5), one readily obtains

$$\ddot{x} + c_1(1+x^2)\dot{x} - (\xi_0 + \mu\cos(2w\tau))x + c_2 x^3 = 0. \tag{7}$$

3.1. Periodic Oscillations

In this subsection, we aim to study how a local bifurcation of the magnetic flux dynamo of Eq.(7) is modified when the control parameter due to the parametric excitation fluctuates. This is an important question regarding the problem of the self-generation of a magnetic field by means of a conducting fluid, known as the dynamo instability and also to understand the oscillatory properties of the Bullard-type disk dynamo. As the amplitude μ of the parametric force increases, an oscillation develops in which higher harmonics are present but is ignored in comparison with the fundamental component in this studies. Since the system is symmetrical, we assume, for the time being, that this fundamental component is of odd order; hence a periodic solution for Eq. (7) may be written as

$$x(\tau) = A_1(\tau)\cos(w\tau) + A_2(\tau)\sin(w\tau) \tag{8}$$

The coefficients in the right side of Eq. (8) may be found by the method of harmonic balance [14, 15], where $A_1(\tau)$ and $A_2(\tau)$ are the amplitude of the oscillations. The expression of the vibration amplitude and phase are given by

$$\rho = \sqrt{A_1^2 + A_2^2} \quad , \quad \psi = arc\tan(\frac{A_1}{A_2}). \tag{9}$$

Substituting Eq. (8) into (7) and equating the coefficients of the terms containing only $\sin(\omega\tau)$ and $\cos(\omega\tau)$ separately to zero yields

$$\left\{\begin{array}{l} (c_1 + \frac{c_1}{4}\rho^2 + \frac{c_1}{2}A_1^2)\dot{A}_1 + (2\omega + \frac{c_1}{2}A_1A_2)\dot{A}_2 \\ +(-\xi_0 - \omega^2 - \frac{\mu}{2} + \frac{3}{4}c_2\rho^2)A_1 + (c_1\omega + \frac{c_1}{4}\omega\rho^2)A_2 = 0 \\ (-2\omega + \frac{c_1}{2}A_1A_2)\dot{A}_1 + (c_1 + \frac{c_1}{4}\rho^2 + \frac{c_1}{2}A_2^2)\dot{A}_2 \\ -(c_1\omega + \frac{c_1}{4}\omega\rho^2)A_1 + (-\xi_0 - \omega^2 + \frac{\mu}{2} + \frac{3}{4}c_2\rho^2)A_2 = 0 \end{array}\right. \tag{10}$$

The steady-state vibrations are obtained for $\dot{A}_1 = 0$ and $\dot{A}_2 = 0$. The new system is given by

$$\left\{\begin{array}{l} (-\xi_0 - \omega^2 - \frac{\mu}{2} + \frac{3}{4}c_2\rho^2)A_1 + (c_1\omega + \frac{c_1}{4}\omega\rho^2)A_2 = 0 \\ (-c_1\omega - \frac{c_1}{4}\omega\rho^2)A_1 + (-\xi_0 - \omega^2 + \frac{\mu}{2} + \frac{3}{4}c_2\rho^2)A_2 = 0 \end{array}\right. \tag{11}$$

after some simple algebra we get simple equation

$$(-\xi_0 - \omega^2 + \frac{3}{4}c_2\rho^2)^2 + (c_1\omega + \frac{c_1}{4}\omega\rho^2)^2 = (\frac{\mu}{2})^2. \tag{12}$$

From this relation the component ρ of the periodic solution is determined. To better understand the role of the parametric excitation on the periodic oscillations, we plot the forced response (see Figure 2) and the frequency response (see Figures 3 and 4) with the fixed parameters $c_1 = 0.11$, $\xi_0 = 0.5$ and $c_2 = 0.0108$.

From Figure 2, one can said that the radial magnetic field oscillations shows the phenomenon of hysteresis, in certain range of the amplitude of the parametric excitation, successively one, two and three types of solutions are possible. This is important since from this figure, for a fixed frequency excitation, one can defined the range of the amplitude μ where the dynamo effect is possible. From figure 2, it is clear that for smaller value of the parametric excitation amplitude, there is no periodic oscillation. Another phenomenon observed in Fig. 2 is the synchronization effects on the amplitude of oscillations. We also, note a domain of unfolding amplitude, we observe one minimum amplitude ρ_{min} and one maximum ρ_{max} amplitude.

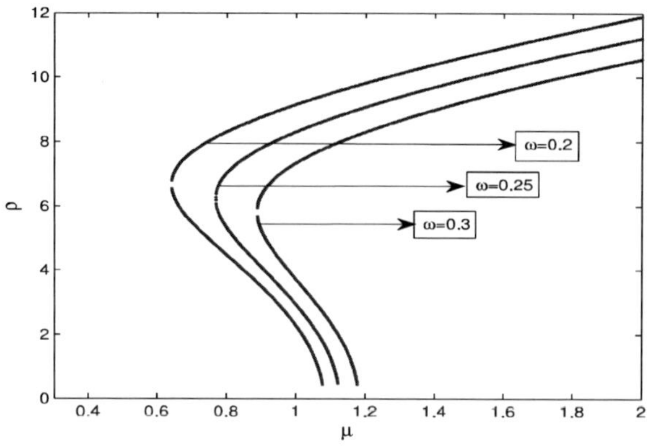

Figure 2. Amplitude response curves ρ versus the parametric μ.

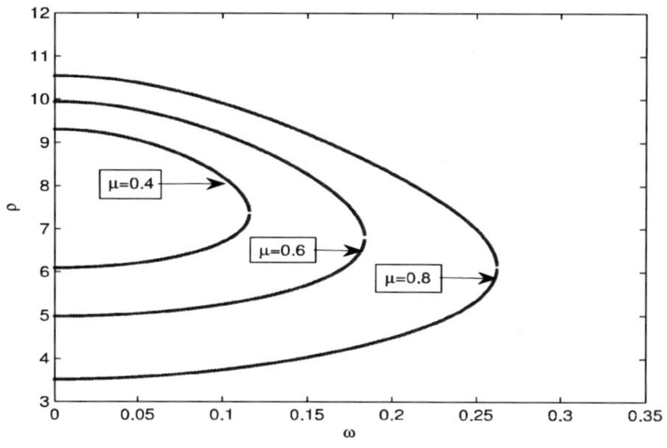

Figure 3. Amplitude response curves ρ versus the frequency ω.

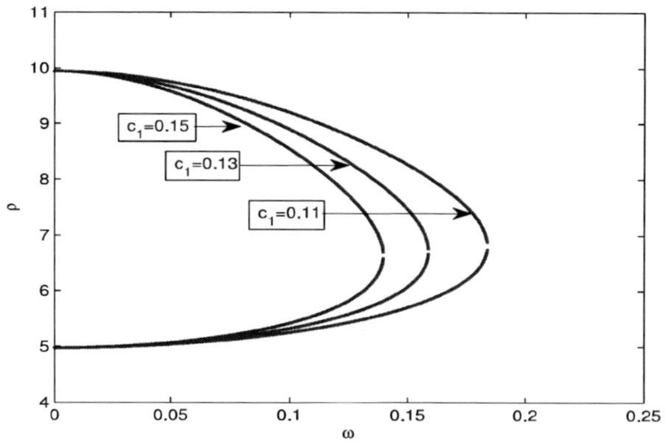

Figure 4. Amplitude response curves ρ versus the damping coefficient c_1 for $\mu = 0.6$.

Fig. 3 represent the amplitude ρ versus frequency ω. We note that the response amplitude has a multi-valued curve that leads to a semi-oval and all solutions are unstable. As the amplitude μ increases, we observe that the semi-oval shifts to the right. For a given frequency, we observe two values of the amplitude of the oscillations. This figure also presents saddle-node bifurcation. It is also clear that for fixed amplitude of the parametric excitation, as the frequency of the parametric excitation increases, no periodic solution appears. Fig. 4 shows the effect of the damping factor c_1. As the damping factor c_1 increases, we observe that the semi-oval shifts to the left and is contracted. The main difference between this figure and figure 4 is that when the frequency is zero, the starting points of the periodic solutions are the same. Another phenomenon observed in Figs. 3 and 4 is the synchronization effects on the amplitude of oscillations.

3.2. Stability Analysis of the Stationary States

The stability of each stationary state can be determined by using the method of Andronov and Vitt [16]. To this purpose let us assume that singular point is subjected to small perturbations characterized by slight variation of A_1 and A_2. That is,

$$A_1 = a_{01} + a_1 \quad and \quad A_2 = a_{02} + a_2 \tag{13}$$

where a_{01} and a_{02} are the values of A_1 and A_2 obtained by solving system of Eq.(11) and a_1 and a_2 are the perturbations. Substituting Eq.(13) into the system of Eq.(10), and keeping only the linear terms in a_1 and a_2, we obtain

$$\begin{cases} (c_1 + \frac{c_1}{4}\rho^2 + \frac{c_1}{2}a_{01}^2)\dot{a}_1 + (2\omega + \frac{c_1}{2}a_{01}a_{02})\dot{a}_2 \\ +(-\xi_0 - \omega^2 - \frac{\mu}{2} + \frac{3}{4}c_2\rho^2)a_1 + (c_1\omega + \frac{c_1}{4}\omega\rho^2)a_2 = 0 \\ \\ (-2\omega + \frac{c_1}{2}a_{01}a_{02})\dot{a}_1 + (c_1 + \frac{c_1}{4}\rho^2 + \frac{c_1}{2}a_{02}^2)\dot{a}_2 \\ -(c_1\omega + \frac{c_1}{4}\omega\rho^2)a_1 + (-\xi_0 - \omega^2 + \frac{\mu}{2} + \frac{3}{4}c_2\rho^2)a_2 = 0 \end{cases} \tag{14}$$

by solving this system, we obtain

$$\begin{cases} \dot{a}_1 = \dfrac{f_2f_7 - f_3f_6}{f_1f_6 - f_2f_5}a_1 + \dfrac{f_2f_8 - f_4f_6}{f_1f_6 - f_2f_5}a_2 \\ \\ \dot{a}_2 = \dfrac{f_1f_7 - f_3f_5}{f_2f_5 - f_1f_6}a_1 + \dfrac{f_1f_8 - f_4f_5}{f_2f_5 - f_1f_6}a_2 \end{cases} \tag{15}$$

where the expressions of f_i are given in Appendix. Thus the steady-state solutions are stable if the eigenvalues (λ) of the coefficient matrix of the right-hand sides of Eq. (15) all have their real parts negative. These eigenvalues are characterized by the algebraic equation

$$\lambda^2 - 2H_1\lambda + H_0 = 0 \tag{16}$$

where

$$H_1 = \frac{1}{2}\left(\frac{f_2f_7 - f_3f_6}{f_1f_6 - f_2f_5} + \frac{f_1f_8 - f_4f_6}{f_1f_6 - f_2f_5}\right) \tag{17}$$

and

$$
\begin{aligned}
H_0 &= \left(\frac{f_2 f_7 - f_3 f_6}{f_1 f_6 - f_2 f_5}\right)\left(\frac{f_1 f_8 - f_4 f_6}{f_1 f_6 - f_2 f_5}\right) \\
&- \left(\frac{f_1 f_7 - f_3 f_5}{f_2 f_5 - f_1 f_6}\right)\left(\frac{f_2 f_8 - f_4 f_6}{f_1 f_6 - f_2 f_5}\right)
\end{aligned} \tag{18}
$$

According to the Routh-Hurrwitz criterion, equilibrium states will be unstable if the characteristic equation Eq.(16) has at least one root with a positive real part. Otherwise, they are stable. Thus, the stability domain is given by the equations $H_1 \prec 0 \quad and \quad H_0 \succ 0$.

Fig. 5 presents the amplitudes response curves for different values of the external excitation and the unstable and stable domain for the parameters $c_1 = 0.11$, $\xi_0 = 0.5$ and $c_2 = 0.0108$.

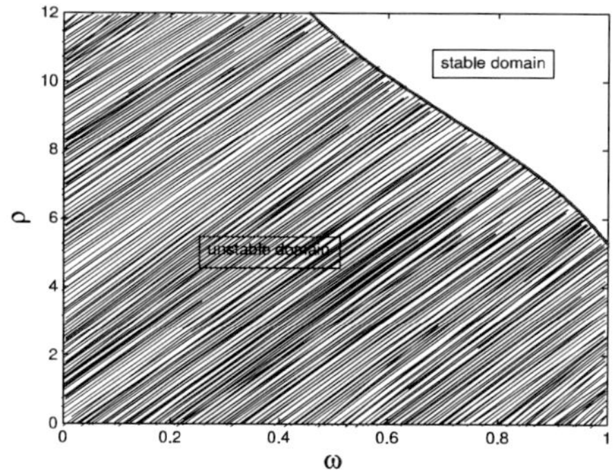

(a)

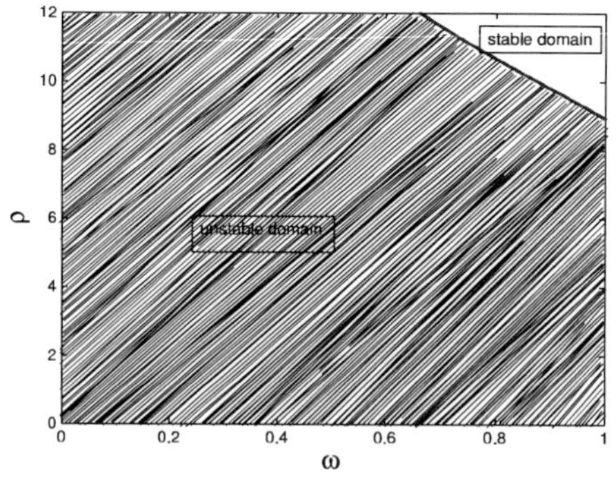

(b)

Figure 5. Frequency response curve and stable domain. (6.a) Stable and unstable domain for $\mu = 3.00$, (6.b) Stable and unstable domain for $\mu = 4.50$.

Figures 5(a) and 6(b) are obtained respectively for the amplitude $\mu = 3.00$ and $\mu = 4.50$. These figures illustrate the above results and the effect of the amplitude μ on the stability of the system. The white domain corresponds to stable solutions while the shaded domain represents the unstable solutions domain. This figures shows that as μ increases the unstable domain increase. The unstable domain corresponds to the dynamo state.

4. Detection of the Instability Threshold in the Model

4.1. Melnikov Analysis

The generalized Melnikov method developed by Wiggins [17, 18] consists of studying a system in which the unperturbed problem is an integrable Hamiltonian system having a normally hyperbolic invariant set whose stable and unstable manifold intersect non-transversally. In this section, we consider the case $\xi_0 > 0$ and $\mu > 0$ and we studies the chaotic behaviors of Eq. (7) in which c_1, and μ are assumed to be small parameters. A transformation of $c_1 \to \varepsilon c_1$, $\mu \to \varepsilon\mu$ is done in order to apply the ε first-order perturbation scheme of the Melnikov theory. Hence, Eq. (7) may be rewritten as

$$\begin{cases} \dot{x} = y \\ \dot{y} = \xi_0 x - c_2 x^3 + \varepsilon(\mu x \cos(2\omega t) - c_1(1 + x^2)y) \end{cases} \tag{19}$$

The unperturbed system is obtain by $\varepsilon = 0$, equation (19) become

$$\begin{cases} \dot{x} = y \\ \dot{y} = \xi_0 x - c_2 x^3 \end{cases} \tag{20}$$

System of Eq.(20) is an Hamiltonian system with potential function given by

$$H(x, y) = \frac{1}{2}y^2 - \frac{\xi_0}{2}x^2 + \frac{c_2}{4}x^4. \tag{21}$$

In which the potential function

$$V(x) = -\frac{\xi_0}{2}x^2 + \frac{c_2}{4}x^4. \tag{22}$$

We notice that in the case of double-well potential, the system has three equilibrium points. We have two stable fixed points and one unstable saddle point at $\left(\pm\sqrt{\frac{\xi_0}{c_2}}, 0\right)$ *and* $(0, 0)$.

Fig. 6 represents the configurations of the potential.

Figs. 6(a) and 6(b) show respectively the double-well and single-well potential corresponding to the Eq.(22). Hence, the no-dynamo states are stable for $Z_0 < 1$ and the dynamo states are stable for $Z_0 > 1$. In all the next section, only the second case is considered. Eq.(21) has a symmetric of homoclinic trajectories connected each unstable point to itself given by

$$\begin{cases} X_S = \pm\sqrt{\dfrac{2\xi_0}{c_2}} \sec h(\sqrt{\xi_0}t) \\ Y_S = \pm\xi_0\sqrt{\dfrac{2}{c_2}} \tanh(\sqrt{\xi_0}t) \sec h(\sqrt{\xi_0}t) \end{cases} \tag{23}$$

The signs on the right hand side of system equation (23) refer to the right and left half planes. Both solutions determine the separatrix orbit, since it separates two types of orbits in

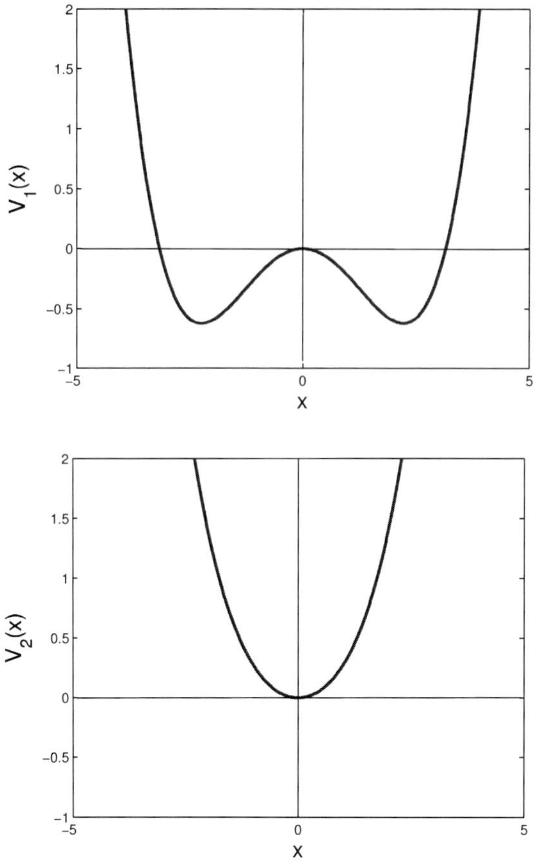

(a)

(b)

Figure 6. The two configurations of the potential.(2.a) two well: $\xi = 0.5$, (2b) single well: $\xi' = -0.5$ and $c_2 = 0.1$.

space. According to Melnikov theory we need the previous ingredients in order to calculate the Melnikov function defined by

$$M(t_0) = -c_1 \int_{-\infty}^{+\infty} Y_s^2 dt - c_1 \int_{-\infty}^{+\infty} Y_s^2 X_s^2 dt + \mu \int_{-\infty}^{+\infty} Y_s X_s \cos(\omega(t + t_0) dt \tag{24}$$

After substitution of the homoclinic solutions given in Eq. (24), the Melnikov function needs to be evaluated. It appears as

$$M(t_0) = -\frac{2c_1\xi_0^2}{c_2} I_0 - \frac{4c_1\xi_0^3}{c_2^2} I_1 - \frac{2\mu\xi_0^{\frac{3}{2}}}{c_2} I_2 \sin(2\omega t_0) \tag{25}$$

where,

$$\begin{cases} I_0 = \int_{-\infty}^{+\infty} \tanh^2(\sqrt{\xi_0}t)\sec h^2(\sqrt{\xi_0}t)dt \\ I_1 = \int_{-\infty}^{+\infty} \tanh^2(\sqrt{\xi_0}t)\sec h^4(\sqrt{\xi_0}t)dt \\ I_2 = \int_{-\infty}^{+\infty} \tanh(\sqrt{\xi_0}t)\sec h^2(\sqrt{\xi_0}t)\sin(2\omega t)dt \end{cases} \tag{26}$$

This needs to be evaluated of the last integrals, appears that the Melnikov function given by

$$M(t_0) = -\frac{48c_1\xi_0^{\frac{5}{2}} + 60c_1c_2\xi_0^{\frac{3}{2}}}{45c_2^2} - \frac{4\mu\pi\omega^2}{c_2\sinh(\frac{\pi\omega}{\sqrt{\xi_0}})}\sin(2\omega t_0). \tag{27}$$

According to Melnikov theory, this function is related to the distance between this function has simple zeros, it implies that there is a critical parameter μ_c corresponding to the parameterized excitations, for which homoclinic tangles intersect. For a certain frequency ω the critical amplitude depends on the damping coefficient c_1 and the nonlinear coefficient c_2. This critical amplitude may be written as

$$\mu_c \geq \frac{c_1}{45c_2\pi\omega^2}\left(12\xi_0^{\frac{5}{2}} + 15c_2\xi_0^{\frac{3}{2}}\right)\sinh(\frac{\pi\omega}{\sqrt{\xi_0}}). \tag{28}$$

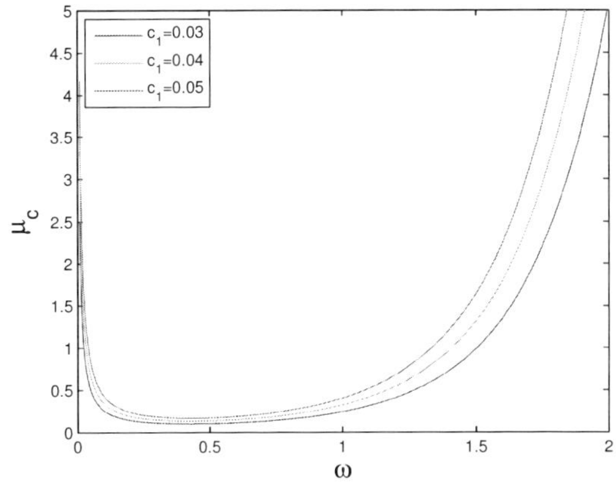

Figure 7. Critical amplitude μ_c for three different damping coefficient values. $c_2 = 0.1$ and $\xi_0 = 0.5$.

Figure 7 presents the critical value of the amplitude μ_c, this figure shows that for a three fixed values of damping coefficient c_1, the threshold for homoclinic chaos decreases as the damping coefficient decrease. Thus, the transverse orbits intersections appear in the system when the amplitude of the parametric excitation exceeds a breaking value. The analytic threshold is $\mu_s = 0.1845$ For a value of c_1, the line corresponds to the no dynamo (underneath) and an effect dynamo (to the over).

4.2. Phase Portraits and Power Spectra

We analyze the behavior of the magnetic flux over the parametric instability threshold according to amplitude of the parametric excitation for a given frequency. We use the portrait of phase, the temporal set and the power spectra as the indicators. The system is solved numerically by using the fourth-order Runge-Kutta method and is shown in Figs. 8-10. The following parameters are fixed $c_1 = 0.03$, $c_2 = 0.1$, $\xi_0 = 0.5$ and $\omega = 0.5$.

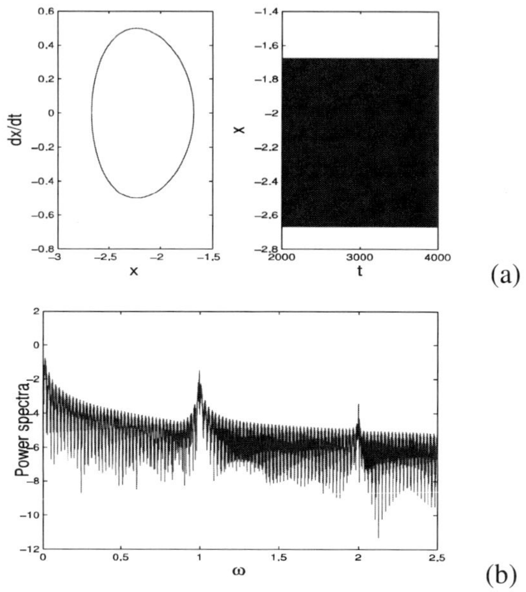

Figure 8. (a) Phase portrait and time series, (b) power spectra corresponding for the parameters $\varepsilon = 0.2$ and $\mu = 0.1$.

Phase portrait of the limit cycle and power spectra corresponding of Eq. (19) are shown in figures 8(a) and 8(b). The period-one spectrum consists of a primary peak. This behavior corresponds to no dynamo. This cascade ends with a saddle-node bifurcation catastrophe with intermittency [1] leading to the attractor shown in Figs. 9(a) and 10(a). The power spectrum Fig. 9(b) has an additional subharmonic at one value of the frequency and additional combination tones. When the system becomes chaotic Fig. 10(b), the power spectrum becomes quite broadband with significantly increased energy levels at low frequencies.

4.3. Basin of Attraction

A basin of attraction is an important means of checking whether a regulation of the fractal boundaries of basin of attraction has been realized [20]. We want to study how is the effect

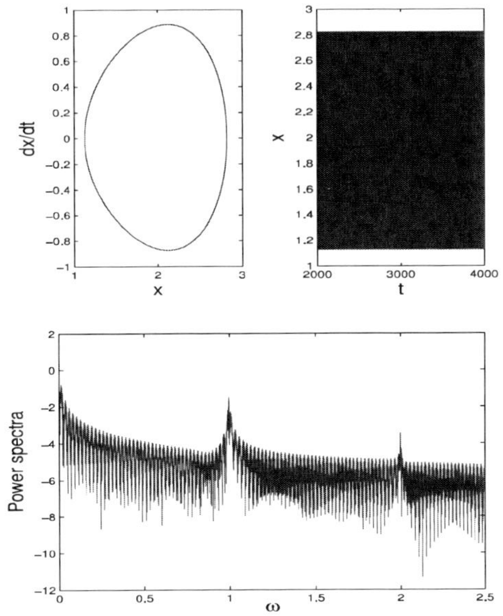

(a)

(b)

Figure 9. (a) Phase portrait and time series, (b) power spectra corresponding for the parameters $\varepsilon = 0.2$ and $\mu = 0.7$.

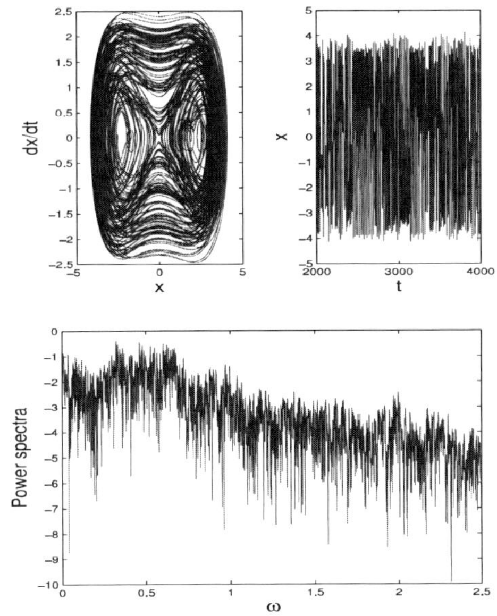

(a)

(b)

Figure 10. (a) Phase portrait and time series, (b) power spectra corresponding for the parameters $\varepsilon = 0.2$ and $\mu = 1.2$.

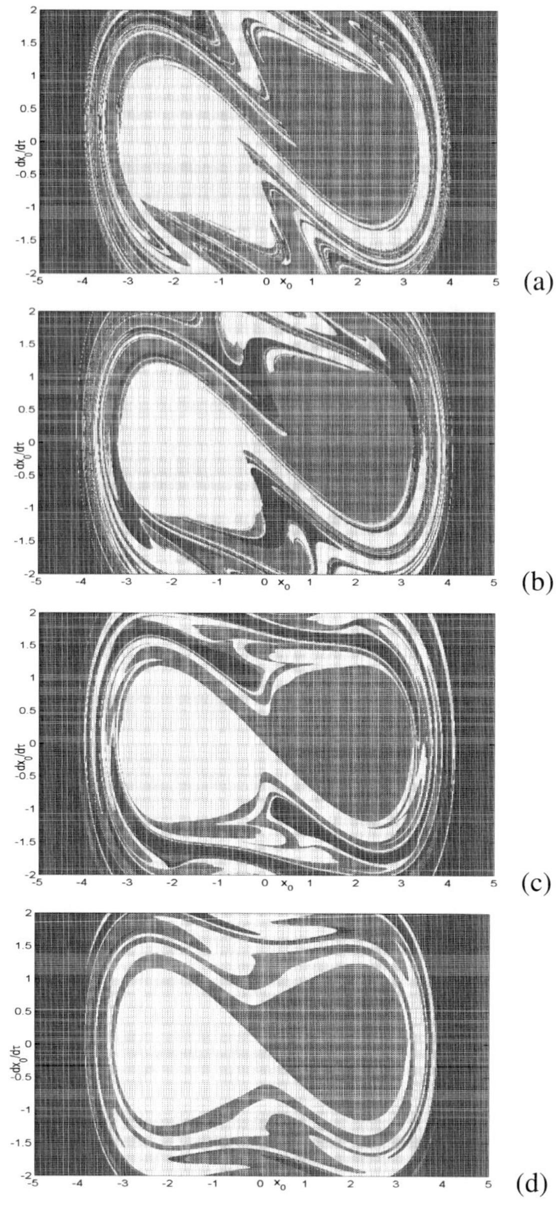

Figure 11. Basin of attraction for (a) $\mu = 1.0$, (b) $\mu = 0.7$, (c) $\mu = 0.3$, (d) $\mu = 0.1$ for $\varepsilon = 0.2$, $c_1 = 0.03$ $c_2 = 0.1$ and $\xi_0 = 0.5$ and $\omega = 0.5$.

of using parametric excitation on the equation of the oscillator and how the basins of attraction are affect as the amplitude μ of the parametric excitation is varied. To show the fractal structure, we consider the case of the bifurcation of motion close to the frequency $\omega = 0.5$. Hence we fix $\varepsilon = 0.2$, $c_1 = 0.03$, $c_2 = 0.1$ and $\xi_0 = 0.5$, it has been shown that the Melnikov threshold value is given analytically by $\mu_{critical} \simeq 0.2$. Figures 11 show the basins of three coexisting attractors of the parametric Duffing oscillator with nonlinear damping term. The picture was made by starting with 200×20 grid of points in $[-5, 5] \times [-2, 2]$ as initial values, and by solving the differential equation for initial condition representing each pixel, and coloring the pixel blue, black, and green depending on which sink orbit attracts the orbit. The three attractors in Fig. 11 (with green color, blue color and black color) are two stable fixed points and one unstable fixed point. In fact the green color (respectively the blue color) correspond to the left well (respectively the right well) potential given by Fig. 2(a). These two colors may display two periodic attractors in the basin of attraction while the black color which corresponds to the unstable point (middle of potential well given by Fig. 2(a)) may display chaotic attractor in the basin of attraction. From Figs. 11(a) and 11(b), the fractal shape of the basin boundaries of attraction is observed. From Fig. 11(c), it is shown that under the Melnikov control, the shape of the basin is smooth and the boundary of the basin consists of the smooth curve. From figure 11, it is shown that the magnetic flux exceeding a given threshold forms a fractal set. The dimension of these fractal basin boundaries depends on the value of the threshold magnetic field (multifractality).

Conclusion

We have studied parametric oscillations and transition to chaos that can arise when radial magnetic field is subjected to parametric excitation in a reduced disk dynamo of the Bullard-type. First, it is shown that the radial magnetic field instability displays hysteresis behavior in the case of the forced response. Also, the frequencies curves display the growth and the saturation of the time periodic magnetic field at high enough parametric frequency. This phenomenon predicts a Hopf bifurcation of convective nature at onset. The stability of the steady state oscillations has been presented showing the boundary between stable and unstable (dynamo states) solutions.

Special attention has been paid also to the analytical estimation of the threshold of the appearance of chaotic motion (i.e., dynamo states). This has been derived using Melnikov method. We have proved that as the parametric amplitude exceeds certain threshold, dynamo states occur. Finally, this parametric excitation plays an important role Bullard-type dynamo with various nonlinear phenomena.

Appendix

Expressions of the f_i and the b_i are given by the following expressions

$$\begin{cases} f_1 = c_1 + \dfrac{c_1}{4}\rho^2 + \dfrac{c_1}{2}a_{01}^2 \\ f_2 = 2\omega + \dfrac{c_1}{2}a_{01}a_{02} \\ f_3 = -\xi_0 - \omega^2 - \dfrac{\mu}{2} - \dfrac{3}{4}c_2\rho^2 \\ f_4 = c_1\omega + \dfrac{c_1}{4}\omega\rho^2 \end{cases}$$

$$\begin{cases} f_5 = 2\omega + \dfrac{c_1}{2} a_{01} a_{02} \\ f_6 = c_1 + \dfrac{c_1^2}{4}\rho^2 + \dfrac{c_1}{2} a_{02}^2 \\ f_7 = -c_1\omega - \dfrac{c_1}{4}\omega\rho^2 \\ f_8 = -\xi_0 - \omega^2 + \dfrac{\mu}{2} - \dfrac{3}{4} c_2 \rho^2 \end{cases}$$

The solutions of the stationary system are given by

$$\begin{cases} a_{01}^2 = \dfrac{\rho^2}{2} - \dfrac{(-1 - \omega^2 + \frac{3}{4} c_2 \rho^2)\rho^2}{\mu} \\ a_{02}^2 = \dfrac{\rho^2}{2} + \dfrac{(-1 - \omega^2 + \frac{3}{4} c_2 \rho^2)\rho^2}{\mu} \end{cases}$$

All the integrals used for the computation of Melnikov functions are:

$$\int_{-\infty}^{+\infty} \frac{\sinh^p(t)}{\cosh^q(t)} dt = B(\frac{p+1}{2}, \frac{q-p}{2}) \tag{29}$$

$$\int_{-\infty}^{+\infty} \tanh At \sec hAt \sin Btdt = \frac{\pi}{A^2} \sec h(\frac{\pi B}{2A}) \tag{30}$$

$$\int_{0}^{+\infty} \frac{\cos At}{\cosh^2 Bt} dt = \frac{\pi A}{2B^2 \sinh(\frac{\pi A}{2B})} \tag{31}$$

The Euler Beta functions are related to the Gamma Euler Function through the expression:

$$B(a, b) = \frac{\Gamma(a)\Gamma(b)}{\Gamma(a+b)} \tag{32}$$

Remember that the Euler Gamma function has the following property: $\Gamma(n+1) = n\Gamma(n)$, where $\Gamma(n+1) = n!$

Acknowledgments

Martin Siewe Siewe is indebted to the "Institut des Hautes Études Scientifiques" (IHES) for hosting him for a short visit research and where part of this work was finalized.

References

[1] N. Leprovost, B. Dubrulle, F. Plunian. Instability of the homopolar disk-dynamo in presence of white noise, *Magnetohydrodynamics*, 40 (2004) 1-12.

[2] Janis Priede, Raul Avalos-Zuniga, Franck Plunian. Homopolar oscillating-disc dynamo driven by parametric resonance, *Physics Letters A* 374 (2010) 584-587.

[3] M. Chiba and M. Tosa. Swing Excitation of Galactic Magnetic Fields Induced by Spiral Density Waves, *Mon. Not. R. Astr. Soc.* 244 (1990) 714-726.

[4] D. Schmitt and G. Rudiger. Dynamos in resonance, *Astron. Astrophys.* 264 (1992) 319-326.

[5] E.N. Lorentz. Deterministic non-periodic flow, *Journal of Atmospheric Science*, 20 (1963) 130-141.

[6] H. Haken. Analogy between higher instabilities in fluids and lasers, *Physics Letters*, 53A (1975) 77-78.

[7] E. Knobloch. Chaos in segmented disc dynamo, *Physics Letters*, A 82 (1981) 439-440.

[8] E. C. Bullard. The stability of a homopolar dynamo, *Proc. Camb. Philos. Soc.* 51 (1995) 744-760.

[9] H. K. Moffatt. A self-consistent treatment of simple dynamo systems, *Geophys. Astrophys. Fluid Dyn.* 14 (1979) 147-166.

[10] R. Hide. Structural instability of the Rikitake disk dynamo, G*eophys. Res. Let.* 22 (1995) 1057-1059.

[11] J. -S. Chen and D. B. Bogy. *J. Appl. Mech.*, 59 (1992) 390.

[12] E. Knobloch. Chaos in the segmented disc dynamo, *Phys. Lett.* A82 (1981) 439-440.

[13] F. Plunian, F. Marty, and A. Alemany. Chaotic behaviour of the Rikitake dynamo with symmetric mechanical frictions and azimuthal currents, *Proc. R. Soc.* London, Ser. A 454 (1998) 1-8.

[14] A. H. Nayfeh, D. T. Mook. *Nonlinear Oscillations*, Wiley, New York (1979).

[15] M. Siewe, F. M. Moukam, C. Tchawoua, P. Woafo, Nonlinear Response and Suppression of Chaos by Weak Harmonic Perturbation Inside a Triple Well Φ^6-Rayleigh Oscillator Combined to Parametric Excitations, *J. Comput. Non. Dynam.* 1 (2006) 196.

[16] Andronov A, Vitt A. Zur Theorie des Mitnehmens von van der Pol, *Zur Arch Electrotech* 24 (1930) 99-110.

[17] S. Wiggins, *Introduction to Applied Nonlinear Dynamical Systems and Chaos*, Spinger-Verlag, Berlin (1990).

[18] Guckenheimer, P. Holmes, Nonlinear Oscillations, *Dynamical Systems and Bifurcations of Vector Field*, Springer, New York (1983).

[19] A. J. Lichtenberg and M. A. Lieberman, *Regular and Stochastic Motion*, Springer, New York, (1983) .

[20] H. Cao, G. Chen. Global and local control of homoclinic and heteroclinic bifurcations, *International Journal of Bifurcation and Chaos*, 15 (2005), pp. 2411-2432.

In: Dynamical Systems
Editor: Jean Michel Tchuenche, pp. 201-233
ISBN: 978-1-62808-001-8

Chapter 8

BIRHYTHMICITY, SYNCHRONIZATION AND CHAOS IN AN ENZYME-SUBSTRATE REACTION WITH A FERROELECTRIC BEHAVIOR

H. G. Enjieu Kadji[1,2] *and P. Woafo*[2]
[1]Monell Chemical Senses Center, Philadelphia, Pennsylvania, US
[2]Laboratory of Modelling and Simulation in Engineering and Biomimetics and Prototypes, University of Yaoundé I, Yaoundé, Cameroon

Abstract

This chapter addresses the issues of birhythmicity, synchronization and chaos in an enzyme-substrate reaction with ferroelectric behavior modeled by a multi-limit cycles van der Pol oscillator, a self-sustained system that possesses mechanisms to damp oscillations that grow too large and a source of energy to pump those that become too small. Theoretical analyses based on the method of perturbation are utilized to determine the amplitudes and frequencies of limit cycles for autonomous oscillatory states. The unidirectional and bi-directional synchronization processes in the model are addressed and their stability boundaries are determined using the Floquet theory and Whittaker's method. Routes leading to chaos and strange attractors are investigated via the bifurcation diagrams and maximum Lyapunov exponent. Chaos synchronization is also investigated for unidirectional and bi-directional coupling schemes. The fourth order Runge-Kutta algorithm coded in Pascal is used to perform the numerical simulations.

1. Introduction

Nonlinear oscillations have been considered as growing sources of interest for decades in several complex systems [1-7]. These systems have extremely diverse dynamics and are able to exhibit regular oscillatory states or/and chaotic oscillations. These complex systems comprise a particular family whose oscillations are self-sustained in autonomous states. An example of such systems is the classical van der Pol oscillator that serves as a paradigm for smoothly single limit cycle or relaxation oscillations [8]. In the presence of an external excitation, the van der Pol oscillator exhibits various interesting phenomena such as harmonic, subharmonic and superharmonic oscillations [9-14], frequency entrainment [15,16], devils staircase [17], chaotic behavior in a small range of control parameters [18,19]. The generalization of the single limit cycle van der Pol oscillator including cubic nonlinear term (so called Duffing-van der Pol or van der Pol-Duffing oscillator) represents another case of self-sustained system whose dynamics has been widely investigated [20-23]. However, in modeling processes comprising more than a stable limit cycle, there is a need to go beyond the classical van der Pol oscillator and Duffing-van der Pol oscillator. In that regards, previous attempts have considered nonlinear dynamics and synchronization of extended van der Pol oscillator that possesses more than a single stable limit cycle [9,10,13,14,24]. This chapter aims at reviewing the phenomena of birhythmicity, bifurcation, synchronization and chaos in an enzyme-reaction which exhibits multi-limit cycles oscillations. The chapter is organized as follows: Section 2 introduces the model of enzyme-substrate reaction under consideration while section 3 addresses the amplitudes and frequencies of limit cycles. Section 4 tackles the issue of unidirectional coupling synchronization while section 5 investigates the bi-directional coupling synchronization. Section 6 deals with bifurcation, strange attractors and routes to chaos in the model. Section 7 investigates chaos synchronization of the model and conclusions are given in section 8.

2. Model of Enzyme-Substrate Reaction

Coherent oscillations in biological systems are considered here through the case of an enzymatic substrate reaction with ferroelectric behavior in brain wave model [25]. The following suggestions made by **Frohlich** [26,27] are taken as a physical basis for a theoretical investigation:

- When metabolic energy is available, long-wavelength electric vibrations are very strongly and coherently excited in active biological system.
- Biological systems have metastable states with a very high electric polarization.

These long range interactions may lead to a selective transport of enzymes and thus, rather specific chemical reactions may become possible. For this survey, let us consider a population of enzyme molecules of which N are in the excited polar state while R remains not excited. We assume that S is the number of substrate molecules. Both the enzyme and substrate show long range selective interactions which tend to increase their level by influx. Each transition from the non-polar (or weakly polar) ground state of enzyme to the highly polar excited state leads to the chemical destruction of a substrate molecule. Additionally, there are also spontaneous transitions from the excited one to the ground (or weakly polar) state and it is assumed that the rate of increase of the activated enzymes is proportional to their own concentration N, to the rate of the unexcited enzymes R and to the number of the substrate molecules S. In such a context, the model can be described by following set of nonlinear differential equations:

$$\frac{dN}{d\tau} = \nu NRS - \xi N, \tag{1}$$

$$\frac{dS}{d\tau} = \gamma S - \nu NRS, \tag{2}$$

$$\frac{dR}{d\tau} = \xi N - \nu NRS - \lambda(R - C). \tag{3}$$

ν represents the strength of the nonlinear enzyme-substrate reaction, ξ the decay rate of excited enzymes to the ground (or weakly polar) state and γ the range attraction of the substrate particles due to the autocatalytic reactions. C is the equilibrium concentration of the unexcited enzymes molecules in the absence of the excited enzyme and substrate *i.e* when $N = S = 0$. One supposes that the equilibrium of the unexcited enzyme concentration is quickly reached in order to simplify the above nonlinear equations. This is also called an adiabatic elimination of the fast variable. Thus the equations above are reduced to a well-known LotkaVoltera equations [28]

$$\frac{dN}{d\tau} = \nu NCS - \xi N, \tag{4}$$

$$\frac{dS}{d\tau} = \gamma S - \nu NCS. \tag{5}$$

Here, the number of activated enzyme molecules N can be considered as a population of predator while the substrate molecules S can be viewed as population of prey. From Eqs. (4) and (5), two steady states $(N_0, S_0) = (0, 0)$ and $(N_0, S_0) = (\frac{\gamma}{\nu C}, \frac{\xi}{\nu C})$ are obtained. By perturbing these activated enzymes and substrate molecules around the nontrivial steady state, the following variational equations are derived:

$$\begin{aligned} \frac{d\varepsilon}{d\tau} &= \gamma\eta + \nu C\eta\varepsilon, \\ \frac{d\eta}{d\tau} &= -\xi\varepsilon - \nu C\eta\varepsilon. \end{aligned} \tag{6}$$

ε and η are the excess concentrations of activated enzymes and substrate molecules beyond their equilibrium values N_0 and S_0 respectively. Based on **Frohlich's** ideas, one may assume that in large regions of the system of proteins, substrates, ions and structured water are activated by the chemical energy available from substrate enzyme reactions. These regions are expected to oscillate coherently with a frequency of about 10^{11} Hertz and via an interaction arising from it, chemical oscillations in the number of substrate and activated enzyme molecules with a very low frequency $\sqrt{\xi\gamma}$ (possibly in the 20 Hertz region which has been found in the brain) might be carried out around the equilibrium state [12]. This oscillation also represents an electric oscillation through the high dipole moment of excited enzymes. The electric dipole moment of excited enzyme is partially screened by the ions and the remaining polarization causes the system to display a tendency towards a ferroelectric instability. On the other hand, electric resistances against the system's tendency to become ferroelectric also have to be considered through their contribution defined as $-\sigma^2 P$ and which is viewed as a relaxation term. Assuming the macroscopic polarization P to be proportional to the time dependent number ε of the excited enzyme molecules, a nonlinear dielectric contribution is obtained and given as follows

$$\frac{d\varepsilon}{d\tau} = (\kappa^2 e^{-r^2\varepsilon^2} - \sigma^2)\varepsilon. \tag{7}$$

Since the electrical field E interacts with the polarization, it is also important to include its contribution which consists of an internal field due to thermal fluctuation and an externally applied field on excited enzymes. E does not need to be an electrical field necessary and can also represents for example external chemical influences (*e.g* an input or an output of enzyme molecules through the transport phenomena). Therefore, Adding both the chemical and the dielectric contribution finally lead us to the set of equations system

$$\frac{d\varepsilon}{d\tau} = \gamma\eta + \nu C\eta\varepsilon + (\kappa^2 e^{-r^2\varepsilon^2} - \sigma^2)\varepsilon + E(\tau),$$
$$\frac{d\eta}{d\tau} = -\xi\varepsilon - \nu C\eta\varepsilon. \tag{8}$$

For small values of ε and η, one can neglect the nonlinear term $\nu C\eta\varepsilon$ and expand the exponential function if $r^2\varepsilon^2 << 1$. Therefore, when using a Taylor expansion of the exponential function and the following rescalings

$$\omega_0 = \sqrt{\xi\gamma}, \quad t = \omega_0\tau, \quad x = \Xi\varepsilon, \quad \Xi = \sqrt{\frac{3}{\kappa^2 - \sigma^2}}\kappa r, \quad \zeta = \frac{\kappa^2 - \sigma^2}{\omega_0},$$
$$F(t) = \frac{\Xi}{\omega_0^2}\frac{d}{dt}E(\frac{t}{\omega_0}), \quad \alpha = \frac{5}{18\kappa^2}(\kappa^2 - \sigma^2), \quad \beta = \frac{7}{162\kappa}(\kappa^2 - \sigma^2)^2,$$

one can conclude that this enzyme-substrate reaction is modeled by the following non-dimensional nonlinear differential equation:

$$\ddot{x} - \zeta(1 - x^2 + \alpha x^4 - \beta x^6)\dot{x} + x = F\cos\Omega t. \tag{9}$$

This equation describes a modified van der Pol oscillator. Here, an overdot denotes the derivative with respect to time while $\dot{x}$ physiologically represents the rate of change of the amount of excited enzyme molecules. α and β are positive parameters that measure the degree of tendency of the system to a ferroelectric instability compared to its electric resistance with such a behavior while ζ is the parameter of nonlinearity. This equation had been proposed by Kaiser [12] as a model to simulate specific biophysical processes involving the coexistence of two stable oscillatory states (*so called birhytmicity*).

3. Amplitudes and Frequencies of Limit Cycles

We consider the situation where the oscillator is unforced ($F = 0$). In order to find amplitudes and frequencies of the limit cycles, we utilized the **Lindsted's** perturbation method [6]. For that purpose, it is interesting to set $\tau = \omega t$ where ω is an unknown frequency. This permits the frequency and the amplitude to interact. We assume that the periodic solution of equation (9) can be approximated as follows:

$$x(\tau) = x_0(\tau) + \zeta x_1(\tau) + \zeta^2 x_2(\tau) + ..., \tag{10}$$

where the functions $x_i(\tau)$ $(i = 0, 1, 2, ...)$ are periodic functions of τ of period 2π. Moreover, the frequency ω can be represented by the following expression

$$\omega = \omega_0 + \zeta\omega_1 + \zeta^2\omega_2 + ..., \tag{11}$$

where ω_i $(i = 0, 1, 2, ...)$ are unknown constants at this point. Substituting the expressions (10) and (11) in equation (9) and equating the coefficients of ζ^0, ζ^1 and ζ^2 to zero, we obtain the following equations at different order of ζ:
order ζ^0

$$\omega_0^2 \ddot{x}_0 + x_0 = 0. \tag{12}$$

order ζ^1

$$\omega_0^2 \ddot{x}_1 + x_1 = \omega_0(1 - x_0^2 + \alpha x_0^4 - \beta x_0^6)\dot{x}_0 - 2\omega_0\omega_1\ddot{x}_0. \tag{13}$$

order ζ^2

$$\begin{aligned} \omega_0^2 \ddot{x}_2 + x_2 &= \omega_0[(1 - x_0^2)\dot{x}_1 - 2x_0\dot{x}_0 x_1] - 2\omega_0\omega_1\ddot{x}_1 - (\omega_1^2 + 2\omega_0\omega_2)\ddot{x}_0 \\ &+\omega_1(1 - x_0^2)\dot{x}_0 + \omega_1(\alpha - \beta x_0^2)x_0^4\dot{x}_1 \\ &+\omega_0[(\alpha - \beta x_0^2)x_0^4\dot{x}_1 + (4\alpha - 6\beta x_0^2)x_0^3\dot{x}_0 x_1]. \end{aligned} \tag{14}$$

The property $x(\tau + 2\pi) = x(\tau)$ and the initial condition $\dot{x}(0) = 0$ are used to determine unknown quantities in the above equations as follows:

$$x_i(\tau + 2\pi) = x_i(\tau); \quad \dot{x}_i(0) = 0; \quad i = 0, 1, 2. \tag{15}$$

Solving equation (12) and using conditions (15) allow to obtain the following solutions:

$$x_0 = A\cos\tau, \tag{16}$$

$$\omega_0 = 1, \tag{17}$$

where A is the amplitude of the limit cycle. In virtue of the solutions (16) and (17), equation (13) is rewritten as

$$\begin{aligned} \ddot{x}_1 + x_1 &= \left(\frac{5\beta}{64}A^6 - \frac{\alpha}{8}A^4 + \frac{1}{4}A^4 - 1\right) A\sin\tau + 2\omega_1 A\cos\tau \\ &+ \left(\frac{9\beta}{64}A^7 - \frac{3\alpha}{16}A^5 + \frac{1}{4}A^3\right)\sin 3\tau + \left(\frac{5\beta}{64}A^7 - \frac{\alpha}{16}A^5\right)\sin 5\tau + \frac{\beta}{64}A^7\sin 7\tau. \end{aligned}$$

From this latter equation, the secularity conditions (so called the solvability conditions) allows to obtain the following equation:

$$\frac{5\beta}{64}A^6 - \frac{\alpha}{8}A^4 + \frac{1}{4}A^2 - 1 = 0, \tag{18}$$

$$\omega_1 = 0. \tag{19}$$

Eq(18) provides the amplitudes of limit cycles as function of the parameters α and β. A general expression for a periodic solution of equation (18) can then be written as follows:

$$x_1 = \Psi_1\cos\tau + \Psi_2\sin\tau + \Psi_3\sin 3\tau + \Psi_4\sin 5\tau + \Psi_5\sin 7\tau, \tag{20}$$

where

$$\Psi_2 = \frac{73}{1024}\beta A^7 - \frac{1}{12}\alpha A^5 + \frac{3}{32}A^3, \quad \Psi_3 = -\frac{1}{32}\left(\frac{9\beta}{16}A^7 - \frac{3\alpha}{4} + A^3\right),$$

$$\Psi_4 = -\frac{1}{384}\left(\frac{5\beta}{4} - \alpha A^5\right), \quad \Psi_5 = -\frac{\beta}{3072}A^7.$$

The value of Ψ_1 remains unknown for the moment and will be determined in the next step. The secularity condition for the solution $x_2(\tau)$ yields to the following solutions

$$\Psi_1 = 0, \tag{21}$$

$$\begin{aligned}\omega_2 &= \frac{93\beta}{65536}A^{12} - \frac{69\alpha\beta}{16384}A^{10} + \left(\frac{3\alpha^2}{1024} + \frac{67\alpha\beta}{8192}\right)A^8 - \left(\frac{73\beta}{2048} + \frac{\alpha}{96}\right)A^6 \\ &+ \left(\frac{1}{128} + \frac{\alpha}{24}\right)A^4 - \frac{3}{64}A^2.\end{aligned} \tag{22}$$

Therefore, the solution of equation (9) can be approximated by

$$x(t) = A\cos\omega t + \zeta(\Psi_2 \sin\omega t + \Psi_3 \sin 3\omega t + \Psi_4 \sin 5\omega t + \Psi_5 \sin 7\omega t) + O(\zeta^2), \tag{23}$$

where the frequency ω is given by

$$\omega = 1 + \zeta^2\omega_2 + O(\zeta^3). \tag{24}$$

When considering the particular case of $\alpha = \beta = 0$, the amplitude of the autonomous state for Eq. (18) is $A = 2$ and that corresponds to the limit cycle's amplitude of the classical van der Pol oscillator. The amplitudes of limit cycles A_i and their corresponding frequencies ω_i $(1 \leq i \leq 3)$ are obtained respectively from Eqs. (18) and (24) using the Dichotomie algorithm. According to the values of the parameters α and β, Eq. (18) can lead to one or three positive roots which correspond to the amplitudes of limit cycles. When three limit cycles are obtained, two of them are stable while one is unstable [29]. These domains of α and β for which the system possesses one or three limit cycles have been summarized in a chart plotted in the plane (β, α) as shown in Fig.1. Such a coexistence of two stable limit cycles with different amplitudes and frequencies (or periods) separated by an unstable limit cycle for a given set of parameters is referred to as birhythmicity [30]. The unstable limit cycle represents the separatrix between the basins of attraction of the two stable limit cycles. Birhythmicity provides the capability of switching back and forth, upon appropriate perturbation or parameter change, between two distinct types of stable oscillations characterized by markedly different periods (or frequencies) and amplitudes. Such a phenomenon is encountered in many areas of science. In biochemistry for instance, consider a simple two variables for periodic oscillations of the limit cycle provided by the case of a product activated enzyme reaction as shown in Fig. 2. Such a reaction forms the core of models proposed for glycotic oscillations in yeast and muscle [31,32]. The model represented in Fig. 2 mimics an enzyme-substrate reaction that possesses either one or two stable limit cycles

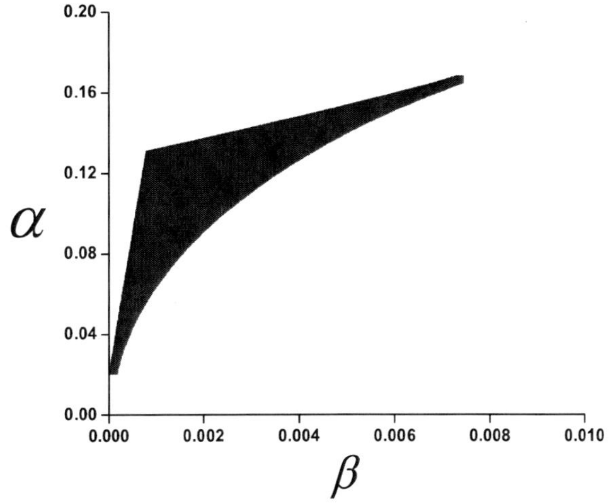

Figure 1. Chart providing values of α and β for which the model possesses a single limit cycle (non-shaded zone) and three limit cycles (shaded zone).

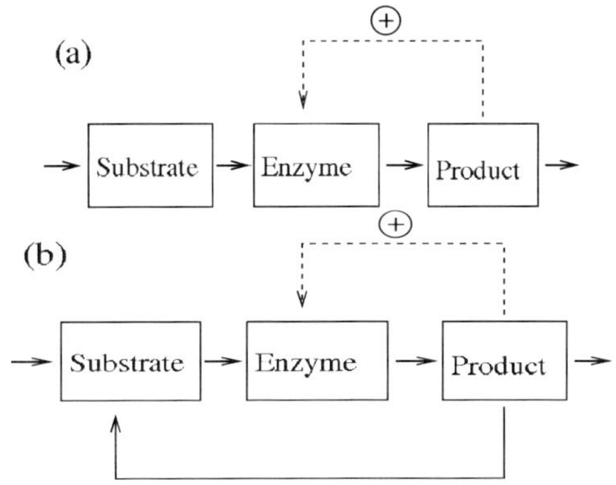

Figure 2. Models for oscillatory enzyme-substrate reactions. (a): The product activates only the enzyme and the reaction leads to a single limit cycle, (b) The product activates both the enzyme and substrate, generating two stable limit cycles.

as described by Eq. (9). For the specific case of Phosphofructokinase (**PFK**) which is an allosteric enzyme, Fig. 2 represents the reaction of catalysis of the Adenosine-TriPhosphate (**ATP**) in the form of $Mg^{-}ATP^{2-}$ (substrate) into a product named Adenosine Diphosphate (**ADP**) in the form of $Mg^{-}ADP^{-}$. The use of particular initial conditions for a defined set of the parameters α and β, one or two stable limit cycles can be obtained as for those derived from Eq. (9). During that process whose purpose is the activation of the Fructose 6-phosphate to the fructose 1,6-diphosphate, the activity of the **PFK** is controlled both by the concentration of the substrate and the concentration of the product. Thus, a more important presence of concentration of the **ATP** which indicates a state of energy load

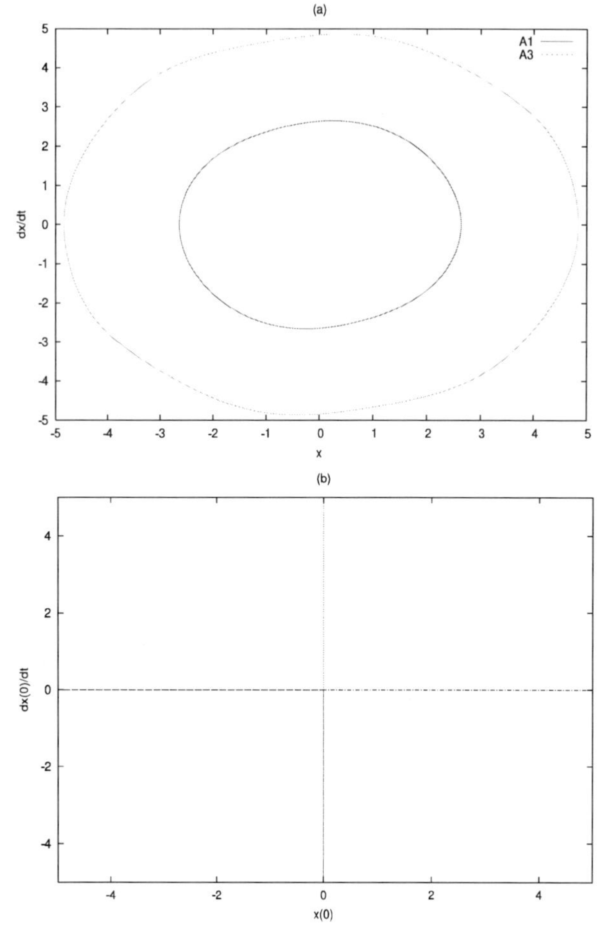

Figure 3. Phases portrait of the two stable limit cycles (a) and their corresponding basins of attraction (b) computed for $\zeta = 0.1$, $\alpha = 0.144$ and $\beta = 0.005$.

higher of the cell will support an action of the **ADP** (allosteric activator) on the **PFK** to reduce its affinity with the **ATP** and also to prevent with the **ATP** from having its inhibiting effect. Beyond a critical parameter value, the steady state admitted by the system becomes unstable and the system then evolves towards a stable limit cycle (see Fig. 2(a)). Additionally, when the concentration of the product becomes more important than that of the substrate, part of the **APD** is recycled into **ATP**, thus giving the possibility of existence of a second limit cycle as displayed in Fig. 2(b). Birhythmicity can also originate from circadian rhythms based on genetic regulation [33-35] or calcium oscillations [36]. The coexistence of two stable limit cycles can also be observed in the model of two isozymes activated by their reaction product [37]. When $\alpha = 0.144$ and $\beta = 0.005$ in the model under investigation, we obtain from Eqs. (18) and (24) two stable limit cycle's amplitude given by $A_1 = 2.6390$ and $A_3 = 4.8395$ with their related frequencies $\omega(A_1) = 1.0011$ and $\omega(A_3) = 1.0545$ respectively, while the unstable limit cycle's amplitude is $A_2 = 3.9616$

with the frequency $\omega(A_2) = 1.0114$. The stable limit cycles and their corresponding basins of attraction can be determined from a direct numerical simulation of Eq. (9) using the fourth-order Runge-Kutta algorithm as it is shown in Fig. 3.

Table 1. Comparison between analytical and numerical characteristics of the multi-limit cycles with the parameters of Figure 3

Analytical Amplitude	Numerical Amplitude	Analytical Frequency	Numerical Frequency
A_1=2.63901	A_1=2.6391	$\omega(A_1)$=1.0011	$\omega(A_1)$=0.99971
A_2=3.9616	Unstable	$\omega(A_2)$=1.0114	Unstable
A_3=4.83952	A_3=4.83955	$\omega(A_3)$=1.0545	$\omega(A_3)$=0.99987

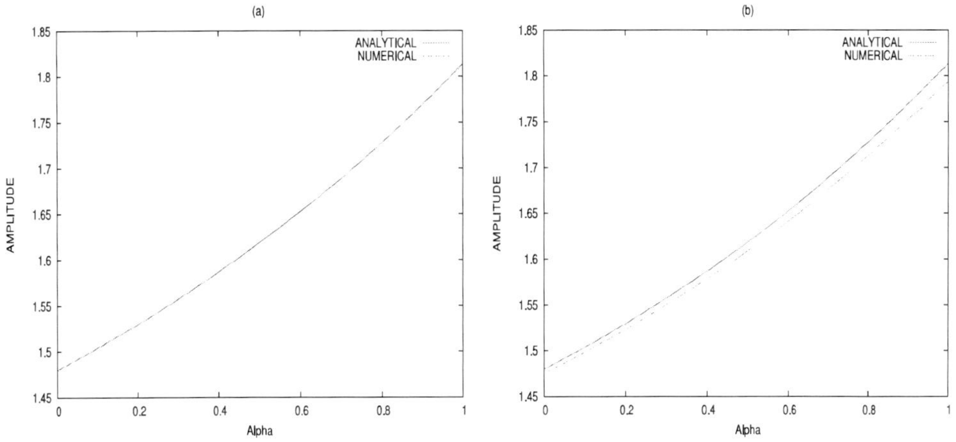

Figure 4. Comparison between analytical and numerical amplitude-response curves of the limit-cycle $A(\alpha)$ for $\beta = 0.55$. (a): $\zeta = 0.005$, (b): $\zeta = 2$.

In Fig. 3(a), both inner and outer regular trajectories correspond respectively to the limit cycle amplitude A_1 and A_3. In Fig. 3(b), the shaded area represents attraction to A_1 and the non-shaded the attraction to A_3. Therefore, any defined set of initial conditions selected within the shaded region will lead to the limit-cycle of amplitude A_1 while the limit-cycle of amplitude A_3 will be obtained from any choice of initial conditions selected elsewhere. The comparison between amplitudes and frequencies characteristics derived both from the analytical and numerical treatments are given in Table 1. In order to check the accuracy of the analytical approach used, the comparison between analytical and numerical amplitude responses-curves $A(\alpha)$ and $A(\beta)$ are plotted respectively in Fig. 4 and Fig. 5 both for ζ very small and ζ large. For both ranges of ζ, there is a good fitting between the results of analytical investigations and those obtained through numerical simulations.

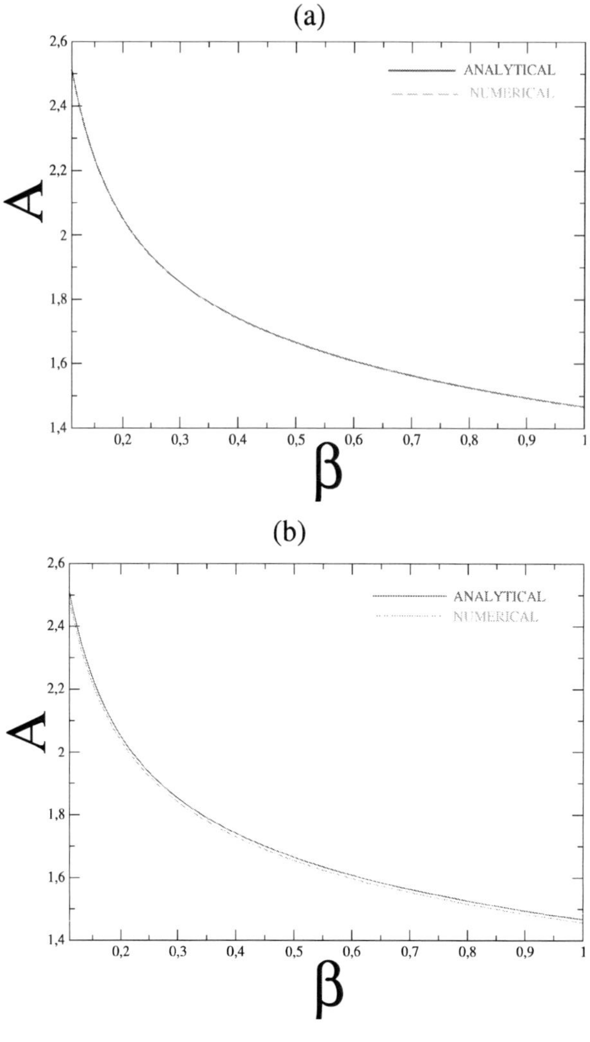

Figure 5. Comparison between analytical and numerical amplitude-response curves of the limit-cycle $A(\beta)$ for $\alpha = 0.55$. (a): $\zeta = 0.005$, (b): $\zeta = 2$.

4. Unidirectional Coupling Synchronization

4.1. Problem Statement

The model under consideration can give birth to one limit cycle or to three limit cycles, based upon the values of α, β and initial conditions. Consequently, if two of such models are launched with two different initial conditions, they will lastly circulate either on the different limit cycles or on the same limit cycle but with different phases. The objective of the synchronization is therefore to phase-lock (phase synchronization) both oscillators or to force one of the oscillators from its limit cycle to that of the other oscillator.

We investigate in this section the unidirectional coupling synchronization. The master system (**M**) is described by the component x while the slave system (**S**) has the corresponding component u. The enslavement is carried out by coupling the slave to the master as follows:

$$\begin{aligned} &\ddot{x} - \zeta(1 - x^2 + \alpha x^4 - \beta x^6)\dot{x} + x = 0, \\ &\ddot{u} - \zeta(1 - u^2 + \alpha u^4 - \beta u^6)\dot{u} + u = K(u - x)H(t - T_0), \end{aligned} \tag{25}$$

where K is the feedback coupling coefficient or strength, T_0 the onset time of the synchronization process and H the Heaviside function defined as

$$H(l) = \begin{cases} 0 & \text{if } l < 0, \\ 1 & \text{if } l \geq 0. \end{cases}$$

The role of the feedback is to force the convergence of the slave towards the master orbit. Throughout the study, we use $T_0 = 600$. It should be emphasized that other values of T_0 can be used and will lead to the same results. Practically, such type of coupling occurs in cells or systems when an external stimulus substrate coming from the outside of the cell is selected, amplified as a signal or transformed through an enzyme and transported as a product. In such a situation, the reaction process is unidirectional and irreversible. The coupling of two such a system is described by the diffusion of their solute concentration. The coupling is written as the difference in the concentration . Since within each system there are inhibitors (negative direction of solute flow) and promoters (positive direction of solute flow) that determine the direction of the solute flow, it follows that the coupling coefficient can be negative or positive.

4.2. Stability Boundaries of the Synchronization Process

When the synchronization starts at T_0, one must make sure that the process is stable to avoid instability and the lost of synchronization during the process. For example, time during which enzymes are not stable in the model under investigation corresponds to the period during which they are not able to take part in a reaction process because of the modification of their structures. Thus, the product of reaction cannot be obtained or are not readily available. Such a situation is for instance catastrophic in biochemistry for glycolic oscillations in muscle. Consequently, we have to know the closeness between the master and the slave at each time through the new variable

$$z(t) = u(t) - x(t). \tag{26}$$

The stability of the synchronization process is therefore examined by the boundlessness of $z(t)$ which in the linear regime obeys to the following equation:

$$\ddot{z} - \zeta(1 - x^2 + \alpha x^4 - \beta x^6)\dot{z} + [1 - K + \zeta(2x - 4\alpha x^3 + 6\beta x^5)\dot{x}]z = 0. \tag{27}$$

Let us first consider the first component of the solution $x(t)$ given by

$$x(t) = A\cos\omega t. \tag{28}$$

With this approximation, an error of order ζ is made (see Eq. (23)). If we set $\tau = \omega t$ into Eq. (28), the variational Eq. (27) becomes

$$\ddot{z} + [2\lambda + F(\tau)]\dot{z} + G(\tau)z = 0, \tag{29}$$

where the parameters $\lambda, F(\tau), G(\tau)$ are defined as follows:

$$\begin{aligned}
\lambda &= -\frac{\zeta H}{2\omega}, \\
F(\tau) &= -\frac{\zeta}{\omega}[I\cos 2\tau + J\cos 4\tau + L\cos 6\tau], \\
G(\tau) &= \frac{1}{\omega^2}[1 - K + M\sin 2\tau + N\sin 4\tau + P\sin 6\tau],
\end{aligned}$$

with

$$\begin{aligned}
&H = 1 - \frac{1}{2}A^2 + \frac{3\alpha}{8}A^4 - \frac{5\beta}{16}A^6, \\
&I = -\frac{1}{2}A^2 + \frac{\alpha}{2}A^4 - \frac{15\beta}{32}A^6, \\
&J = \frac{\alpha}{8}A^4 - \frac{3\beta}{16}A^6, \\
&L = -\frac{\beta}{32}A^6, \\
&M = -\zeta\omega A^2(2 + 2\alpha A^2 - \frac{15\beta}{8}A^4), \\
&N = \frac{\zeta\omega}{2}A^4(\alpha - \frac{3\beta}{2}A^2), \\
&P = -\frac{3\zeta\beta}{16}A^6.
\end{aligned}$$

From the expression of $G(\tau)$, we find that for $K > 1$, $G(\tau)$ has a permanent negative part $(1-K)/\omega^2$, thus inducing a continuous increase of the deviation $z(t)$ (see Eq. (29)). In such a case, the slave continuously drifts away from its original limit cycle and no synchronization with the master is possible even if both of them are on the same or different limit cycles as it is shown respectively in Figs. 6 and 7. To find other stability boundaries of the process, Eq. (29) is transformed into its standard form by introducing a new variable Γ defined as follows:

$$z = \Gamma\exp(-\lambda\tau)\exp\left[-\frac{1}{2}\int F(\tau)d\tau\right]. \tag{30}$$

Consequently, we demonstrate that Γ satisfies the following Hill equation [1,2]:

$$\begin{aligned}
&\ddot{\Gamma} + [a_0 + 2a_{1s}\sin 2\tau + 2a_{1c}\cos 2\tau + 2a_{2s}\sin 4\tau + 2a_{2c}\cos 4\tau + 2a_{3s}\sin 6\tau \\
&+2a_{3c}\cos 6\tau + 2a_{4c}\cos 8\tau + 2a_{5c}\cos 10\tau + 2a_{6c}\cos 12\tau]\Gamma = 0,
\end{aligned} \tag{31}$$

where the parameters a_0, a_{ns} and a_{nc} $(1 \leq n \leq 6)$ are given below:

$$a_0 = \frac{1}{\omega^2}\left[1 - K - \frac{\zeta^2}{8}(2H^2 + I^2 + J^2 + L^2)\right],$$

$$a_{1s} = -\frac{1}{2\omega}(\zeta I - \frac{M}{\omega}), \quad a_{1c} = -\frac{\zeta^2}{8\omega^2}(2HI + IJ + JL),$$

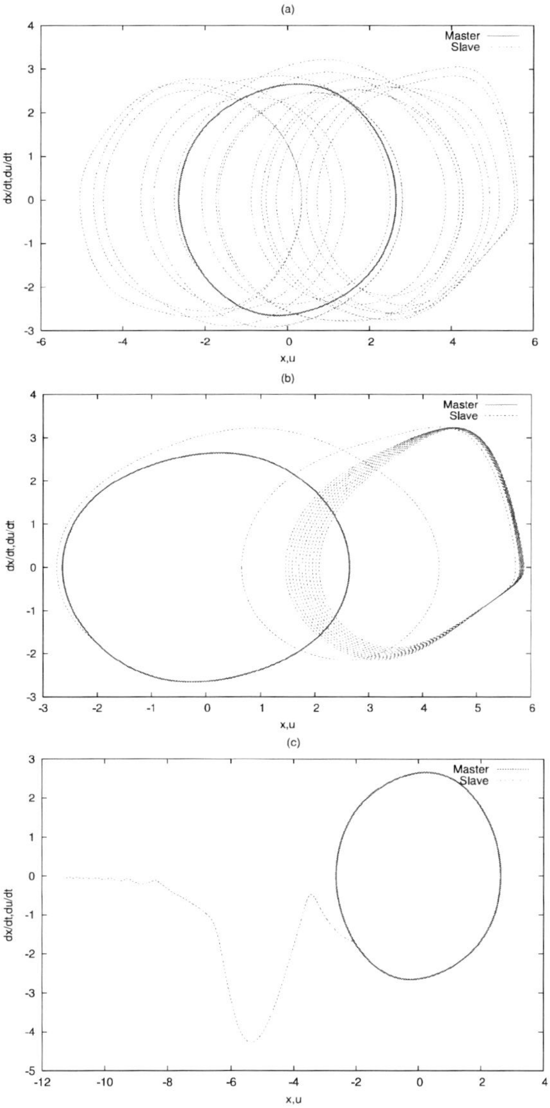

Figure 6. Phase portraits showing how the slave continuously drifts away from the master's limit cycle when both master and slave evolve on the same limit cycle's amplitude A_1 with the parameters of Fig. 3 and the initial conditions $(x(0), \dot{x}(0)) = (1, 1)$ and $(u(0), \dot{u}(0)) = (2, 1)$. (a): $K = 0.99$, (b): $K = 1.001$, (c): $K = 3.50$.

$$a_{2s} = -\frac{1}{2\omega}(2\zeta J - \frac{N}{\omega}), \quad a_{2c} = -\frac{\zeta^2}{16\omega^2}(4HJ + I^2 + 2IL),$$

$$a_{3s} = -\frac{1}{2\omega}(3\zeta L - \frac{P}{\omega}), \quad a_{3c} = -\frac{\zeta^2}{8\omega^2}(2HL + IJ),$$

$$a_{4c} = -\frac{\zeta^2}{16\omega^2}(2IL + J^2), \quad a_{5c} = -\frac{\zeta^2}{8\omega^2}JL, \quad a_{6c} = -\frac{\zeta^2}{16\omega^2}L^2.$$

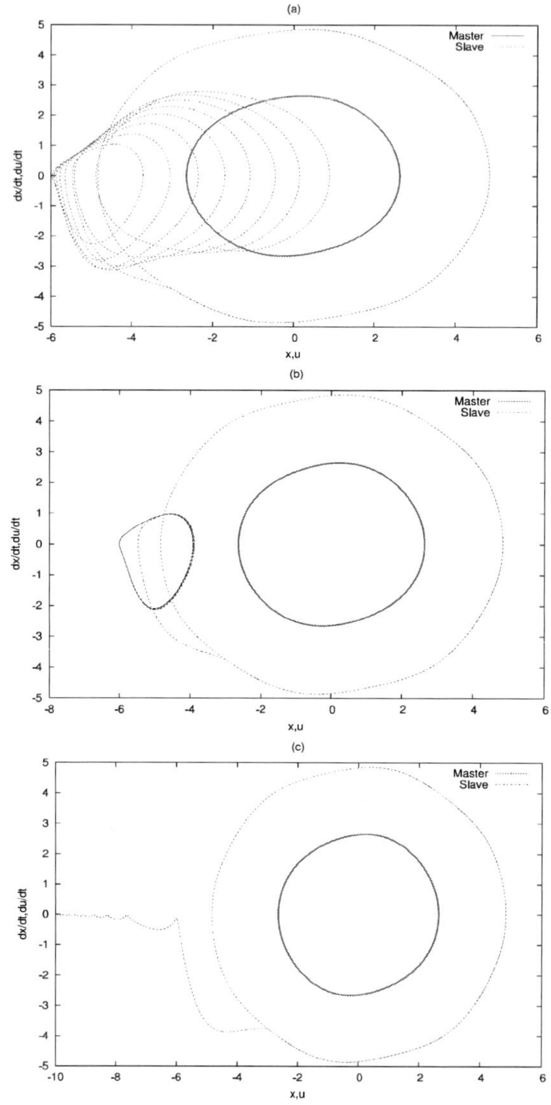

Figure 7. Phase portraits showing how the slave continuously drifts away from the master's limit cycle with the parameters of Fig. 3 when both master and slave evolve on two different limit cycles with the initial conditions $(x(0), \dot{x}(0)) = (1, 1)$ and $(u(0), \dot{u}(0)) = (4, 4)$. (a): $K = 0.99$, (b): $K = 1.0001$, (c): $K = 2.0$.

The Whittaker method [2] is utilized to determine the stability boundaries of the synchronization process around the six main parametric resonances revealed by Eq. (31). Its solution in the nth unstable region has the form

$$\Gamma = e^{\varphi\tau} \sin(n\tau - \varrho), \tag{32}$$

where φ is the characteristic exponent, ϱ a constant while n stands for the order of the parametric resonances. From Eqs. (31) and (32), the expression of the characteristic exponent is obtained and given as follows:

$$\varphi^2 = -(a_0 + n^2) + \sqrt{4n^2 a_0 + a_{ns}^2 + a_{nc}^2}. \tag{33}$$

Eq. (30) implies that for $z(t)$ to tend to zero as time increases, the real part $-\lambda \pm \varphi$ should be negative. Under these circumstances, the stability of the synchronization process is guaranteed if the following criteria are satisfied:

$$Q_n = (a_0 - n^2)^2 + 2(a_0 + n^2)\lambda^2 + \lambda^4 - a_{ns}^2 - a_{nc}^2 > 0; \quad 1 \leq n \leq 6. \tag{34}$$

Saying the process is stable here means the synchronization is achieved. To confirm the validity of these criteria, Eqs. (34) is integrated numerically . The values of parameters α and β are those defined for the limit cycles A_1, A_2 and A_3 in the previous section when $\zeta = 0.1$. For the stable limit cycles, two applications are to be considered here.
We first consider that the master and the slave are on the same stable limit cycle with the amplitude A_1 or A_3. For instance with the amplitude A_3, we obtain from Eqs. (34) that the synchronization process is unstable for $K \in D_{Anal}[\mathbf{S}(A_3) \to \mathbf{M}(A_3)] = [-0.034, 0) \cup (0, 0.003] \cup (1, +\infty)$ while from the numerical simulation of Eqs (25), we find that the synchronization is unstable for $K \in D_{Num}[\mathbf{S}(A_3) \to \mathbf{M}(A_3)] = [-0.020, 0) \cup (0, 0.015] \cup (1, +\infty)$ (the sets of initial conditions for numerical simulations are $(x(0), \dot{x}(0)) = (5, 5)$ and $(u(0), \dot{u}(0)) = (4, 4)$, both leading to the final state whose amplitude is A_3). We note here that the synchronization process is unstable means that $z(t)$ never goes to zero, but has a bounded oscillatory behavior or goes to infinity. Thus we find that the analytical procedure gives good results comparatively to the numerical ones.

For the second application, we assume that the master and the slave are on different limit cycles. It is found from our analytical investigation that the synchronization process is unstable for $K \in D_{Anal}[\mathbf{S}(A_1) \to \mathbf{M}(A_3)] = [-0.034, 0) \cup (0, 0.003] \cup (1, +\infty)$ if the slave comes from the limit cycle whose amplitude is A_1 to follow the master at the limit cycle with the amplitude A_3. For numerical simulations, the master and the slave are now initially launched respectively with the conditions $(x(0), \dot{x}(0)) = (5, 5)$ leading to A_3 and $(u(0), \dot{u}(0)) = (1, 1)$ leading to A_1. We find from this numerical procedure that the slave transition from A_1 to A_3 is not achieved if $K \in D_{Num}[\mathbf{S}(A_1) \to \mathbf{M}(A_3)] = [-0.019, 0) \cup (0, 0.015] \cup (1, +\infty)$.

The fitting between the analytical and numerical results is good. To illustrate these results, we have represented graphically the temporal evolution of the deviation between the master and the slave for a feedback coupling strength K chosen both in the unstable and stable domains of synchronization. In the unstable regions, we find that the deviation $z(t)$ grows infinitely or behaves bounded while in the stable areas, $z(t)$ decreases to zero with

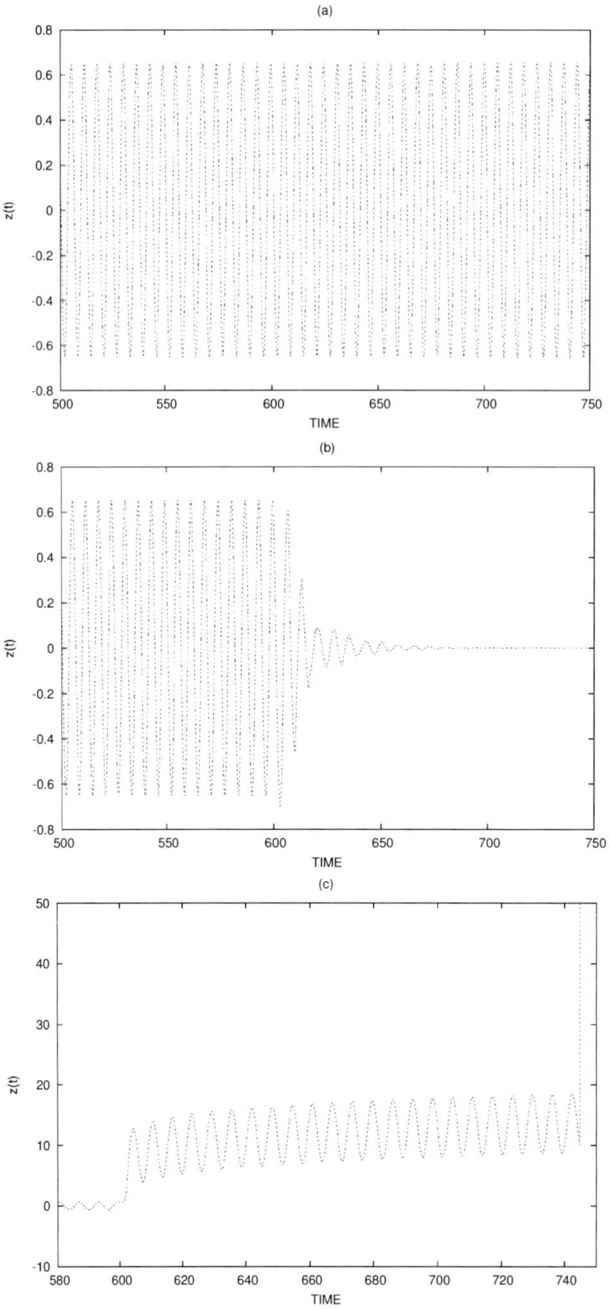

Figure 8. Temporal variation of the deviation $z(t)$ for the first application with the parameters of Fig. 3. (a): $K = -0.015$, (b): $K = 0.30$, (c): $K = 4$.

increasing time as displayed in Figs. 8 and 9 respectively for the first and the second applications. In the case where the slave leaves A_3 to A_1, we just have to invert the initial

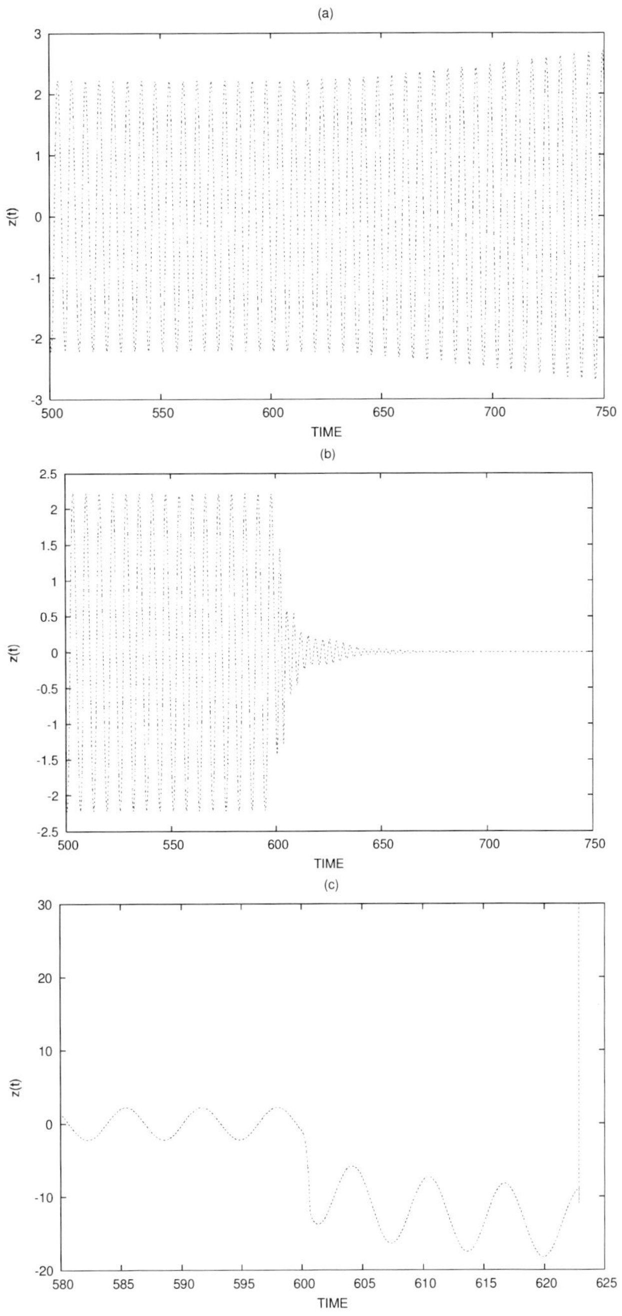

Figure 9. Temporal variation of the deviation $z(t)$ for the second application with the parameters of Fig. 3. (a): $K = 0.010$, (b): $K = -2.50$, (c): $K = 20$.

conditions for the numerical simulation. It is found from the investigations that the synchronization process is unstable for $K \in D_{Num}[\mathbf{S}(A_3) \rightarrow \mathbf{M}(A_1)] = [-0.252, 0) \cup (0, +\infty)$

numerically while from Eq. (34), it is found that $K \in D_{Anal}[\mathbf{S}(A_3) \rightarrow \mathbf{M}(A_1)] = (0, 0.001] \cup (1, +\infty)$. It appears from this latter transition a lack of agreement between numerical and analytical results. Thus, the transition from the orbit with small amplitude to another orbit with higher amplitude is easier than in the reverse situation. Such a situation is understandable since moving from a small to a large limit cycle arises whenever the perturbation brings the system across the boundary set by the unstable limit cycle. Also, the basin of attraction of the larger amplitude oscillatory state A_3 is much more extended than that of the small amplitude oscillatory region A_1. Similar results have been obtained for the cyclic Adenosine Monophosphate (**cAMP**) system of D. discoideum by Goldbeter and Martiel [38] who have reported that the transition from the large to small limit cycle necessitates much finer tuning of the perturbation with respect to both the magnitude and the phase at which it is applied. The reason for such asymmetry in the sensitivity of two stable cycles with respect to the perturbation has been reported in the model of an allosteric enzyme activated by its reaction product [39]. The instability boundaries also depend on the choice of the initial conditions because they influence the speed of reaction of the biological process and consequently, the rate at which the product is obtained (see an example hereafter). To show how the approximation in $x(t)$ affects the stability domains, we now consider the full expression (23) in the variational Eq. (27). Therefore, a Hill equation containing harmonics till $\cos 42\tau$ is obtained with thus the possibility of having 21 parametric resonances. We solved analytically that Hill equation around the first main parametric resonance using the criteria (34) and found that the process is now analytically unstable for $K \in \widehat{D}_{Anal}[\mathbf{S}(A_3) \rightarrow \mathbf{M}(A_3)] = [-0.033, 0) \cup (0, 0.020] \cup (1, +\infty)$ when both master and slave are on the limit cycle whose amplitude is A_3. Considering the second application, the unstable areas are now defined as $K \in \widehat{D}_{Anal}[\mathbf{S}(A_1) \rightarrow \mathbf{M}(A_3)] = [-0.033, 0) \cup (0, 0.020] \cup (1, +\infty)$ analytically when the slave comes from A_1 to follow the master at A_3. But when the slave comes from A_3 to follow the master at A_1, the synchronization is unstable for $K \in \widehat{D}_{Anal}[\mathbf{S}(A_3) \rightarrow \mathbf{M}(A_1)] = [-0.001, 0) \cup (0, 0.003] \cup (1, +\infty)$. Thus, it comes that using the full expression of the solution $x(t)$ has improved the accuracy between analytical and numerical stability boundaries of the synchronization process.
After the stability has been analyzed through the linear variational Eq. (27) and linear stability domains have been derived, we also aim to search the effects of the nonlinearities on the stability boundaries of the synchronization process for the slave transition from A_3 to A_1. Then the deviation $z(t)$ is now governed by the following nonlinear variational equation

$$\begin{aligned}
&\ddot{z} - \zeta[1 - x^2 + \alpha x^4 - \beta x^6 + \alpha(z^4 + 6x^2 z^2 + 4xz^3 + 4x^3 z) - \beta(z^6 + 6xz^5 \\
&+15x^2 z^4 + 20x^3 z^3 + 15x^4 z^2 + 6x^5 z) - z^2 - 2xz]\dot{z} + [1 - K + \zeta(2x - 4\alpha x^3 \\
&+6\beta x^5)\dot{x}]z + \zeta[\beta(z^6 + 15x^2 z^4 + 6xz^5 + 20x^3 z^3 + 15x^4 z^2) + z^2 \\
&-\alpha(z^4 + 6x^2 z^2 + 4xz^3)]\dot{x} = 0. \qquad (35)
\end{aligned}$$

Eq. (35) is integrated only numerically due to the mathematical difficulties in handling such an equation analytically. This is achieved by using the complete expression of the solution $x(t)$ to obtain the results for stability boundaries. It should be emphasized that when the slave comes from the limit cycle whose amplitude is A_3 to follow the master at the limit cycle's amplitude A_1, we conclude from these numerical investigations that the synchronization domains depend on the set of initial conditions used to launch the slave. For instance, by using $(u(0), \dot{u}(0)) = (3.5, 3.5)$, we find that the synchronization is unstable for $K \in \widehat{D}^{\dagger}_{Num}[\mathbf{S}(A_3) \rightarrow \mathbf{M}(A_1)] = [-0.240, 0) \cup (0, +\infty)$ while for $(u(0), \dot{u}(0)) = (2.8, 2.8)$, instability during the process appears rather for $K \in \widehat{D}^{\ddagger}_{Num}[\mathbf{S}(A_3) \rightarrow \mathbf{M}(A_1)] = [-0.009, 0) \cup (0, 0.010) \cup (0.840, +\infty)$. This is understandable since in the case of multiple limit cycles, the synchronization boundaries depend on what can be termed the stability basins for synchronization. From a biological point of view, one could conclude that there are situations for which it is difficult, even impossible to find a physiological behavior (limit cycle A_1) starting from a pathological behavior (limit cycle A_3).

5. Bi-Directional Coupling Synchronization

The bi-directional coupling synchronization (so-called the mutual synchronization) occurs when the individual units or oscillators in a system alter their individual rhythms just enough so they all act in unison. Such phenomenon appears in many populations of biological oscillators. It is the case for instance with crickets that chirp in unison, electrically synchronous pacemaker cells and groups of women whose menstrual cycles become mutually synchronize (see Refs. [4,40] and references therein). Two types of bi-directional coupling configurations are to be considered: Symmetric and antisymmetric coupling. But before addressing this issue, it is useful to emphasize that only the case of oscillators on the same limit cycle but that were launched with different initial conditions will be considered. This case is analytically tractable by the formalism developed for the unidirectional coupling in the previous section.
For the symmetric coupling scheme, the enslavement is carried out as follows:

$$\begin{aligned} \ddot{x} - \zeta(1 - x^2 + \alpha x^4 - \beta x^6)\dot{x} + x &= K(x - u)H(t - T_0), \\ \ddot{u} - \zeta(1 - u^2 + \alpha u^4 - \beta u^6)\dot{u} + u &= K(u - x)H(t - T_0). \end{aligned} \tag{36}$$

In such a case, the coupling is of the same type (either excitatory or inhibitory) but in both directions [41]. Setting

$$\begin{aligned} x &= x_s + z_1, \\ u &= x_s + z_2, \end{aligned} \tag{37}$$

allows obtaining the following coupled variational equations

$$\ddot{\delta} - \zeta(1 - x_s^2 + \alpha x_s^4 - \beta x_s^6)\dot{\delta} + [1 + \zeta(2x_s - 4\alpha x_s^3 + 6\beta x_s^5)\dot{x}_s]\delta = 0, \tag{38}$$

$$\ddot{\nu} - \zeta(1 - x_s^2 + \alpha x_s^4 - \beta x_s^6)\dot{\nu} + [1 - 2K + \zeta(2x_s - 4\alpha x_s^3 + 6\beta x_s^5)\dot{x}_s]\nu = 0, \tag{39}$$

where $\delta = z_1 + z_2$, $\nu = z_1 - z_2$ and x_s is a stable limit cycle. Eq. (38) is just the stability equation of x_s, thus $\delta = 0$. From Eq. (39), it is found that when $K > 0.5$, ν will grow indefinitely and thus, trajectories of the biological systems gradually diverge one from the other as it is shown in Fig. 10. When only the first component of the solution $x(t)$ and the analytical procedure used in the previous section are taken into account, we find that

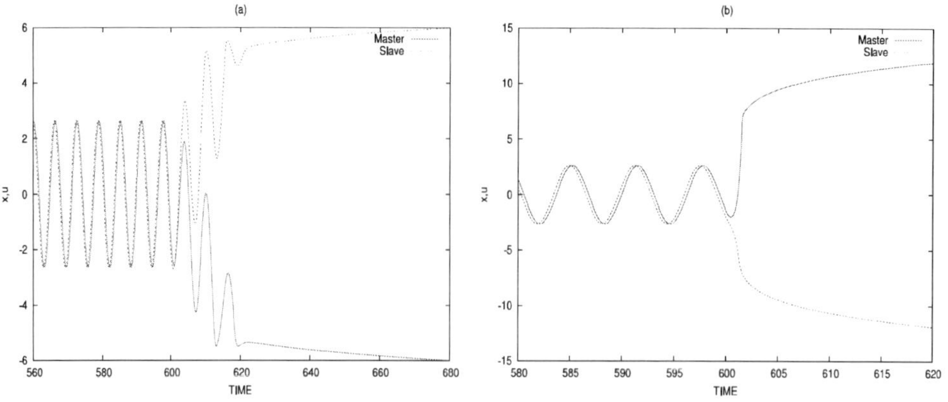

Figure 10. Time histories showing how both Master and Slave trajectories continuously diverge in the symmetric bi-directional coupling with the parameters of Fig. 3 and the initial conditions $(x(0), \dot{x}(0)) = (1, 1)$ and $(u(0), \dot{u}(0)) = (2, 1)$. (a): $K = 0.505$, (b): $K = 5.0$.

the phase-locking process cannot be obtained for $K = 0$ and $K \in (0.5, +\infty)$ when both biological systems are on the limit cycle whose amplitude is A_1, while numerically the synchronization is not achieved for $K \in [-0.004, 0) \cup (0, 0.004] \cup (0.5, +\infty)$. But, when both the oscillators are on the limit cycle whose amplitude is A_3, the synchronization process is analytically unstable for $K \in [-0.017, 0) \cup (0, 0.0015] \cup (0.5, +\infty)$ while from numerical simulations, instability occurs for $K \in [-0.009, 0) \cup (0, 0.007] \cup (0.5, +\infty)$. The agreement between our analytical and numerical results is good and to illustrate such a situation, we have plotted in Fig. 11 the time histories of the deviation $z(t)$ when both the oscillators are on the limit cycle amplitude A_1. Similar time histories configuration are also obtained when the oscillators are on the limit cycle's amplitude A_3. When considering the entire expression of the solution $x(t)$ (see Eq. (23)), the process is not achieved analytically if $K \in (0, 0.001] \cup (0.5, +\infty)$ in the case where the master and the slave evolve on the limit cycle A_1. In the case where the two oscillators are on the orbit A_3, the phase-locking process is not realized for $K \in [-0.016, 0) \cup (0, 0.001] \cup (0.5, +\infty)$ following the analytical treatment. As in the case of the one-way coupling synchronization, considering the full expression of the solution $x(t)$ significantly improved the stability boundaries of the synchronization process. Therefore the fitting becomes very good between numerical and analytical results. For the antisymmetric coupling, the enslavement is defined as follows:

$$\begin{aligned} \ddot{x} - \zeta(1 - x^2 + \alpha x^4 - \beta x^6)\dot{x} + x &= K(u - x)H(t - T_0), \\ \ddot{u} - \zeta(1 - u^2 + \alpha u^4 - \beta u^6)\dot{u} + u &= K(u - x)H(t - T_0). \end{aligned} \tag{40}$$

Here, the coupling is such that the inhibitory is in one direction and the excitatory in the other [41]. The stability of the synchronization process is tackled through the following coupled variational equations:

$$\ddot{\delta} - \zeta(1 - x_s^2 + \alpha x_s^4 - \beta x_s^6)\dot{\delta} + [1 + \zeta(2x_s - 4\alpha x_s^3 + 6\beta x_s^5)\dot{x}_s]\delta + 2K\nu = 0, \tag{41}$$

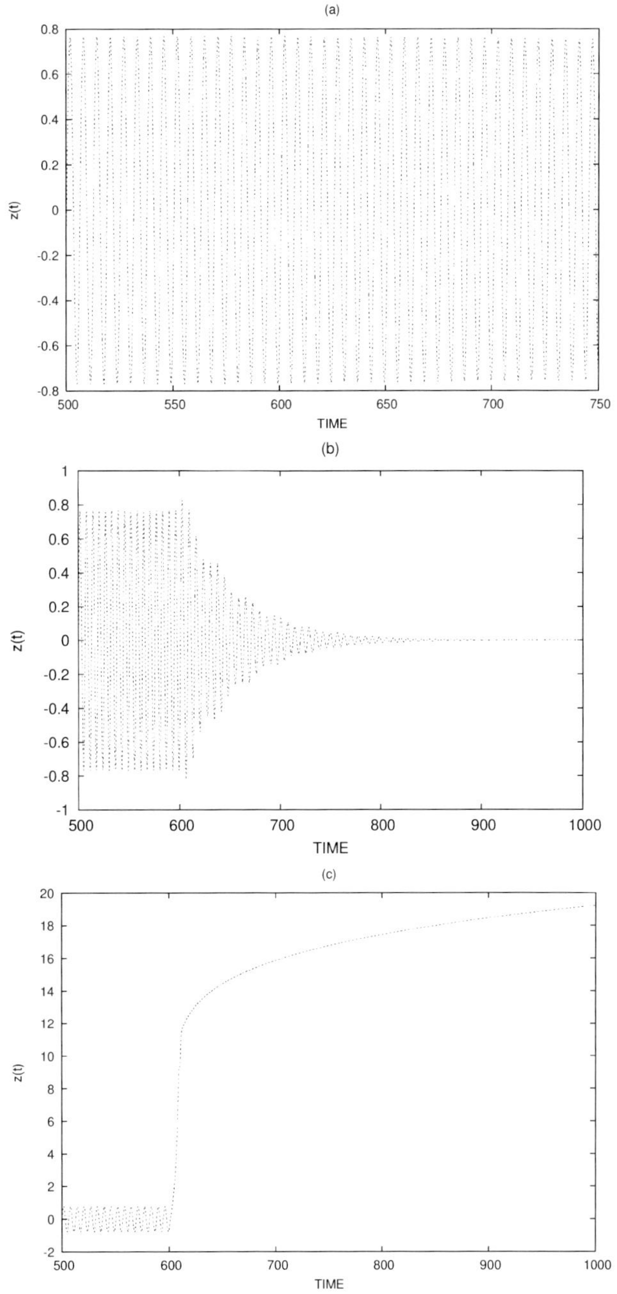

Figure 11. Temporal variation of the deviation $z(t)$ in the two-way coupling when the oscillators evolve on the limit cycle A_1: (a): $K = -0.002$, (b): $K = 0.10$, (c): $K = 0.55$.

$$\ddot{\nu} - \zeta(1 - x_s^2 + \alpha x_s^4 - \beta x_s^6)\dot{\nu} + [1 + \zeta(2x_s - 4\alpha x_s^3 + 6\beta x_s^5)\dot{x}_s]\nu = 0. \tag{42}$$

From Eq. (42), we have asymptotically $\nu = 0$ since it is just the variational equation around the stable limit cycle x_s. Therefore, Eq. (41) asymptotically reduces to Eq. (42) and also gives $\delta = 0$ for any K. Thus, for the antisymmetric coupling, the synchronization is possible for all K, except obviously the value $K = 0$ for which the coupling can not be realized. By setting $\alpha = \beta = 0$, this analytical approach confirms the results obtained numerically by Leung [42] for the synchronization of two classical van der Pol oscillators with a two-way coupling scheme.

6. Bifurcations and Transitions to Chaos

As most nonlinear systems, the biological systems are able to display regular and chaotic motions depending on some perturbation of their initial states, internal and external parameters. Thus, it is of interest to find ways through which strange attractors arise in biological systems. This is because chaotic oscillations that have been earlier thought as inconvenient have become sources of liability because of their potential applications. Indeed, chaotic motions are of interest in executing activity adaptation and state transitions in response to environmental changes, and consequently creates a rich repertory of responses [43]. The quenching of chaos in enzyme-substrate reactions is important in medical science since chaos control techniques are expected to bring about new diagnostic tools and therapies for certain types of diseases, including cardiac arrhythmias [44], human eye [45] and epilepsy [46]. On the other hand, the existence of chaos is sometimes needed. Thus, the interests have been devoted to the idea that the brain may utilize transition in and out of chaos in its processing to form a complex system displaying self-organization [47-50] capable of generating, new types of structure through bifurcations, a sudden qualitative change in structure occurring at a critical value of a continuously varying parameter. In this section, we explore ways through which chaos arises in the model described by Eq. (9)). For this purpose, Eq. (9)) is computed numerically using the fourth order RungeKutta algorithm. The resulting bifurcation diagram and variation of the maximum Lyapunov exponent are plotted as functions of the amplitude of the external excitation F. The maximum Lyapunov exponent is defined as follows:

$$\mathbf{Lya} = \lim_{t\to\infty} ln \frac{\left(\sqrt{dx^2 + dv_x^2}\right)}{t}, \tag{43}$$

where dx and dv_x are the variations of x and $\dot{x}$ respectively. The time period of the periodic stroboscopic bifurcation diagram used to map the transition is $T = \frac{2\pi}{\Omega}$. Fig. 12 shows the bifurcation diagram and its corresponding maximum Lyapunov exponent respectively when the amplitude of external excitation F varies and the following transitions are observed. When F continuously increases from zero, the system moves from a quasi-periodic states to a period-5 orbit at F= 1.34. That period-5 orbit remains in the system until F = 8.224 where chaos occurs with some sporadic windows of quasi-periodic orbits and this persists until F = 9.302 where only quasi-periodic oscillations continue to be displayed. When F = 9.50, there is a transition from quasi-periodic behavior to a period-8 orbit. The period-8 orbit exists until F = 9.55 where a period-5 orbit occurs and evolves to a period-4 orbit. At F = 9.86, a transition from period-4 orbit to small windows of chaotic oscillations and quasi-periodic

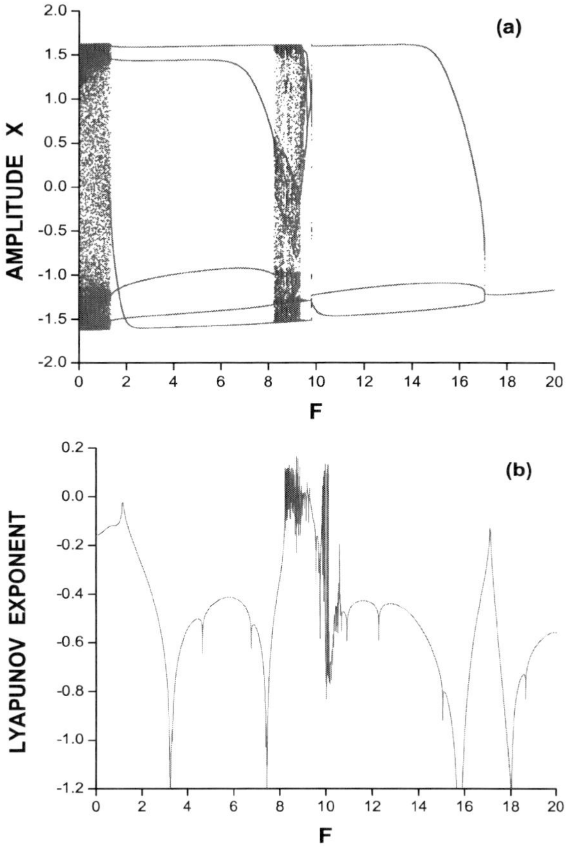

Figure 12. Bifurcation diagram (a) and the corresponding maximum Lyapunov exponent (b) for $\alpha = 2.55, \beta = 1.50, \zeta = 2, \Omega = 3.465$.

orbit. Those windows continue to be displayed until F = 9.96 where a period-3 orbit takes place before leading to a period-1 orbit (harmonic oscillations) at F = 17.056. A phase portrait showing chaotic behavior is plotted in Fig. 13 for $F = 8.27$ with the parameters of Fig.12. For another set of parameter $\alpha = 0.144, \beta = 0.05, \zeta = 3.5, \Omega = 3.465, F = 11.40$ but with two different initial conditions, the model exhibits two particular types of chaotic attractors (see Fig. 14) which are complementary since one of the attractors can evolve towards the other through symmetry inversion. Such type of attractors have also been reported by Leung in the study of the synchronization of two classical van der der Pol oscillator [18].

7. Chaos Synchronization

Chaos synchronization has increasingly become a key source of interest since the pioneer works of Pecora [51] who had used a common drive signal to achieve the synchronization of two coupled chaotic oscillators. Since then, many other works about the applicability

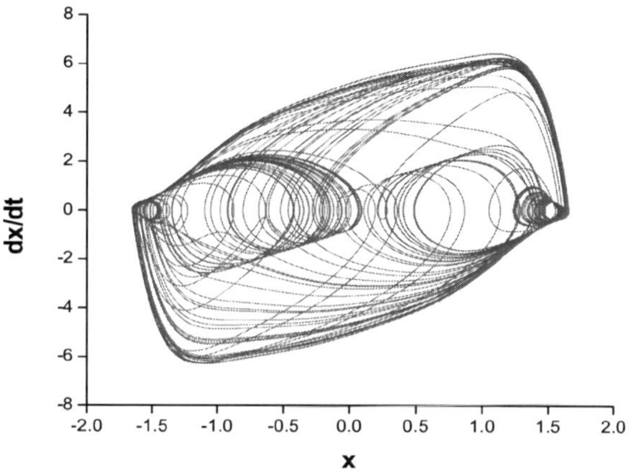

Figure 13. Chaotic phase portrait with the parameters of Fig. 12 for $F = 8.27$.

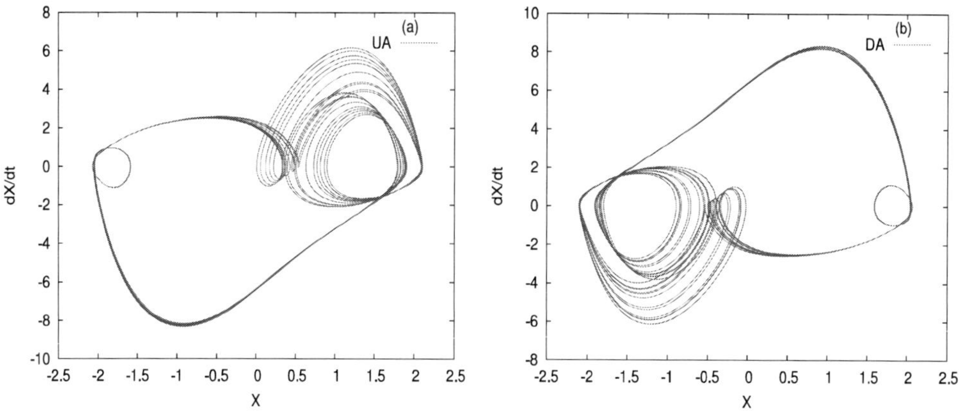

Figure 14. (a): Symmetrical chaotic orbits for $\alpha = 0.144, \beta = 0.05, \zeta = 3.5, \Omega = 3.465, F = 11.40$). (a): Upward attractor (**UA**) $((x(0), \dot{x}(0)) = (3, 3))$; (b): Downward attractor (**DA**) $(x(0), \dot{x}(0)) = (2.5, 2.5)$.

of chaos have been carried out. For instance chaos synchronization is used in engineering communication to, mask the information bearing signals [52-58], help understanding the pressure and flow regulation of neighboring functional units of the kidney [59] and, allows good interpretation in data assimilation for numerical weather prediction models [60].

7.1. Unidirectional Chaos Synchronization

This section aims to investigate synchronization between two coupled chaotic enzyme-substrate reaction. If two of such systems are launched with two different initial conditions, they will lastly circulate either on the different or same chaotic orbit. The enslavement is carried out by coupling the slave to the master as follows:

$$\ddot{x} - \zeta(1 - x^2 + \alpha x^4 - \beta x^6)\dot{x} + x = F\cos\Omega t,$$
$$\ddot{u} - \zeta(1 - u^2 + \alpha u^4 - \beta u^6)\dot{u} + u = F\cos\Omega t + K(u - x)H(t - T_0), \quad (44)$$

The role of the feedback is to force the convergence of the slave towards the master orbit. Two cases are to be considered depending on the Master and Slave positions. Thus, the

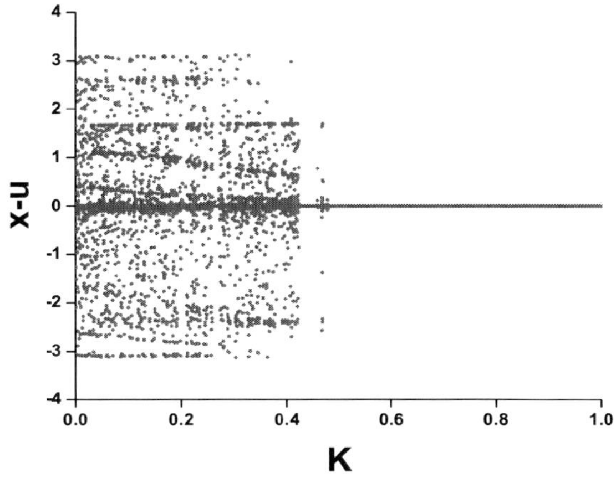

Figure 15. Bifurcation diagram x-u versus the coupling parameter coefficient in an unidirectional coupling configuration with the parameters of figure 13.

chaos synchronization is investigated when both Master and Slave evolve on the same or on different chaotic orbits. In that regards, the bifurcation diagram $x - u$ versus the coupling parameter K and correlated states variables $\frac{dx}{dt}$, $\frac{du}{dt}$ will be evaluated in order to figure out conditions of chaos synchronization.

Thus, if the difference $x - u < 10^{-5}$ or if the plot of correlated variables is parallel to the first diagonal, chaos synchronization takes place in the system.

In the case of two chaotic attractors derived from Fig. 12 with the parameters of Fig. 13, the bifurcation diagram $x - u$ is plotted for $F = 8.27$ in Fig. 15 and it reveals chaos synchronization for $K \in [0.479, 1)$.

As for symmetric chaotic attractors shown in Fig. 14, transitory processes leading to synchronization when the Master and Slave are on the same or different orbits are addresses via the plots of correlated variables. Thus, when both strange attractors are on same orbit, synchronisation occurs slowly on the **DAs** than when they are on the **UAs** as reported in Figs. 16 and 17 respectively.

We now consider the case where phase trajectories of enzyme-substrate reaction belong to two symmetrical chaotic orbits. In the case where the Master is launched on a **UA** and the Slave on the **DA**, Fig. 18 shows the transitory processes leading to their synchronization for $K = 0.98$. The master is launched on the **DA** and the Slave on the **UA** with the set of initial conditions $x(0), \dot{x}(0) = (3.0, 3.0)$ and $u(0), \dot{u}(0) = (2.0, 2.5)$ respectively. Figure 18(a) is plotted for $t \in [400, 500]$ $i.e$ before the coupling takes place in the process and it can be observed that both Master and slave are evolving along the downward and upward chaotic

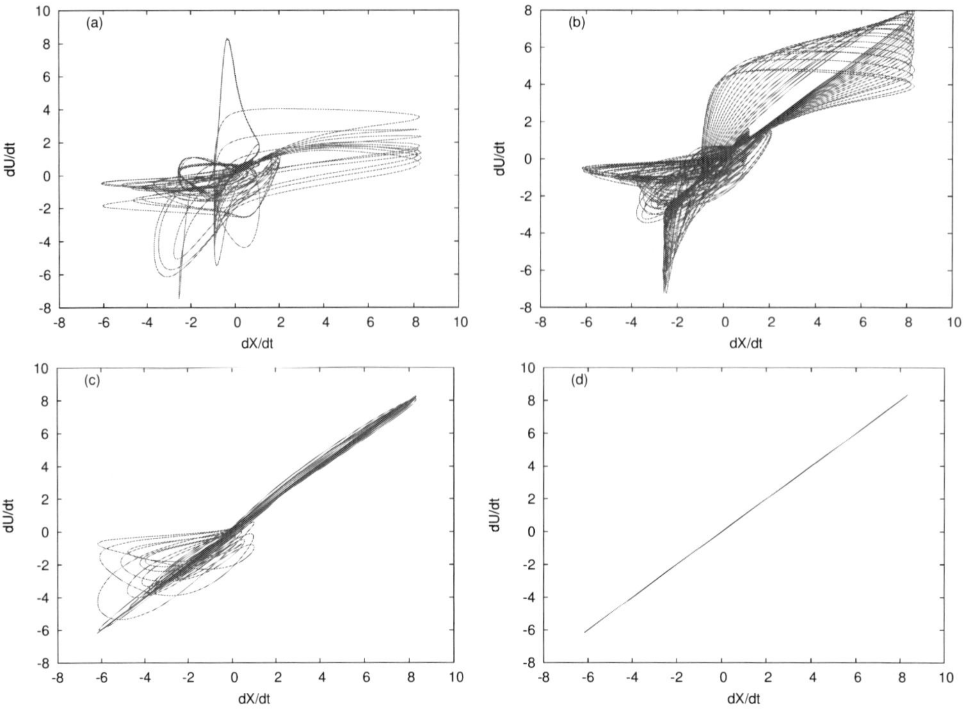

Figure 16. Transitory processes leading to chaos synchronization between two **DAs** for $K = 0.98$. (a): $t \in [580, 650]$; (b): $t \in [650, 800]$; (c): $t \in [800, 1000]$; (d): $t \in [1000, 1200]$.

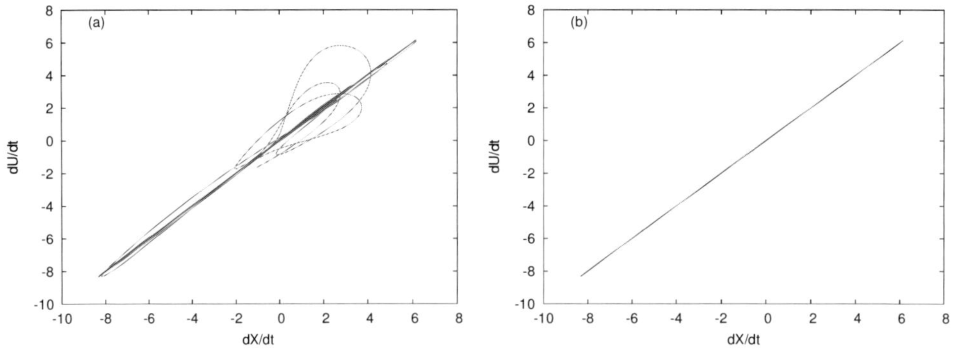

Figure 17. Transitory processes leading to chaos synchronization between two **UAs** for $K = -1.5$. (a): $t \in [580, 650]$; (b): $t \in [650, 800]$.

orbit respectively. When the coupling starts at $T_0 = 600$, the Slave's trajectory begin to be attracted by one of the Master as shown in Figure 18(b) plotted for $t \in [600, 650]$. In Figure 18(c) plotted for $t \in [650, 700]$, the master and slave are on the verge of synchronization.

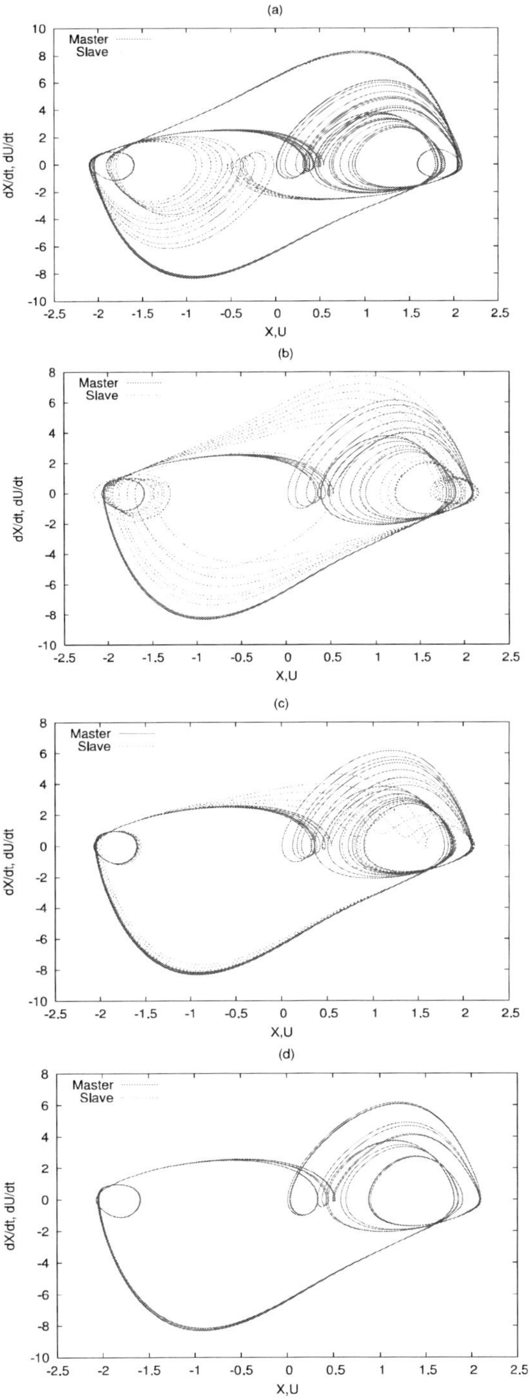

Figure 18. Transitory processes leading to chaos synchronization between two symmetrical attractors in an unidirectional coupling scheme for $K = 0.98$.

The full synchronization finally occurs in Figure 18(d) plotted for $t \in [750, 800]$ where the trajectories and phases of both Master and Slave are fully identical.
The K-dependency of synchronization noticed for non-chaotic oscillations still true even when oscillations are chaotic. Specifically, for any $K > 1$, coupled chaotic enzymatic-substrate reactions can not be synchronized since the Slave will be continuously driven away from the Master as shown in Figure 19.

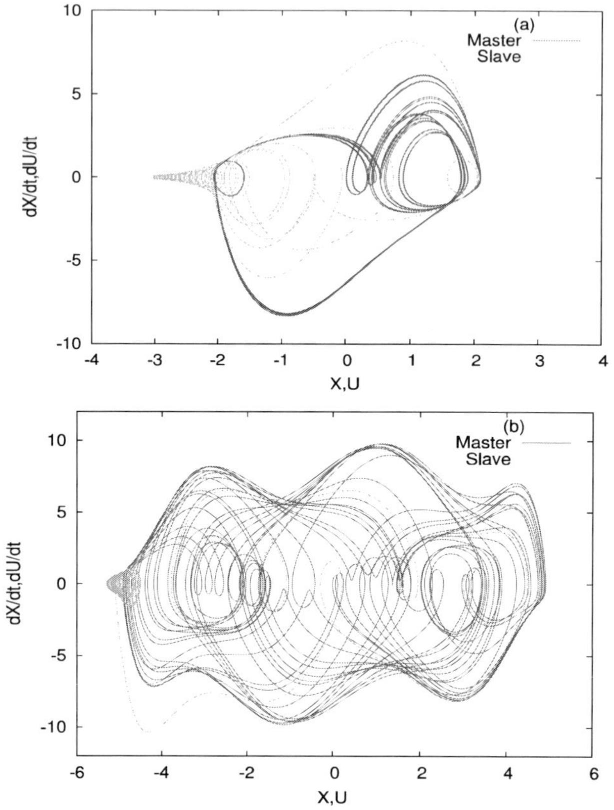

Figure 19. Phase portraits showing the slave continuously drifting away from the master's chaotic orbit. (a): $K = 1.2$; (b): $K = 1.0$.

7.2. Bi-directional Chaos Synchronization

Two different coupling schemes will be considered, namely the symmetric and anti-symmetric configurations. As for the symmetric coupling, enslavement is described by the following equation:

$$\begin{aligned}\ddot{x} - \zeta(1 - x^2 + \alpha x^4 - \beta x^6)\dot{x} + x &= F\cos\Omega t + K(x-u)H(t-T_0),\\ \ddot{u} - \zeta(1 - u^2 + \alpha u^4 - \beta u^6)\dot{u} + u &= F\cos\Omega t + K(u-x)H(t-T_0). \quad (45)\end{aligned}$$

In this case, the bifurcation diagram $x - u$ reveals the existence of chaos synchronization as shown in Fig. 20 for $K \in [0.273, 1)$.

However, for the antisymmetric coupling which is described by the following equation:

$$\begin{aligned}
&\ddot{x} - \zeta(1 - x^2 + \alpha x^4 - \beta x^6)\dot{x} + x = F\cos\Omega t + K(u - x)H(t - T_0),\\
&\ddot{u} - \zeta(1 - u^2 + \alpha u^4 - \beta u^6)\dot{u} + u = F\cos\Omega t + K(u - x)H(t - T_0),
\end{aligned} \tag{46}$$

the bifurcation diagram $x - u$ reveals the lack chaos synchronization (see Fig. 21).

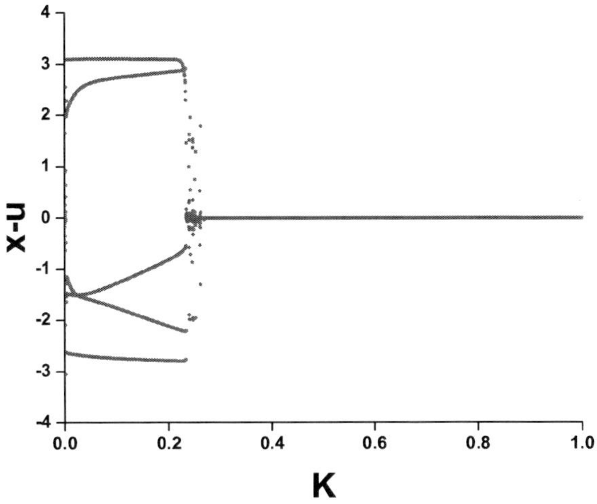

Figure 20. Bifurcation diagram x-u versus the coupling parameter coefficient in the symmetric bi-directional coupling configuration with the parameters of figure 13.

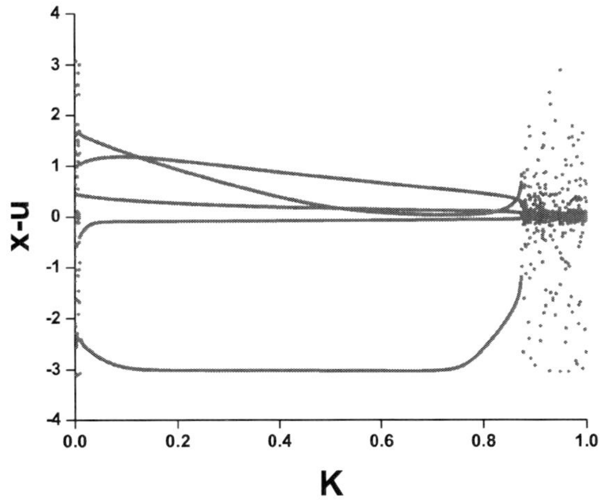

Figure 21. Bifurcation diagram x-u versus the coupling parameter coefficient in the antisymmetric bi-directional coupling configuration with the parameters of figure 13.

8. Conclusion

This chapter has reviewed some key issues occurring in an enzyme-substrate reaction that possesses ferroelectric behavior. The reaction has been modeled by a modified van der Pol oscillator which has the capability of exhibiting two coexisting stable limit cycles and one unstable limit cycle. The equations for amplitudes and frequencies of those cycles have been determined analytically and validated by numerical simulations. The conditions for which two of such reactions can display stable synchronization under both unidirectional and bi-directional coupling schemes have been revealed in autonomous states. When subjected to an external periodic force, transitions to multi-periodic and chaos orbits have been investigated based upon the bifurcation diagram and maximum Lyapunov exponent. In spite of chaos occurrence in the model, synchronization remains achievable for the unidirectional and symmetric bi-directional couplings, but not during antisymmetric bi-directional coupling. This provides some insights regarding coupling configuration and chaos synchronization. Overall, the results obtained here can be of implications in biology, neuroscience and biochemistry. Indeed in biology, the master-slave coupling situation is of interests for example in the case of displacement from one limit cycle to the other. In fact, one supposes that both physiological and pathological systems possess each one a stable limit cycle called respectively the physiological limit cycle and the pathological limit cycle. The purpose of particular controls in biology, with therapeutic goal, is to move a limit cycle, representing a pathological homeostasis until reaching the limit cycle which corresponds to physiological homeostasis. This situation is also found in the case of displacement between two physiological cycles. In a master-slave relationship configuration, the critical values of the coupling strength for which the synchronization process is stable are paramount for providing some leverage to a master. This study is of great interest and needs further investigations such as considering the effects of delay in the enzyme-substrate reaction synchronization since the brain may utilize such a reaction transitions in and out of chaos [48] in its processing to form a complex system [50], displaying self-organization [49], capable of generating new types of structures through bifurcation, a sudden qualitative change in structure occurring at a critical value of a continuous varying parameter. It has been shown that chaotic behavior of enzymes lead to the chaotic inter-beat interval of the human heart and that a regular heartbeat is a sign of disease and a strong predictor of imminent cardiac arrest [62,63]. Moreover, many physiological systems are not normally in a stable state of homeostatic balance characterized by regular motions of enzymes, but in a dynamical stable state characterized by chaotic fluctuation within a certain area of the phase space [62,64].

References

[1] A. H. Nayfeh, D. T. Mook. *Nonlinear Oscillations*, New York: Wiley, (1979).

[2] C. Hayashi, *Nonlinear Oscillations in Physical Systems*, McGraw Hill, New York, (1964).

[3] Y. Kuramoto, *Chemical Oscillations, Waves and Turbulence*, Springer-Verlag, Berlin, (1984).

[4] A. Pikovsky, M. Rosemblum, J. Kurths, *Synchronization: A Universal Concept in Nonlinear Systems*, Cambridge: University Press, (2001).

[5] J. Guckenheimer, P. J. Holmes, *Nonlinear oscillations, dynamical systems and bifurcations of vectors fields*, Berlin: Springer-Verlag, (1984).

[6] P. Hagedorn, *Non-linear Oscillations*, Second edition, Clarendon Press, Oxford, (1988).

[7] A. Goldbeter, *Biochemical oscillations and cellular rhythms: the molecular bases of periodic and chaotic behavior*, Cambridge: Cambridge University Press, (1996).

[8] van der Pol, B. Philos Mag 700, 43, (1922), 978, 72, (1926), 65, 7-3 (1927), *Proc. IRE* 1051, 22 (1934).

[9] H. G. Enjieu Kadji, R. Yamapi, J. B. Chabi Orou, *Chaos* 033113 (2007).

[10] H. G Enjieu Kadji, J. B. Chabi Orou, R. Yamapi, P. Woafo, *Chaos Solitons and Fractals* 32, 862 (2007).

[11] J. C. Chedjou, H. B. Fotsin, P. Woafo, *Phys Scr* 390, 393 (1997).

[12] F. Kaiser, *Coherent excitations in biological systems: specific effects in externally driven self-sustained oscillating biophysical systems*, Berlin, Heidelberg: Springer-Verlag, (1983).

[13] F. Kaiser, C. Eichwald, *Int J Bifurcat Chaos* 485, 491 (1991).

[14] F. Kaiser, C. Eichwald, *Int J Bifurcat Chaos* 711, 715 (1991).

[15] D. Storti, R. H. Hand, *Int. J. Non-Linear Mechanics* 23 (3), 231 (1988).

[16] C. S. Manohar, R. N. Iyengar, *Int. J. Non-Linear Mechanics* 26 (5), 679 (1991).

[17] U. Parlitz, W. Lauterborn, *Phys Rev* A 1428, 1434 (1987).

[18] P. Woafo, R. Kraenkel, *Phys. Rev.* E 65, 036225 (2002).

[19] B. Ryals, L. S. Young, *Chaos* 22, 043140 (2012).

[20] A. Venkatesan, M. Lakshmanan, *Phys. Rev.* E 56, 6324 (1997).

[21] A. P. Kuznetsov, N. V. Stankevich, L. V. Turukina, *Physica* D 238(14), 1203 (2009).

[22] A. Chudzik, P. Perlikowski, A. Stefanski, T. Kapitaniak, *Int. J. Bifurcation Chaos* 21, 1907 (2011).

[23] H. G. Enjieu Kadji, R. Yamapi, *Physica* A 370 (2), 316 (2006).

[24] A. Cheage Chamgoue, R. Yamapi, P. Woafo, *Eur. Phys. J. Plus* 127, 59 (2012).

[25] F. Kaiser, Z *Naturforsch* A 294, 304 (1978).

[26] H. Frohlich, *Int J Quantum Chem* 641, 649 (1988).

[27] H. Frohlich, *Quantum mechanical concepts in biology, in theoretical physics and biology*, p. 1322, (1969).

[28] V. Voltera, *Lecons sur la theorie mathenatique de la lutte pour la vie*, Paris, Gauthier-Villars, (1931).

[29] F. Kaiser, Coherent Modes in Biological Systems": In *Biological Effects of Nonionizing Radiation*, ed. Illinger, K. H., A.C.S Symp. Series 157, 219 (1981).

[30] O. Decroley and A. Goldbeter, *Proc. Natl. Acad. Sci.* U.S.A 79, 6917 (1982).

[31] A. Goldbeter, D. Gonze, G. Houart, J. C. Leloup, J. Halloy, G. Dupont, *Chaos* 11, 1, 247 (2001).

[32] A. K. Gustavsson, D. D. van Niekerk, C. B. Adiels, F. B. du Preez, M. Goksor, J. L. Snoep, *FEBS Journal* 279 (16), 2837 (2012).

[33] J. C. Leloup, D. Gonze, A. Goldbeter, *J. Biol. Rhytms* 14, 433 (1999).

[34] J. C. Leloup, A. Goldbeter, *J. Theor. Biol.* 198, 445 (1999).

[35] W. Abou-Jaoude, M. Chaves, J. L. Gouze, *PLoS One* 6 (2), e17075 (2011).

[36] T. Haberichter, M. Marhl, R. Heinrich, *Biophys*. Chem. 90, 17 (2001).

[37] Y-X. Li, A. Goldberter, *J. Theor. Biol.* 138, 149 (1989).

[38] A. Goldbeter, J. L. Martiel, *FEBS Lett.* 191, 149 (1985).

[39] A. Goldbeter, F. Moran, Complex pattern of excitability and oscillations in biochemical system, From *The Organization of Cell Metabolism,* Edited by G. Rickey Welch and James S. Clegg, Plenum Publishing Corporation, (1987).

[40] R. E. Mirollo, S. H. Strogatz, *SIAM J. Appl. Math.* 50, 1645 (1990).

[41] S. A. Campbell, R. Edwards, P. V. den Driessche, *SIAM J. Appl. Math.* 65, 316 (2004).

[42] H. K. Leung, *Phys. Rev.* E 58, 5704 (1998).

[43] M. I. Rabinovich, H. D. I Abarbanel, *Neuroscience* 87 (1), 5 (1998).

[44] G. E. Louridas, A. G. Louridas, *Int. J. Applied Science and Technology* 2 (7), 24 (2012).

[45] K. M. Hampson, E. A. H. Mallen, *Biomedical Optics Express* 3 (5), 863 (2012)

[46] S. J. Schiff, K. Jerger, D. H. Duong, T. Chang, M. L. Spano, W. L. Ditto, *Nature* 370, 615 (1994).

[47] H. J. Schuster, *Deterministic chaos*, Berlin, Springer-Verlag (1986).

[48] C. C. King, *Progr Neurobiol* 279, 308 (1991).

[49] S. Barton, *Am Psychol* 5, 14 (1994).

[59] R. Ruthen, *Sci Am* 110, 117 (1993).

[51] L. M. Pecora, T. L. Carroll, *Phys. Rev. Lett.* 64, 821 (1990).

[52] L. Kocarev, U. Parlitz, *Phys. Rev. Lett.* 5028, 5031 (1995).

[53] B. Nana, P. Woafo, S. Domngang, *Commun Nonlinear Sci Numer Simulat* 2266, 2276 (2009).

[54] W. Kinzel, A. Englert, I. Kanter, *Phil. Trans. R. Soc.* A 376, 389 (2010).

[55] J. Ohtsubo, *Semiconductor Lasers* 111, 463 (2013).

[56] S. Petit-Watelot, J. V. Kim, A. Ruotolo, R. M. Otxoa, K. Bouzehouane, J. Grollier, A. Vansteenkiste, B. V. de Wiele, V. Cros, T. Devolder, *Nature Physics* 8, 682 (2012).

[57] J. Pan, Q. Ding, B. Du, *Int. J. Bifurcation Chaos* 22, 1250125 (2012).

[58] N. Jiang, C. Zhang, K. Qiu, *Optics Letters* 37 (21), 4501 (2012).

[59] E. Mosekilde, J. L. Laugesen, Zh. T. Zhusubaliyev, *AIP Conf. Proc.* 1468, 276 (2012).

[60] Y. H. Zhao, X. C. Zhao, W. Cheng, *Advanced Materials Research* 614-615, 866 (2012).

[61] D. A. Hsieh, *The Journal of Finance*, DOI: 10.1111/j.1540-6261.1991.tb04646.x (2012).

[62] A. L. Goldberger, D. R. Rigney, B. J. West, *Sci Am* 42, 49 (1990).

[63] C. K. Peng, J. Mietus, J. M. Hausdorff, S. Havlin, H. E. Stanley, A. L. Goldberger, *Phys. Rev. Lett.* 1343, 1346 (1993).

[64] A. L. Goldberger, *Perspect Biol Med* 40, 543 (1997).

In: Dynamical Systems
Editor: Jean Michel Tchuenche, pp. 235-259
ISBN: 978-1-62808-001-8

Chapter 9

Dynamics of Low and High Pathogenic Avian Influenza in Wild Bird Population

***Necibe Tuncer*[1,*], *Juan Torres*[2,†] *and Maia Martcheva*[2,‡]**
[1]Department of Mathematics, University of Tulsa, Oklahoma, USA
[2]Department of Mathematics, University of Florida, Gainesville, Florida, US

Abstract

This chapter introduces an avian influenza model which includes the dynamics of low pathogenic avian influenza (LPAI) and high pathogenic avian influenza (HPAI). The model structures the LPAI-recovered individuals by time-since-recovery and involves the cross-immunity that LPAI infection generates toward the HPAI. Reproduction numbers ($\mathcal{R}_0^{L_w}, \mathcal{R}_0^{H_w}$) and invasion reproduction numbers ($\hat{\mathcal{R}}_{H_w}, \hat{\mathcal{R}}_{L_w}$) of LPAI and HPAI are computed. It is shown that the system has a unique disease-free equilibrium that is locally and globally stable if $\mathcal{R}_0^{L_w} < 1$ and $\mathcal{R}_0^{H_w} < 1$. If $\mathcal{R}_0^{L_w} > 1$ a unique LPAI dominance equilibrium exists. Similarly, if $\mathcal{R}_0^{H_w} > 1$ a unique HPAI dominance equilibrium exists. The equilibria are locally stable if $\hat{\mathcal{R}}_{H_w} < 1$ ($\hat{\mathcal{R}}_{L_w} < 1$ correspondingly). A unique coexistence equilibrium is present if both invasion numbers are larger than one. Simulations show that this coexistence equilibrium can lose stability and coexistence in the form of sustained oscillations is possible. Cross-immunity and duration of protection increase the probability of coexistence. Simulations also show that increasing LPAI transmission increases LPAI prevalence and decreases HPAI prevalence. This observation in part may explain why wild birds which have much higher transmission of LPAI compared to domestic birds also have much lower prevalence of HPAI.

Keywords: mathematical models, age-structured differential equations, reproduction number, invasion number, LPAI, HPAI, H5N1, avian influenza

*E-mail address: necibe-tuncer@utulsa.edu
†E-mail address: torres@math.ufl.edu
‡E-mail address: maia@ufl.edu

1. Introduction

Avian (bird) Influenza (flu) viruses belong to a group of viruses called Influenza A. There are three types of influenza viruses: Influenza A, B and C. Influenza A viruses infect many different avian and mammalian species including humans [13]. Humans can be infected with all three types of influenza viruses while birds can be only infected with Influenza A virus [45]. Influenza A strains are classified by their surface proteins: *haemagglutinin* (HA) and *neuraminidase* (NA). Majority of all HA/NA combinations have been isolated in wild birds especially waterfowls and shorebirds [1, 45]. The subtype HA has 16 distinct molecules (H1-H16) and NA has 9 distinct molecules (N1-N9). The H and N combination names the subtype, for instance the virus that caused one of the deadliest pandemics in the history, which is known as the "Spanish Flu," was H1N1 subtype. Two other major influenza pandemics have occurred in the 20th century : "Asian Influenza" caused by H2N2 subtype in 1957 and "Hong Kong Influenza" caused by H3N2 subtype in 1968.

An influenza pandemic happens when a new influenza subtype spreads in the human population. Influenza viruses have the capability of evolving rapidly and jumping between species. They evolve through two evolutionary mechanisms: *drift* and *shift*. Drift is small and gradual changes on the surface proteins (antigens) which occur both in Influenza A and Influenza B viruses. Shift evolution occurs through reassortment which is the mixing of two influenza strains into a new strain with capabilities of both strains. Since Influenza A viruses infect many different species, shift occurs only in Influenza A type viruses. The research on influenza subtypes H2N2 and H3N2 which caused Asian and Hong Kong influenza respectively, recovered that these subtypes had surface protein genes almost certainly from both influenza virus of avian and human origins [4, 37]. This evidence indicates that antigenic shift occurred through reassortment of both avian and human influenza viruses. On the other hand, influenza subtype H1N1 that caused the "Spanish Flu" was more related to avian influenza than to influenza from any other species which suggested that an avian influenza virus was adapted to human-to-human transmissible pathogen and caused a pandemic [2]. It is highly possible that the next pandemic influenza can be caused by avian influenza.

The pandemic potential of highly pathogenic avian influenza of subtype H5N1 leads to continued concerns. The first human avian influenza case of subtype H5N1 appeared in Hong Kong in 1997, which caused the death of a boy. Since then, as of September 2012, there are total of 608 reported H5N1 human cases worldwide, out of which 359 resulted in death. Even though, the number of H5N1 infected human cases is small, the case fatality rate, which is approximately 60% is very high. Highly pathogenic avian influenza of subtype H5N1 is currently at the top of the list for a pandemic threat. The main reason for the high pandemic potential is that the virus is capable of rapid evolution and at some point might emerge as an effective human-to-human transmissible pathogen and cause a pandemic.

Avian influenza viruses are further classified into two groups based on their ability to cause disease: low pathogenic avian influenza (LPAI) and high pathogenic avian influenza (HPAI). Viruses of most subtypes persist in low pathogenic form, typically producing asymptomatic or mild illness in wild or domestic birds. Studies suggest that LPAI strains from the H5 and H7 subtypes circulating in wild birds can evolve into strains of Highly

Pathogenic Avian Influenza (HPAI), after spillover infection to domestic birds [1, 8, 38, 42]. The World Organization for Animal Health (OIE) defines a virus of the H5 or H7 subtype as a *highly pathogenic avian influenza virus* if it can cause at least 75% mortality in 4-week to 8-week old chickens infected intravenously [47, 42].

1.1. Low Pathogenic Avian Influenza

The host range of AI in wild birds is not known but some species of aquatic birds (such as ducks, geese, and shorebirds), serve as natural reservoirs of AI viruses [8, 41]. All subtypes of AI viruses isolated this far persist in wild birds as low pathogenic forms. LPAI viruses of the H5 subtype, which is of primary interest here, are generally reported at very low prevalence rates in ducks [23, 33], but the prevalence may vary by location or year. LPAI viruses have been reported in domestic birds, most frequently turkeys, ducks and chickens. Although LPAI viruses of many subtypes can be found in poultry, viruses from the H5 (and H7) subtypes are most frequently reported there [43].

1.2. High Pathogenic Avian Influenza

HPAI was first described in poultry in Italy in 1878 [29]. Unlike the low pathogenic influenza, which mostly affects the respiratory tract, the highly pathogenic form infects multiple organs and systems of infected birds. In poultry, HPAI is characterized by high mortality rate, often over 80% within 48 hours. Before 2002, HPAI was rarely found in wild birds. For this reason, it is still an open question whether HPAI viruses are endemic in wild birds. Since 2002, HPAI has been isolated from multiple species of wild birds [41]. The HP H5N1, which is now the main cause of concern, is believed to have emerged from an LPAI virus circulating in chickens sometime in 1996 [46]. After potentially undergoing additional mutation, the HPAI H5N1 virus infected 18 people in Hong Kong in 1997, six of whom died [9]. Since 2003 HPAI H5N1 has been regularly infecting humans primarily though bird-to-human infection. The emergence of a virus that can pass directly from birds to humans showed that pigs are not a necessary link in the chain. Since then, viruses of other subtypes have also made the transition LPAI→HPAI and can pose a significant threat [37].

1.3. Cross Immunity

Infection with one strain of Influenza A type viruses provides cross protection against infection with antigenically similar strains. The partial protection offered by the cross immunity is high only if the new strain infecting the host and the strain responsible for immunity are closely related. The protection is ineffective against viruses with major antigenic divergence [6, 7]. However, the unique evolutionary capabilities of Influenza type A viruses (drift and shift) still causes a concern, since prior infection may only reduce the virus spread but can not prevent the infection [10]. Studies in [40] and [10] suggest that prior infection with LPAI provides cross protection against infection with HPAI.

Just before the first H5N1 human case in 1997, there was H5N1 poultry outbreak in chicken farms in Hong Kong. Even though, evidence showed the presence of H5N1 virus, most of the chickens in the poultry markets did not show any symptoms and appeared healthy [39]. Seo and Webster, motivated by this puzzling situation, set up an experiment to test the hypothesis that prior infection with H9N1 (which is LPAI) provides partial protection against highly pathogenic avian H5N1 infections [40]. There are mainly two types of immunity caused by prior infection: cell mediated immunity and antibody mediated immunity. The prior infection with LPAI provides cell mediated immunity toward HPAI [40, 10]. The protection mediated by cellular immune response is established by $CD8^+$ T cells. These findings of the research done by the authors in [40, 10] indicate that it is possible to induce cell mediated immunity toward high pathogenic avian influenza viruses by low pathogenic avian influenza. In terms of control measures, it is promising to develop vaccines that emphasize cell mediated cellular immunity.

1.4. Mathematical Modeling of Avian Influenza

Avian influenza is perhaps the most dangerous disease linking humans and animals at present. Because of its deadly pandemic potential, even in its current pre-pandemic stage it has caused significant hardship and economic loss [35]. In the last 5-10 years avian influenza has enjoyed significant attention from mathematical modelers, largely due to its key position among infectious diseases.

Early models of avian influenza focused on humans, investigating the potential impact of a hypothetical pandemic and exploring strategies for its possible mitigation [12, 15, 14, 27, 32]. Other models focused on the present status quo centered on infection of domestic birds, and current control strategies which primarily target poultry. For example, a spatial farm-based model treating poultry-farms as units [24] and SIR models for within-flock transmission of H5N1 [44] were developed. Despite the importance of a number of emergent diseases, many of which arise from spillover infections from animals, few models to date involve both animals and humans linked by a pathogen. This situation has been changing recently, particularly in relation to AI. The simplest model that captures a bird-to-human transmission pathway of HPAI involves domestic birds and humans [18]. The dynamics of HPAI in poultry are given by a simple SI model, as infected domestic birds either die from the infection or are culled to prevent further spread. To introduce the model, we denote the number of susceptible domestic birds by $S_d(t)$, and the number of domestic birds infected by HPAI by $I_{H_d}(t)$. Furthermore, we denote susceptible humans by $S(t)$, and humans infected with HPAI by $I(t)$. The AI model consists of the following two systems:

$$\textbf{Domestic birds:} \qquad \begin{cases} \dfrac{dS_d}{dt} = \Lambda_d - \beta_{H_d} I_{H_d} S_d - \mu_d S_d, \\ \dfrac{dI_{H_d}}{dt} = \beta_{H_d} I_{H_d} S_d - (\mu_d + \nu_{H_d}) I_{H_d} \end{cases} \tag{1.1}$$

where the parameters for the domestic bird population are: Λ_d – the recruitment rate of domestic birds, μ_d – the natural death rate of domestic birds, β_{H_d} – the transmission coefficient of HPAI among domestic birds, and ν_{H_d} – the death rate of domestic birds due to HPAI.

$$\textbf{Humans:} \qquad \begin{cases} \dfrac{dS}{dt} = \Lambda - \beta I_{H_d} S - \mu S \\ \dfrac{dI}{dt} = \beta I_{H_d} S - (\nu + \alpha + \mu) I \end{cases} \tag{1.2}$$

The parameters related to human dynamics are: Λ – the birth/recruitment rate of humans, μ – the natural death rate of humans, β – the transmission coefficient of HPAI from birds to humans, ν – the death rate of humans due to HPAI, and α – the recovery rate. Model (1.1-1.2) has very simple dynamics. However, it fails to capture most of the complexity and characteristic features of AI transmission and evolution. Recognizing the importance of both birds and humans in the transmission and evolution of AI, a number of models involving domestic birds and humans [18, 20, 19, 22] were developed. Some more elaborate models even involve wild birds, as well as domestic birds and humans [28, 26, 16, 3]. These models typically incorporate the hypothetical scenario in which HP H5N1 avian influenza becomes adapted to humans and starts transmitting efficiently from human-to-human. This models the shift evolutionary mechanism. The drift evolutionary mechanism was originally modeled by Pease [36]. Pease's drift model has recently been extensively studied [17, 30]. A novel model combining effects of both drift and shift evolution of influenza A was discussed by Martcheva [31], where it was found that drift evolution may be responsible for the 365-day oscillation of human flu, as well as 365-day oscillations of the number of humans infected with HP H5N1, as observed in data.

In this chapter we investigate the dynamics of LPAI and HPAI in birds. The interaction between LPAI and HPAI in birds has been priorly studied by several authors [28, 5] but the effect of cross-immunity that LPAI gives to HPAI has not been investigated.

2. The Model

Influenza strains can compete in two ways: depletion of susceptible hosts, and cross-immunity [6, 7, 34]. Cross-immunity is a mechanism by which infection with one strain provides partial protection against infection with another. A number of references in the biological literature suggest that infection with LPAI can provide partial protection against HP H5N1 both in poultry and in wild birds. Seo and Webster [40] infected a group of chickens with LPAI H9N2 influenza virus. They found that the group infected by H5N1 within 30 days of inoculation by LPAI had a 100% survival rate and reduced clinical signs. As time between the two infections grew, the protection started to fade and the morbidity of infection with HPAI grew. This study suggests that the immunity provided by prior infection with LPAI is temporary, declining with time-since-recovery from LPAI. To incorporate this variability of cross-immunity in wild birds, we let τ be the time-since-recovery from low pathogenic influenza and $q_w(\tau)$ be the variable cross-immunity (high cross-immunity is modeled by low q_w) imparted from LPAI to infection with HPAI. To introduce the model, let the birth/recruitment rate of wild birds be denoted by Λ_w and the natural mortality rate for wild birds by μ_w. The number of susceptible wild birds is $S_w(t)$, the number of LPAI infected wild birds is $I_{L_w}(t)$, and the number of HPAI infected wild birds is $I_{H_w}(t)$. The density of individuals that have recovered from LPAI is $r_{L_w}(\tau, t)$ and the number of individuals that have recovered from HPAI is $R_{H_w}(t)$. The dynamics of LPAI and HPAI in wild birds are captured by the following model

$$\begin{cases} \dfrac{dS_w}{dt} &= \Lambda_w - \beta_{11}^L I_{L_w} S_w - \beta_{11}^H I_{H_w} S_w - \mu_w S_w, \\ \dfrac{dI_{L_w}}{dt} &= \beta_{11}^L I_{L_w} S_w - (\mu_w + \alpha_{L_w}) I_{L_w} \\ \dfrac{\partial r_{L_w}}{\partial t} + \dfrac{\partial r_{L_w}}{\partial \tau} &= -q_w(\tau)\beta_{11}^H I_{H_w} r_{L_w} - \mu_w r_{L_w} \\ r_{L_w}(0,t) &= \alpha_{L_w} I_{L_w} \\ \dfrac{dI_{H_w}}{dt} &= \beta_{11}^H I_{H_w} S_w \\ & \quad + \beta_{11}^H I_{H_w} \displaystyle\int_0^\infty q_w(\tau) r_{L_w}(\tau,t)d\tau - (\mu_w + \alpha_{H_w} + \nu_{H_w}) I_{H_w} \\ \dfrac{dR_{H_w}}{dt} &= \alpha_{H_w} I_{H_w} - \mu_w R_{H_w}. \end{cases} \tag{2.1}$$

where α_{L_w} and α_{H_w} are the recovery rates for wild birds, and ν_{H_w} is the HPAI-induced mortality. A simplifying assumption is that most wild birds that recover from HPAI do so because they may have had prior exposure to LPAI. This assumption is justified based on the experimental studies that suggest that LPAI strains of the same subtype or different subtype induce partial cross-immunity toward infection with HPAI strains [11, 21]. We make the simplifying assumption that recruitment does not depend upon focal bird numbers. This could be relaxed in future, for instance by introducing a logistic growth term.

In the following section we analyze the dynamics of low pathogenic and high pathogenic avian influenza within the wild bird population. The model that captures the interaction of LPAI and HPAI in poultry is quite similar. Therefore, symmetrical results are true for the domestic bird population.

3. LPAI-HPAI Dynamics in Wild Birds

We study the existence and stability of equilibria of the wild bird system (2.1). To determine the equilibria, we solve the following system which is obtained by setting the time derivatives in (2.1) equal to zero.

$$\begin{cases} 0 &= \Lambda_w - \beta_{11}^L I_{L_w} S_w - \beta_{11}^H I_{H_w} S_w - \mu_w S_w, \\ 0 &= \beta_{11}^L I_{L_w} S_w - (\mu_w + \alpha_{L_w}) I_{L_w} \\ \dfrac{\partial r_{L_w}}{\partial \tau} &= -q_w(\tau)\beta_{11}^H I_{H_w} r_{L_w} - \mu_w r_{L_w} \\ r_{L_w}(0) &= \alpha_{L_w} I_{L_w} \\ 0 &= \beta_{11}^H I_{H_w} S_w \\ & \quad + \beta_{11}^H I_{H_w} \displaystyle\int_0^\infty q_w(\tau) r_{L_w}(\tau,t)d\tau - (\mu_w + \alpha_{H_w} + \nu_{H_w}) I_{H_w} \\ 0 &= \alpha_{H_w} I_{H_w} - \mu_w R_{H_w}. \end{cases} \tag{3.1}$$

The wild bird system has 4 equilibria. The first is the disease-free equilibrium. The second and the third are the LPAI-only and HPAI-only equilibria (i.e. boundary equilibria). The fourth one is the coexistence equilibrium (i.e. interior equilibrium) which represents the state in which both LPAI and HPAI are present in the wild bird population.

3.1. Disease-Free Equilibrium

The wild-bird-model (2.1) has a disease-free equilibrium (DFE) ε_0 given by

$$\varepsilon_0 = (S_w^*, 0, 0, 0, 0)\,,$$

where $S_w^* = \frac{\Lambda_w}{\mu_w}$. The basic reproduction number for LPAI in wild birds, denoted by $\mathcal{R}_0^{L_w}$, is given by

$$\mathcal{R}_0^{L_w} = \frac{\beta_{11}^L \Lambda_w}{\mu_w(\mu_w + \alpha_{L_w})}\,,$$

and the basic reproduction number for HPAI in wild birds, denoted by $\mathcal{R}_0^{H_w}$, is given by

$$\mathcal{R}_0^{H_w} = \frac{\beta_{11}^H \Lambda_w}{\mu_w(\mu_w + \alpha_{H_w} + \nu_{H_w})}\,.$$

These basic reproduction numbers are threshold values which determine whether LPAI or HPAI can invade the disease-free equilibrium.

Theorem 1. *If $\mathcal{R}_0^{L_w} < 1$ and $\mathcal{R}_0^{H_w} < 1$ then the DFE, ε_0, is locally asymptotically stable.*

Proof. Let (u, v, x, y, z) be the perturbations around the steady state. Expressing the perturbations as

$$S_w(t) = S_w^* + u(t)\,, \quad I_{L_w}(t) = v(t)\,, \quad r_{L_w}(\tau, t) = x(\tau, t)\,, \quad I_{H_w}(t) = y(t)\,, \quad R_{H_w}(t) = z(t)\,,$$

we substitute into the PDE system (2.1). Using the equation for the disease-free equilibrium and dropping the quadratic terms in the perturbations, we obtain the following linear system involving only perturbations.

$$\begin{aligned}
\frac{du}{dt} &= \beta_{11}^L S_w^* v - \beta_{11}^H S_w^* y - \mu_w u,\\
\frac{dv}{dt} &= \beta_{11}^L S_w^* v - \beta_{11}^H S_w^* y - (\mu_w + \alpha_{L_w})v\\
\frac{\partial x}{\partial t} + \frac{\partial x}{\partial \tau} &= -\mu_w x\\
x(0,t) &= \alpha_{L_w} v\\
\frac{dy}{dt} &= \beta_{11}^H S_w^* y - (\mu_w + \alpha_{H_w} + \nu_{H_w})y\\
\frac{dz}{dt} &= \alpha_{H_w} y - \mu_w z
\end{aligned} \tag{3.2}$$

To investigate the local stability of the DFE, we study the solutions of the system (3.2). Suppose that the linear system (3.2) has exponential solutions, that is we look for solutions of the following form:

$$u = \bar{u}e^{\lambda t}, \quad v = \bar{v}e^{\lambda t}, \quad x = \bar{x}(\tau)e^{\lambda t}, \quad y = \bar{y}e^{\lambda t}, \quad z = \bar{z}e^{\lambda t}. \tag{3.3}$$

Substituting the above solutions into linearized system (3.2), we get the following eigenvalue problem;

$$\begin{aligned}
\lambda\bar{u} &= -\beta_{11}^L S_w^* \bar{v} - \beta_{11}^H S_w^* \bar{y} - \mu_w \bar{u}, \\
\lambda\bar{v} &= \beta_{11}^L S_w^* \bar{v} - (\mu_w + \alpha_{L_w})\bar{v} \\
\lambda\bar{y} &= \beta_{11}^H S_w^* \bar{y} - (\mu_w + \alpha_{H_w} + \nu_{H_w})\bar{y} \\
\lambda\bar{z} &= \alpha_{H_w}\bar{y} - \mu_w \bar{z}
\end{aligned} \tag{3.4}$$

which is combined with the following first order ODE;

$$\lambda\bar{x} + \frac{d\bar{x}}{d\tau} = -\mu_w \bar{x}, \quad \bar{x}(0) = \alpha_{L_w}\bar{v}.$$

Solving the above differential equation, we obtain;

$$\bar{x} = \alpha_{L_w}\bar{v}e^{-(\lambda+\mu_w)\tau}.$$

Solutions of (3.4) gives the eigenvalues $\{\lambda_i\}_{i=1}^4$ of the linearized differential operator in (3.2). The eigenvalue problem (3.4) is linear. The second equation involves only $\bar{v}$ and is independent from $\bar{u}$, $\bar{y}$ and $\bar{z}$. Similar statement is true for the third equation involving only $\bar{y}$. Solving the second equation we get

$$\lambda_2 = \beta_{11}^L S_w^* - (\mu_w + \alpha_{L_w}),$$

which is negative since $\mathcal{R}_0^{L_w} < 1$. Solving the third equation we get $\lambda_3 = \beta_{11}^L S_w^* - (\mu_w + \alpha_{H_w} + \nu_{H_w})$. Clearly, $\lambda_3 < 0$, since $\mathcal{R}_0^{H_w} < 1$. The other two eigenvalues are $\lambda_1 = \lambda_4 = -\mu_w < 0$. □

The basic reproduction number $\mathcal{R}_0^{L_w}$ ($\mathcal{R}_0^{H_w}$) measures the average number of new low (high) pathogenic infections generated by a single wild bird infected with low (high) pathogenic avian influenza in a completely susceptible wild bird population. Thus, Theorem 1 implies that the LPAI and HPAI can be eliminated from the wild bird population if $\mathcal{R}_0^{L_w} < 1$ and $\mathcal{R}_0^{H_w} < 1$ and if initially the number of wild birds infected with LPAI and HPAI are in the the basin of attraction of the DFE.

3.1.1. Global Stability of the Disease-Free Equilibrium

Now, we prove the global asymptotic stability of the disease-free equilibrium ε_0.

Theorem 2. *If $\mathcal{R}_0^{L_w} < 1$ and $\mathcal{R}_0^{H_w} < 1$ then the DFE, ε_0, is globally asymptotically stable.*

Proof. From the first equation in the system (2.1) we obtain the following inequality

$$S'_w \leq \Lambda_w - \mu_w S_w.$$

From this inequality we have

$$\limsup_t S_w(t) \leq \frac{\Lambda_w}{\mu_w}. \tag{3.5}$$

Integrating the second equality in system (2.1) we have

$$I_{L_w}(t) = e^{-(\mu_w+\alpha_{L_w})t} I_{L_w}(0) + \beta_{11}^L \int_0^t e^{-(\mu_w+\alpha_{L_w})(t-\sigma)} S_w(\sigma) I_{L_w}(\sigma) d\sigma.$$

Changing the variable of integration in the integral, we have

$$I_{L_w}(t) = e^{-(\mu_w+\alpha_{L_w})t} I_{L_w}(0) + \beta_{11}^L \int_0^t e^{-(\mu_w+\alpha_{L_w})\sigma} S_w(t-\sigma) I_{L_w}(t-\sigma) d\sigma.$$

Taking a $\limsup$ of both sides of this equality, we obtain the following inequality:

$$\limsup_t I_{L_w}(t) \leq \frac{\beta_{11}^L \Lambda_w}{\mu_w(\mu_w+\alpha_{L_w})} \limsup_t I_{L_w}(t)$$

where the coefficient infront the $\limsup$ on the right hand side is exactly $\mathcal{R}_0^{L_w}$. Since $\mathcal{R}_0^{L_w} < 1$, this inequality is only possible if $\limsup_t I_{L_w}(t) = 0$. Hence, $I_{L_w}(t) \to 0$ as $t \to \infty$. Next, we integrate the partial differential equation along the characteristic lines. We obtain,

$$r_{L_w}(\tau,t) = \begin{cases} \alpha_{L_w} I_{L_w}(t-\tau) e^{-\int_0^\tau q_w(\sigma) I_{H_w}(t-\tau+\sigma)d\sigma - \mu_w \tau} & \tau < t \\ r^0(\tau-t) e^{-\int_0^t q_w(\tau-t+\sigma) I_{H_w}(\sigma)d\sigma - \mu_w t} & \tau > t \end{cases} \tag{3.6}$$

where $r^0(\tau) = r_{L_w}(\tau,0)$. Consider the term $\int_0^\infty q_w(\tau) r_{L_w}(\tau,t)d\tau$. We claim

$$\limsup_t \int_0^\infty q_w(\tau) r_{L_w}(\tau,t) d\tau = 0.$$

Indeed, using (3.6) we have

$$\begin{aligned} \limsup_t & \int_0^\infty q_w(\tau) r_{L_w}(\tau,t) d\tau \\ & \leq \alpha_{L_w} \int_0^t q_w(\tau) I_{L_w}(t-\tau) e^{-\mu_w \tau} d\tau + \int_t^\infty q_w(\tau) r^0(\tau-t) e^{-\mu_w t} d\tau \qquad (3.7) \\ & \leq \alpha_{L_w} \int_0^t q_w(\tau) I_{L_w}(t-\tau) e^{-\mu_w \tau} d\tau + e^{-\mu_w t} \int_0^\infty r^0(\tau) d\tau. \end{aligned}$$

Taking the $\limsup$ from both sides of the above inequality gives the claim. We denote by

$$Q(t) = \int_0^\infty q_w(\tau) r_{L_w}(\tau,t) d\tau.$$

To conclude the proof, we integrate the equation for high pathogenic influenza and change the variable of integration.

$$I_{H_w}(t) = e^{-(\mu_w+\alpha_{H_w}+\nu_{H_w})t} I_{H_w}(0) + \beta_{11}^H \int_0^t e^{-(\mu_w+\alpha_{H_w}+\nu_{H_w})\sigma} S_w(t-\sigma) I_{H_w}(t-\sigma) d\sigma$$
$$+\beta_{11}^H \int_0^t e^{-(\mu_w+\alpha_{H_w}+\nu_{H_w})\sigma} Q(t-\sigma) I_{H_w}(t-\sigma) d\sigma. \tag{3.8}$$

Taking $\limsup$ of both sides of the above equality, we have

$$\limsup_t I_{H_w}(t) \leq \frac{\beta_{11}^H \Lambda_w}{\mu_w(\mu_w + \alpha_{H_w} + \nu_{H_w})} \limsup_t I_{H_w}(t).$$

The coefficient on the right-hand side infront $\limsup_t I_{H_w}(t)$ is exactly $\mathcal{R}_0^{H_w}$. Since $\mathcal{R}_0^{H_w} < 1$, the only way the above inequality can hold is if $\limsup_t I_{H_w}(t) = 0$, that is if $I_{H_w}(t) \to 0$ as $t \to \infty$. That completes the proof.

□

3.2. LPAI-only and HPAI-only Equilibria

We study the competition of the low and high pathogenic avian influenza in wild bird population. Like the basic reproduction number, the invasion number is a threshold quantity that determines if one pathogen can invade the other pathogen's equilibrium. The invasion numbers are very useful in understanding the dynamics between low and high pathogens in wild bird population. We denote by $\hat{\mathcal{R}}_{L_w}$ the invasion number of LPAI when the system is at HPAI-only equilibrium. The invasion number of LPAI is

$$\hat{\mathcal{R}}_{L_w} = \frac{\beta_{11}^L(\mu_w + \alpha_{H_w} + \nu_{H_w})}{\beta_{11}^H(\mu_w + \alpha_{L_w})} = \frac{\mathcal{R}_0^{L_w}}{\mathcal{R}_0^{H_w}}.$$

The invasion number $\hat{\mathcal{R}}_{L_w}$ gives the ability of LPAI to invade the HPAI-only equilibrium which is measured as the secondary infections one LPAI-infected wild bird can produce in a wild bird population where HPAI is at equilibrium. We denote by $\hat{\mathcal{R}}_{H_w}$ the invasion number of HPAI. Similar definition is true for the invasion number of HPAI. The invasion number of HPAI is

$$\begin{aligned} \hat{\mathcal{R}}_{H_w} &= \frac{\beta_{11}^H \hat{S}_w + \beta_{11}^H \int_0^\infty q_w(\tau)\hat{r}_w d\tau}{\mu_w + \alpha_{H_w} + \nu_{H_w}} \\ &= \frac{\mathcal{R}_0^{H_w}}{\mathcal{R}_0^{L_w}}\left(1 + \frac{\alpha_{L_w}\mu_w}{\alpha_{L_w}+\mu_w}(\mathcal{R}_0^{L_w} - 1)\int_0^\infty q_w(\tau)e^{-\mu_w\tau} d\tau\right). \end{aligned} \tag{3.9}$$

The wild bird system (2.1) has two boundary equilibria: LPAI-only and HPAI-only equilibria. We denote the LPAI-only equilibrium by $(\hat{S}_w, \hat{I}_{L_w}\hat{r}_{L_w}, 0, 0)$ and the HPAI-only equilibrium by $(\tilde{S}_w, 0, 0, \tilde{I}_{H_w}\tilde{R}_{H_w})$. In the following two theorems we prove the existence of boundary equilibria.

Theorem 3. *If* $\mathcal{R}_0^{L_w} > 1$, *then there exists a unique LPAI-only equilibrium* $(\hat{S}_w, \hat{I}_{L_w}\hat{r}_{L_w}, 0, 0)$ *in which* $\hat{S}_w = \frac{\mu_w+\alpha_{L_w}}{\beta_{11}^L}$, $\hat{I}_{L_w} = \frac{\mu_w}{\beta_{11}^L}\left(\mathcal{R}_0^{L_w} - 1\right)$ *and* $\hat{r}_{L_w} = \alpha_{L_w}\hat{I}_{L_w}e^{-\mu_w\tau}$.

Proof. To find the LPAI-only equilibrium, we look for time-independent solutions of the form $(\hat{S}_w, \hat{I}_{L_w}\hat{r}_{L_w}(\tau), 0, 0)$. We set the time derivatives in wild bird system (2.1) equal to zero and obtain the following system for the LPAI-only equilibrium.

$$\begin{aligned} 0 &= \Lambda_w - \beta_{11}^L \hat{I}_{L_w}\hat{S}_w - \mu_w\hat{S}_w \\ 0 &= \beta_{11}^L \hat{I}_{L_w}\hat{S}_w - (\mu_w + \alpha_w)\hat{I}_{L_w} \\ \frac{d\hat{r}_{L_w}}{d\tau} &= -\mu_w\hat{r}_{L_w} \\ \hat{r}_{L_w}(0) &= \alpha_{L_w}\hat{I}_{L_w} \end{aligned} \tag{3.10}$$

The system consists of one first order ODE whose initial condition depends on the solution $\hat{I}_{L_w}$ and two algebraic equations. The second equation can be rewritten as

$$0 = \left(\beta_{11}^L\hat{S}_w - (\mu_w + \alpha_w)\right)\hat{I}_{L_w}.$$

Solving for $\hat{S}_w$ we get,

$$\hat{S}_w = \frac{\mu_w + \alpha_w}{\beta_{11}^L}.$$

Substituting the expression for $\hat{S}_w$ into the first equation and solving it for $\hat{I}_{L_w}$ we get,

$$\hat{I}_{L_w} = \frac{\Lambda_w}{\mu_w + \alpha_{L_w}} - \frac{\mu_w}{\beta_{11}^L}.$$

Since the basic reproduction number for LPAI is $\mathcal{R}_0^{L_w} = \frac{\beta_{11}^L\Lambda_w}{\mu_w(\mu_w+\alpha_{L_w})}$, we rearrange the terms in $\hat{I}_{L_w}$ by factoring out $\frac{\mu_w}{\beta_{11}^L}$ to obtain

$$\hat{I}_{L_w} = \frac{\mu_w}{\beta_{11}^L}\left(\mathcal{R}_0^{L_w} - 1\right).$$

Finally we solve the differential equation $\frac{d\hat{r}_{L_w}}{d\tau} = -\mu_w\hat{r}_{L_w}$ with $\hat{r}_{L_w}(0) = \alpha_{L_w}\hat{I}_{L_w}$ whose solution is

$$\hat{r}_{L_w}(\tau) = \alpha_{L_w}\hat{I}_{L_w}e^{-\mu_w\tau}$$

□

Theorem 4. *If* $\mathcal{R}_0^{H_w} > 1$, *then there exists a unique HPAI-only equilibrium* $(\tilde{S}_w, 0, 0, \tilde{I}_{H_w}\tilde{R}_{H_w})$ *where* $\tilde{S}_w = \frac{\mu_w+\alpha_{L_w}+\nu_{H_w}}{\beta_{11}^H}$, $\hat{I}_{H_w} = \frac{\mu_w}{\beta_{11}^H}\left(\mathcal{R}_0^{H_w} - 1\right)$ *and* $\tilde{R}_{H_w} = \frac{\alpha_{H_w}}{\beta_{11}^H}\left(\mathcal{R}_0^{H_w} - 1\right)$.

Proof. Proof is very similar to the proof of Theorem 3, and will not be repeated. □

Theorem 5. *The LPAI-only equilibrium is locally asymptotically stable if* $\hat{\mathcal{R}}_{H_w} < 1$ *and unstable if* $\hat{\mathcal{R}}_{H_w} > 1$.

Proof. As before, we start by linearizing the system. We denote by (u, v, x, y, z) the perturbations around the steady state and set

$$S_w(t) = \hat{S}_w + u(t)\,, I_{L_w}(t) = \hat{I}_{L_w} + v(t)\,, r_{L_w}(\tau, t) = \hat{r}_{L_w}(\tau) + x(\tau, t)\,, I_{H_w}(t) = y(t)\,, R_{H_w}(t) = z(t)\,.$$

Substituting the above expressions onto (2.1), we obtain the following linear system for perturbations.

$$\begin{aligned}
\frac{du}{dt} &= -\beta_{11}^L \hat{I}_{L_w} u - \beta_{11}^L \hat{S}_w v - \beta_{11}^H \hat{S}_w y - \mu_w u, \\
\frac{dv}{dt} &= \beta_{11}^L \hat{I}_{L_w} u + \beta_{11}^L \hat{S}_w v - (\mu_w + \alpha_{L_w}) v \\
\frac{\partial x}{\partial t} + \frac{\partial x}{\partial \tau} &= -\beta_{11}^H \hat{r}_{L_w} q_w(\tau) y - \mu_w x \\
x(0, t) &= \alpha_{L_w} v \\
\frac{dy}{dt} &= \beta_{11}^H \hat{S}_w y + \beta_{11}^H y \int_0^\infty q_w(\tau) \hat{r}_{L_w} d\tau - (\mu_w + \alpha_{H_w} + \nu_{H_w}) y \\
\frac{dz}{dt} &= \alpha_{H_w} y - \mu_w z
\end{aligned} \tag{3.11}$$

We expect that the solutions are exponential and seek for solutions of the form (3.3). Substituting the these solutions (3.3) into linearized system (3.11), we get the following linear eigenvalue problem which consists of algebraic equations and a differential equation. The algebraic equations can be represented in the following matrix form

$$A\omega = \lambda \omega \tag{3.12}$$

where $\omega = (\bar{u}\,, \bar{v}\,, \bar{y}\,, \bar{z})^T$ and the matrix A is;

$$A = \begin{pmatrix} -\beta_{11}^L \hat{I}_{L_w} - \mu_w & -\beta_{11}^L \hat{S}_w & -\beta_{11}^H \hat{S}_w & 0 \\ \beta_{11}^L \hat{I}_{L_w} & \beta_{11}^L \hat{S}_w - (\mu_w + \alpha_{L_w}) & 0 & 0 \\ 0 & 0 & D & 0 \\ 0 & 0 & \alpha_{H_w} & -\mu_w \end{pmatrix}$$

where $D = \beta_{11}^H \hat{S}_w + \beta_{11}^H \int_0^\infty q_w(\tau) \hat{r}_{L_w} d\tau - (\mu_w + \alpha_{H_w} + \nu_{H_w})$. The differential equation is:

$$\lambda \bar{x} + \frac{d\bar{x}}{d\tau} = -\beta_{11}^H \hat{r}_{L_w} q_w(\tau) \bar{y} - \mu_w \bar{x}\,, \quad \bar{x}(0) = \alpha_{L_w} \bar{v}\,.$$

We solve the non-homogeneous differential equation by first multiplying with the integral factor $e^{(\lambda+\mu_w)\tau}$. We then integrate and obtain the following solution;

$$\bar{x}(\tau) = \alpha_{L_w} \bar{v} e^{-(\lambda+\mu_w)\tau} - \int_0^\tau \beta_{11}^H \hat{r}_{L_w} q_w(s) \bar{y} e^{-(\lambda-\mu_w)(\tau-s)} ds\,.$$

The LPAI-only equilibrium is stable, if and only if the eigenvalues $\{\lambda_i\}_{i=1}^4$ of the algebraic eigenvalue problem (3.12) are all with negative real parts. The third equation in (3.12) involves only $\bar{y}$. We solve the third equation and obtain

$$\lambda_3 = \beta_{11}^H \hat{S}_w + \beta_{11}^H \int_0^\infty q_w(\tau)\hat{r}_{L_w} d\tau - (\mu_w + \alpha_{H_w} + \nu_{H_w}).$$

The eigenvalue $\lambda_3 < 0$ if $\hat{R}_{H_w} < 1$. We see that eigenvalue $\lambda_4 = -\mu_w$. Note that $\beta_{11}^L \hat{S}_w - (\mu_w + \alpha_{L_w}) = 0$ Thus, the characteristic equation for eigenvalues λ_1 and λ_2 is;

$$\lambda^2 + \lambda\left(\beta_{11}^H \hat{I}_{L_w} + \mu_w\right) + \beta_{11}^L \hat{S}_w \beta_{11}^L \hat{I}_{L_w} = 0$$

Since $\lambda_1 + \lambda_2 < 0$ and $\lambda_1\lambda_2 > 0$, there are either $\lambda_1 < 0$ and $\lambda_2 < 0$ or two complex conjugate eigenvalues that satisfy $\Re\lambda_1 < 0$, $\Re\lambda_2 < 0$. □

Theorem 6. *The HPAI-only equilibrium is locally asymptotically stable if* $\hat{\mathcal{R}}_{L_w} < 1$ *and unstable if* $\hat{\mathcal{R}}_{L_w} > 1$.

Proof. Proof of Theorem 6 is very similar to the proof of Theorem 5, and will be omitted. □

3.3. Coexistence Equilibrium

In this subsection, we investigate the existence, uniqueness and the stability of the coexistence equilibrium (i.e. interior equilibrium). We denote the coexistence equilibrium by $(S_w^{**}, I_{L_w}^{**}, r_{L_w}^{**}, I_{H_w}^{**}, R_{H_w}^{**})$. Coexistence equilibrium represents the state for which both low pathogenic and high pathogenic avian influenza are endemic in the wild bird population. We first show the existence and uniqueness of the coexistence equilibrium by the following theorem.

Theorem 7. *There exists a unique coexistence equilibrium* $(S_w^{**}, I_{L_w}^{**}, r_{L_w}^{**}, I_{H_w}^{**}, R_{H_w}^{**})$ *iff* $\hat{\mathcal{R}}_{L_w} > 1$, *and* $\hat{\mathcal{R}}_{H_w} > 1$.

Proof. The coexistence equilibrium satisfies the following steady state equation

$$\begin{cases} 0 &= \Lambda_w - \beta_{11}^L I_{L_w}^{**} S_w^{**} - \beta_{11}^H I_{H_w}^{**} S_w^{**} - \mu_w S_w^{**}, \\ 0 &= \beta_{11}^L I_{L_w}^{**} S_w^{**} - (\mu_w + \alpha_{L_w}) I_{L_w}^{**} \\ \dfrac{dr_{L_w}^{**}}{d\tau} &= -q_w(\tau)\beta_{11}^H I_{H_w}^{**} r_{L_w}^{**} - \mu_w r_{L_w}^{**} \\ r_{L_w}^{**}(0) &= \alpha_{L_w} I_{L_w}^{**} \\ 0 &= \beta_{11}^H I_{H_w}^{**} S_w^{**} + \beta_{11}^H I_{H_w}^{**} \displaystyle\int_0^\infty q_w(\tau) r_{L_w}^{**}(\tau) d\tau - (\mu_w + \alpha_{H_w} + \nu_{H_w}) I_{H_w}^{**} \\ 0 &= \alpha_{H_w} I_{H_w}^{**} - \mu_w R_{H_w}^{**} \end{cases} \tag{3.13}$$

Solving the second equation for S_w^{**}, we get;

$$S_w^{**} = \frac{\mu_w + \alpha_{L_w}}{\beta_{11}^L}.$$

We then substitute the expression for S_w^{**} to the first equation and solve for $I_{L_w}^{**}$ and obtain:

$$I_{L_w}^{**} = \mu_w \left(\mathcal{R}_0^{L_w} - 1\right) - \frac{\beta_{11}^H}{\beta_{11}^L} I_{H_w}^{**}.$$

To determine $r_{L_w}^{**}$, we solve the differential equation

$$\frac{dr_{L_w}^{**}}{d\tau} = -q_w(\tau)\beta_{11}^H I_{H_w}^{**} r_{L_w}^{**} - \mu_w r_{L_w}^{**}, \quad r_{L_w}^{**}(0) = \alpha_{L_w} I_{L_w}^{**}$$

and obtain

$$r_{L_w}^{**} = \alpha_{L_w} I_{L_w}^{**} e^{-\mu_w \tau - \beta_{11}^H I_{H_w}^{**} \int_0^\tau q_w(s)ds}.$$

Substituting the expressions for S_w^{**} and $r_{L_w}^{**}$ into the fourth equation, we see that $I_{H_w}^{**}$ satisfies the equation $F(I_{H_w}^{**}) = 0$ where $F(x)$ is the following monotone decreasing function

$$F(x) = \frac{\beta_{11}^H}{\beta_{11}^L}(\mu_w + \alpha_{L_w}) + \frac{\beta_{11}^H}{\beta_{11}^L}\alpha_{L_w}\left(\mu_w(\mathcal{R}_0^{L_w} - 1) - \beta_{11}^H x\right)\int_0^\infty q_w(\tau)e^{-\mu_w\tau - \beta_{11}^H x \int_0^\tau q_w(s)ds} d\tau$$
$$- (\mu_w + \alpha_{H_w} + \nu_{H_w}). \quad (3.14)$$

There exists a unique positive $I_{H_w}^{**}$ in the interval $(0, \hat{I}_{L_w})$ such that $F(I_{H_w}^{**}) = 0$, since

$$\begin{aligned}
F(0) &= \frac{\beta_{11}^H}{\beta_{11}^L}(\mu_w + \alpha_{L_w}) + \frac{\beta_{11}^H}{\beta_{11}^L}\left(\mu_w(\mathcal{R}_0^{L_w} - 1)\right)\int_0^\infty q_w(\tau)e^{-\mu_w\tau}d\tau - (\mu_w + \alpha_{H_w} + \nu_{H_w}) \\
&= \beta_{11}^H \hat{S}_w + \beta_{11}^H \int_0^\infty q_w(\tau)\hat{r}_{L_w}(\tau)d\tau - (\mu_w + \alpha_{H_w} + \nu_{H_w}) \\
&= (\mu_w + \alpha_{H_w} + \nu_{H_w})(\hat{\mathcal{R}}_{H_w} - 1) \\
&> 0
\end{aligned}$$

and

$$\begin{aligned}
F(\hat{I}_{L_w}) &= \frac{\beta_{11}^H}{\beta_{11}^L}(\mu_w + \alpha_{L_w}) - (\mu_w + \alpha_{H_w} + \nu_{H_w}) \\
&= (\mu_w + \alpha_{H_w} + \nu_{H_w})\left(\frac{1}{\hat{\mathcal{R}}_{L_w}} - 1\right) \\
&< 0.
\end{aligned}$$

□

3.3.1. Stability of the Coexistence Equilibrium:

To investigate the stability of coexistence equilibrium $(S_w^{**}, I_{L_w}^{**}, r_{L_w}^{**}, I_{H_w}^{**}, R_{H_w}^{**})$ we start by linearizing the system (2.1). We set

$$S_w(t) = S_w^{**} + u(t), \quad I_{L_w}(t) = I_{L_w}^{**} + v(t), \quad r_{L_w}(\tau, t) = r_{L_w}^{**} + x(\tau, t), \quad I_{H_w}(t) = I_{H_w}^{**} + y(t),$$

$R_{H_w}(t) = R_{H_w}^{**} + z(t)$. Substituting into the equations of the system (2.1) and using the equation (3.13) for the coexistence equilibrium, we get the following linear system after dropping the quadratic terms in perturbations:

$$\frac{du}{dt} = -\beta_{11}^L S_w^{**} v - \beta_{11}^L I_{L_w}^{**} u - \beta_{11}^H I_{H_w}^{**} u - \beta_{11}^H S_w^{**} y - \mu_w u \tag{3.15}$$

$$\frac{dv}{dt} = \beta_{11}^L S_w^{**} v + \beta_{11}^L I_{L_w}^{**} u - (\mu_w + \alpha_{L_w}) v \tag{3.16}$$

$$\frac{\partial x}{\partial t} + \frac{\partial x}{\partial \tau} = -q_w(\tau)\beta_{11}^H r_{L_w}^{**} y - q_w(\tau)\beta_{11}^H I_{H_w}^{**} x - \mu_w x \tag{3.17}$$

$$x(0,t) = \alpha_{L_w} v$$

$$\frac{dy}{dt} = \beta_{11}^H I_{H_w}^{**} u + \beta_{11}^H S_w^{**} y + \beta_{11}^H I_{H_w}^{**} \int_0^\infty q_w(\tau) x(\tau,t) d\tau \tag{3.18}$$

$$+ \beta_{11}^H y \int_0^\infty q_w(\tau) r_{L_w}^{**} d\tau - (\mu_w + \alpha_{H_w} + \nu_{H_w}) y$$

$$\frac{dz}{dt} = \alpha_{H_w} y - \mu_w z \tag{3.19}$$

We look for exponential solutions of the system (3.15)-(3.19). Substituting (3.3) into (3.15)-(3.19), we get the following eigenvalue problem:

$$\lambda \bar{u} = -\beta_{11}^L S_w^{**} \bar{v} - \beta_{11}^L I_{L_w}^{**} \bar{u} - \beta_{11}^H I_{H_w}^{**} \bar{u} - \beta_{11}^H S_w^{**} \bar{y} - \mu_w \bar{u} \tag{3.20}$$

$$\lambda \bar{v} = \beta_{11}^L S_w^{**} \bar{v} + \beta_{11}^L I_{L_w}^{**} \bar{u} - (\mu_w + \alpha_{L_w}) \bar{v} \tag{3.21}$$

$$\lambda \bar{x} + \frac{d\bar{x}}{d\tau} = -q_w(\tau)\beta_{11}^H r_{L_w}^{**} \bar{y} - q_w(\tau)\beta_{11}^H I_{H_w}^{**} \bar{x} - \mu_w \bar{x} \tag{3.22}$$

$$\bar{x}(0) = \alpha_{L_w} \bar{v}$$

$$\lambda \bar{y} = \beta_{11}^H I_{H_w}^{**} \bar{u} + \beta_{11}^H S_w^{**} \bar{y} + \beta_{11}^H I_{H_w}^{**} \int_0^\infty q_w(\tau) \bar{x}(\tau) d\tau \tag{3.23}$$

$$+ \beta_{11}^H \bar{y} \int_0^\infty q_w(\tau) r_{L_w}^{**} d\tau - (\mu_w + \alpha_{H_w} + \nu_{H_w}) \bar{y}$$

$$\lambda \bar{z} = \alpha_{H_w} \bar{y} - \mu_w \bar{z} \tag{3.24}$$

We solve the non-homogeneous differential equation (3.22), by first multiplying with the integrating factor;

$$e^{(\lambda+\mu_w)\tau + \beta_{11}^H I_{H_w}^{**} \int_0^\tau q_w(s) ds}.$$

We then integrate both sides from 0 to τ and use the initial condition $\bar{x}(0) = \alpha_{L_w} \bar{v}$ to obtain:

$$\bar{x}(\tau) = \alpha_{L_w} \bar{v} e^{-(\lambda+\mu_w)\tau - \beta_{11}^H I_{H_w}^{**} \int_0^\tau q_w(s) ds} - \beta_{11}^H \bar{y} f(\tau), \tag{3.25}$$

where

$$f(\tau) = \int_0^\tau q_w(s) r^{**}_{L_w}(s) e^{-(\lambda+\mu_w)(\tau-s) - \beta^H_{11} I^{**}_{H_w} \int_s^\tau q_w(\sigma)d\sigma} ds. \tag{3.26}$$

Since $\beta^L_{11} S^{**}_w - (\mu_w + \alpha_{L_w}) = 0$, equation (3.21) reduces to

$$\lambda \bar{v} = \beta^L_{11} I^{**}_{L_w} \bar{u}\,. \tag{3.27}$$

Similarly using the equation of coexistence equilibrium (3.13), from (3.23) we see that $\beta^H_{11} S^{**}_w + \beta^H_{11} \int_0^\infty q_w(\tau) r^{**}_{L_w} d\tau - (\mu_w + \alpha_{H_w} + \nu_{H_w}) = 0$. Hence, (3.23) reduces to

$$\lambda \bar{y} = \beta^H_{11} I^{**}_{H_w} \bar{u} + \beta^H_{11} I^{**}_{H_w} \int_0^\infty q_w(\tau) \bar{x}(\tau) d\tau\,. \tag{3.28}$$

Substituting the solution for $\bar{x}(\tau)$, (3.25), into the reduced equation for $\bar{y}$ (3.28), we obtain the following equation for $\bar{y}$ which involves only $\bar{u}$ and $\bar{v}$

$$\lambda \bar{y} = \beta^H_{11} I^{**}_{H_w} \bar{u} + A(\lambda) \bar{v} - \beta^H_{11} I^{**}_{H_w} \beta^H_{11} B(\lambda) \bar{y} \tag{3.29}$$

where

$$A(\lambda) = \alpha_{L_w} \beta^H_{11} I^{**}_{H_w} \int_0^\infty \left(q_w(\tau) e^{-(\lambda+\mu_w)\tau - \beta^H_{11} I^{**}_{H_w} \int_0^\tau q_w(s) ds} \right) d\tau\,. \tag{3.30}$$

and

$$B(\lambda) = \int_0^\infty q_w(\tau) f(\tau) d\tau\,. \tag{3.31}$$

Combining these reduced equations for (3.27) and (3.29) together with (3.20), we get the following eigenvalue problem;

$$\begin{aligned}
\lambda \bar{u} &= -\beta^L_{11} S^{**}_w \bar{v} - \beta^L_{11} I^{**}_{L_w} \bar{u} - \beta^H_{11} I^{**}_{H_w} \bar{u} - \beta^H_{11} S^{**}_w \bar{y} - \mu_w \bar{u} \\
\lambda \bar{v} &= \beta^L_{11} I^{**}_{L_w} \bar{u} \\
\lambda \bar{y} &= \beta^H_{11} I^{**}_{H_w} \bar{u} + A(\lambda) \bar{v} - \beta^H_{11} I^{**}_{H_w} \beta^H_{11} B(\lambda) \bar{y}\,.
\end{aligned} \tag{3.32}$$

This system will have non-zero solution for $(\bar{u}, \bar{v}, \bar{y})$ if the determinant of this system is zero, that is, if

$$\begin{vmatrix}
-\lambda - \beta^L_{11} I^{**}_{L_w} - \beta^H_{11} I^{**}_{H_w} - \mu_w & -\beta^L_{11} S^{**}_w & \beta^H_{11} S^{**}_w \\
\beta^L_{11} I^{**}_{L_w} & -\lambda & 0 \\
\beta^H_{11} I^{**}_{H_w} & A(\lambda) & -\lambda - \beta^H_{11} I^{**}_{H_w} B(\lambda)
\end{vmatrix} = 0. \tag{3.33}$$

From the determinant, we obtain the characteristic equation for the eigenvalue problem (3.32):

$$\lambda^3 + \lambda^2 \left(\beta_{11}^H I_{H_w}^{**} \beta_{11}^H B(\lambda) + \beta_{11}^L I_{L_w}^{**} + \beta_{11}^H I_{H_w}^{**} + \mu_w\right)$$

$$+\lambda \left[\left(\beta_{11}^L I_{L_w}^{**} + \beta_{11}^H I_{H_w}^{**} + \mu_w\right)\left(\beta_{11}^H I_{H_w}^{**} \beta_{11}^H B(\lambda)\right) + \beta_{11}^H S_w^{**} \beta_{11}^H I_{H_w}^{**} + \beta_{11}^L S_w^{**} \beta_{11}^L I_{L_w}^{**}\right]$$

$$+\beta_{11}^H S_w^{**} \beta_{11}^L I_{L_w}^{**} A(\lambda) + \beta_{11}^H I_{H_w}^{**} \beta_{11}^H B(\lambda) \beta_{11}^L S_w^{**} \beta_{11}^L I_{L_w}^{**} = 0.$$

This characteristic equation has roots with positive real parts. Thus instability and oscillations occur. We show that sustained oscillations are possible in the system (2.1) by demonstrating it with a specific example in the next section.

4. Numerical Simulations

In this section, we perform several numerical simulations of the model (2.1). We obtain the approximate solution of the system (2.1) by constructing an implicit finite difference method. We discretize the domain $D = \{(\tau, t) : 0 \leq \tau \leq A\,,\ 0 \leq t \leq T\}$ by taking equal step sizes in both t and τ direction. Thus, $\Delta t = \Delta \tau$.

In all simulations, we choose the cross-immunity function $q_w(\tau)$ to be the following step function;

$$q_w(\tau) = \begin{cases} 0 & \text{if } \tau \leq a \\ q & \text{if } \tau > a \end{cases} \tag{4.1}$$

where a is an arbitrary constant. Since the variable τ is the time-since-recovery from low pathogenic avian influenza, this cross immunity function $q_w(\tau)$ means that the wild birds are fully protected from HPAI for a period of time a. After that period a, the protection wanes. This is a reasonable assumption for low pathogenic and high pathogenic influenza in the avian population, and it is in agreement with the results of the study by [40].

The first question that we address with simulations is whether system can (2.1) can exhibit oscillations in which both the LPAI and the HPAI oscillate at non-zero values. To address this question, we analyze the characteristic equation. With the above choice of $q_w(\tau)$, we compute the integrals in $f(\tau)$, $A(\lambda)$, and $B(\lambda)$ and get;

$$f(\tau) = q\alpha_{L_w} I_{L_w}^{**} \frac{1}{\lambda}\left(e^{\lambda\tau} - e^{\lambda a}\right) e^{-(\lambda+\mu_w)\tau - \beta_{11}^H I_{H_w}^{**} q(\tau - a)}$$

and

$$A(\lambda) = \frac{\beta_{11}^H I_{H_w}^{**} \alpha_{L_w} q e^{-(\lambda+\mu_w)a}}{\lambda + \mu_w + \beta_{11}^H I_{L_w}^{**} q}$$

and

$$B(\lambda) = \frac{q^2 \alpha_{L_w} I_{L_w}^{**} e^{-\mu_w a}}{(\lambda + \mu_w + \beta_{11}^H I_{L_w}^{**} q)(\mu_w + \beta_{11}^H I_{L_w}^{**} q)}.$$

After analyzing the characteristics equation we find parameters of the model (2.1), whose solutions exhibits oscillations. These parameters are given in Table 1. To illustrate the oscillations, we simulate the solutions of the wild bird system (2.1) using an implicit

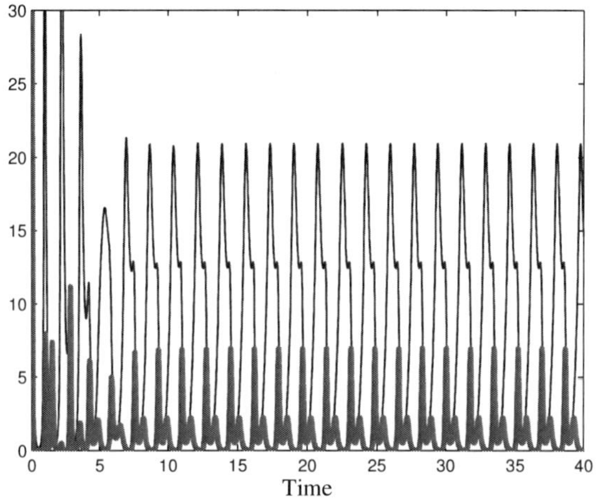

Figure 1. The number of wild birds infected with LPAI (thin line) and the number of wild birds infected with HPAI (thick line) exhibit oscillations.

finite difference method. Time is measured in years, and the final time for the simulations are $T = 40$ and $A = 30$. The number of wild birds infected with LPAI and the number of wild birds infected with HPAI, which exhibit sustained oscillations, are plotted in Figure 1. The Figure shows that the oscillations in LPAI have a much larger amplitude than the oscillations of HPAI suggesting that the instability of the dynamics of LPAI is more pronounced. Furthermore, the oscillations in HPAI follow the oscillations of LPAI with the peak in HPAI occurring right after the drop in LPAI. This behavior partly reminds of a Lotka-Volterra predator-prey dynamics where the oscillations in the predator follow the oscillations of the prey with 1/4 of a turn. This analogy is perhaps not surprising since HPAI infects individuals recovered from LPAI.

Table 1. Parameter values of the wild birds model (2.1) **which exhibits oscillations**

Parameter	Value	Parameter	Value	Parameter	Value
Λ_w	1020	β_{11}^L	0.1278	$r_{L_w}(\tau, 0)$	$28/T$
μ_w	$1/2$	β_{11}^H	0.7140	$I_{H_w}(0)$	2
ν_{H_w}	460.925	$S_w(0)$	2000	q	1
α_{H_w}	$365/7$	$I_{L_w}(0)$	10	a	0.25
α_{L_w}	$365/7$				

The next question we address is how changes in the transmission of LPAI affect the competition of the strains. To investigate the impact of β_{11}^L, we obtain the approximate solutions of the system (2.1) for three different values of β_{11}^L. We fix all other parameters and change only β_{11}^L. The results show that increasing β_{11}^L increases the LPAI prevalence and decreases the HPAI prevalence. We plot $I_{L_w}(t)$ and $I_{H_w}(t)$ in Figure 2.

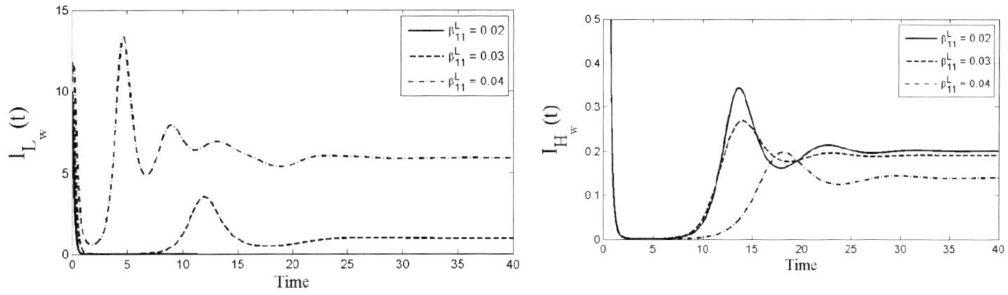

Figure 2. LPAI and HPAI prevalence for three different values of β_{11}^{L}.

System (2.1) can model the dynamics of LPAI and HPAI in both wild and domestic birds. The observation that increasing transmission increases prevalence of LPAI and decreases prevalence of HPAI may shed light on why wild birds have lower prevalence of HPAI compared to domestic birds. The immunity created by circulating LPAI protects wild birds from HPAI. In contrast, domestic birds are protected from LPAI and consequently, they experience much more serious symptoms from HPAI and HPAI has much higher prevalence in domestic birds, particularly in the poultry population of some countries.

Another question of interest is how the duration of cross-immunity affects the LPAI and HPAI prevalence. Since we assume that LPAI provides full protection against HPAI for $\tau \in [0, a]$, it is interesting to know how the length of this protection affects the dynamics. We take three different values of a and plot $I_{L_w}(t)$ and $I_{H_w}(t)$ in Figure 3. We observe that increasing duration of cross-immunity decreases the prevalence of HPAI among wild bird populations. What is unexpected is that cross-immunity also increases the prevalence of LPAI, even though it does not directly affect LPAI. The change in LPAI is at least as pronounced as the change in HPAI.

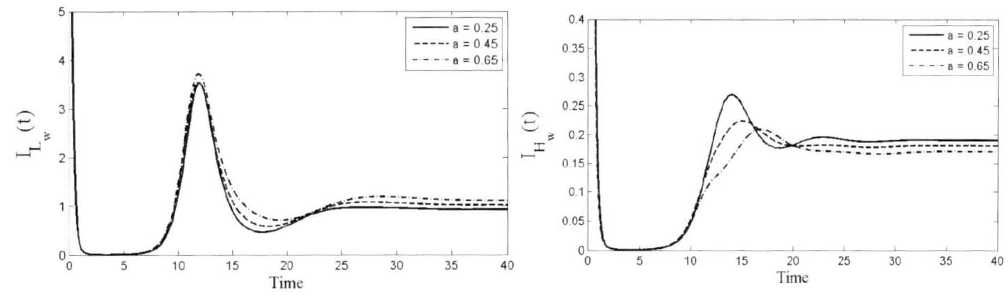

Figure 3. Effect of cross-immunity on HPAI prevalence.

Another question that we consider is the impact of LPAI and HPAI transmission rates β_{11}^{H} and β_{11}^{L} on the HPAI prevalence at coexistence equilibrium. To investigate this topic, we consider the equation for the $I_{H_w}^{**}$. The HPAI prevalence in the coexistence equilibrium satisfies the equation $F(I^{**}_{H_w}) = 0$ given in (3.14) which involves the following integral

$$C = \int_0^\infty q_w(\tau) e^{-\mu_w \tau - \beta_{11}^H I^{**}_{H_w} \int_0^\tau q_w(s)ds} d\tau \,.$$

We compute the integral $C(\lambda)$ by taking $q_w(\tau)$ to be the step function in (4.1) and obtain

$$C = \frac{q e^{-\mu_w a}}{\mu_w + \beta_{11}^H I^{**}_{H_w} q}.$$

Thus, the HPAI coexistence equilibrium, $I^{**}_{H_w}$ satisfies

$$\frac{\beta_{11}^H}{\beta_{11}^L}(\mu_w + \alpha_{L_w}) + \frac{\beta_{11}^H}{\beta_{11}^L}\alpha_{L_w}\left(\mu_w(\mathcal{R}_0^{L_w} - 1) - \beta_{11}^H I^{**}_{H_w}\right)\frac{q e^{-\mu_w a}}{\mu_w + \beta_{11}^H I^{**}_{H_w} q} - \\ - (\mu_w + \alpha_{H_w} + \nu_{H_w}) = 0 \quad (4.2)$$

We plot (4.2) for several values of β_{11}^L in Figure 4 . We consider how the two transmis-

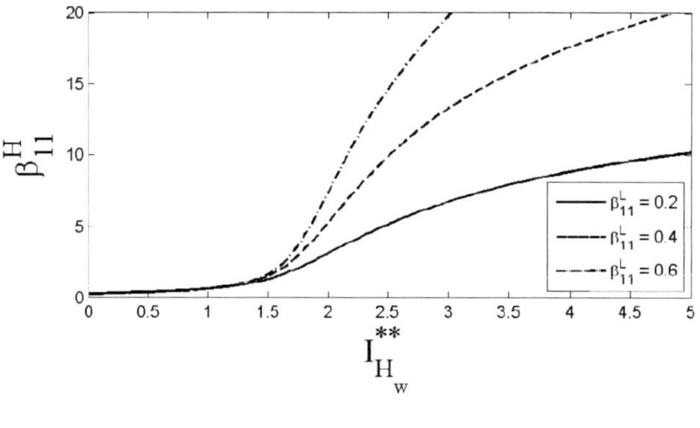

Figure 4. HPAI coexistence equilibrium.

sion rates impact $I^{**}_{H_w}$. We notice that for low values of HPAI transmission rate and low HPAI prevalence, the HPAI prevalence does not depend on the transmission rate of LPAI. However, as the HPAI prevalence increases it becomes more and more sensitive to the LPAI transmission rate. A surprising conclusion is that for higher HPAI prevalences, the HPAI prevalence depends significantly on the LPAI transmission rate β_{11}^L. It is further clear that increase in β_{11}^H increases the HPAI prevalence $I^{**}_{H_w}$ while increase in the LPAI transmission rate β_{11}^L decreases HPAI prevalence $I^{**}_{H_w}$. This observation again suggests that higher transmission of LPAI in wild birds may be responsible for lower prevalence of HPAI.

Finally, we recall that HPAI is locally stable if $\hat{\mathcal{R}}_{L_w} < 1$, and LPAI is locally stable if $\hat{\mathcal{R}}_{H_w} < 1$. We notice that the two invasion numbers can be written as functions of the two reproduction numbers:

$$\hat{\mathcal{R}}_{L_w} = h(\mathcal{R}_0^{L_w}, \mathcal{R}_0^{H_w}) \qquad \hat{\mathcal{R}}_{H_w} = g(\mathcal{R}_0^{L_w}, \mathcal{R}_0^{H_w}).$$

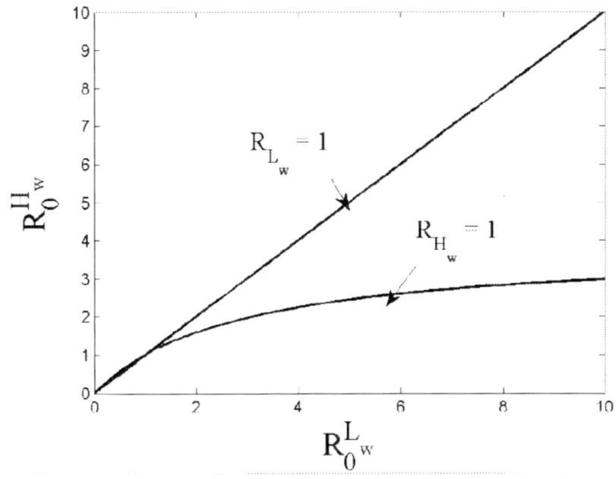

Figure 5. The regions of coexistence and dominance.

Therefore, we can plot the curves $h(\mathcal{R}_0^{L_w}, \mathcal{R}_0^{H_w}) = 1$ and $g(\mathcal{R}_0^{L_w}, \mathcal{R}_0^{H_w}) = 1$ in the $(\mathcal{R}_0^{L_w}, \mathcal{R}_0^{H_w})$ plane. The figure we obtain is given in Figure 5. The figure shows that the area in the $(\mathcal{R}_0^{L_w}, \mathcal{R}_0^{H_w})$ plane where HPAI dominates is the largest. HPAI dominates whenever $\mathcal{R}_0^{H_w} > \mathcal{R}_0^{L_w}$. The area where LPAI dominates is below the lower curve. The area where LPAI and HPAI coexist is the area between the curves. This area is larger if q is larger and when a is smaller. That is to say that coexistence is more likely if the full immunity provided by LPAI is shorter.

5. Discussion

This chapter introduces a LPAI-HPAI avian influenza model. HPAI-infected individuals can infect individuals recovered from LPAI at some reduced infectivity. However, we assume that LPAI-infected individuals cannot infect HPAI-recovered individuals. Furthermore, LPAI-recovered individuals are structured by time-since-recovery and the immunity to HAPI created by prior infection by LPAI wanes as the time-since-recovery increases. This model describes the HPAI-LPAI dynamics in both domestic and wild birds.

We find that the system has a unique disease-free equilibrium. We define the reproduction numbers for the LPAI and HPAI. If both reproduction number are smaller than one, then we show that the disease-free equilibrium is both locally and globally stable. If one of the reproduction numbers is greater than one, then the disease-free equilibrium is unstable. Furthermore, we find that if the reproduction number of LPAI is greater than one, there is a unique LPAI-only equilibrium. Similarly, if the reproduction number of HPAI is greater than one, there is a unique HPAI-only equilibrium. We define the invasion reproduction numbers of LPAI and HPAI. The invasion reproduction number of LPAI is greater than one if and only if the reproduction number of LPAI is greater then the reproduction number of HPAI. We prove that the LPAI-only equilibrium is locally asymptotically stable if the HPAI invasion number is smaller than one, that is if the HPAI cannot invade the equilibrium of the LPAI. Similarly we prove that the HPAI-only equilibrium is locally asymptotically stable if the LPAI invasion number is smaller than one, that is the LPAI cannot invade the equilibrium of the HPAI. This is to say that HPAI dominates in the population and drives

LPAI to extinction if and only if HPAI reproduction number is larger than the LPAI reproduction number. HPAI dominates for more values of the reproduction numbers than LPAI and every time the reproduction number of HPAI is larger than the reproduction number of LPAI; however, we believe that for realistic parameter values LPAI has a higher reproduction number, particularly in wild birds.

Finally we show that if both invasion numbers are larger than one coexistence equilibrium is present. The mechanism of coexistence is cross-immunity. We show that the coexistence equilibrium is unique. In addition, we show through simulations that the coexistence equilibrium may lose stability and coexistence in the form of sustained oscillations is possible. In this case HPAI's peak follows right after the LPAI's peak.

Simulations also suggest that increasing the transmission coefficient of LPAI increases the prevalence of LPAI and decreases the prevalence of HPAI. Furthermore, increasing the transmission coefficient of HPAI and decreasing the transmission coefficient of LPAI both increase the HPAI prevalence. Based on these simulations we conclude that higher transmission of LPAI in wild birds may be responsible for the lower prevalence of HPAI compared to poultry. Finally, assuming that LPAI protects against HPAI completely for a period of time, we investigate the effect the duration of protection has on the LPAI and HPAI prevalence. We find that increasing the duration of protection has an impact on both the LPAI prevalence and the HPAI prevalence. In particular, it increases the LPAI prevalence and decreases the HPAI prevalence.

Several articles have investigated the dynamics of LPAI and HPAI before [28, 5]. However, the effects of cross-immunity and duration of protection have not been studied. In this article we investigate these effects as well as how transmission impacts the prevalence of LPAI and HPAI. Furthermore, prior studies have found that ODE models with cross-immunity have note been able to exhibit sustain oscillations [6, 7]. To generate oscillations in the multi-strain influenza dynamics, a quarantine state was introduced in [34]. Here, we have used a different approach to model the oscillations in avian influenza dynamics, namely, we have introduced a time-since-recovery independent variable. We show via simulations that the coexistence equilibrium can be destabilized and coexistence in the form of sustained oscillation is possible.

Acknowledgment

The authors acknowledge support from the NSF under grant DMS-1220342.

References

[1] D.J. Alexander, An overview of epidemiology of avian influenza, *Vaccine*, 25 (2006), p. 5637-5644.

[2] A. Apisornthanarak, M.D., L. Mundy, M.D. Infection control for Avian Influenza (H5N1) in Healtcare Settings. Chapter 3. Avian Influenza REsearch Progress. E.P. Allegra (Editor), (2008) p.73-78.

[3] F.B. Agusto, A.B. Gummel, Theoretical assessment of avian influenza vaccine, *Dis. Cont. D. Sys. B* 13(1) (2010), p. 1-25.

[4] R.B. Belshe. The origins of pandemic influenza-lessons demo the 1918 virus, *N. Engl. J. Med.* 353(2005), p. 2209-11.

[5] L. Bourouiba, A. Teslya, J. Wu, Highly pathogenic avian influenza outbreak mitigated by seasonal low pathogenic strains: Insights from dynamic modeling, *JTB* 271 (2011), p. 181-201.

[6] C. Castillo-Chavez, H. Hethcote, V. Andreasen, S. Levin, W.M. Liu, Epidemiological models with age structure, proportionate mixing and cross-immunity, *J. Math. Biol.* 27 (1989), p. 159-165.

[7] C. Castillo-Chavez, H. Hethcote, V. Andreasen, S. Levin, W.M. Liu, Cross-immunity in the dynamics of homogeneous and heterogeneous populations, *Mathematical Ecology (Trieste, 1986)*, World Sci. Publishing, Teaneck, NJ, (1988), p. 303-316.

[8] L. Clark, J. Hall, Avian influenza in wild birds: status as reservoirs, and risks to humans and agriculture, *Ornithol. Monogr.* 60 (2006), p. 3-29.

[9] E. Class, *et al.*, Human influenza A H5N1 virus related to highly pathogenic avian influenza virus, *Lancet* 351 (1998), p. 472-477.

[10] E. O'Neill, J.M. Riberdy, R.G. Webster, D.L. Woodland, Heterologous protection against lethal A/Hong Kong/156/97 (H5N1) influenza virus infection in C57BL/6 mice, *J. Gen. Virol.* 81 (2000), p. 2689-96.

[11] S.R. Fereidouni, E. Starick, M. Beer, D. Kalthoff *et al.*, Highly pathogenic avian influenza virus infection of mallards with homo- and heterosubtypic immunity induced by LPAI viruses, *PLoS One* 4 (8), e6705.

[12] N.M. Ferguson, D.A.T. Cummings, S. Cauchemez, C. Fraser, S. Riley, A. Meeyai, S. Iamsirithaworn, D.S. Burke, Strategies for containing an emerging influenza pandemic in Southeast Asia, *Nature* 437 (2005), p. 209-214.

[13] B.N. Fields, D.M. Knipe, P.M. Howley(eds), *Fields Virology*, 3rd Ed., Lippincott–Raven, Philadelphia, 1996.

[14] T. C. Germann, K. Kadau, I. M. Longini, Jr., C. A. Macken, Mitigation strategies for pandemic influenza in the United States, *PNAS* 103(15) (2006), p. 5935-5940.

[15] R. F. Grais, J. H. Ellis, G. E. Glass, Assessing the impact of airline travel on the geographic spread of pandemic influenza, *Eur. J. Epidem.* 18 (2003), p. 1065-1072.

[16] A.B. Gummel, Global dynamics of a two-strain avian influenza model, *Int.J. Comp. Math.* 86(1) (2009), p.85-108.

[17] H. Inaba, Endemic threshold and stability in an evolutionary epidemic model, in Mathematical Approaches for Emerging and Reemerging Infectious Diseases: Models, Methods and Theory, *IMA Vol. Math. Appl.* 126, Springer, New York, (2002), p. 337-359.

[18] S. Iwami, Y. Takeuchi, X. Liu, Avian-human influenza epidemic model, *Math. Biosci.* 207 (2007), p. 1-25.

[19] S. Iwami, Y. Takeuchi, X. Liu, Avian flu pandemic: Can we prevent it?, *JTB* 257 (2009), p. 181-190.

[20] S. Iwami, Y. Takeuchi, A. Korobeinikov, X. Liu, Prevention of avian influenza epidemic: What policy should we choose?, *JTB* 252 (2008), p. 732-741.

[21] D. Kalthoff, A. Breithaupt, J.P. Teifke, A. Globig, *et al.* Pathogenicity of highly pathogenic avian influenza virus (H5N1) in adult mute swans, *EID* 14 (2008), p. 1267-1270.

[22] K.I. Kim, Z. Lin, L. Zhang, Avian-human influenza epidemic model with diffusion, *Nonlin. Analysis: RWA* 11 (2010), p. 313-322.

[23] S. Krauss, D. Walker, S.P. Pryor, L. Niles, L. Chenghong, V.S. Hinshaw, R.G. Webster, Influenza A viruses of migrating wild aquatic birds in North America, *Vector Borne Zoonotic Dis.* 4(3) (2004), p.177-189.

[24] A. Le Menach, E. Vargu, R. F. Grais, D. L. Smith, A. Flahault, Key strategies for reducing spread of avian influenza among commercial poultry holdings: lessons for transmission to humans, *Proc. R. Soc. B* 273 (2006), p. 2467-2475.

[25] X.Z. Li, J.X. Liu, M. Martcheva, An age-structured two-strain model with superinfection, *Math. Biosci. Eng.* 7 (1) (2010), p. 125-149.

[26] R. Liu, V.R.S.K. Duvvuri, J. Wu, Spread pattern formation and its implications for control strategies, *Math. Model. Nat. Phenom.* 3(7) (2008), p. 161-179.

[27] I.M. Longini, A. Nizam, S. Xu, K. Ungchusak, W. Hanshaoworakul, D.A.T. Cummings, M.E. Halloran, Containing pandemic influenza at the source, *Science* 309 (2005), p. 1083-1087.

[28] J. Lucchetti, M. Roy, M. Martcheva, An avian influenza model and its fit to human avian influenza cases, in *"Advances in Disease Epidemiology"* (J.M. Tchuenche, Z. Mukandavire, eds.), Nova Science Publishers, New York, NY, 2009, p. 1-30.

[29] B. Lupiani, S. Reddy, The history of avian influenza, *Comp. Immunol. Microbiol. Infect. Dis.* 32(4) (2009), p.311-323.

[30] P. Magal, S. Ruan Sustained oscillations in an evolutionary epidemiological model of influenza A drift, *Proc. R. Soc. A* 466 (2116) (2010), p. 965-992.

[31] M. Martcheva, An evolutionary model of influenza A with drift and shift, *J. Biol. Dynamics* 6 (2) (2012), p.299-332.

[32] C.E. Mills, J.M. Robins, C.T. Bergstrom, M. Lipsitch, Pandemic influenza: risk of multiple introductions and the need to prepare for them, *PLoS Medicine* 3 (6) (2006), p. 1-5.

[33] V.J. Munster, A. Wallensten, C. Baas, G.F. Rimmelzwaan, M. Schutten, B. Olsen, A.D. Osterhaus, R.A. Fouchier, Mallards and highly pathogenic avian influenza ancestral viruses, Northern Europe, *EID* 11(10) (2005), p. 1545-1551.

[34] M. Nuño, Z. Feng, M. Martcheva, C. Castillo-Chavez, Dynamics of two-strain influenza with isolation and partial cross-immunity, *SIAM J. Appl. Math.* 65 (3) (2005), p. 964-982.

[35] J. Otte, J. Hinrichs, J. Rushton, D. Roland-Holst, D. Zilberman, Impacts of avian influenza virus on animal production in developing countries, *Perspectives in Agr., Vet. Sci., Nutrition, Nat. Resources* 3 (2008), No. 080.

[36] C.M. Pease, An Evolutionary epidemiological mechanism with applications to type A influenza, *Theor. Pop. Biol.* 31 (1987), p. 422-452.

[37] R.S. Schrijver, G. Koch (eds.), *Avian Influenza: Prevention and Control*, Springer, The Netherlands, 2005.

[38] I. Scones, The international response to avian influenza: science, policy, politics, *Avian Influenza: Science, Policy and Politics*, (I. Scoones, ed.), EarthScan, London, (2010), p. 1-18.

[39] K. F. Shortridge, Poultry and influenza H5N1 outbreaks in Hong Kong, 1997: abridged chronology and virus isolation, *Vaccine* 17(1999), S26-S29.

[40] S.H. Seo, R.G. Webster, Cross-reactive, cell-mediated immunity and protection of chickens from lethal H5N1 influenza virus infection in Hong Kong poultry markets, *J.Virol.* 75 (2001), p. 2516-2525.

[41] D.E. Stallknecht, J.D. Brown, Ecology of avian influenza in wild birds, *Avian Influenza*, (D.E. Swayne, ed.), p.43-58, 2008.

[42] D.L. Suarez, Influenza A virus, in *Avian Influenza*, (D.E. Swayne, ed.), p.3-17, 2008.

[43] D.E. Swayne, Epidemiology of avian influenza in agricultural and other man-made system, in *Avian Influenza*, (D.E. Swayne, ed.), p.59-85, 2008.

[44] T. Tiensin, M. Nielen, H. Vernooij, *et. al*, Transmission of the highly pathogenic avian influenza virus H5N1 within flocks during the 2004 epidemic in Thailand, *JID* 196 (2007), p. 1679-1684.

[45] R.G. Webster, W.J. Bean, O.T. Gorman, T.M. Chambers, Y. Kawaoka, Evolution and ecology of influenza A viruses, *Microbiol. Rev.* 56 (1) (1992) p. 152-179.

[46] USDA, *An Early Detection System for Highly Pathogenic H5N1 Avian Influenza in Wild Migratory Birds*, www.usda.gov/documents/wildbirdstrategicplanpdf.pdf.

[47] World Organization for Animal Health, *Avian influenza* 2.7.12, Terrestrial animal health code-2006, World Organization for Animal Health, Paris, France, 2006.

In: Dynamical Systems
Editor: Jean Michel Tchuenche, pp. 261-292
ISBN: 978-1-62808-001-8

Chapter 10

ANALYSIS OF A PREDATOR-PREY MODEL WITH SIGMOIDAL RESPONSE FUNCTION AND SIS DISEASE

J. J. Tewa[1,5,*], *P. Tchinda Mouofo*[2,5], *V. Kamla*[3,5] *and S. Bowong*[4,5]
[1]Department of Mathematics and Physics,
National Advanced School of Engineering (Polytechnic),
University of Yaounde I, Yaounde, Cameroon
[2]Department of Mathematics, Yaounde, Cameroon
University of Yaounde I, Faculty of Science,
[3]Department of Mathematics and Computer Science, Cameroon
University of Ngaoundere, ENSAI,
[4]Department of Mathematics and Computer Science, Cameroon
University of Douala, Faculty of Science,
[5]GRIMCAPE team project, LIRIMA, Yaounde, Cameroon
UMI 209 IRD/UPMC UMMISCO, University of Yaounde I, Cameroon

Abstract

A predator-prey system with prey harvesting, sigmoidal response function is considered. Three derived models with SIS disease only among the prey population, the predator population and the two populations respectively are obtained. Global qualitative and bifurcation analysis are combined to determine the global dynamics of these models. The bifurcation analysis of the model depending on all parameters indicates that it exhibits bifurcations phenomena, including the saddle-node bifurcation and the transcritical bifurcation. There is no limit cycle for these models. Mathematical analysis of the models regarding stability and influence of disease has been performed. The disease modifies considerably the qualitative analysis of our systems since it can disappear or persist inside population. Numerical simulations are given in order to illustrate analytical results.

Keywords: Predator-Prey model; Infectious disease; Sigmoidal function; Harvesting; Global Stability; Bifurcation

*E-mail address: tewajules@gmail.com, Tel: +(237) 77 71 13 69 (Corresponding author).

1. Introduction

For predator-prey interaction, a functional response of predator to the prey density refers to the change in the density of prey attached per unit time per predator as the prey density changes. The Lotka-Volterra response function $\phi_1(x) = r\,x$, is a straight line through the origin and is unbounded. In 1913, Michaelis and Menten proposed the response function $\phi_2(x) = \dfrac{\omega\,x}{a+x}$ in studying enzymatic reactions, where $\omega > 0$ denotes the maximal growth rate of the species and $a > 0$ the half-saturation constant. C. S. Holling [2] also used in 1959 this function as one of the predator functional responses. Another response function is $\phi_3(x) = \dfrac{\omega_0 x^2}{x^2 + \omega_1\,x + \omega_2}$ which is called a sigmoidal response function, while the simplified form $\dfrac{\omega\,x^2}{a_0 + x^2}$ is known as Holling type-III function.

Many epidemiological models [14, 15, 16, 17, 18, 19, 20, 21, 30, 37, 38, 39, 40, 41, 42] and ecological models [22, 23, 24, 25, 26, 27, 28, 29, 31, 32, 33, 34, 35, 36, 43, 44] has been analyzed in the literature. There are also many models in the literature that combine the two thematics (epidemiology and ecology) [45, 3, 4, 5, 6, 12, 11, 13].

Generally, there are more macroparasitic infections which can affect only preys, only predators, or both preys and predators. The predators can be infected when swallowing the infected preys, and the preys can be infected by contacting the excrement of the infected predators.

Dynamical systems of infectious diseases are mostly based on compartment structures that were initially proposed by Kermack and McKendrick (1927,1932) and developed later by many other researchers. In order to formulate a dynamic model for the transmission of an epidemic disease, the population in a given region is often divided into several different groups or compartments. Such a model describing the dynamic relations among these compartments is called a compartment model.

Mathematical modeling has been playing an important role in epidemiology, the study of spread and control of infectious diseases. Mathematical models contributes in two major aspects of epidemiology : (i) to build and test epidemiological theories and (ii) to plan, implement and evaluate detection, control and prevention programmes. Formally, "epidemiological modeling" refers to dynamic modeling, where sometime, the population is divided into compartments based on their health status, namely, susceptible, exposed, infectious or recovered. The movement between compartments by becoming infected, progressing, recovering or migrating can be specified by differential equations. The advantage of mathematical modeling of infectious diseases is the economy, clarity and precision of mathematical formulation. Moreover, there are many mathematical techniques available for determining the threshold, equilibrium, bifurcation, periodic solutions and stability behaviour of the system.

Concerning predator-prey models with infectious disease, Anderson and May (1986) in [3] analyzed the invasion, persistence and spread of infectious diseases within animal and plant communities. In that model, they assume that the disease is transmitted only within the prey species, with bilinear incidence. Since their results, there has been many predator-prey models with infectious diseases. Mainul Haque and Ezio Venturino (2006) in [44] analyzed the role of transmissible diseases in the Holling-Tanner predator-prey model;

Bairagi, Roy and Chattopadhyay (2007) in [33] analyzed the role of infection on the stability of a predator-prey system with several response functions. Pada Das, Kundu and Chattopadhyay (2011) in [34] analyzed a predator-prey mathematical model without harvesting, with both the populations affected by diseases but they used Holling type-I functional response in their paper. There has been many studies on the effects of harvesting on population growth [22, 29, 46]. Makinde [46] thinks that from the point of view of human needs, the exploitation of biological resources and harvesting of populations are commonly practiced in fishery, forestry, and wildlife management. There is a wide range of interest in the use of bioeconomic models to gain insight into the scientific management of renewable resources like fisheries and forestries. These models considered in the literature were all different from the model considered in this paper with the sigmoidal response function of type III, prey's harvesting and bilinear incidence. Our goal is to analyze the influence of a Susceptible-Infectious-Susceptible (SIS) infectious disease in the dynamics of the two populations. Despite numerous works on predator-prey system incorporating the harvesting, to this day, still no scholar studies System (1), System (3) and System (5) with the assumption of the harvesting on prey species, on predator species and prey and predator species. Although, there are some similarities in our models considered here with the others in the literature, concerning some assumptions.

2. Description of the Model

The models considered here consist of two populations. The variables $x(t)$ and $y(t)$ denote the densities of preys and predators respectively. γ is the recover rate of infectious preys to become susceptible and λ is the adequate contact rate between susceptible and infected prey. The following assumptions hold for our models.

1. In the absence of disease and predation, the prey population grows according to the logistic law with intrinsic growth rate $r > 0$ and carrying capacity $K > 0$.

2. In the presence of disease in prey, predator or both prey and predator populations respectively, the prey population is divided into two classes namely, susceptible prey $x(t)$ and infectious prey $z(t)$, at any time t, the predator population is divided into two classes namely, susceptible predator $y(t)$ and infectious predator $w(t)$, at any time t.

3. Only the susceptible preys are assumed to be capable of reproducing with logistic law. The infected preys in System (1) and System (5) are removed by recover with rate γ, death with rate q_1 or predation. However, the infected and susceptible also contribute to the population growth toward the carrying capacity. The infected predators in System (3) and System (5) are removed by recover with rate μ, ν or death with rate $\bar{d}$.

4. There is a constant death rate $\bar{d} > 0$ for the predators and the predation coefficients are $m > 0$, $m_1 > 0$. The coefficient of conversing prey into predator is c, with $0 < c \leq 1$. The response functional of the predator is assumed to be sigmoidal :
$$\phi_3(x) = \frac{\omega_0\, x^2}{x^2 + \omega_1\, x + \omega_2} = \frac{m\, x^2}{ax^2 + bx + 1}, \text{ where } m = \frac{\omega_0}{\omega_2}, a = \frac{1}{\omega_2}, b = \frac{\omega_1}{\omega_2}.$$

5. A susceptible or infected predator cannot distinguish an infected or healthy prey.

6. The disease is not genetically inherited. The infected population do not become immune.

2.1. The Model when Disease Spreads among Prey Population

The model obtained with the simple mass action when the disease spreads only among the prey population is defined by :

$$\begin{cases} \dot{x}(t) = rx(t)\left(1 - \dfrac{x(t)}{K}\right) - \dfrac{m\,x^2(t)y(t)}{a\,x^2(t) + b\,x(t) + 1} - \lambda\,x(t)\,z(t) - h\,x(t) + \gamma\,z(t), \\ \dot{z}(t) = \lambda\,x(t)\,z(t) - \dfrac{m_1\,z^2(t)\,y(t)}{a\,z^2(t) + b\,z(t) + 1} - (\gamma + q_1)z(t), \\ \dot{y}(t) = \dfrac{c\,m\,x^2(t)\,y(t)}{a\,x^2(t) + b\,x(t) + 1} - \dfrac{m_2\,z^2(t)\,y(t)}{a\,z^2(t) + b\,z(t) + 1} - d\,y(t), \\ x(t) \geq 0,\ z(t) \geq 0,\ y(t) \geq 0, \end{cases} \tag{1}$$

where the constant h denotes the rate of preys's harvesting and is also called the intensity of harvesting, d is the death rate of infected predators, the parameter m_2 which can be written as $m_2 = c_1\,m_1$ can be positive or negative. When $m_2 < 0$, the consumption of infected preys has positive effect for predator dynamics and when $m_2 > 0$, this consumption of infected preys has bad effect.

The parameters r, k, a, m, c, d, h are strictly positive and $b \geq 0$. Trough the linear transformation and time scaling $(X, Z, Y, T) = \left(\dfrac{x}{K}, \dfrac{z}{K}, \dfrac{y}{cK}, c\,m\,K^2\,t\right)$, System (1) becomes

$$\begin{cases} \dot{x}(T) = \rho\,x(T)(1 - x(T)) - y(T)\,p(x) - \delta\,x(T)\,z(T) - h\,x(T) + \theta\,z(T), \\ \dot{z}(T) = \delta\,x(T)\,z(T) - \bar{m}_1\,y(T)\,p(z) - (\theta + q)z(T), \\ \dot{y}(T) = y(T)\,p(x) - \bar{m}_2\,y(T)\,p(z) - \eta\,y(T), \\ x(T) \geq 0,\ z(T) \geq 0,\ y(T) \geq 0, \end{cases} \tag{2}$$

where the function p is given by $p(x) = \dfrac{x^2}{a\,x^2 + b\,x + 1}$, and the other parameters are $\rho = \dfrac{r}{c\,m\,k^2}$, $\alpha = a\,K^2$, $\beta = b\,K$, $\delta = \dfrac{\lambda}{c\,m\,K}$, $h = \dfrac{h_1}{c\,m\,K^2}$, $\theta = \dfrac{\gamma}{c\,m\,K^2}$, $\bar{m}_1 = \dfrac{m_1}{m}$, $\bar{m}_2 = \dfrac{m_2}{cm}$, $\eta = \dfrac{d}{c\,m\,K^2}$, $q = \dfrac{q_1}{c\,m\,K^2}$.

2.2. The Model when Disease Spreads among Predator Population

We assume here that infected predators still can catch preys at a different rate η than sound ones. The parameter η can be thought to be less than m if the disease affects the ability of hunting of the predators or larger than m, if we want to emphasize that the interactions with infected predators cause the preys to die because of disease even if they are not caught. ν is the recovered rate of infected predators, δ is the adequate contact rate between susceptible and infected predators. All these parameters are positives. The model obtained with the simple mass action is

$$\begin{cases} \dot{x} = r\,x\left(1 - \dfrac{x}{K}\right) - \dfrac{m\,x^2 y}{a\,x^2 + b\,x + 1} - \dfrac{\eta\,x^2 w}{a\,x^2 + b\,x + 1} - h_1\,x, \\ \dot{y} = \dfrac{c\,m\,x^2 y}{a\,x^2 + b\,x + 1} - \delta\,y\,w + \nu\,w - d\,y, \\ \dot{w} = \delta\,y\,w + \dfrac{e\,m\,x^2\,w}{a\,x^2 + b\,x + 1} - (\nu + \bar{d})w, \\ x \geq 0, \quad y \geq 0, \quad w \geq 0. \end{cases} \tag{3}$$

Through the following linear transformation and time scaling $(X, Y, W, T) = \left(\dfrac{x}{K}, \dfrac{y}{cK}, \dfrac{w}{cK}, cmK^2 t\right)$, we can reduce the number of parameters and then obtain the following system

$$\begin{cases} \dot{x}(T) = \rho x(T)(1 - x(T)) - y(T)\,p(x) - hx(T) - \eta_1 w(T)p(x), \\ \dot{y}(T) = y(T)p(x) - \delta_1 y(T)w(T) + \nu_1 w(T) - d_1 y(T), \\ \dot{w}(T) = \delta_1 y(T)w(T) + e_1 w(T)p(x) - (\nu_1 + \bar{d}_1)w(T), \\ x \geq 0, \quad y \geq 0, \quad w \geq 0, \end{cases} \tag{4}$$

where $p(x) = \dfrac{x^2}{\alpha x^2 + \beta x + 1}$, and the parameters are $\rho = \dfrac{r}{cmk^2}$, $h = \dfrac{h_1}{cmK^2}$, $\eta_1 = \dfrac{\eta}{m}$, $\alpha = a\,K^2$, $\beta = b\,K$, $\delta_1 = \dfrac{\delta}{m\,K}$, $\nu_1 = \dfrac{\nu}{cmK^2}$, $d_1 = \dfrac{d}{cmK^2}$, $e_1 = \dfrac{e\eta}{c\,m}$, $\bar{d}_1 = \dfrac{\bar{d}}{cmK^2}$.

2.3. The Model when Disease Spreads among Both Prey and Predator Populations

The following system is obtained with disease in prey and predator populations:

$$\begin{cases} \dot{x} = r\,x\left(1 - \dfrac{x}{k}\right) - \lambda\,x\,z - h_1\,x + \gamma\,z - \dfrac{m_1\,x^2\,y}{a\,x^2 + b\,x + 1} - \dfrac{\eta_1\,x^2\,w}{a\,x^2 + b\,x + 1}, \\ \dot{z} = \lambda\,x\,z - (\gamma + q_1)z - \dfrac{m_1\,z^2\,y}{a\,z^2 + b\,z + 1} - \dfrac{\eta_2\,z^2\,w}{a\,z^2 + b\,z + 1}, \\ \dot{y} = \mu\,w - \delta\,y\,w - d\,y + \dfrac{c\,m\,x^2\,y}{a\,x^2 + b\,x + 1} - \dfrac{c_1\,m_1\,z^2\,y}{a\,z^2 + b\,z + 1}, \\ \dot{w} = \delta\,y\,w - \mu\,w - \bar{d}\,w + \dfrac{e\,\eta_1\,x^2\,w}{a\,x^2 + b\,x + 1} - \dfrac{c_2\,\eta_2\,z^2\,w}{a\,z^2 + b\,z + 1}, \\ x \geq 0,\ z \geq 0,\ y \geq 0,\ w \geq 0, \end{cases} \tag{5}$$

where μ is the recover rate of infected predators, δ is the adequate contact rate between susceptible and infected predators, the rest of parameters are given as in the previous cases. In order to reduce the number of parameters, we consider the following transformation $X = \dfrac{x}{k}$, $Z = \dfrac{z}{k}$, $Y = \dfrac{y}{c\,k}$ $W = \dfrac{w}{c\,k}$, $T = c\,m\,k^2\,t$, $\rho = \dfrac{r}{c\,m\,k^2}$, $\bar{m}_1 = \dfrac{m_1}{m}$, $h = \dfrac{h_1}{c\,m\,k^2}$, $\bar{\lambda} = \dfrac{\lambda}{c\,m\,k}$, $\bar{\gamma} = \dfrac{\gamma}{c\,m\,k^2}$, $\bar{\eta}_1 = \dfrac{\eta_1}{m}$, $\bar{d} = \dfrac{d}{c\,m\,k^2}$, $\bar{m}_2 = \dfrac{c_1\,m_1}{c\,m}$, $\bar{\delta} = \dfrac{\delta}{m\,k}$, $\bar{\mu} = \dfrac{\mu}{c\,m\,k^2}$, $e_1 = \dfrac{e\,\eta_1}{c\,m}$, $\bar{m}_3 = \dfrac{m_3}{c\,m}$, $\bar{\eta}_2 = \dfrac{\eta_2}{m}$, $\bar{\eta}_3 = \dfrac{c_2\,\eta_2}{c\,m}$, $\bar{q} = \dfrac{q_1}{c\,m\,k^2}$, $\bar{d}_2 = \dfrac{d}{c\,m\,k^2}$. System (5) becomes

$$\begin{cases} \dot{x} = \rho\, x(1-x) - \bar{\lambda} x\, z - h\, x + \bar{\gamma} z - \,y\, p(x) - \bar{\eta}_1\, w\, p(x), \\ \dot{z} = \bar{\lambda}\, x\, z - (\bar{\gamma} + q) z - \bar{m}_1\, y\, p(z) - \bar{\eta}_2\, w\, p(z), \\ \dot{y} = \bar{\mu}\, w - \bar{\delta}\, y\, w - \bar{d}_1\, y + y\, p(x) - \bar{m}_2\, y\, p(z), \\ \dot{w} = \bar{\delta}\, y\, w - (\bar{\mu} + \bar{d}_2)\, w + e_1\, w\, p(x) - \bar{\eta}_3\, w\, p(z), \\ x \geq 0,\ z \geq 0,\ y \geq 0,\ w \geq 0, \end{cases} \tag{6}$$

where $p(x)$ is defined as previously.

3. Mathematical Analysis

Lemma 1. *Let us consider System (2), System (4) and System (6).*

1. *The set* $\Omega = \left\{ (x,\, z,\, y) \in \mathbb{R}_3^+ : 0 < S = x + z + y \leq \dfrac{\chi}{\xi} \right\}$ *is the attracting region for all solutions initiating in the interior of the positive orthant, where ξ is a constant such that $\xi < \min\{\eta;\, q\}$ and* $\chi = \dfrac{(\rho - h + \xi)^2}{4\rho}$.

2. *Every solution of System (4) is nonnegative for $t > 0$. Furthermore, for sufficiently large value of t, these solutions are bounded.*

3. *Every solution of System (6) is nonnegative for $t > 0$. Furthermore, for sufficiently large value of t, these solutions are bounded.*

Proof.

1. For System (2), let $S(t) = x(t) + z(t) + y(t)$ and $\xi > 0$ be a constant. Then

$$\frac{dS}{dt} + \xi S = (\rho - h + \xi)x - \rho\, x^2 + (\xi - q)z + (\xi - \eta)y - (\bar{m}_1 + \bar{m}_2) y\, p(z). \tag{7}$$

We choose ξ such that $0 < \xi < \min\{\eta; q\}$. Then Eq (7) can be written as

$$\begin{aligned} \frac{dS}{dt} + \xi S \ &\leq\ (\rho - h + \xi)x - \rho x^2 \leq -\rho \left[\left(x - \frac{\rho - h + \xi}{2\rho} \right)^2 - \left(\frac{\rho - h + \xi}{2\rho} \right)^2 \right], \\ &\leq\ \frac{(\rho - h + \xi)^2}{4\rho} - \rho \left(x - \frac{\rho - h + \xi}{2\rho} \right)^2 \leq \frac{(\rho - h + \xi)^2}{4\rho} = \chi. \end{aligned}$$

By using the differential inequality [47], we obtain $0 < w(t) \leq \dfrac{\chi}{\xi}(1 - e^{-\xi t}) + w(0)e^{-\xi t}$. The limit when $t \to \infty$ gives $0 < w(t) \leq \dfrac{\chi}{\xi}$. This end the proof of the first part.

2. We define the function $A : \mathbb{R}_+ \to \mathbb{R}_+$ by $A(t) = d_1 x + y + w$. Then,

$$\frac{dA}{dt} + d_1 A = -\psi \rho x^2 + \psi x(\rho - h + \psi) + y(-\psi p(x) + p(x)) + w(-\psi \eta_1 p(x) + e_1 p(x) - \bar{d}_1 + d_1).$$

Setting $\psi = \max\left\{1; \dfrac{e_1}{\eta_1}\right\}$. Therefore,

$$\frac{dA}{dt} + d_1 A \leq -\psi\rho x^2 + \psi x(\rho - h + d_1) \leq \frac{\psi(\rho - h + d_1)^2}{4\rho}.$$

Using again the differential inequality [47], gives $0 < A(t) \leq \dfrac{\psi(\rho - h + d_1)}{4\rho d_1}(1 - e^{-d_1 t}) + A(0)e^{-d_1 t}$. The limit when $t \to \infty$, gives $0 < A(t) \leq \dfrac{\psi(\rho - h + d_1)}{4\rho d_1}$. This end the proof of the second part of the lemma.

3. Let's consider the function B define by $B(t) = u\,x + u\,z + y + w$, where u is a positive constant. Let $s \in \mathbb{R}_+$. Then,

$$\begin{aligned}
\frac{dB}{dt} + sB &= u\rho x(1-x) - uh\,x - uyp(x) - u\bar{\eta}_1 wp(x) - u\,q\,z - u\bar{m}_1\,y\,p(z) - u\bar{\eta}_2 wp(z) \\
&- \bar{d}_1 + y\,p(x) - \bar{m}_2\,y\,p(z) - \bar{d}_2 w + e_1 wp(x) - \bar{\eta}_3 wp(z) + s\,u\,x + s\,u\,z + s\,y + s\,w, \\
&\leq -u\rho x^2 + u\,x(\rho - h + s) + uz(s - q) + y(-up(x) - \bar{d}_1 + p(x) + s) \\
&+ w(-u\bar{\eta}_1 p(x) - \bar{d}_2 + e_1 p(x) + s).
\end{aligned}$$

Setting $u = \max\left\{1, \dfrac{e_1}{\eta_1}\right\}$. Therefore,

$$\frac{dB}{dt} + sB \leq -u\rho x^2 + u\,x(\rho - h + s) + uz(s - q) + y(-\bar{d}_1 + s) + w(-\bar{d}_2 + s).$$

Taking $s = \min\{\bar{d}_1, q\}$ gives

$$\frac{dB}{dt} + sB \leq -u\rho x^2 + u\,x(\rho - h + s) \leq \frac{u(\rho - h + s)^2}{4\rho}.$$

This end the proof of the third part of the lemma.

3.1. Existence of Equilibria

We are only interested in analyzing the solutions in the nonnegative closed first quadrant because the solutions outside of the first quadrant are not biologically interpretable. The following theorems hold concerning possible equilibria of the models.

Theorem 1. *Let us consider System (2). Setting* $\Delta = \eta^2\beta^2 + 4\eta(1 - \eta\alpha)$. *The possible equilibria are given by* $E_0 = (0,0,0)$ *which always exists;* $E_1 = (\tilde{x}_1, 0, 0)$ *where* $\tilde{x}_1 = \dfrac{\rho - h}{\rho}$ *exists if* $\rho - h > 0$; $E_2 = (\tilde{x}_2, \tilde{z}_2, 0)$ *where* $\tilde{x}_2 = \dfrac{\theta + q}{\delta}$, $\tilde{z}_2 = \dfrac{\tilde{x}_2(\rho - h - \rho\tilde{x}_2)}{q}$ *exists if* $\rho - h > 0$ *and* $\tilde{x}_2 < \tilde{x}_1$; $E_2 = (\tilde{x}_3, 0, \tilde{y}_3)$ *where* $\tilde{x}_2 = \dfrac{\eta\beta + \sqrt{\Delta}}{2(1 - \eta\alpha)}$, $\tilde{y}_3 = \dfrac{\tilde{x}_3(\rho - h - \rho\tilde{x}_3)}{\eta}$ *exists if* $\rho - h > 0$, $1 - \eta\alpha > 0$ *and* $\tilde{x}_3 < \tilde{x}_1$; *a possible coexisting equilibrium* $E_4 = (\tilde{x}_4, \tilde{z}_4, \tilde{y}_4)$, *where* $\tilde{x}_4$ *is any positive solution of* $p(\tilde{x}_4) - \bar{m}_2 p(\tilde{z}_4) - \eta = 0$ *which exists when* $\tilde{x}_4 \in]\tilde{x}_2; \tilde{x}_1[$ *and* $\rho - h > 0$, $\tilde{y}_4 = \dfrac{(\delta\tilde{x}_4 - \theta - q)\tilde{z}_4}{\bar{m}_1\,p(\tilde{z}_4)}$, $\tilde{z}_4 = \dfrac{\bar{m}_1\tilde{x}_4(\rho - h - \rho\tilde{x}_4)(p(\tilde{x}_4) - \eta)}{\bar{m}_1(p(\tilde{x}_4) - \eta)(-\theta + \delta\tilde{x}_4) + \bar{m}_2 p(\tilde{x}_4)(\delta\tilde{x}_4 - \theta - q)}$.

Theorem 2. *Let us consider System (4). Setting* $\Delta = \beta^2 d_1^2 + 4d_1(1-d_1\alpha)$, $\bar{\Delta} = \dfrac{\beta^2 \bar{d}_1^2}{e_1^2} + \dfrac{4\bar{d}_1}{e_1}\left(1-\dfrac{\alpha\bar{d}_1}{e_1}\right)$, $b = \dfrac{\dfrac{\beta\bar{d}_1}{e_1}+\sqrt{\bar{\Delta}}}{2\left(1-\dfrac{\alpha\bar{d}_1}{e_1}\right)}$. *The possible equilibria are given by* $P_0 = (0,0,0)$ *which always exists;* $P_1 = (\hat{x}_1, 0, 0)$ *which exists if* $\rho - h > 0$*;* $P_2(\hat{x}_2, \hat{y}_2, 0)$ *where* $\hat{x}_2 = \dfrac{d_1\beta + \sqrt{\Delta}}{2(1-\alpha d_1)}$, $\hat{y}_2 = \dfrac{\hat{x}_2(\rho - h - \rho\hat{x}_2)}{d_1}$ *which exists if* $\rho - h > 0$, $1 - \alpha d_1 > 0$ *and* $\hat{x}_2 < \hat{x}_1$*;* *a possible coexisting equilibrium* $P_3(\hat{x}_3, \hat{y}_3, \hat{w}_3)$ *where* $\hat{y}_3 = \dfrac{-e_1\left(p(\hat{x}_3) - \dfrac{\nu_1 + \bar{d}_1}{e_1}\right)}{\delta_1}$, $\hat{w}_3 = -\dfrac{-\hat{y}_3(p(x_3) - d_1)}{p(x_3) - \dfrac{\bar{d}_1}{e_1}}$ *which exists if* $1 - \dfrac{\alpha\bar{d}_1}{e_1} > 0$ *and* $\hat{x}_3 \in]\hat{x}_2, b[$.

Theorem 3. *Let consider System (6). Setting* $\Delta_a = \beta^2 \bar{d}_1^2 + 4\bar{d}_1(1-\bar{d}_1\alpha)$, $\Delta_b = \dfrac{\beta^2\bar{d}_2^2}{e_1^2} + \dfrac{4\bar{d}_2}{e_1}(1-\dfrac{\alpha\bar{d}_2}{e_1})$, $a = \dfrac{\beta\bar{d}_1 + \sqrt{\Delta_a}}{2(1-\alpha\bar{d}_1)}$, $b = \dfrac{\dfrac{\beta\bar{d}_2}{e_1} + \sqrt{\Delta_b}}{2\left(1-\dfrac{\alpha\bar{d}_2}{e_1}\right)}$. *There are many possible equilibria.*

- $M_0 = (0,0,0,0)$ *which always exists;* $M_1 = (x_1, 0, 0, 0)$ *with* $x_1 = \dfrac{\rho - h}{\rho}$ *which exists if* $\rho - h > 0$*;*
- $M_2 = (x_2, z_2, 0, 0)$ *with* $x_2 = \dfrac{\bar{\gamma} + q}{\bar{\lambda}}$ *and* $z_2 = \dfrac{x_2(\rho - h - \rho x_2)}{q}$, *which exists if* $\rho - h > 0$ *and* $x_1 > x_2$*;*
- $M_3 = (x_3, 0, y_3, 0)$ *with* $x_3 = \dfrac{\beta\bar{d}_1 + \sqrt{\Delta_a}}{2(1-\alpha\bar{d}_1)}$ *and* $y_3 = \dfrac{x_3(\rho - h - \rho x_3)}{\bar{d}_1}$, *which exists if* $\rho - h > 0$, $1 - \alpha\bar{d}_1 > 0$ *and* $x_3 < x_1$*;*
- $M_4 = (x_4, z_4, y_4, 0)$ *where*

$$z_4 = \frac{\bar{m}_1 x_4(\rho - h - \rho x_4)(p(x_4) - \bar{d}_1)}{\bar{m}_1(p(x_4) - \bar{d}_1)(\bar{\lambda}x_4 - \bar{\gamma}) + \bar{m}_2 p(x_4)(\bar{\lambda}x_4 - \bar{\gamma} - q)}, \qquad y_4 = \frac{z_4(\bar{\lambda}x_4 - \bar{\gamma} - q)}{\bar{m}_1 p(z_4)} \tag{8}$$

and x_4 *is a solution of equation* $p(x_4) - \bar{m}_2 p(z_4) - \bar{d}_1 = 0$, *which exists if*

$$\rho - h > 0 \quad \text{and} \quad x_4 \in]x_2, x_1[. \tag{9}$$

- $M_5 = (x_5, 0, y_5, w_5)$ *where*

$$y_5 = \frac{-e_1\left(p(x_5) - \dfrac{\bar{\mu} + \bar{d}_2}{e_1}\right)}{\bar{\delta}}, \qquad w_5 = \frac{-y_5(p(x_5) - \bar{d}_1)}{e_1\left(p(x_5) - \dfrac{\bar{d}_2}{e_1}\right)} \tag{10}$$

and x_5 satisfies $\rho x_5(1 - x_5) - hx_5 - y_5 p(x_5) - \eta_1 w_5 p(x_5) = 0$ which exists if

$$\rho - h > 0, \qquad 1 - \frac{\alpha \bar{d}_1}{e_1} > 0 \quad \textit{and} \quad x_5 \in]a, b[. \tag{11}$$

- *A possible coexisting equilibrium $M_6(x_6, z_6, y_6, w_6)$ with nonnegative and nonzero components: $x_6 \neq 0$, $z_6 \neq 0$, $y_6 \neq 0$, $w_6 \neq 0$.*

Proof of Theorem 1. In Appendix A.

Proof of Theorem 2. In Appendix B.

Proof of Theorem 3. In Appendix C.

3.2. Stability Analysis

Theorem 4. *Let's consider System (2). The following properties hold.*

1. *The equilibrium point E_0 is a saddle point when $\rho > h$. E_0 is the unique equilibrium of model and is globally asymptotically stable in Ω when $\rho - h \leq 0$.*

2. *When $\rho - h > 0$, then the equilibrium point E_1 exists in $\mathbb{R}^3_+$ and is locally asymptotically stable if the following conditions are satisfied : $p(\tilde{x}_1) < \eta$, $\tilde{x}_1 < \dfrac{\theta + q}{\delta}$.*

3. *The equilibrium E_2 when it exists is locally asymptotically stable if $p(\tilde{x}_2) - m_2 p(\tilde{z}_2) - \eta < 0$ and $\rho - h - 2\rho\tilde{x}_2 - \delta\,\tilde{z}_2 < 0$.*

4. *When the equilibrium point E_3 exists, it is locally asymptotically stable if $\tilde{x}_3 < \tilde{x}_2$ and $\rho - h - 2\rho\tilde{x}_3 - \tilde{y}_3\, p'(\tilde{x}_3) < 0$.*

5. *When the equilibrium point E_4 exists, it is locally asymptotically stable if $a_1 > 0$, $a_3 > 0$ and $a_1 a_2 > a_3$, where $a_1 = -\lambda_1 - \lambda_2$,*

$$a_2 = \lambda_1\lambda_2 + p(\tilde{x}_4)p'(\tilde{x}_4)\tilde{y}_4 - \delta\tilde{z}_4(-\delta\tilde{x}_4 + \theta) - \bar{m}_1 p(\tilde{z}_4)\bar{m}_2\tilde{y}_4 p'(\tilde{z}_4),$$

$$a_3 = \bar{m}_1 p(\tilde{z}_4)(-\delta\tilde{x}_4 + \theta)p'(\tilde{x}_4)\tilde{y}_4 - p(\tilde{x}_4)\bar{m}_2\tilde{y}_4\delta\tilde{z}_4 p'(\tilde{z}_4) - \lambda_2\tilde{y}_4 p(\tilde{x}_4)p'(\tilde{x}_4) + \bar{m}_1\bar{m}_2\lambda_1\tilde{y}_4 p(\tilde{z}_4)p'(\tilde{z}_4),$$

$$\lambda_1 = \rho - h - \tilde{y}_4 p'(\tilde{x}_4) - \delta\tilde{z}_4 - 2\rho\tilde{x}_4 \textit{ and } \lambda_2 = \delta\tilde{x}_4 - m_1\tilde{y}_4 p'(\tilde{z}_4) - (\theta + q).$$

Theorem 5. *Let's consider System (4). The following properties hold.*

1. *The equilibrium point P_0 is a saddle point when $\rho > h$. P_0 is the unique equilibrium of model and is globally asymptotically stable when $\rho - h \leq 0$.*

2. *When the equilibrium point P_1 exists, it is locally asymptotically stable if and only if $p(\hat{x}_1) < d_1$.*

3. *When the equilibrium point P_2 exists, it is locally asymptotically stable if only if $\delta_1\hat{y}_2 + e_1 p(\hat{x}_2) - \nu_1 - \bar{d}_1 < 0$ and $\rho - h - 2\rho\hat{x}_2 - \hat{y}_2 p'(\hat{x}_2) < 0$.*

4. *The equilibrium point* P_3 *when it exists is locally asymptotically stable if* $\alpha_1 > 0$, $\alpha_3 > 0$ *and* $\alpha_1\alpha_2 > \alpha_3$, *where* $\alpha_1 = -u_1 - u_2$, $u_1 = \rho - h - 2\rho\hat{x}_3 - \hat{y}_3 p'(\hat{x}_3) - \eta_1\hat{w}_3 p'(\hat{x}_3)$, $u_2 = p(\hat{x}_3) - \delta_1\hat{w}_3 - d_1$,

$$\alpha_2 = u_1u_2 + \eta_1 p(\hat{x}_3)e_1\hat{w}_3 p'(\hat{x}_3) + p(\hat{x}_3)\hat{y}_3 p'(\hat{x}_3) - \delta_1\hat{w}_3(-\delta_1\hat{y}_3 + \nu_1),$$

$$\alpha_3 = e_1 p(\hat{x}_3)\hat{w}_3 p'(\hat{x}_3)(-\delta_1\hat{y}_3+\nu_1)+\eta_1 p(\hat{x}_3)\hat{y}_3 p'(\hat{x}_3)\delta_1\hat{w}_3-\eta_1 p(\hat{x}_3)e_1\hat{w}_3 p'(\hat{x}_3)u_2+\delta_1\hat{w}_3u_1(-\delta_1\hat{y}_3+\nu_1).$$

Theorem 6. *Let's consider System (6). The following properties hold.*

1. *The equilibrium* M_0 *is locally asymptotically stable if* $\rho \leq h$.

2. *When* $\rho > h$, *the equilibrium point* M_1 *exists in* $\mathbb{R}_4^+$ *and is locally asymptotically stable if* $x_1 < x_2$ *and* $p(x_1) < \frac{\bar{\mu}+\bar{d}_2}{e_1}$.

3. *The equilibrium* M_2 *when it exists is locally asymptotically stable if* $e_1p(x_2) - \eta_3 p(z_2) - (\bar{\mu} + d_2) < 0$, $p(x_2) - \bar{m}_2 p(z_2) - \bar{d}_1$ *and* $\rho - h - 2\rho x_2 - \bar{\lambda} z_2 < 0$.

4. *The equilibrium* M_3 *when it exists, is locally asymptotically stable if* $\bar{\delta} y_3 + e_1 p(x_3) - \bar{\mu} - \bar{d}_2 < 0$, $x_3 < x_2$ *and* $\rho - h - 2\rho x_3 - y_3 p'(x_3) < 0$.

5. *Equilibrium* M_4 *is locally asymptotically stable if* $\bar{\delta} y_4 + e_1 p(x_4) - \eta_3 p(z_4) - \bar{\mu} - \bar{d}_2 < 0$, $a_1 > 0$, $a_3 > 0$ *and* $a_1a_2 > a_2$, *where* $a_1 = -\lambda_1 - \lambda_2$,

$$a_2 = \lambda_1\lambda_2 + p(x_4)y_4p'(x_4)\lambda_2 - \bar{\lambda}z_4(-\bar{\lambda}x_4 + \bar{\gamma}) - \bar{m}_1 p(z_4)\bar{m}_2 y_4 p'(z_4),$$

$$a_3 = \bar{m}_1 p(z_4)y_4 p'(x_4)(-\bar{\lambda}x_4 + \bar{\gamma}) - p(x_4)\bar{\lambda}z_4\bar{m}_2 y_4 p'(z_4) - p(x_4)y_4p'(x_4)\lambda_2 + \bar{m}_1 p(z_4)\bar{m}_2 y_4 p'(z_4)\lambda_1,$$

with $\lambda_1 = \rho - h - 2\rho x_4 - \bar{\lambda} z_4 - y_4 p'(x_4)$, $\lambda_2 = \bar{\lambda} x_4 - \bar{m}_1 y_4 p'(z_4) - \bar{\gamma} - q$ *and* $\lambda_3 = \bar{\delta} y_4 + e_1 p(x_4) - \eta_3 p(z_4) - \bar{\mu} - \bar{d}_2$.

6. *Let* $\lambda_1 = \rho - h - 2\rho x_5 - y_5 p'(x_5) - \eta_1 w_5 p'(x_5)$ *and* $\lambda_3 = -\bar{\delta} w_5 - \bar{d}_1 + p(x_5)$. *The equilibrium* M_5 *is locally asymptotically stable if* $x_5 < x_2$, $\alpha_1 > 0$, $\alpha_3 > 0$ *and* $\alpha_1\alpha_2 > \alpha_3$ *where* $\alpha_1 = -\lambda_1 - \lambda_3$,

$$\alpha_2 = \lambda_1\lambda_3 + \eta_1 p(x_5)e_1 w_5 p'(x_5) + y_5 p'(x_5)p(x_5) - \bar{\delta} w_5(\bar{\mu} - \bar{\delta} y_5),$$

$$\alpha_3 = p(x_5)e_1 w_5 p'(x_5)(\bar{\mu} - \bar{\delta} y_5) + \eta_1 p(x_5)y_5 p'(x_5)\bar{\delta} w_5 - \eta_1 p(x_5)e_1 w_5 p'(x_5)\lambda_3 + \bar{\delta} w_5(\bar{\mu} - \bar{\delta} y_5)\lambda_1.$$

Proof of Theorem 4. In Appendix D.

Remark 1. *The global stability of* E_0 *allow us to conclude that if the harvesting of the prey population exceeds it growth rate, then the populations of preys and predators will disappear.*

Proof of Theorem 5. In Appendix E.

Proof of Theorem 6. In Appendix F.

4. Existence of Periodic Solutions

Theorem 7. *There is no periodic solution for System (2) in the interior of the positive quadrant of the xz-plane.*

Proof. Let $H(x,z) = \frac{1}{xz}$. It is obvious that $H(x,z)$ is positive in the interior of the nonnegative quadrant of the xz-plane. Let $h_1(x,z) = \rho x(1_x) - \delta xz - hx + \theta z$ and $h_2(x,z) = \delta xz - (\theta + q)z$. Then

$$\Delta(x,z) = \frac{\partial}{\partial x}(h_1 H) + \frac{\partial}{\partial y}(h_2 H) = -\left(\frac{\rho}{z} + \frac{\theta}{x^2}\right) < 0.$$

From the above equation, we note that $\Delta(x,z)$ does not change sign and is not identically zero in the interior of the positive quadrant of the xz-plane. By Dulac-Bendixon criteria, it follows that there is no closed trajectory in the interior of the positive quadrant of the xz-plane, and hence the theorem follows.

Theorem 8. *There is no periodic solution for System (4) in the interior of the positive quadrant of the xy-plane.*

Proof. Let $H(x,y) = \frac{1}{xy}$. It is obvious that $H(x,y)$ is positive in the interior of the nonnegative quadrant of the xy-plane. Let $f_1(x,y) = -\delta_1 yw + \nu_1 w - d_1 y$ and $f_2(x,y) = \delta_1 yw - (\nu_1 + \bar{d}_1)w$. Then

$$\Delta(x,y) = \frac{\partial}{\partial x}(f_1 H) + \frac{\partial}{\partial y}(f_2 H) = -\frac{\nu_1}{y^2} < 0.$$

From the above equation, we note that $\Delta(x,y)$ does not change sign and is not identically zero in the interior of the positive quadrant of the xy-plane. By Dulac-Bendixon criteria, it follows that there is no closed trajectory in the interior of the positive quadrant of the xy-plane, and hence the theorem follows.

5. Conclusion

Three predator-prey models incorporating harvesting have been studied in this work. A suitable harvesting can have influence on the persistence property of our harvesting systems. Concerning the analysis of harvesting on the dynamic behaviors, our systems exhibit some interesting phenomenon such as possibility of using in order to help human improving the scientific management of renewable resources such as fisheries and forest trees. One can also consider the economic profit of the harvesting. The presence of disease can destabilize these properties if the disease persists in the population. Something should be done in order to control and prevent the disease. The biological interpretation of bifurcations is that the predator coexists with the susceptible preys and the infected preys, exhibiting oscillatory balance behavior. To avoid extinction of the species, one should look carefully some parameters, namely, the rate of infection, the death rate of the infective population and the growth rate of the susceptible population.

6. Numerical Simulations

6.1. For Predator-Prey Model with Disease Spreads among the Preys

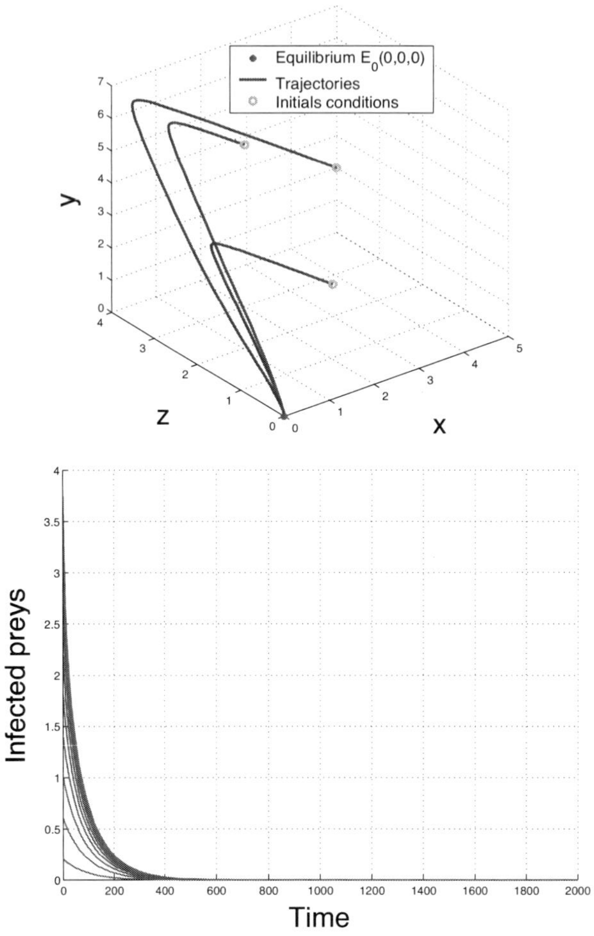

Figure 1. Time evolution of each variable for System (2) when $\rho - h < 0$ and with many initials conditions values. The population converges to the equilibrium $E_0(0, 0, 0)$.

Figure 1 illustrates the stability of equilibrium $E_0(0, 0, 0)$ and Figure 2 illustrates the stability of equilibrium $E_1(0.6, 0, 0)$. Figure 3 illustrates the stability of $E_2(0.25, 0, 0.33)$, Figure 4 confirms the stability of E_3, Figure 5 confirms the stability of E_4.

6.2. For Predator-Prey Model with Disease Spreads among the Predators

Figure 6 illustrates the stability of equilibrium $P_0(0, 0, 0)$ and Figure 7 illustrates the stability of equilibrium $P_1(0.27, 0, 0)$. Figure 8 illustrates the stability of $P_2(0.089, 0.364, 0)$, Figure 9 confirms the stability of $P_3(0.25; 0.19; 0.64)$.

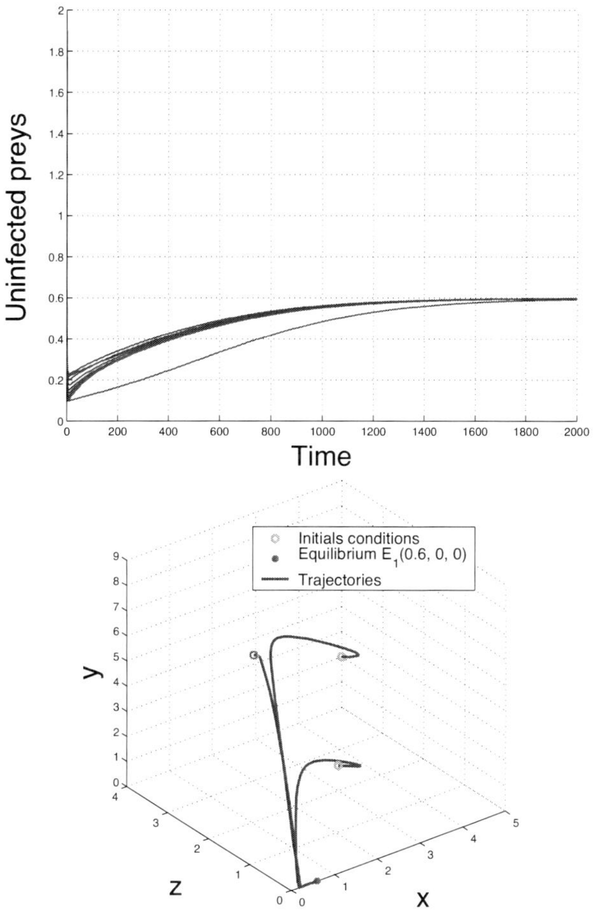

Figure 2. Behaviors of the uninfected preys, infected preys and predators in time of System (2) for $\rho = 0.0051$, $h = 0.002$, $\eta = 0.5$. We have $\tilde{x}_1 < \dfrac{\theta + q}{\delta}$, $p(\tilde{x}_1) < \eta$ and population converges to the equilibrium $E_1(0.6, 0, 0)$.

6.3. For Predator-Prey Model with Disease Spreads among Both the Predator and Prey Populations

Figure 10 illustrate the stability of equilibrium $M_4(0.14, 0.14, 0.38, 0)$. Figure 11 illustrates the stability of equilibrium $M_5(0.11, 0, 0.55, 0.57)$.

Appendix A. The Proof of Theorem 1

Equilibria of System (2) can be obtained by setting the right hand side of all equations equal to zero. It can be checked that System (2) has five possible nonnegative equilibria. When y and z are equal to zero in the system at equilibrium, we obtain the following equation:

$$\rho x(\rho - h + \rho x) = 0. \tag{12}$$

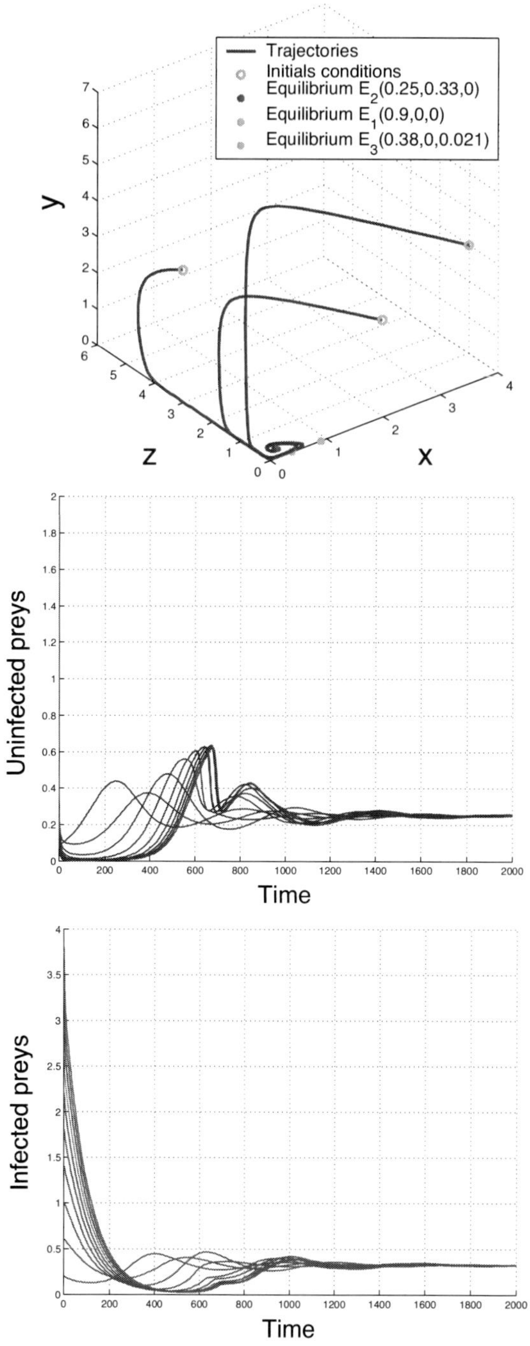

Figure 3. The equilibrium $E_2(0.25, 0, 0.33)$ of System (2) is locally asymptotically stable when $p(\tilde{x}_2) - \bar{m}_2 p(\tilde{z}_2) - \eta < 0$ and $\rho - h - 2\rho\tilde{x}_2 - \delta z_2 < 0$.

From equation (12), one has $x = 0$ or $x = \dfrac{\rho - h}{\rho}$. So, the equilibrium namely $E_0 = (0, 0, 0)$ always exists and $E_1(\tilde{x}_1, 0, 0)$ exists if $\rho - h > 0$ where $\tilde{x}_1 = \dfrac{\rho - h}{\rho}$. When $y = 0$

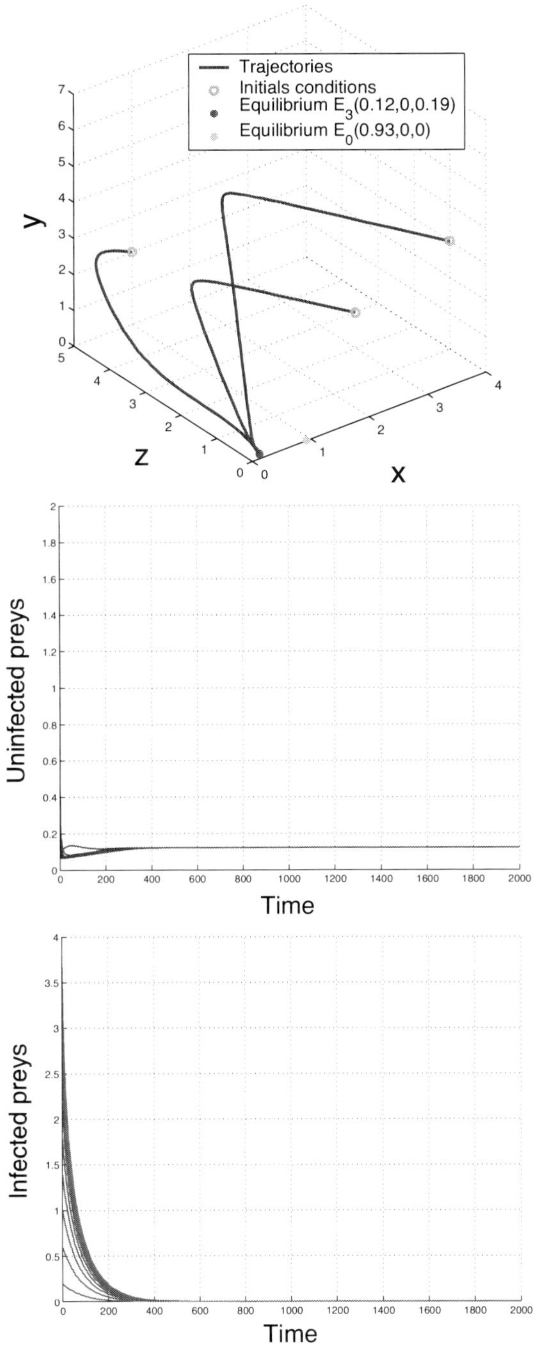

Figure 4. Behaviors of the uninfected preys, infected preys and predators of System (2) in time when Equilibrium E_3 is asymptotically stable.

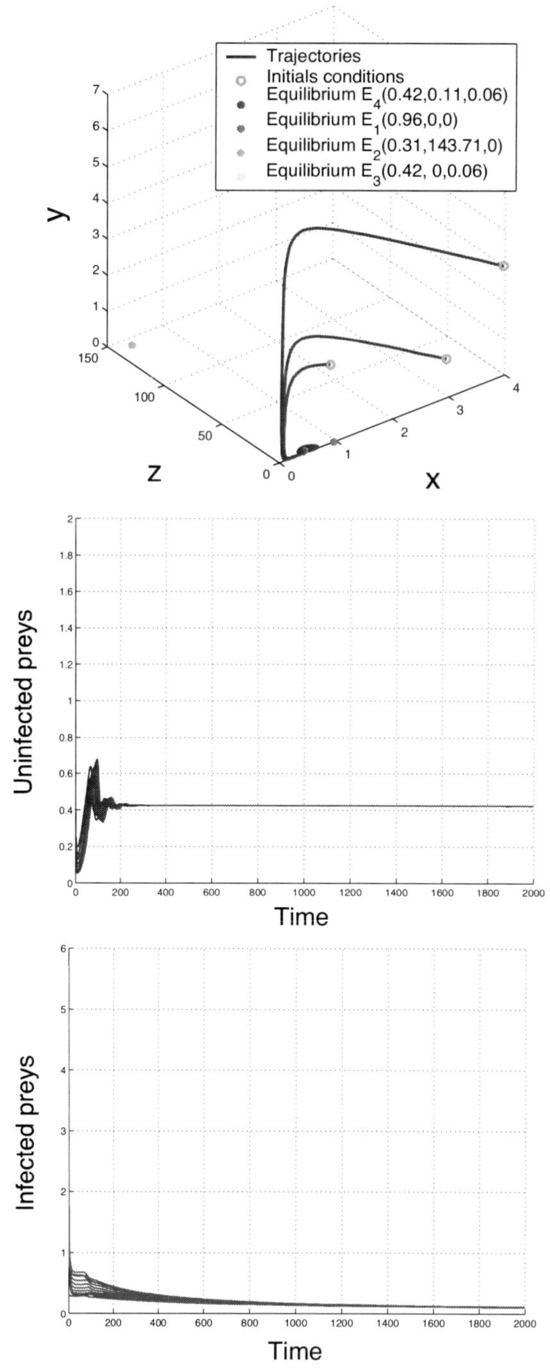

Figure 5. Behaviors of the uninfected preys, infected preys and predators of System (2) in time when Equilibrium E_4 is asymptotically stable.

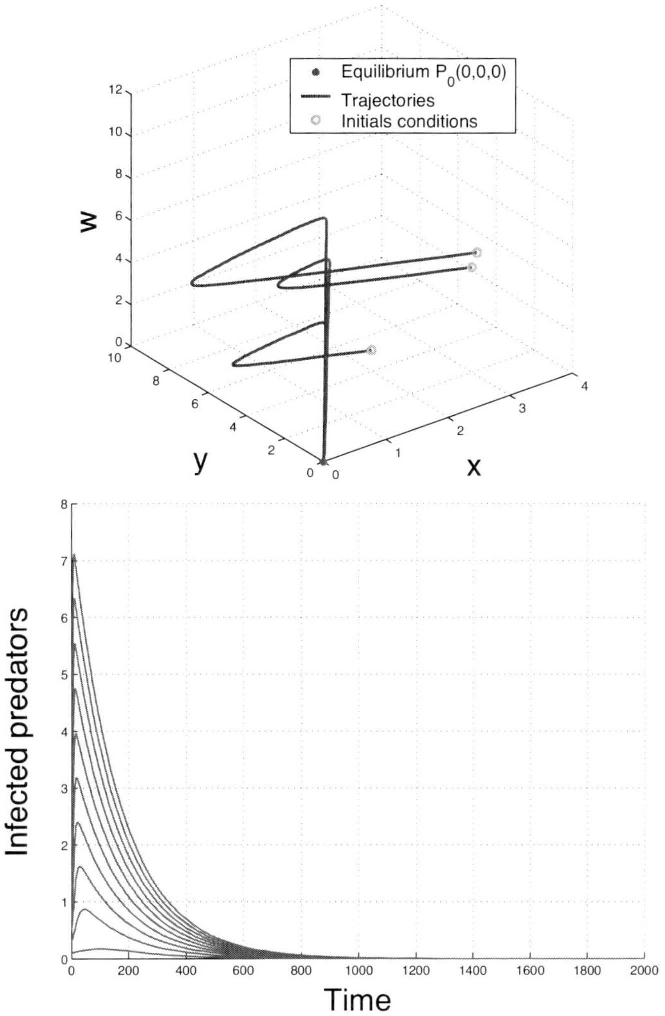

Figure 6. Time evolution of all the population for System (4) when $\rho - h < 0$ and with many initials conditions values. Population converges to the equilibrium $P_0(0, 0, 0)$.

and $z \neq 0$, we have $x = \dfrac{\theta + q}{\delta}$ and $z = \dfrac{x(\rho - h - \rho x)}{q}$. So, equilibrium $E_2(\tilde{x}_2, \tilde{z}_2, 0)$ exists if $\rho - h > 0$ and $\tilde{x}_2 < \tilde{x}_1$. When $z = 0$ and $y \neq 0$, we have

$$\begin{cases} \rho x(1 - x) - yp(x) - hx = 0, \\ p(x) - \eta = 0. \end{cases} \tag{13}$$

From the second equation of (13), we have $p(x) = \eta$. Substituting the value of $p(x)$ in the first equation of (13), we obtain

$$\tilde{y} = \frac{x(\rho - h - \rho x)}{\eta}. \tag{14}$$

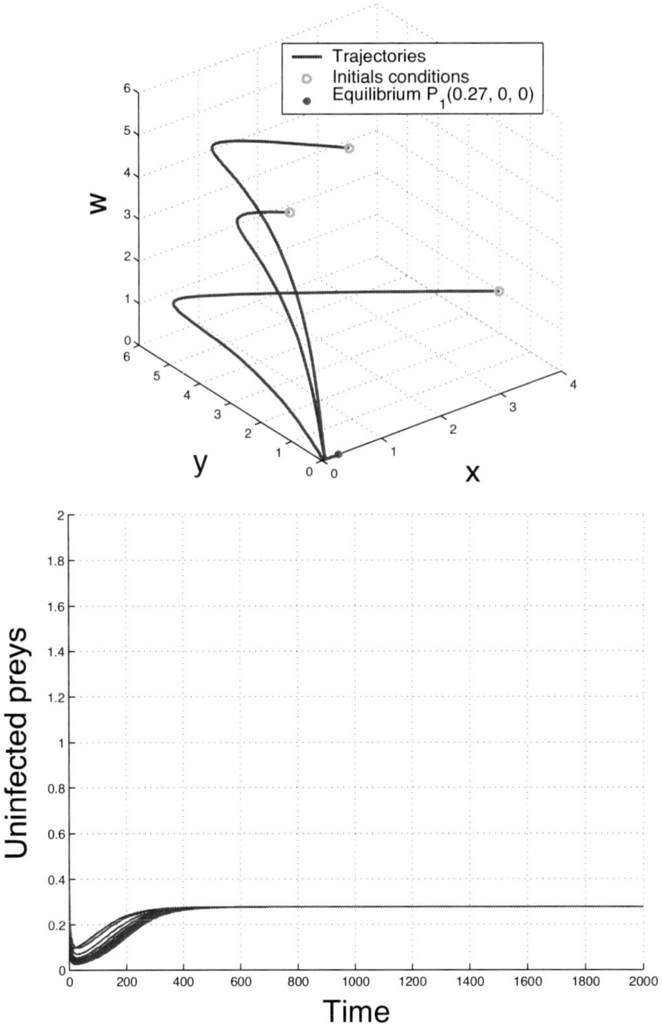

Figure 7. Time evolution of all the population for System (4) when equilibrium P_1 is asymptotically stable. Population converges to the equilibrium $P_1(0.27, 0, 0)$.

Since x and y are positive, we have $x \in]\,0\,; \dfrac{\rho - h}{\rho}[$, with $\rho - h > 0$. From the second equation of (13), a little algebraic manipulation yields

$$(1 - \eta\alpha)x^2 - \eta\beta x - \eta = 0. \tag{15}$$

- If $1 - \eta\alpha = 0$, then we have $x = \dfrac{-1}{\beta} < 0$ and is not admissible.

- If $1 - \eta\alpha \neq 0$, then, the discriminant of equation (15) is given by $\Delta = \eta^2\beta^2 + 4\eta(1 - \eta\alpha)$. If $\Delta < 0$, then equation (15) has no solution. If $\Delta = 0$, then the unique solution of equation (15) is $\tilde{x} = \dfrac{\eta\beta}{2(1-\eta\alpha)}$. But, $\Delta = 0$ yields to $(1 - \eta\alpha) = -\dfrac{\eta\beta^2}{4} < 0$.

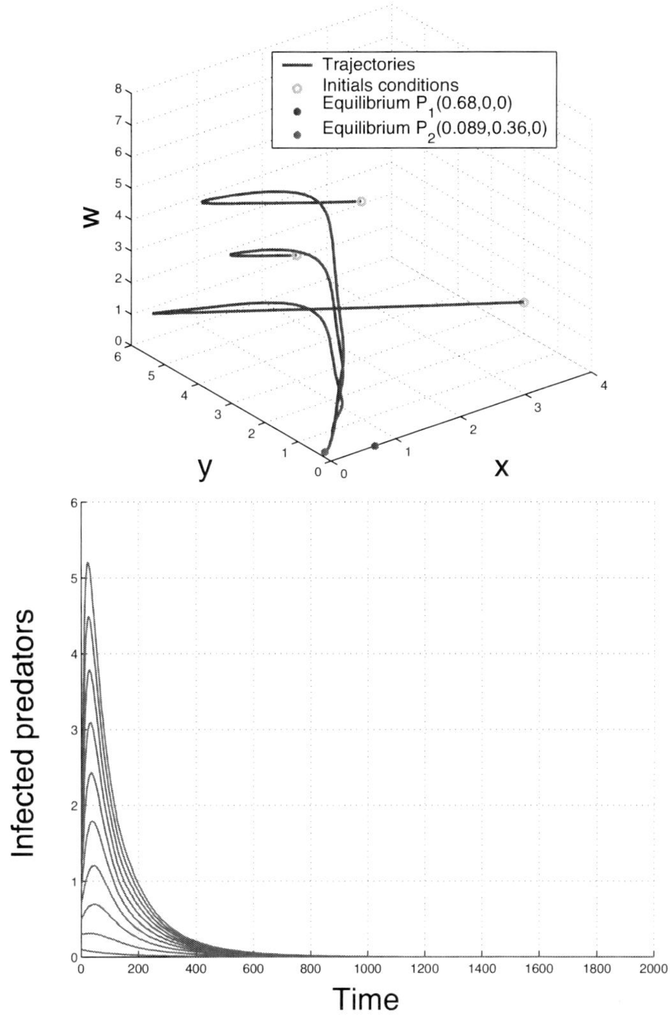

Figure 8. Time evolution of all the population for System (4) when equilibrium P_2 is asymptotically stable. Equilibrium P_1 exist and is unstable. Population converges to the equilibrium $P_2(0.089, 0.364, 0)$.

Then $\tilde{x} < 0$ and is not admissible. If $\Delta > 0$, then equation (15) has two solutions given by: $\tilde{x}_{01} = \dfrac{\eta\beta - \sqrt{\Delta}}{2(1-\eta\alpha)}$ and $\tilde{x}_{02} = \dfrac{\eta\beta + \sqrt{\Delta}}{2(1-\eta\alpha)}$.

If $1 - \eta\alpha < 0$ then $\tilde{x}_{01} < 0$ and $\tilde{x}_{02} < 0$ which are not admissible. If $1 - \eta\alpha > 0$ then $\Delta > 0$, $\tilde{x}_{01} < 0$ and $\tilde{x}_3 = \tilde{x}_{02} > 0$. It may be note that since $\tilde{x}_3 \in]\,0\,; \tilde{x}_1[$, we have the following inequality: $\tilde{x}_3 < \tilde{x}_1$. Hence, if

$$\rho - h > 0, \quad 1 - \eta\alpha > 0, \quad \tilde{x}_3 < \tilde{x}_1, \tag{16}$$

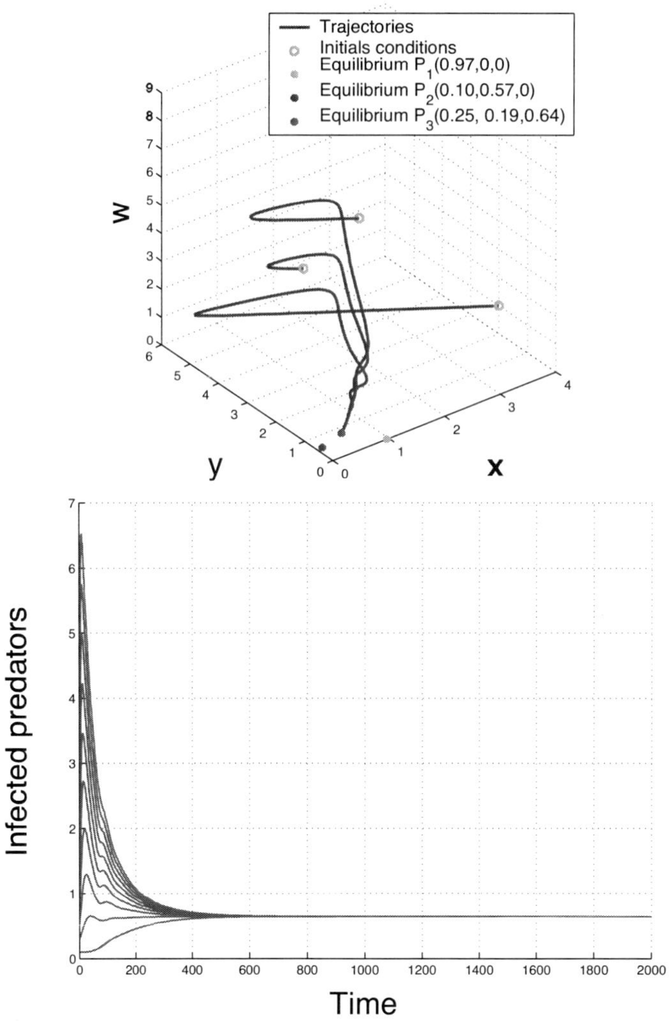

Figure 9. Time evolution of all the population for System (4) when equilibrium P_3 is asymptotically stable. Equilibria P_0, P_1 and P_2 exist and are unstable. Population converges to the equilibrium $P_3(0.25, 0.19, 0.64)$.

then the model has a positive equilibrium $(\tilde{x}_3, 0, \tilde{y}_3)$ where $\tilde{x}_3 = \dfrac{\eta\beta + \sqrt{\Delta}}{2(1-\eta\alpha)}$ and $\tilde{y}_3 = \dfrac{\tilde{x}_3(\rho - h - \rho\tilde{x}_3)}{\eta}$. This proves the existence of equilibrium $E_3 = (\tilde{x}_3, 0, \tilde{y}_3)$. Let us consider now x_4, z_4 and y_4 the positive solution of algebraic system of equations

$$\rho\tilde{x}(1-\tilde{x}) - \tilde{y}\,p(\tilde{x}) - \delta\tilde{x}\tilde{z} - h\,\tilde{x} + \theta\,\tilde{z} = 0, \tag{17}$$

$$\delta\tilde{x}\tilde{z} - \overline{m}_1\tilde{y}\,p(\tilde{z}) - (\theta + q)\,\tilde{z} = 0, \tag{18}$$

$$p(\tilde{x}) - \overline{m}_2 p(\tilde{z}) - \eta = 0. \tag{19}$$

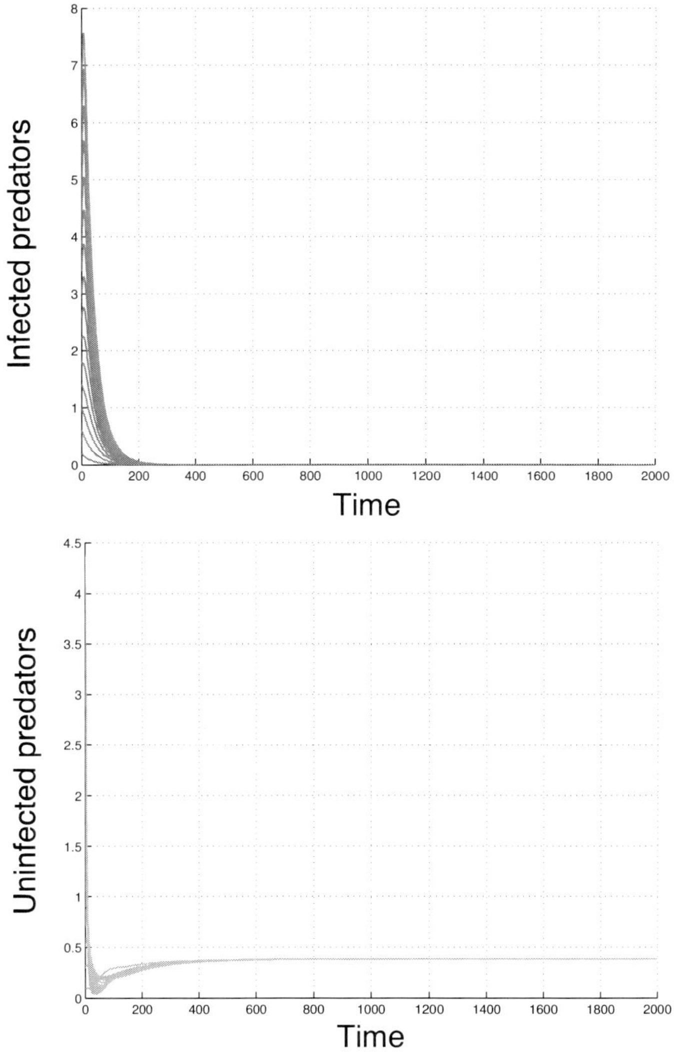

Figure 10. Time evolution of all the population for System (6) when equilibrium $M_4(0.14, 0.14, 0.38, 0)$ is asymptotically stable. The population converges to the equilibrium M_4.

From these equations, we have

$$\tilde{z} = \frac{\tilde{x}(\rho - h - \rho\tilde{x})\bar{m}_1(p(\tilde{x}) - \eta)}{m_2(\delta\tilde{x} - \theta - q)p(\tilde{x}) + \bar{m}_1(\delta\tilde{x} - \theta)(p(\tilde{x}) - \eta)}, \quad \tilde{y} = \frac{(\delta\tilde{x} - \theta - q)\tilde{z}}{\bar{m}_1 p(\tilde{z})} \tag{20}$$

So, if $\tilde{x} \in]\tilde{x}_2, \tilde{x}_1[$, then $\tilde{z} > 0$ and $\tilde{y} > 0$ and $\tilde{x}_4$ is the positive solution of

$$p(\tilde{x}) - \bar{m}_2 p\left(\frac{\tilde{x}(\rho - h - \rho\tilde{x})\bar{m}_1(p(\tilde{x}) - \eta)}{m_2(\delta\tilde{x} - \theta - q)p(\tilde{x}) + \bar{m}_1(\delta\tilde{x} - \theta)(p(\tilde{x}) - \eta)}\right) - \eta = 0. \tag{21}$$

This ends the proof.

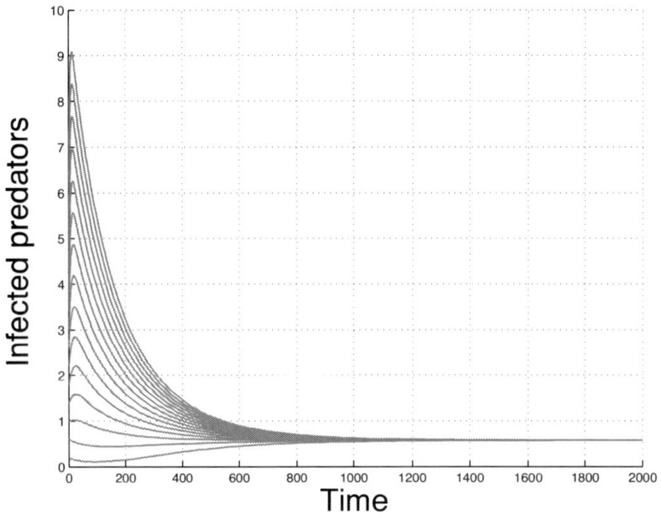

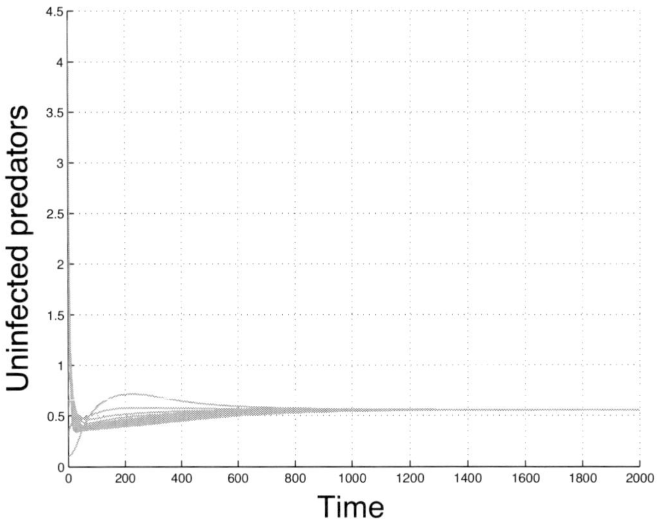

Figure 11. Time evolution of all the population for System (6) when equilibrium $M_5(0.11, 0, 0.55, 0.57)$ is asymptotically stable. The population converges to the equilibrium M_5.

Appendix B. The Proof of Theorem 2

Equilibria of System (4) can be obtained by setting the right hand side of all equations equal to zero. It can be check that the model has an equilibrium $P_0(0, 0, 0)$ which always exists and a possible equilibrium $P_1(\bar{x}_1, 0, 0)$ where $x_1 = \dfrac{\rho - h}{\rho}$ which exists when $\rho - h > 0$.

When w=0 and $y \neq 0$, we have the following system

$$\begin{cases} \rho\bar{x}_2(1-\bar{x}_2) - \bar{y}_2\, p(\bar{x}_2) - h\bar{x}_2 = 0, \\ p(\bar{x}_2) - d_1 = 0. \end{cases} \tag{22}$$

The first equation of (22) gives $\bar{y}_2 = \dfrac{\bar{x}_2(\rho - h - \rho\bar{x}_2)}{p(\bar{x}_2)}$. Since $\bar{x}_2 > 0$ and $\bar{y}_2 > 0$, we must have $\rho - h > 0$ and $\bar{x}_2 < \bar{x}_1$. With these conditions, the positive solution $(\bar{x}_2, \bar{y}_2)$ of (22) is $\bar{x}_2 = \dfrac{d_1\beta + \sqrt{d_1^2\beta^2 + 4d_1(1-\alpha d_1)}}{2(1-\alpha d_1)}$, $\bar{y}_2 = \dfrac{\bar{x}_2(\rho - h - \rho\bar{x}_2)}{p(\bar{x}_2)}$, which exists if $1 - \alpha d_1 > 0$. We then have the equilibrium $E_2(\bar{x}_2, \bar{y}_2, 0)$.

Let us consider the nonnegative solution $(\bar{x}_3, \bar{y}_3, \bar{w}_3)$ of the system

$$\begin{cases} \rho\bar{x}_3(1-\bar{x}_3) - \bar{y}_3\, p(\bar{x}_3) - h\bar{x}_3 - \eta_1\bar{w}_3 p(\bar{x}_3) = 0, \\ \bar{y}_3 p(\bar{x}_3) - \delta_1\bar{y}_3\bar{w}_3 + \nu_1\bar{w}_3 - d_1\bar{y}_3 = 0, \\ \delta_1\bar{y}_3 + e_1 p(\bar{x}_3) - (\nu_1 + \bar{d}_1) = 0, \\ \bar{x}_3 > 0, \quad \bar{y}_3 > 0, \quad \bar{w}_3 > 0. \end{cases} \tag{23}$$

The third equation of (23) gives $\bar{y}_3 = \dfrac{-e_1\left(p(\bar{x}_3) - \frac{\nu_1+\bar{d}_1}{e_1}\right)}{\delta_1}$. So, $\bar{y}_3 > 0$ if $p(\bar{x}_3) - \frac{\nu_1+\bar{d}_1}{e_1} < 0$. From the second equation of (23), we have $\bar{w}_3 = -\dfrac{\bar{y}_3(p(\bar{\bar{x}}_3) - d_1)}{e_1\left(p(\bar{x}_3) - \frac{\bar{d}_1}{e_1}\right)}$. So, $\bar{w}_3 > 0$ if $\dfrac{p(\bar{x}_3 - d_1)}{p(\bar{x}_3) - \frac{\bar{d}_1}{e_1}} < 0$. Since $e_1 < 1$, we have $d_1 < \bar{d}_1 < \dfrac{\bar{d}_1}{e_1} < \dfrac{\bar{d}_1 + \nu_1}{e_1}$. So, if $p(\bar{x}_3) - d_1 > 0$ and $p(\bar{x}_3) - \dfrac{\bar{d}_1}{e_1} < 0$, then $\bar{y}_3 > 0$ and $\bar{w}_3 > 0$. Note that $p(\bar{x}_3) - d_1 > 0$ if $1 - \alpha d_1 > 0$ and $\bar{x}_3 \in]\bar{x}_2, +\infty[$; $p(\bar{x}_3) - \dfrac{\bar{d}_1}{e_1} < 0$ if $1 - \alpha\frac{\bar{d}_1}{e_1} > 0$ and $x_3 \in]0, b[$ with $b = \dfrac{\beta\frac{\bar{d}_1}{e_1} + \sqrt{\beta^2\frac{\bar{d}_1^2}{e_1^2} + 4\frac{\bar{d}_1}{e_1}(1 - \alpha\frac{\bar{d}_1}{e_1})}}{2(1 - \alpha\frac{\bar{d}_1}{e_1})}$. Note that $\bar{x}_2 < b$ since $p(\bar{x}_2) < p(b)$ and p is an increasing function. since $1 - \alpha\frac{\bar{d}_1}{e_1} > 0 \Rightarrow 1 - \alpha d_1 > 0$. We can conclude that if $1 - \alpha\frac{\bar{d}_1}{e_1} > 0$ and $\bar{x}_3 \in]\bar{x}_2, b[$, then $E_3(\bar{x}_3, \bar{y}_3, \bar{w}_3)$ exists where $\bar{x}_3$ is the solution of $\rho\bar{x}_3(1-\bar{x}_3) - \bar{y}_3\, p(\bar{x}_3) - h\bar{x}_3 - \eta_1\bar{w}_3 p(\bar{x}_3) = 0$. This end the proof.

Appendix C. The Proof of Theorem 3

Equilibria of System (6) can be obtained by setting the right hand side of all equations equal to zero. One can observe that M_0 is a solution. For $z = y = w = 0$, equilibrium M_1 exists if $\rho - h > 0$. When $y = w = 0$ and $z \neq 0$, we obtain the equilibrium $M_2 = (x_2, z_2, 0, 0)$

with $x_2 = \dfrac{\bar{\gamma}+q}{\bar{\lambda}}$ and $z_2 = \dfrac{x_2(\rho - h - \rho x_2)}{q}$. This equilibrium exists when $\rho - h > 0$ and $x_2 < x_1$. If $z = w = 0$, the right hand side of all equations is reduced to

$$\begin{cases} \rho x(1-x) - hx - yp(x) = 0, \\ \bar{m}_1 p(x) - \bar{d}_1 = 0. \end{cases} \tag{24}$$

Solving the system and using the fact that $x > 0$ and $y > 0$, our previous condition hold for the existence of M_3. When $w = 0$, the steady states of the system 6 are solution of the following algebraic equations:

$$\begin{cases} \rho x(1-x) - \bar{\lambda} xz - hx + \bar{\gamma} z - yp(x) = 0, \\ \bar{\lambda} xz - (\bar{\gamma}+q)z - \bar{m}_1 yp(z) = 0, \\ p(x) - \bar{m}_2 p(z) - \bar{d}_1 = 0. \end{cases} \tag{25}$$

Solving System (25) yields (8) and since $x > 0$, $z > 0$ and $y > 0$, we must have the condition (9). When $z = 0$, the steady states of model are solution of the following system

$$\begin{cases} \rho x(1-x) - hx - yp(x) - \bar{\eta}_1 wp(x) = 0, \\ \bar{\mu} w - \bar{\delta} yw - \bar{d}_1\, y + yp(x) = 0, \\ \bar{\delta} y + e_1 p(x) - \bar{\mu} - \bar{d}_2 = 0 \end{cases} \tag{26}$$

Solving the second and third equation of (26), we obtain (10). Using the fact that $\bar{d}_1 < \bar{d}_2 < \dfrac{\bar{d}_2}{e_1} < \dfrac{\bar{d}_2 + \bar{\mu}}{e_1}$, we conclude that if $p(x_5) - \dfrac{\bar{d}_2}{e_1} < 0$ and $p(x_5) - \bar{d}_1 > 0$ then $w_5 > 0$ and $y_5 > 0$. Note that $p(x_5) - \dfrac{\bar{d}_2}{e_1} < 0$ if $1 - \alpha\frac{\bar{d}_2}{e_1} < 0$ and $x_5 \in]0, b[$ where b is defined as in the theorem. $p(x_5) - \bar{d}_1 > 0$ if $1 - \alpha\bar{d}_1 > 0$ and $x \in]a, +\infty[$. Since $a < b$ and $1 - \alpha\bar{d}_1 > 1 - \alpha\dfrac{\bar{d}_2}{e_1}$, then equilibrium M_5 exists if equation (11) is satisfied.

When $x \neq 0$, $z \neq 0$, $\neq y \neq 0$ and $w \neq 0$, we have the following system:

$$\begin{cases} \rho x(1-x) - \bar{\lambda} xz - hx + \bar{\gamma} z - yp(x) - \bar{\eta}_1 wp(x) = 0, \\ \bar{\lambda} xz - (\bar{\gamma}+q)z - \bar{m}_1 yp(z) - \bar{\eta}_2 wp(z) = 0, \\ \bar{\mu} w - \bar{\delta} yw - \bar{d}_1\, y + yp(x) - \bar{m}_2 yp(z) = 0, \\ \bar{\delta} y - (\bar{\mu} + \bar{d}_2) + e_1 p(x) - \bar{\eta}_3 p(z) = 0. \end{cases} \tag{27}$$

From the third and fourth equation of the previous system, we have

$$y = \frac{\eta_3 p(z) + \bar{\mu} + d_2 - e_1 p(x)}{\bar{\delta}} \text{ and } w = \frac{y(\bar{d}_1 - p(x) + \bar{m}_2 p(z))}{e_1 p(x) - \eta_3 p(z) - \bar{d}_2}.$$

Thus, using the first and the second equation of our system, we obtain

$$z = \frac{[\eta_3 p(z) + \bar{\mu} + d_2 - e_1 p(x)][\bar{m}_1 p(z)(e_1 p(x) - \eta_3 p(z) - \bar{d}_2) + \bar{\eta}_2 p(z)(d_1 - p(x) + \bar{m}_2 p(z)]}{(\bar{\lambda} x - \bar{\gamma} - q)} \tag{28}$$

$$x = \frac{yp(x) + \eta_1 wp(x) - \bar{\gamma} z}{\rho - h - \rho x - \bar{\lambda} z} \tag{29}$$

When the equilibrium M_6 exists, it is the positive solution of the fixe point equation (28) and (29).

Appendix D. The Proof of Theorem 4

The jacobian matrix of System (2) at any point $(\tilde{x}, \tilde{z}, \tilde{y})$ is given by

$$J_{(\tilde{x},\tilde{z},\tilde{y})} = \begin{pmatrix} \rho - h - 2\rho\tilde{x} - \delta\tilde{z} - \tilde{y}p'(\tilde{x}) & -\delta\tilde{x} + \theta & -p(\tilde{x}) \\ \delta\tilde{z} & \delta x - \bar{m}_1 \tilde{y} p'(\tilde{z}) - \theta - q & -\bar{m}_1 p(\tilde{z}) \\ \tilde{y}p'(\tilde{x}) & -\bar{m}_2 \tilde{y} p'(\tilde{z}) & p(\tilde{x}) - \bar{m}_2 p(\tilde{z}) - \eta \end{pmatrix}.$$

1. The characteristic equation at $E_0(0,0,0)$ is $(\rho - h - \lambda)(-\theta - q - \lambda)(-\eta - \lambda) = 0$. Then the three eigenvalues of the jacobian evaluated at E_0 are $\lambda_1 = \rho - h$, $\lambda_2 = -\theta - q$ and $\lambda_3 = -\eta$. Therefore, when $\rho > h$, E_0 is a saddle point and when $\rho < h$, E_0 is locally asymptotically stable. When $\rho = h$, zero is an eigenvalue; so, in order to analyze the stability of E_0 when $\rho = h$, we use the central manifold theorem. The central manifold equation is $W^c(0) = \{(x, v) \in \mathbb{R} \times \mathbb{R}^2,\ v = h(x),\ |\ x\ |< \vartheta,\ h(0) = 0,\ Dh(0) = 0\}$, where $\vartheta > 0$ is a small constant and $v = (z, y)$. Let $z = a_k\, x^k$ and $y = b_l x^l$, $k, l \geq 2$. In the neighborhood of zero, we have $p(x) \equiv 0$. So, we have

   $$\dot{z} = k a_k x^{k-1} \dot{x} = k a_k x^{k-1}(-\rho x^2 - yp(x) - \delta x z + \theta z) = -\rho k a_k x^{k+1} - \delta a_k^2 k x^{2k} + k a_k^2 \theta x^{2k-1}.$$

   Moreover, $\dot{z} = \delta a_k x^{k+1} - (\theta + q) a_k x^k$. By identification, we obtain $a_k = 0$ $\forall k \geq 2$ and $z \equiv 0$. We also have $\dot{y} = -\rho l b_l x^{l+1}$ and $\dot{y} = -\eta b_l x^l$. So, $b_l = 0$ $\forall l \geq 2$ and $y \equiv 0$. So, we have $W^c(0) = \{(x, 0, 0)\}$. On the central manifold, $\dot{x} = -\rho x^2$. Solving the above differential equation gives $x = \dfrac{1}{\rho t + \frac{1}{x_0}}$. Therefore, the equilibrium point E_0 is stable.

2. The characteristic equation at $E_1 = (\tilde{x}_1, 0, 0)$ is $(-\rho - h - \lambda)(\delta x_1 - \theta - q - \lambda)(p(\tilde{x}_1) - \eta - \lambda) = 0$. Therefore, when $\rho - h > 0$, the equilibrium E_1 lies in $\mathbb{R}_3^+$ and is locally asymptotically stable if $\tilde{x}_1 < \tilde{x}_2$ and $p(\tilde{x}_1) < \eta$.

3. The characteristic equation at point point E_2 is $\Big(p(\tilde{x}_2) - \bar{m}_2 p(\tilde{z}_2) - \eta - \lambda\Big)\Big(\lambda^2 - \lambda(\rho - h - \delta z_2 - 2\rho x_2) + \delta q z_2\Big) = 0$. Then $\lambda = p(\tilde{x}_2) - \bar{m}_2 p(\tilde{z}_2) - \eta$ is an eigenvalue and the other eigenvalues are the solution of equation $\lambda^2 - \lambda(\rho - h - \delta z_2 - 2\rho x_2) + \delta q z_2 = 0$. Using the Hurwitz criterion gives the condition for local stability of E_2.

4. Studying the jacobian matrix corresponding to $E_3 = (x_3, 0, y_3)$, one can observe that the eigenvalues are solution of the following equation $\Big(\delta\tilde{x}_3 - \theta - q - \lambda\Big)\Big(\lambda^2 - \lambda(\rho - h - \tilde{y}_3 p'(\tilde{x}_3) - 2\rho\tilde{x}_3) + p(\tilde{x}_3)p'(\tilde{x}_3)\tilde{y}_3\Big) = 0$. Using the Hurwitz criterion, E_3 is locally asymptotically stable if $\tilde{x}_3 < \tilde{x}_2$ and $\rho - h - \tilde{y}_3 p'(\tilde{x}_3) - 2\rho\tilde{x}_3 < 0$.

5. The characteristic equation evaluate at $E_4 = (\tilde{x}_4, \tilde{z}_4, \tilde{y}_4)$ is given by $\lambda^3 + a_1\lambda^2 + a_2\lambda + a_3 = 0$, where a_1, a_2 and a_3 are given in the theorem. Using the Hurwitz criterion, we have the result.

Appendix E. The Proof of Theorem 5

The jacobian matrix of System (4) at any point (x, y, w) is

$$J(x, y, w) = \begin{pmatrix} \rho - h - 2\rho x - (y + \eta_1 w)p'(x) & -p(x) & -\eta_1 p(x) \\ yp'(x) & p(x) - \delta_1 w - d_1 & \nu_1 - \delta_1 y \\ e_1 w p'(x) & \delta_1 w & \delta_1 y + e_1 p(x) - (\nu_1 + \bar{d}_1) \end{pmatrix}.$$

1. The eigenvalues of the jacobian matrix evaluate at P_0 are $\rho - h$, $-d_1$ and $-\nu_1 - \bar{d}_1$. Therefore, when $\rho > h$, P is a saddle point and when $\rho < h$, P_0 is locally stable. When $\rho = h$, zero is an eigenvalue of the jacobian matrix at E_0; so, in order to analyze the stability of E_0 when $\rho = h$, we use the central manifold theorem. The central manifold equation is $W^c(0) = \{(x, v) \in \mathbb{R} \times \mathbb{R}^2,\ v = h(x),\ | x | < \vartheta,\ h(0) = 0,\ Dh(0) = 0\}$, where $\vartheta > 0$ is a small constant and $v = (y, w)$. Let $y = a_k x^k$ and $w = b_l x^l$, $k, l \geq 2$. In the neighborhood of zero, we have $p(x) \equiv 0$. Thus,

$$\dot{w} = l b_l x^{l-1}\dot{x} = l b_l x^{l-1}(-\rho x^2 - yp(x) - \eta_1 w p(x)) = -\rho l b_l x^{l+1}.$$

We also have $\dot{w} = \delta_1 a_k b_l x^{k+l} - (\mu_1 + \bar{d}_1)b_l x^l$. By identification, we have $b_l = 0$ $\forall l \geq 2$ and $w \equiv 0$. Furthermore, $\dot{y} = -k\rho a_k x^{k+1}$ and $\dot{y} = -d_1 a_k x^k$ and we obtain $a_k = 0\ \forall k \geq 2$, $y \equiv 0$. So, $h(x) = 0$ and the first equation of System (4) can be written as $\dot{x} = -\rho x^2$. Thus, the equilibrium P_0 is locally asymptotically stable.

2. The jacobian matrix at $P_1 = (\bar{x}_1, 0, 0)$ becomes

$$J(P_1) = \begin{pmatrix} -(\rho - h) & -p(\bar{x}_1) & -\eta_1 p(\bar{x}_1) \\ 0 & p(\bar{x}_1) - d_1 & \nu_1 \\ 0 & 0 & e_1 p(\bar{x}_1) - (\nu_1 + \bar{d}_1) \end{pmatrix}.$$

The eigenvalues are $-(\rho - h)$, $p(\bar{x}_1) - d_1$ and $e_1 p(\bar{x}_1) - (\nu_1 + \bar{d}_1)$. Note that P_1 exists if $\rho - h > 0$. Since $p(\bar{x}_1) - d_1 < 0 \Rightarrow p(\bar{x}_1) - \dfrac{\bar{d}_1 + \nu_1}{e_1} < 0$, we conclude that P_1 is locally stable if $p(\bar{x}_1) - d_1 < 0$.

3. The Jacobian matrix at $P_2(\bar{x}_2, \bar{y}_2, 0)$ is given by $J_{P_2} = \begin{pmatrix} \lambda_1 & -p(\bar{x}_2) & -\eta_1 p(\bar{x}_2) \\ \bar{y}_2 p'(\bar{x}_2) & 0 & \nu_1 - \delta_1 \bar{y}_2 \\ 0 & 0 & \lambda_2 \end{pmatrix}$ where $\lambda_1 = \rho - h - 2\rho\bar{x}_2 - \bar{y}_2 p'(\bar{x}_2)$ and $\lambda_2 = \delta_1 \bar{y}_2 + e_1 p(\bar{x}_2) - \nu_1 - \bar{d}_1$. Thus, the characteristic equation is $(\lambda_2 - \lambda)(\lambda^2 - \lambda_1 \lambda + \bar{y}_2 p(\bar{x}_2) p'(\bar{x}_2)) = 0$. Using the Hurwitz criterion, we can conclude that the equilibrium point P_2 is locally asymptotically stable if $\lambda_1 < 0$ and $\lambda_2 < 0$.

4. The characteristic equation of the jacobian matrix at equilibrium $P_3 = (\bar{x}_3, \bar{y}_3, \bar{w}_3)$ is $\lambda^3 + \alpha_1 \lambda^2 + \alpha_2 \lambda + \lambda_3 = 0$, where the coefficient α_i are given as previously. Using the Hurwitz criterion, the equilibrium P_3 when it exists is locally asymptotically stable if only if $\alpha_1 > 0$, $\alpha_3 > 0$ and $\alpha_1 \alpha_2 > \alpha_3$.

Appendix F. The Proof of Theorem 6

Let's consider System (6).

1. Using the jacobian matrix of System (6) at M_0 gives the following characteristic equation
$$(\rho - h - \lambda)(-\bar{\gamma} - \lambda)(-\bar{d} - \lambda)(-\bar{\mu} - \bar{d} - \lambda) = 0.$$
The eigenvalues are $\lambda_1 = \rho - h$, $\lambda_2 = -\bar{\gamma}$, $\lambda_3 = -\bar{d}$ and $\lambda_4 = -\bar{\mu} - \bar{d}$. Then, M_0 is locally asymptotically stable if $\rho - h < 0$ and unstable if $\rho - h > 0$.

   When $\rho = h$, the system can be written as
$$\begin{cases} \dot{x} = Ax + f(x, v), \\ \dot{v} = Bv + g(x, v), \end{cases} \tag{30}$$
where $A = 0$, $v = \begin{pmatrix} z \\ y \\ w \end{pmatrix}$, $g(x, v) = \begin{pmatrix} \bar{\lambda} x\, z - \bar{m}_1\, y\, p(z) - \bar{\eta}_2 w p(z) \\ -\bar{\delta} y\, w + \bar{m}_1\, y\, p(x) - \bar{m}_2\, y\, p(z) \\ \bar{\delta} y\, w + e_1 \bar{m}_1\, w\, p(x) - \bar{m}_3\, w\, p(z) \end{pmatrix}$, $f(x, v) = -\rho x^2 - \bar{\lambda} x\, z + \bar{\gamma} z - \bar{m}_1\, y\, p(x) - \bar{\eta}_1 w p(x)$ and $B = \begin{pmatrix} -\bar{\gamma} & 0 & 0 \\ 0 & -\bar{d} & \bar{\mu} \\ 0 & 0 & -\bar{\mu} - \bar{d} \end{pmatrix}$. We want to use the center manifold theorem. Setting $v = h(x)$ implies $\dot{v} = h'(x) f(x, h(x))$. From equation (30), $h(x) = B^{-1}\left(h'(x) f(x, h(x)) - g(x, h(x))\right)$. In order to obtain the approximative expression of the center manifold, we set
$$\begin{cases} h_0(x) = 0, \\ h_{n+1}(x) = B^{-1}\left(h'_n(x) f(x, h_n(x)) - g(x, h_n(x))\right), \end{cases} \tag{31}$$
where $B^{-1} = \begin{pmatrix} -\dfrac{1}{\bar{\gamma}} & 0 & 0 \\ 0 & -\dfrac{1}{\bar{d}} & -\dfrac{\bar{\mu}}{\bar{d}(\bar{\mu} + \bar{d})} \\ 0 & 0 & -\dfrac{1}{\bar{\mu} + \bar{d}} \end{pmatrix}$. It is easy to show that $h_n(x) = 0\ \forall n$.

Thus, we obtain $\dot{x} = -\rho x^2$ and this prove that M_0 is locally asymptotically stable.

2. The jacobian matrix at M_1 gives the characteristic equation

$$(-\rho + h - \lambda)(\bar{\lambda}\tilde{x}_1 - \bar{\gamma} - \lambda)(\bar{m}_1 p(\tilde{x}_1) - \bar{d} - \lambda)(e_1 \bar{m}_1 p(\tilde{x}_1) - (\bar{\mu} + \bar{d}) - \lambda) = 0.$$

The eigenvalues are $\lambda_1 = -(\rho - h)$, $\lambda_2 = \bar{\lambda}\tilde{x}_1 - \bar{\gamma}$, $\lambda_3 = \bar{m}_1 p(\tilde{x}_1) - \bar{d}$ and $\lambda_4 = e_1 \bar{m}_1 p(\tilde{x}_1) - (\bar{\mu} + \bar{d})$. Thus, P_1 is locally asymptotically stable if $\bar{\lambda}\tilde{x}_1 < \bar{\gamma}$ and $p(\tilde{x}_1) < \min\left\{\dfrac{\bar{d}}{\bar{m}_1}, \dfrac{\bar{\mu} + \bar{d}}{e_1 \bar{m}_1}\right\}$.

3. The characteristic equation of the jacobian matrix at M_2 is

$$(\lambda_2 - \lambda)(\bar{\lambda}\tilde{x}_2 - \bar{\gamma} - \lambda)(\lambda^2 - \lambda_1 \lambda + \bar{m}_1^2 \tilde{y}_2 p(\tilde{x}_2) p'(\tilde{x}_2)) = 0,$$

where the parameters $\lambda_1 = \rho - h - 2\rho\tilde{x}_2 - \bar{m}_1 \tilde{y}_2 p'(\tilde{x}_2)$ and $\lambda_2 = \bar{\delta}\tilde{y}_2 + e_1 \bar{m}_1 p(\tilde{x}_2) - (\bar{\mu} + \bar{d})$. Using the Hurwitz criterion, we can conclude that the equilibrium M_2 is locally asymptotically stable if $\lambda_1 < 0$, $\lambda_2 < 0$ and $\bar{\lambda}\tilde{x}_2 < \bar{\gamma}$.

4. Let $s_1 = \rho - h - 2\rho x_3 - \bar{m}_1 y_3 p'(x_3) - \bar{\lambda} z_3$, $s_2 = \bar{\lambda} x_3 - \bar{\gamma} - \bar{m}_1 y_3 p'(z_3)$ and $s_3 = -\bar{\delta} y_3 + e_1 \bar{m}_1 p(x_3) - (\bar{\mu} + \bar{d}) - \bar{m}_3 p(z_3)$. The characteristic equation at M_3 is

$$(-s_3 + \lambda)(\lambda^3 + a_1 \lambda^2 + a_2 \lambda + a_3) = 0,$$

where $a_1 = -s_1 - s_2$, $a_2 = s_1 s_2 + \bar{\lambda} z_3 x_3 - \bar{\lambda}\bar{\gamma} z_3 + \bar{m}_1^2 p(x_3) y_3 p'(x_3) - \bar{m}_1 p(z_3) \bar{m}_2 y_3 p'(z_3)$, $a_3 = -\bar{m}_1^2 p(z_3) y_3 p'(x_3) \bar{\lambda} x_3 + \bar{m}_1 p(z_3) y_3 p'(x_3) \bar{\gamma} - \bar{m}_1 p(x_3) \bar{\lambda} z_3 \bar{m}_2 y_3 p'(z_3) - \bar{m}_1^2 p(x_3) y_3 s_2 p'(x_3) + \bar{m}_1 p(z_3) \bar{m}_2 y_3 p'(z_3) s_1$. Then, we can conclude that when the equilibrium M_3 exists, it is locally asymptotically stable if $a_1 > 0$, $a_3 > 0$, $a_1 a_2 > a_3$ and $s_3 < 0$.

5. Let $q_1 = \rho - h - 2\rho x_4 - \bar{m}_1 y_4 p'(x_4) - \bar{\eta}_1 w_4 p'(x_4)$ and $q_2 = \bar{m}_1 x_4 - \bar{d}$. The characteristic equation of the jacobian matrix at equilibrium M_4 is

$$(-\bar{\lambda} x_4 + \bar{\gamma} + \xi)(\xi^3 + a_1 \xi + a_2 \xi + a_3) = 0,$$

where $a_1 = -q_1 - q_2$, $a_2 = q_1 q_2 + \bar{\eta}_1 p(x_4) e_1 \bar{m}_1 w_4 p'(x_4) + \bar{m}_1 p(x_4) m_1 y_4 p'(x_4) - \bar{\delta} w_4 (-\bar{\delta} y_4 + \bar{\mu})$, $a_3 = \bar{m}_1^2 p(x_4) e_1 w_4 p'(x_4)(-\bar{\delta} y_4 + \bar{\mu}) + \bar{\eta}_1 p(x_4) \bar{m}_1 y_4 p'(x_4) \bar{\delta} w_4 - \bar{\eta}_1 p(x_4) \lambda_2 e_1 \bar{m}_1 w_4 p'(x_4) + \lambda_1 \bar{\delta} w_4 (-\bar{\delta} y_4 + \bar{\mu})$. Then, using the Hurwitz criterion, we can conclude that the equilibrium M_4 is locally asymptotically stable if $a_1 > 0$, $a_3 > 0$, $a_1 a_2 > a_3$ and $\bar{\lambda} x_4 - \bar{\gamma} < 0$.

6. One can consider the right hand side of the system at equilibrium with $x \neq 0$, $y \neq 0$, $z \neq 0$, $w \neq 0$ and apply the implicit function theorem. Then, since the right hand side of the system is consisted of polynomial equations, there exists at least one solution in $\mathbb{R}_4^+$.

References

[1] A. D. Bazykin, Aleksandr Iosifovich Khibnik, Bernd Krauskopf; Nonlinear Dynamics of Interacting Populations, *World Scientific*, 1998, 193 pages.

[2] C. S. Holling, The functional response of predators to prey density and its role in mimicry and population regulation, *Mem. Entomolog. Soc. Can.*, 45 (1965), pp. 3-60.

[3] R.M. Anderson and R.M. May, The invasion, persistence and spread of infectious diseases within animal and plant communities, *Phil. Trans. R. Sot.* London B314, (1986) 533-570.

[4] K.P. Hadeler and H.I. Freedman, Predator-prey populations with parasitic infection, *J. Math. Biol.* 27, (1989) 609-631.

[5] E. Venturino, The influence of diseases on Lotka-Volterra systems, *Rocky Mt. J. Math.* 24, (1994) 381402.

[6] E. Venturino, Epidemics in predator-prey models: Disease in the prey, *In Mathematical Population Dynamics : Analysis of Heterogeneity*, Volume One : Theory of Epidemics, (Edited by O. Arino, D. Axelrod, M. Kimmel and M. Langlais), pp. 381-393, Wuerz, Winnipeg, Canada, (1995).

[7] B. Hellriegel, Modelling the immune response to malaria with ecological concepts: short-term behaviour against long term equilibrium., *Proc R Soc Lond B Biol Sci*, 250(1992), pp. 97105.

[8] G. Birkoff , G.C. Rota, *Ordinary Differential Equations.*, Ginn, 1982.

[9] N. P. Bhatia and G. P. Szegö, *Stability theory of dynamical systems.*, Springer-Verlag; 1970.

[10] R. Anguelov, Y. Dumont, J.M-S. Lubuma and M. Shillor, Comparison of some standard and nonstandard numerical methods for the MSEIR epidemiological model, *Proceedings of the International Conference of Numerical Analysis and Applied Mathematics*, Crete, Greece, 18- 22 September 2009, American Institute of Physics Conference Proceedings-AIP 1168, Volume 2, 2009, 1209-1212.

[11] P.J. Hudson, A.P. Dobson and D. Newborn, Do parasites make prey more vulnerable to predation? Red grouse and parasites, *J. Anim. Ecol.* 61, (1992) 681-692.

[12] Hadeler, K. P. and Freedman, H. I. (1989). Predator-prey populations with parasitic infection. *J Math Biol* 27: 609-631.

[13] Ying-Hen Hsieh and Chin-Kuei Hsiao, Predator-prey model with disease infection in both populations, *Mathematical Medicine and Biology* (2008)25, 247-266

[14] S. Bowong, J.J. Tewa, Mathematical analysis of a tuberculosis model with differential infectivity, *Commun. Nonlinear Sci. Numer. Simul.* 14 (2009), 4010-4021

[15] A. Iggidr, J.C. Kamgang, G. Sallet, J.J. TEWA, Global analysis of new malaria intrahost models with a competitive exclusion principle, *SIAM. J. Appl. Math.* 67 (N1) (2006) 260-278.

[16] A. Fall, A. Iggidr, G. Sallet, J.J. Tewa, Epidemiological models and Lyapunov functions, *Math. Model. Nat. Phenom.* 2 (1) (2006) 55-71

[17] P. Adda, A. Iggidr, J.C. Kamgang, G. Sallet, J.J. Tewa, General models of host-parasite systems. *Global analysis, Discrete Contin. Dyn. Syst.* B 8 (N1) (2007), 1-17.

[18] A. Iggidr, J. Mbang, G. Sallet, J.J. Tewa, Multi-compartment models, *Discrete Contin. Dyn. Syst.* B (Suppl.) (2007) 506-519

[19] N. Bame, S. Bowong, J. Mbang, G. Sallet, J.J. Tewa, Global stability analysis for SEIS models with n latent classes, *Math. Biosci. Eng.* 5 (N1) (2008) 20-33.

[20] A. Iggidr, J. Mbang, G. Sallet, J.J. Tewa, Multi-compartment models, *Discrete Contin. Dyn. Syst.* (2007) 506-519.

[21] J.J. Tewa, R. Fokouop, B. Mewoli, S. Bowong; Mathematical analysis of a general class of ordinary differential equations coming from within-hosts models of malaria with immune effectors; *Applied Mathematics and Computation* 218 (2012) 7347-7361

[22] T. K. Kar; Selective Harvesting in a Prey-Predator Fishery with Time Delay; *Applied Mathematics and Computation* 388 (2003) 449-458

[23] Shigui Ruan, Dongmei Xiao; Global Analysis in a predator-prey system with nonmonotonic functional response; *SIAM J. APPL. MATH.* (2001), Vol. 61, No. 4, pp. 1445-1472

[24] Yang Kuang, H. I. Freedman; Uniqueness of Limit Cycles in Gause-Type Models of Predator-Prey Systems; *Mathematical Biosciences* 88 : 67-84 (1988)

[25] Kuo-Shung Cheng; Uniqueness of a limit cycle for a predator-prey system; *SIAM J. MATH. ANAL.* Voi. 12, No. 4, July 1981

[26] LII-Perng Liouf and Kuo-Shung Cheng; On the uniqueness of a limit cycle for a predator-prey system; *SIAM J. MATH. ANAL.* Vol. 19, No. 4, July 1988

[27] Sze-Bi Hsu, Tzy-Wei Hwang, Yang Kuang; Global analysis of the Michaelis-Menten-type ratio-dependent predator-prey system; *J. Math. Biol.* 42, 489-506 (2001)

[28] G. J. Butler and G. S. K. Wolkowicz; Predator-mediated competition in the chemostat; *J. Math. Biol.* (1986) 24:167-191

[29] Safar Irandoust, Ahmad Golbabai, Hosein Kheiri and Davood Ahmadian; Homotopy analysis method for solving ratio-dependent predator-prey system with constant effort harvesting by using two parameters H_1 and H_2; *Acta Universitatis Apulensis*, ISSN: 1582-5329, No. 25, (2011), pp. 327-340

[30] Jean Jules Tewa, Samuel Bowong, C. S. Oukouomi Noutchie ; Mathematical analysis of two-patch model of tuberculosis disease with staged progression, *Applied Mathematical Modelling* 36, (2012) 5792-5807.

[31] M. Haque, A predator-prey model with disease in the predator species only, *Nonlinear Anal. Real World Appl.* 11 (2010) 2224-2236

[32] M. Haque, D. Greenhalgh, When predator avoids infected prey : A model based theoretical studies. *IMA J. Math. Med. Biol.* 27 (2009) 75-94.

[33] N. Bairagi, P. K. Roy, J. Chattopadhyay ; Role of infection on the stability of a predator-prey system with several response functions- A comparative study, *Journal of Theoretical Biology*, 248 (1) (2007) 10-25, ISSN 0022-5193

[34] Krishna Padas Das, Kusumika Kundu, J. Chattopadhyay, A predator-prey mathematical model with both populations affected by diseases, *Ecological Complexity* 8 (2011) 68-80

[35] M. Haque, E. Venturino; Modelling disease spreading in symbiotic communities; *Wildlife :Destruction, Conservation and Biodiversity*; Nova Science Publishers, USA (www.novapublishers.com); 2009, ISBN : 978-1-60692-974-2

[36] M. Haque, S. Sarwardi, S. Preston and E. Venturino; Efect of delay in a Lotka-Volterra type predator-prey model with a transmissible disease in the predator species; *Mathematical Biosciences*, 234 (1), (2011) 47-57

[37] D. P. Moualeu, S. Bowong, J. J. Tewa, Y. Emvudu ; Analysis of the Impact of Diabetes on the Dynamical Transmission of Tuberculosis, *Math. Model. Nat. Phenom*, 2012, Vol. 7 (3), (2012) 117-146 DOI: 10.1051/mmnp/20127309;

[38] Jean Jules Tewa, Samuel Bowong, Boulchard Mewoli ; Mathematical analysis of two-patch model for the dynamical transmission of tuberculosis, *Applied Mathematical Modelling*, 36, (2012) 2466-2485;

[39] Samuel Bowong, Jean Jules Tewa, Global analysis of a dynamical model for transmission of tuberculosis with general contact rate, *Communication in Nonlinear Science and Numerical Simulation*, Vol.15, (2010) 3621-3631;

[40] Jean Jules Tewa, Samuel Bowong, Boulchard Mewoli, Jurgen Kurths ; *Two-patch transmission of Tuberculosis, Mathematical Population Studies* (MPS), Volume 18, Issue 2, (2011) 189-205;

[41] Yves Emvudu, Boulchard Mewoli, Jean Jules Tewa, Jean Pierre Kouenkam ; Epidemiological model for the spread of Anti-Tuberculosis resistance; *International Journal of Information and Systems Sciences*, Institute for Scientific Computing and Information; Volume 7, number 4, (2011) 279-301;

[42] Samuel Bowong, Jean Jules Tewa, Jean Claude Kamgang ; Stability analysis of the transmission dynamics of tuberculosis models, *World Journal of Modelling and Simulation* (WJMS), Volume 7, number 2, (2011) 83-100;

[43] M. Haque, J. Zhen, E. Venturino; Rich dynamics of Lotka-Volterra type predator-prey model system with viral disease in prey species; *mathematical methods in the Applied Science* 32 (2009) 875-898

[44] M. Haque, E. Venturino; The role of transmissible diseases in the Holling-Tanner predator-prey model; *Theoretical Population Biology* 70 (2006) 273-288

[45] Jean Jules Tewa, Valaire Yatat Djeumen, Samuel Bowong; Predator-prey model with Holling response function of type II and SIS infectious disease; *Applied Mathematical Modelling* , 2012 (to appear)

[46] O. D. Makinde, Solving ratio-dependent predator-prey system with constant effort harvesting using Adomian decomposition method, *Applied Mathematics and Computation*, vol. 186, no. 1, (2007) pp. 17-22.

[47] J. Hale, *Ordinary Differential Equations*, J. Wiley & Sons (1980).

In: Dynamical Systems
Editor: Jean Michel Tchuenche, pp. 293-314
ISBN: 978-1-62808-001-8

Chapter 11

MODELLING THE EFFECTS OF MULTI-INTERVENTION CAMPAIGNS FOR THE MALARIA EPIDEMIC IN MALAWI

***P. M. Mwamtobe*[1,*], *J. M. Tchuenche*[2,†], *M. Maliyoni*[3] *and S. D. Hove-Musekwa*[4]**
[1]Department of Mathematics and Statistics, University of Malawi, The Malawi Polytechnic, Chichiri, Blantyre, Malawi
[2]Mathematics Department, University of Dar es Salaam, Dar es Salaam, Tanzania
[3]Mathematics Department, DMI-St. John the Baptist University, Mangochi, Malawi
[4]Department of Applied Mathematics, National University of Science and Technology, Ascot, Bulawayo, Zimbabwe

Abstract

Malaria is a public health problem for about 2.4 billion people despite decades of efforts to control the disease. A mathematical model is developed to assess the potential impact of protection (insecticide-treated mosquito bed-nets and indoor residual spraying) and treatment on the transmission dynamics of malaria in Malawi. The model is analyzed qualitatively to determine, the effective reproduction number which is necessary for country-wide control of malaria is computed. The equilibria of the model are determined. The disease-free equilibrium is locally asymptotically stable whenever $R_e < 1$ and unstable when $R_e > 1$, the existence of a unique endemic equilibrium case. Numerical simulations indicate that the two control strategies (protection and treatment) in lowering the number of exposed and infected people. Our results highlight the population-level impact of some model parameters such as the infection and biting rates.

Keywords: Malaria, intermittent preventive treatment, insecticide-treated nets, indoor residual spraying

*E-mail address: pmwamtobe@gmail.com
†E-mail address: jmtchuenche@gmail.com

1. Introduction

Malaria is a vector-borne infectious disease mainly found in tropical regions (Sub-Saharan Africa, Central and South America, the Indian sub-continent, South-East Asia and the Pacific islands). It is a life-threatening disease transmitted through the bites of infected mosquitoes [23]. It is estimated that about 200 million persons are at constant risk of infection. *Plasmodium falciparum* malaria is the major cause of mortality and morbidity in the tropical and sub-tropical areas of the globe (about three-quarter of reported malaria cases). There are four different types of Plasmodium parasites - *Plasmodium falciparum* (the only parasite which causes malignant malaria) - *Plasmodium vivax* (causes benign malaria with less severe symptoms, the vector can stay in the liver for up to three years and can lead to a relapse) - *Plasmodium malariae* also causes benign malaria and is relatively rare - *Plasmodium ovale* (causes benign malaria and can stay in the blood and liver for many years without causing symptoms). This study will focus mainly on the malignant malaria. Severe malaria can affect the patient's brain and central nervous system and can be fatal [4, 23].

Malaria transmission rates can differ depending on local factors such as rainfall patterns (since mosquitoes breed in wet conditions), the proximity of mosquito breeding sites to households and people, and the type of mosquito species in the area. Some regions have a fairly constant number of cases throughout the year; these countries are termed "malaria endemic". In other areas, there are malaria seasons usually coinciding with the rainy season. Large and devastating epidemics can occur when the mosquito-borne parasite is introduced into areas where people have had little prior contact with the infecting parasite and have little or no immunity to malaria, or when people with low immunity move into areas where malaria is endemic. These epidemics can be triggered by wet weather conditions and further aggravated by floods or mass population movements driven by conflicts. The illness can result in high rates of miscarriage and causes over 10% of maternal deaths (soaring to a 50% death rate in cases of severe disease) annually; semi-immune pregnant women risk severe anaemia and impaired foetal growth even if they show no signs of acute disease [23]. The World Health Organization (WHO) recommends the use of insecticide-treated nets for night-time prevention of mosquito bites; for pregnant women in highly endemic areas, preventive doses of sulfadoxine-pyrimethamine (IPT/SP) to periodically clear the placenta of parasites and indoor residual spraying to kill mosquitoes that rest on the walls and ceilings of houses [22]. The combined role of preventative measures such as insecticide-treated mosquito bed nets, indoor residual spraying and treatment on the transmission dynamics of malaria has not been given prominence. It is therefore against this background that this study is carried out.

Beyond the human toll, malaria wreaks significant economic havoc in endemic regions, decreasing gross domestic product (GDP) by as much as 1.3% in countries with high levels of transmission - the disease accounts for up to 40% of public health expenditures, 30%-50% of in-patient hospital admissions and up to 60% of out-patient health clinic visits [23]. The Government of Malawi has put in place several control strategies through the National Malaria Control Programme to reduce and possibly (ultimately) eliminate malaria. The main strategic areas that have been identified for scaling-up of malaria control activities include; malaria case management, intermittent preventive treatment (IPT) of pregnant women using sulfadoxine-pyrimethamine (SP) and malaria prevention with special emphasis on the use of insecticide-treated mosquito nets (ITNs) as well as indoor residual spraying (IRS) [16].

Various compartmental model for the spread of malaria have been proposed [17, 8, 20]. Smith and Musekwa [19] showed that if mosquito birth rate increased by 25% , then, the minimal effective spraying period reduced by half, and if doubled, the period reduced by three quarters. Our goal is to assess the role of preventive measures namely insecticide-treated mosquito bed-nets (ITNs) and indoor residual spraying (IRS) as well as treatment on the transmission dynamics of malaria in Malawi without actually targeting a certain high-risk group. The rest of the paper is organized as follows. In section 2, the model is formulated and its qualitative analysis carried out. Numerical simulations to support the theoretical results of Section 2 are provided in Section 3. Discussion and conclusion conclude the chapter in Section 4.

2. Model Formulation and Analysis

We formulate the model with the population under study being sub-divided into compartments according to individuals disease status. We consider the total population sizes denoted by $N_h(t)$ and $N_v(t)$ for the human hosts and female mosquitoes respectively. We employ the $SEIRS$ framework to describe a disease with temporary immunity on recovery from infection. $SEIRS$ model indicates that the passage of individuals is from the susceptible class, S, to the exposed class, E, then to infective class, I, and finally to the recovery class, R. $S(t)$ represents the number of individuals not yet infected with the malaria parasite at time t. The latent or exposed class $E(t)$ (individuals are infected but not infectious). Individuals in the $I(t)$ class are infected with malaria and are capable of transmitting the disease to susceptible mosquitoes. $R(t)$ represents the class of individuals who have recovered from the disease. It is assumed that they cannot transmit the infection to mosquitoes as they have no (or not enough) plasmodium parasites in their bodies. The protected class $P(t)$ comprises of individuals who rarely come into contact with the vectors (by using ITNs and IRS) or who have develop some level of protection if they have the Sickle-cell trait. Individuals receiving therapy are lumped into the treatment class $T(t)$. These last two classes have been included due to the use of ITNs and IRS as the preventive measures and treatment as a control measure. The susceptible human population is increased by recruitment (birth and immigration) at a constant rate, Λ. All the recruited individuals are assumed to be naive when joining the community. The fraction τ of the susceptible recruited individuals are taken to be under preventive control and join the protected class. The likelihood of infection is assumed to be reduced by a factor of ϑ. We note that protection is effective when $\vartheta = 0$ and ineffective if $\vartheta = 1$. Since we have a protected class, a proportion $g \in [0, 1]$ represents susceptible individuals who migrate to malaria-free areas and thus, become partially protected, but later become exposed once they return to the malaria endemic areas by the force of infection λ_{hp}. This happens when the prevention measures are relaxed or when the presumptive interventions vanish. We assume that $(g + \tau\Lambda) > 0$ in order to ensure that there is a non-zero flow of humans into the protected class. A proportion g uses are protected from mosquitoes bites by using mosquito treated bed nets and indoor residual spraying. The exposed individuals progress to infected class at a constant rate ν. Individuals who have experienced infection may be treated at a constant rate σ and enter the treated class. After a successful treatment, they recover temporarily at an acquired immunity rate ε. Since disease-induced immunity due to malaria is temporary, a fraction ω of individuals leave the recovered state to the susceptible state while the complementary fraction $(1-\omega)$ moves to the protected class due to non-compliance to treatment as a result of fear of drugs and sometimes boredom with the dosage. This means that they do not comply fully and their recovery is temporal. The rate of infection of susceptible individuals is λ_h and the rate at which the infected individuals infect the susceptible mosquitoes is λ_v. The mosquito population is divided into three classes namely; susceptible, $X(t)$; exposed, $Y(t)$; and infectious, $Z(t)$. Female anopheles mosquitoes (male anopheles mosquito is not included in the model because only female mosquito bites animals for blood meals) enter the susceptible class through birth at a rate π. Susceptible mosquitoes become infected by biting infectious humans at a rate α. The parasites (in the form of gametocytes) enter the mosquito with a probability β_{hv}. After some period of time, depending on the ambient temperature and humidity, the parasite develops into sporozoites and enters the mosquito's salivary glands, and the mosquito progresses at a rate θ, from the exposed class to the infectious class. We assume that the infective period of the vector ends with its death, and therefore the vector does not recover from being infective [1]. The mosquitoes leave the population through natural death at a rate, μ_v and infectious mosquitoes suffer an additional death at a rate δ_v due to the presence of infectious parasites in their bodies. The compartmental diagram of the model is shown in Figure 1. The above assumptions lead to the following deterministic system of non-linear ordinary differential equations which describe the progress of malaria with intervention strategies;

$$\left.\begin{aligned}
\frac{dS}{dt} &= (1-\tau)\Lambda - \lambda_h S + \omega R - (g+\mu_h)S \\
\frac{dP}{dt} &= \tau\Lambda + gS + (1-\omega)R - \lambda_{hp}P - \mu_h P \\
\frac{dE}{dt} &= \lambda_h S + \lambda_{hp}P - (\nu+\mu_h)E \\
\frac{dI}{dt} &= \nu E - (\gamma+\sigma+\mu_h+\delta_h)I \\
\frac{dT}{dt} &= \sigma I - (\varepsilon+\mu_h)T \\
\frac{dR}{dt} &= \varepsilon T + \gamma I - (1-\omega)R - (\omega+\mu_h)R \\
\frac{dX}{dt} &= \pi - \lambda_v X - \mu_v X \\
\frac{dY}{dt} &= \lambda_v X - (\theta+\mu_v)Y \\
\frac{dZ}{dt} &= \theta Y - (\mu_v+\delta_v)Z
\end{aligned}\right\}, \tag{1}$$

where $\lambda_h = \frac{\beta_{vh}\alpha Z}{N_h}$, $\lambda_v = \frac{\beta_{hv}\alpha I}{N_h}$, and $\lambda_{hp} = \frac{\beta_{vh}\alpha\vartheta Z}{N_h}$.

The term $\dfrac{\beta_{vh}\alpha\vartheta ZP}{N_h}$ denotes the rate at which the protected individuals P, get infected by infectious mosquitoes Z. It indicates that the rate of infection for the protected individuals is reduced by a factor ϑ.

Adding the first six equations of the model (1), with the case that there is no disease-induced death, that is, $\delta_h = \delta_v = 0$, gives $\dfrac{dN_h}{dt} = \Lambda - \mu_h N_h$, so that $N_h(t) \to \dfrac{\Lambda}{\mu_h}$ as

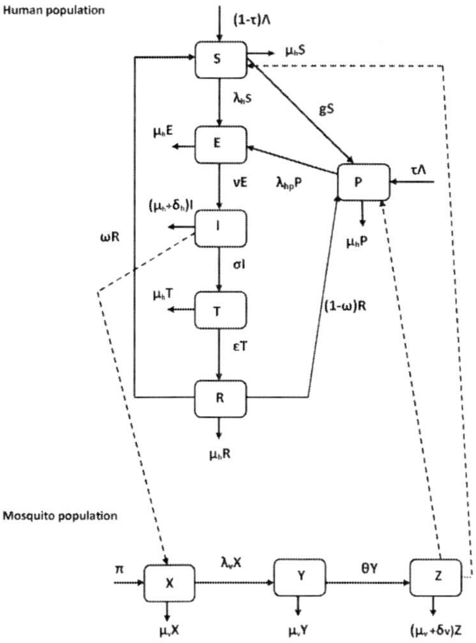

Figure 1. The malaria flow diagram with interventions. The dashed line from infected human class, I, to the susceptible mosquito population, X, shows that the infected human individuals infect the susceptible mosquito population whilst the dashed line from infected mosquito population, Z, to the susceptible human population, S, and the protected human population P shows the transfer of plasmodium parasites from infected mosquito population to susceptible individuals.

$t \to \infty$ [2]. Thus, $\dfrac{\Lambda}{\mu_h}$ is an upper bound of $N_h(t)$ provided that $N_h(0) \leq \dfrac{\Lambda}{\mu_h}$. Further, if $N_h(0) > \dfrac{\Lambda}{\mu_h}$, then $N_h(t)$ will decrease to this level, $\dfrac{\Lambda}{\mu_h}$. Similar calculation for the vector equations shows that $N_v \to \dfrac{\pi}{\mu_v}$ as $t \to \infty$.

The feasible region

$$\Omega = \left\{ (S, P, E, I, T, R, X, Y, Z) \in \mathbb{R}_+^9 : N_h \leq \frac{\Lambda}{\mu_h} = N_h^*, N_v \leq \frac{\pi}{\mu_v} = N_v^* \right\},$$

is positive-invariant and attracting.

Lemma 1. *The region $\Omega \in \mathbb{R}_+^9$ is positively invariant for the model system (1) with initial conditions in $\mathbb{R}_+^9$.*

2.1. Existence and Stability of Equilibrium Points

We analyse system (1) to obtain the equilibrium points of the system and their stability. Let $(S^*, P^*, E^*, I^*, T^*, R^*, X^*, Y^*, Z^*)$ be the equilibrium points of system (1). At an equilibrium point, we have

$$S'(t) = P'(t) = E'(t) = I'(t) = T'(t) = R'(t) = X'(t) = Y'(t) = Z'(t) = 0.$$

2.1.1. Disease-Free Equilibrium, E_0

In the absence of malaria, that is, $E^* = I^* = T^* = R^* = Y^* = Z^* = 0$, model system (1) has an equilibrium point called the disease-free equilibrium, E_0 and is given by

$$\begin{aligned} E_0 &= (S^*, P^*, 0, 0, 0, 0, X^*, 0, 0), \\ &= \left[\frac{(1-\tau)\Lambda}{g+\mu_h}, \frac{\Lambda(\tau\mu_h + g)}{\mu_h(g+\mu_h)}, 0, 0, 0, 0, \frac{\pi}{\mu_v}, 0, 0\right]. \end{aligned} \tag{2}$$

To establish the linear stability of E_0, we employ the Van den Driessche and Watmough's next generation matrix approach [21]. A reproduction number obtained this way determines the local stability of the disease-free equilibrium point for $R_e < 1$ and instability for $R_e > 1$. We do this by first computing the effective reproduction number, R_e of the model. Following [21], the associated next generation matrices F and V of system (1) can be found from $\mathcal{F}_i$ and $\mathcal{V}_i$ respectively, where

$$\mathcal{F}_i = \begin{bmatrix} E' \\ I' \\ Y' \\ Z' \\ T' \end{bmatrix} = \begin{bmatrix} \frac{\beta_{vh}\alpha Z S}{N_h} + \frac{\beta_{vh}\alpha\vartheta Z P}{N_h} \\ 0 \\ \frac{\beta_{hv}\alpha I X}{N_h} \\ 0 \\ 0 \end{bmatrix},$$

and

$$\mathcal{V}_i = \begin{bmatrix} (\mu_h + \nu)E \\ (\gamma + \sigma + \mu_h + \delta_h)I - \nu E \\ (\theta + \mu_v)Y \\ (\mu_v + \delta_v)Z - \theta Y \\ (\varepsilon + \mu_h)T - \sigma I \end{bmatrix}.$$

The Jacobian matrix of $\mathcal{F}_i$ is given by

$$F = \begin{bmatrix} 0 & 0 & 0 & m & 0 \\ 0 & 0 & 0 & 0 & 0 \\ 0 & n & 0 & 0 & 0 \\ 0 & 0 & 0 & 0 & 0 \\ 0 & 0 & 0 & 0 & 0 \end{bmatrix}, \tag{3}$$

where

$m = \beta_{vh}\alpha + \dfrac{\beta_{vh}\alpha\vartheta(\tau\mu_h + g)}{\mu_h(1-\tau)}$, and $n = \dfrac{\beta_{hv}\alpha\pi(g+\mu_h)}{\mu_v\Lambda(1-\tau)}$.

The Jacobian matrix of $\mathcal{V}_i$ is given by

$$V = \begin{bmatrix} a & 0 & 0 & 0 & 0 \\ -b & c & 0 & 0 & 0 \\ 0 & 0 & e & 0 & 0 \\ 0 & 0 & -f & q & 0 \\ 0 & -r & 0 & 0 & j \end{bmatrix}, \tag{4}$$

where
$a = (\mu_h + \nu),\ \ b = \nu,\ \ c = (\gamma + \sigma + \mu_h + \delta_h),\ \ e = (\theta + \mu_v),\ \ f = \theta\ \ q = (\mu_v + \delta_v),$
$r = \sigma,\ \ j = (\varepsilon + \mu_h).$

After some lengthy algebraic manipulations, it can be shown that the effective reproduction number is given by

$$\begin{aligned}\mathcal{R}_e &= \sqrt{\frac{\beta_{vh}\beta_{hv}\alpha^2\theta\pi\mu_h\nu[g(1+\vartheta)+\mu_h(1+\tau\vartheta)]}{\mu_v\Lambda(g+\mu_h)(\mu_h+\nu)(\gamma+\sigma+\mu_h+\delta_h)(\theta+\mu_v)(\mu_v+\delta_v)}}, \\ &= \sqrt{\mathcal{R}_{ev}\mathcal{R}_{eh}},\end{aligned} \tag{5}$$

where

$$\mathcal{R}_{ev} = \frac{\beta_{hv}\theta\alpha\pi}{\mu_v(\theta+\mu_v)(\mu_v+\delta_v)},$$

is the contribution of mosquito population when it infects the humans, and

$$\mathcal{R}_{eh} = \frac{\beta_{vh}\alpha\mu_h\nu(1+\vartheta\mu_h m)}{\Lambda(\mu_h+\nu)(\gamma+\sigma+\mu_h+\delta_h)},$$

is the human contribution when they infect the mosquitoes.
The expression for the effective reproduction number, $\mathcal{R}_e$ has a biological meaning that is readily interpreted from terms under the square root. The term

- $\dfrac{\beta_{hv}\theta\pi\alpha}{\mu_v(\mu_v+\theta)(\mu_v+\delta_v)}$ represents the number of secondary human infections caused by one infected mosquito vector.
- $\dfrac{\beta_{vh}\alpha\mu_h\nu[g(1+\vartheta)+\mu_h(1+\tau\vartheta)]}{\Lambda(g+\mu_h)(\mu_h+\nu)(\gamma+\sigma+\mu_h+\delta_h)}$ represents the number of secondary mosquito infections caused by one infected human host.

The square root represents the geometric mean of the average number of secondary host infections produced by one vector, and the average number of secondary vector infections produced by one host. This effective reproduction number serves as an invasion threshold both for predicting outbreaks and evaluating control strategies that would reduce the spread of the disease. The threshold quantity, $\mathcal{R}_e$, measures the average number of secondary cases generated by a single infected individual in a susceptible human population [9], where a fraction of the susceptible human population is under prevention and the infected class is under treatment. In the absence of any protective measure, the effective reproduction number $\mathcal{R}_e$ with treatment as the only intervention strategy in place becomes

$$\mathcal{R}_{et} = \sqrt{\frac{\beta_{vh}\beta_{hv}\alpha^2\theta\pi\mu_h\nu}{\mu_v\Lambda(\mu_h+\nu)(\gamma+\sigma+\mu_h+\delta_h)(\theta+\mu_v)(\mu_v+\delta_v)}}.$$

Similarly, if the protection is the only intervention strategy being practised, the effective reproduction number becomes

$$\mathcal{R}_{ep} = \sqrt{\frac{\beta_{vh}\beta_{hv}\alpha^2\theta\pi\mu_h\nu[g(1+\vartheta)+\mu_h(1+\tau\vartheta)]}{\mu_v\Lambda(g+\mu_h)(\mu_h+\nu)(\gamma+\mu_h+\delta_h)(\theta+\mu_v)(\mu_v+\delta_v)}}.$$

From the expression for the effective reproduction number $\mathcal{R}_e$, we note that the parameters τ, and ϑ play important roles in the spread of the disease. The following cases can be considered; no disease, that is, the value of the basic reproduction number can be obtained from the value of the effective reproduction number when there are no control measures which are effective. In this case $(\vartheta = 0)$ since this is the reduction of likelihood of infection by protection and $\tau = 0$. Thus the basic reproduction number of the model (1) without control measures is given by

$$\mathcal{R}_0 = \sqrt{\frac{\beta_{vh}\beta_{hv}\alpha^2\nu\theta\pi\mu_h}{\mu_v\Lambda(\mu_h+\nu)(\gamma+\mu_h+\delta_h)(\theta+\mu_v)(\mu_v+\delta_v)}}.$$

From the two reproduction numbers, we notice that

$$\mathcal{R}_e \leq \mathcal{R}_0$$

for $\tau \geq 0$, due to reduction of likelihood of infection by protection. This implies that ITNs and IRS have a positive impact on the malaria dynamics as they contribute in the reduction of secondary infections.

Therefore, from [21] (Theorem 2), the following result holds;

Lemma 2. *The disease-free equilibrium E_0, of the malaria model with intervention strategies (1), given by (2) is locally asymptotically stable if $\mathcal{R}_e < 1$, and unstable if $\mathcal{R}_e > 1$.*

2.1.2. Endemic Equilibrium Point, E_1

When malaria is present, that is, $S^* \neq P^* \neq E^* \neq I^* \neq T^* \neq R^* \neq X^* \neq Y^* \neq Z^* \neq 0$, the system (1) has a steady state, E_1 called the endemic equilibrium. In order to establish the stability of E_1 we express the system (1) in dimensionless variables which are $\frac{S}{N_h} = s$, $\frac{P}{N_h} = p$, $\frac{E}{N_h} = e$, $\frac{I}{N_h} = i$, $\frac{T}{N_h} = \xi$, $\frac{R}{N_h} = r$, $\frac{X}{N_h} = x$, $\frac{Y}{N_h} = y$, $\frac{Z}{N_h} = z$, where $\frac{dN_h}{dt} = \Lambda - \mu_h N_h - \delta_h I$, $\frac{dN_v}{dt} = \pi - \mu_v - \delta_v Z$, and $N_h = N_v = 1$ in the classes $s, p, e, i, \xi, r, x, y, z$ and then differentiating with respect to time, t respectively. This results in the following system.

$$\begin{aligned}
\frac{ds}{dt} &= \frac{1}{N_h}\left[\frac{dS}{dt} - s\frac{dN_h}{dt}\right],\\
&= \frac{1}{N_h}\left[(1-\tau)\Lambda - \beta_{vh}\alpha z s N_h + \omega r N_h - (g+\mu_h)sN_h\right]\\
&\quad - \frac{s}{N_h}\left[\Lambda - \mu_h N_h - \delta_h i N_h\right],\\
&= (1-\tau)\frac{\Lambda}{N_h} - \left[\frac{\Lambda}{N_h} + g - \delta_h i\right] s - \beta_{vh}\alpha z s + \omega r.
\end{aligned}$$

$$\begin{aligned}\frac{dp}{dt} &= \frac{1}{N_h}\left[\frac{dP}{dt} - p\frac{dN_h}{dt}\right],\\ &= \frac{1}{N_h}[\tau\Lambda + gsN_h + (1-\omega)rN_h - \beta_{vh}\alpha\vartheta pN_h - \mu_h pN_h]\\ &\quad - \frac{p}{N_h}[\Lambda - \mu_h N_h - \delta_h iN_h],\\ &= gs + \tau\frac{\Lambda}{N_h} + (1-\omega)r - \beta_{vh}\alpha\vartheta zp - \left[\frac{\Lambda}{N_h} - \delta_h i\right]p.\end{aligned}$$

$$\begin{aligned}\frac{de}{dt} &= \frac{1}{N_h}\left[\frac{dE}{dt} - e\frac{dN_h}{dt}\right],\\ &= \frac{1}{N_h}[\beta_{vh}\alpha zsN_h + \beta_{vh}\alpha\vartheta zpN_h - \nu eN_h - \mu_h eN_h]\\ &\quad - \frac{e}{N_h}[\Lambda - \mu_h N_h - \delta_h iN_h],\\ &= \beta_{vh}\alpha zs + \beta_{vh}\alpha\vartheta zp - \left[\frac{\Lambda}{N_h} + \nu - \delta_h i\right]e.\end{aligned}$$

$$\begin{aligned}\frac{di}{dt} &= \frac{1}{N_h}\left[\frac{dI}{dt} - i\frac{dN_h}{dt}\right],\\ &= \frac{1}{N_h}[\nu eN_h - \mu_h iN_h - \gamma iN_h - \delta_h iN_h - \sigma iN_h] - \frac{i}{N_h}[\Lambda - \mu_h N_h - \delta_h iN_h],\\ &= \nu e - \left[\frac{\Lambda}{N_h} + \sigma + \gamma + \delta_h - \delta_h i\right]i.\end{aligned}$$

$$\begin{aligned}\frac{d\xi}{dt} &= \frac{1}{N_h}\left[\frac{dT}{dt} - t\frac{dN_h}{dt}\right],\\ &= \frac{1}{N_h}[\sigma iN_h - \mu_h tN_h - \varepsilon tN_h] - \frac{t}{N_h}[\Lambda - \mu_h N_h - \delta_h iN_h],\\ &= \sigma i - \left[\frac{\Lambda}{N_h} + \varepsilon - \delta_h i\right]t.\end{aligned}$$

$$\begin{aligned}\frac{dr}{dt} &= \frac{1}{N_h}\left[\frac{dR}{dt} - r\frac{dN_h}{dt}\right],\\ &= \frac{1}{N_h}[\varepsilon tN_h + \gamma iN_h - \omega rN_h - (1-\omega)rN_h - \mu_h rN_h]\\ &\quad - \frac{r}{N_h}[\Lambda - \mu_h N_h - \delta_h iN_h],\\ &= \varepsilon t + \gamma i - \left[\frac{\Lambda}{N_h} + 1 - \delta_h i\right]r.\end{aligned}$$

$$\begin{aligned}\frac{dx}{dt} &= \frac{1}{N_v}\left[\frac{dX}{dt} - x\frac{dN_v}{dt}\right],\\ &= \frac{1}{N_v}[\pi - \beta_{hv}\alpha ixN_v - \mu_h xN_v] - \frac{x}{N_v}[\pi - \mu_v N_v - \delta_v zN_v],\\ &= \frac{\pi}{N_v}(1-x) - \beta_{hv}\alpha ix + \delta_v xz.\end{aligned}$$

$$\begin{aligned}\frac{dy}{dt} &= \frac{1}{N_v}\left[\frac{dY}{dt} - y\frac{dN_v}{dt}\right],\\ &= \frac{1}{N_v}[\beta_{hv}\alpha i x N_v - (\theta+\mu_v)yN_v] - \frac{y}{N_v}[\pi - \mu_v N_v - \delta_v z N_v],\\ &= \beta_{hv}\alpha i x - \left[\frac{\pi}{N_v} + \theta - \delta_v z\right] y.\end{aligned}$$

$$\begin{aligned}\frac{dz}{dt} &= \frac{1}{N_v}\left[\frac{dZ}{dt} - z\frac{dN_v}{dt}\right],\\ &= \frac{1}{N_v}[\theta y N_v - \mu_v z N_v - \delta_v z N_v] - \frac{z}{N_v}[\pi - \mu_v N_v - \delta_v z N_v],\\ &= \theta y - \left[\frac{\pi}{N_v} + \delta_v - \delta_v z\right] z.\end{aligned}$$

$$\frac{dN_h}{dt} = \left[\frac{\Lambda}{N_h} - \mu_h - \delta_h i\right] N_h,$$

and

$$\frac{dN_v}{dt} = \left[\frac{\pi}{N_v} - \mu_v - \delta_v z\right] N_v.$$

Thus, we have the following reduced system of equations;

$$\left.\begin{aligned}\frac{ds}{dt} &= (1-\tau)\frac{\Lambda}{N_h} - \left[\frac{\Lambda}{N_h} + g - \delta_h i\right] s - \beta_{vh}\alpha z s + \omega r\\ \frac{dp}{dt} &= gs + \tau\frac{\Lambda}{N_h} + (1-\omega)r - \beta_{vh}\alpha\vartheta z p - \left[\frac{\Lambda}{N_h} - \delta_h i\right] p\\ \frac{de}{dt} &= \beta_{vh}\alpha z s + \beta_{vh}\alpha\vartheta z p - \left[\frac{\Lambda}{N_h} + \nu - \delta_h i\right] e\\ \frac{di}{dt} &= \nu e - \left[\frac{\Lambda}{N_h} + \sigma + \gamma + \delta_h - \delta_h i\right] i\\ \frac{d\xi}{dt} &= \sigma i - \left[\frac{\Lambda}{N_h} + \varepsilon - \delta_h i\right] t\\ \frac{dr}{dt} &= \varepsilon t + \gamma i - \left[\frac{\Lambda}{N_h} + 1 - \delta_h i\right] r\\ \frac{dx}{dt} &= \frac{\pi}{N_v}(1-x) - \beta_{hv}\alpha i x + \delta_v x z\\ \frac{dy}{dt} &= \beta_{hv}\alpha i x - \left[\frac{\pi}{N_v} + \theta - \delta_v z\right] y\\ \frac{dz}{dt} &= \theta y - \left[\frac{\pi}{N_v} + \delta_v - \delta_v z\right] z\end{aligned}\right\}, \quad (6)$$

with total population sizes N_h and N_v satisfying

$$\frac{dN_h}{dt} = \left[\frac{\Lambda}{N_h} - \mu_h - \delta_h i\right] N_h,$$

$$\frac{dN_v}{dt} = \left[\frac{\pi}{N_v} - \mu_v - \delta_v z\right] N_v.$$

The system can now be reduced to a nine-dimensional system by eliminating s and x since $s = 1-p-e-i-\xi-r$ and $x = 1-y-z$, respectively in the following feasible region Ω where the model makes biological sense

$$\Omega_1 = \left\{ \begin{pmatrix} p \\ e \\ i \\ \xi \\ r \\ N_h \\ y \\ z \\ N_v \end{pmatrix} \in \mathbb{R}^9_+ \middle| \begin{array}{r} p \geq 0, \\ e \geq 0, \\ i \geq 0, \\ \xi \geq 0, \\ r \geq 0, \\ p+e+i+\xi+r \leq 1, \\ N_h \leq \frac{\Lambda}{\mu_h}, \\ y \geq 0, \\ z \geq 0, \\ y+z \leq 1, \\ N_v \leq \frac{\pi}{\mu_v} \end{array} \right\}.$$

v can be shown to be positively invariant with respect to system (7). $\mathbb{R}^9_+$ denotes the non-negative cone of $\mathbb{R}^9$ including its lower dimensional faces. Thus, we have the following system of equations;

$$\left.\begin{aligned}
\frac{dp}{dt} &= g(1-p-e-i-\xi-r) + \tau\frac{\Lambda}{N_h} + (1-\omega)r - \beta_{vh}\alpha\vartheta zp \\
&\quad - \left[\frac{\Lambda}{N_h} - \delta_h i\right] p \\
\frac{de}{dt} &= \beta_{vh}\alpha z(1-p-e-i-\xi-r) + \beta_{vh}\alpha\vartheta zp - \left[\frac{\Lambda}{N_h} + \nu - \delta_h i\right] e \\
\frac{di}{dt} &= \nu e - \left[\frac{\Lambda}{N_h} + \sigma + \gamma + \delta_h - \delta_h i\right] i \\
\frac{d\xi}{dt} &= \sigma i - \left[\frac{\Lambda}{N_h} + \varepsilon - \delta_h i\right] t \\
\frac{dr}{dt} &= \varepsilon t + \gamma i - \left[\frac{\Lambda}{N_h} + 1 - \delta_h i\right] r \\
\frac{dN_h}{dt} &= \Lambda - \mu_h N_h - \delta_h i N_h \\
\frac{dy}{dt} &= \beta_{hv}\alpha i(1-y-z) - \left[\frac{\pi}{N_v} + \theta - \delta_v z\right] y \\
\frac{dz}{dt} &= \theta y - \left[\frac{\pi}{N_v} + \delta_v - \delta_v z\right] z \\
\frac{dN_v}{dt} &= \pi - \mu_v N_v - \delta_v z N_v.
\end{aligned}\right\}. \quad (7)$$

To compute the steady states of system (7), we set the derivatives with respect to time in (7) equal to zero, and then on simplification the following algebraic expressions are obtained;

$$\left.\begin{aligned}
g(1-p-e-i-\xi-r)+\tau\frac{\Lambda}{N_h}+(1-\omega)r &= \beta_{vh}\alpha\vartheta pz+\left[\frac{\Lambda}{N_h}-\delta_h i\right]p\\
\beta_{vh}\alpha z(1-p-e-i-\xi-r)+\beta_{vh}\alpha\vartheta pz &= \left[\frac{\Lambda}{N_h}+\nu-\delta_h i\right]e\\
\nu e &= \left[\frac{\Lambda}{N_h}+\sigma+\gamma+\delta_h-\delta_h i\right]i\\
\sigma i &= \left[\frac{\Lambda}{N_h}+\varepsilon-\delta_h i\right]t\\
\varepsilon t+\gamma i &= \left[\frac{\Lambda}{N_h}+1-\delta_h i\right]r\\
\frac{\Lambda}{N_h} &= \mu_h+\delta_h i\\
\beta_{hv}\alpha i(1-y-z) &= \left[\frac{\pi}{N_v}+\theta-\delta_v z\right]y\\
\theta y &= \left[\frac{\pi}{N_v}+\delta_v-\delta_v z\right]z\\
\frac{\pi}{N_v} &= \mu_v+\delta_v z.
\end{aligned}\right\}. \tag{8}$$

Hence, the dimesionless variables are calculated in terms of i as follows;

$$\left.\begin{aligned}
p &= \frac{d_2 i^2+d_1 i+d_3}{d_4 i+d_5}\\
e &= c_1 i\\
t &= c_2 i\\
r &= c_3 i\\
y &= \frac{\beta_{hv}\alpha i[\beta_{hv}\theta\alpha i+(\beta_{hv}\alpha i+\mu_v+\theta)(\mu_v+\delta_v)]-\theta\beta_{hv}^2\alpha^2 i^2}{(\beta_{hv}\alpha i+\mu_v+\theta)[\theta\beta_{hv}\alpha i+(\beta_{hv}\alpha i+\mu_v+\theta)(\mu_v+\delta_v)]}\\
z &= \frac{k_1 i}{k_6 i+k_5}
\end{aligned}\right\}, \tag{9}$$

for

$c_1=\dfrac{\sigma+\gamma+\mu_h+\delta_h}{\nu},\quad c_2=\dfrac{\sigma}{\mu_h+\varepsilon},\quad c_3=\dfrac{\sigma\varepsilon}{(\mu_h+\varepsilon)(\mu_h+1)},$

$d_1=(g+\tau\mu_h)k_6+(\tau\delta_h+(1-\omega)c_3-c_1g-g-c_2g-c_3g)k_5,$

$d_2=\tau\delta_h+(1-\omega)c_3-c_1g-g-c_2g-c_3g,$

$d_3=[g(1+\vartheta)+\mu_h(1+\tau\vartheta)](\mu_v+\delta_v)(\mu_v+\theta),\quad d_4=\beta_{vh}\alpha\vartheta k_1+(\mu_h+g)k_6,$

$d_5=(\mu_h+g)k_5,\quad k_1=\beta_{hv}\theta\alpha,\quad k_5=(\mu_v+\delta_v)(\mu_v+\theta),\quad k_6=\beta_{hv}\alpha(\theta+\mu_v+\delta_v),$

where i can be obtained by substituting e into the second equation of system (8) which gives

$$c_1 i=\frac{c_4 i+c_4 pi-c_5 i^2}{c_7 i+c_8}.$$

It is clear that either $i=0$ (the disease-free equilibrium point E_0) or

$$p=\frac{d_6 i+d_7}{c_4}\ (endemic\,equilibrium\,point\,E_1) \tag{10}$$

where
$c_4 = \beta_{vh}\beta_{hv}\alpha^2\theta, \;\; c_5 = \beta_{vh}\alpha k_1(1 + c_2 + c_3),$
$c_7 = \beta_{vh}\beta_{hv}\alpha^2\theta + (\mu_h + \nu)(\theta + \mu_v + \delta_v)\beta_{hv}\alpha, \;\; c_8 = (\mu_h + \nu)(\mu_v + \delta_v)(\mu_v + \theta),$
$d_6 = c_1c_7 + c_5$ and $d_7 = c_1c_8 - c_4.$

We seek to establish whether a unique endemic equilibrium exists. This is done by making more realistic assumptions that the protective control measures may not be totally effective, and thus, $0 < \vartheta < 1$.

Existence and Uniqueness of Endemic Equilibrium, E_1

The existence of the endemic equilibrium in Ω, can be determined when $i \neq 0$, that is the equations for p in (9) and (10) will be used. Now, equating (9) and (10) we get,

$$Ai^2 + Bi + C = 0, \tag{11}$$

where

$$\begin{aligned} A &= d_4d_6 - c_4d_2, \\ &= q_1\left[\left(\frac{\sigma + \gamma + \mu_h + \delta_h}{\nu}\right)(\beta_{vh}\beta_{hv}\alpha^2\theta + (\mu_h + \nu)(\theta + \mu_v + \delta_v)\beta_{hv}\alpha)\right] \\ &\quad - q_3\left[\tau\delta_h + \frac{(1-\omega)\sigma\varepsilon}{(\mu_h + \varepsilon)(\mu_h + 1)} - \frac{\sigma + \gamma + \mu_h + \delta_h}{\nu}g - H\right], \\ &> 0, \end{aligned}$$

$$\begin{aligned} B &= d_4d_7 + d_5d_6 - c_4d_1, \\ &= q_1\left[\left(\frac{\sigma + \gamma + \mu_h + \delta_h}{\nu}\right)(\mu_h + \nu)(\mu_v + \delta_v)(\mu_v + \theta) - \beta_{vh}\beta_{hv}\alpha^2\theta\right] \\ &\quad + (\mu_h + g)(\mu_v + \delta_v)(\mu_v + \theta)q_2 \\ &\quad + \left[\beta_{vh}\beta_{hv}\alpha^2\theta\left(\frac{(\mu_h + \varepsilon)(\mu_h + 1) + \sigma(\mu_h + 1) + \sigma\varepsilon}{(\mu_h + \varepsilon)(\mu_h + 1)}\right)\right] \\ &\quad - \beta_{vh}\beta_{hv}\theta\alpha^2[(g + \tau\mu_h)(\theta + \mu_v + \delta_v)\beta_{hv}\alpha + q_4(\mu_v + \delta_v)(\mu_v + \theta)], \\ &> 0, \end{aligned}$$

with
$$H = g - \frac{\sigma g}{(\mu_h + \varepsilon)} - \frac{\sigma\varepsilon g}{(\mu_h + \varepsilon)(\mu_h + 1)},$$
$$q_1 = \beta_{vh}\beta_{hv}\alpha^2\theta\vartheta + (\mu_h + g)(\theta + \mu_v + \delta_v)\beta_{hv}\alpha,$$
$$q_2 = \left(\frac{\sigma + \gamma + \mu_h + \delta_h}{\nu}\right)[\beta_{vh}\beta_{hv}\alpha^2\theta + (\mu_h + \nu)(\theta + \mu_v + \delta_v)\beta_{hv}\alpha],$$
$$q_3 = \beta_{vh}\beta_{hv}\alpha^2\theta,$$
$$q_4 = \tau\delta_h + \frac{(1-\omega)\sigma\varepsilon}{(\mu_h + \varepsilon)(\mu_h + 1)} - \left(\frac{\sigma + \gamma + \mu_h + \delta_h}{\nu}\right)g - g - \frac{\sigma g}{\mu_h + \varepsilon} - \frac{\sigma\varepsilon g}{(\mu_h + \varepsilon)(\mu_h + 1)},$$

$$\begin{aligned} C &= d_5d_7 - c_4d_3, \\ &= (\mu_h + g)(\mu_v + \delta_v)^2(\mu_v + \theta)^2\left(\frac{\sigma + \gamma + \mu_h + \delta_h}{\nu}\right)(\mu_h + \nu) \\ &\quad - \beta_{vh}\beta_{hv}\theta\alpha^2[g(1 + \vartheta) + \mu_h(1 + \tau\vartheta)](\mu_v + \delta_v)(\mu_v + \theta), \\ &= (g + \mu_h)(\mu_v + \delta_v)^2(\mu_v + \theta)^2\left(\frac{\sigma + \gamma + \mu_h + \delta_h}{\nu}\right)(\mu_h + \nu)(1 - \mathcal{R}_e^2). \end{aligned}$$

For $\mathcal{R}_e > 1$, the existence of endemic equilibria is determined by the presence of positive real solutions of the quadratic expression (11).

$$C = (g + \mu_h)(\mu_v + \delta_v)^2(\mu_v + \theta)^2 \left(\frac{\sigma + \gamma + \mu_h + \delta_h}{\nu}\right)(\mu_h + \nu)(1 - \mathcal{R}_e^2) < 0.$$

If x_1, x_2 are the roots of equation (11), then their product

$$x_1 x_2 = \frac{C}{A}.$$

Since $C < 0$ and $A > 0$, then $\frac{C}{A} < 0$. Thus, there exists exactly one positive endemic equilibrium for $i \in (0, 1]$ whenever $\mathcal{R}_e > 1$. This gives the threshold for the endemic persistence. Therefore, we have proved the existence and uniqueness of the endemic equilibrium, E_1 for the system (6). This result is summarized in the following theorem;

Theorem 1. *If $\mathcal{R}_e > 1$, then system (6) has a unique endemic equilibrium E_1.*

The result in Theorem 1 indicates the impossibility of backward bifurcation in the malaria model, since it has no endemic equilibrium when $R_e < 1$. This explains that the model (1) has a globally-asymptotically stable disease-free equilibrium whenever $\mathcal{R}_e \leq 1$. A global stability result for the disease-free equilibrium, E_0 is given below.

2.2. Global Stability of the Disease-free Equilibrium, E_0

Here, the global asymptotic stability property of E_0 of model (1) is explored for the case when $\delta_h = \delta_v = 0$. By considering the disease-free nation in model (1), it follows that, $S = N_h^* - P - E - I - T - R$ and $X = N_v^* - Y - Z$ at the steady-state.

Hence, the global stability of E_0, can be established by considering the following limiting system of model (1);

$$\left.\begin{aligned}
\frac{dP}{dt} &= \tau\Lambda + g(N_h^* - P - E - I - T - R) + (1 - \omega)R - \lambda_{hp}P - \mu_h P \\
\frac{dE}{dt} &= \lambda_h(N_h^* - P - E - I - T - R) + \lambda_{hp}P - (\mu_h + \nu)E \\
\frac{dI}{dt} &= \nu E - (\gamma + \sigma + \mu_h)I \\
\frac{dT}{dt} &= \sigma I - (\varepsilon + \mu_h)T \\
\frac{dR}{dt} &= \varepsilon T + \gamma I - (1 - \omega)R - (\omega + \mu_h)R \\
\frac{dY}{dt} &= \lambda_v(N_v^* - Y - Z) - (\theta + \mu_v)Y \\
\frac{dZ}{dt} &= \theta Y - \mu_v Z
\end{aligned}\right\}, \quad (12)$$

in the positive-invariant region,

$$\Omega_2 = \left\{(P, E, I, T, R, Y, Z) \in \mathbb{R}_+^7 : P + E + I + T + R \leq N_h^*,\ Y + Z \leq N_v^*\right\},$$

where,
$\lambda_h = \dfrac{\beta_{vh}\alpha Z}{N_h}$, $\lambda_{hp} = \dfrac{\beta_{vh}\theta\alpha Z}{N_h}$, and $\lambda_v = \dfrac{\beta_{hv}\alpha I}{N_h}$.
The associated threshold number for the limiting model (12), denoted by $\mathcal{R}_e^{'}$ is given by,

$$\begin{aligned} \mathcal{R}_e^{'} &= \rho(F_3 V_2^{-1}) \\ &= \sqrt{\frac{\beta_{vh}\beta_{hv}N_v^*\alpha^2\theta(A_1+B_1)}{(N_h^*)^2\mu_v(\theta+\mu_v)(\mu_v+\delta_v)(\mu_h+\nu)(\gamma+\sigma+\mu_h+\delta_h)}}, \end{aligned} \tag{13}$$

where, $A_1 = \nu N_h^*$, $B_1 = \vartheta P^*(\nu+\mu_h)$ and the matrix V_2 is as defined in (4). The matrix F_3 is given by

$$F_3 = \begin{bmatrix} 0 & 0 & 0 & \beta_{vh}\alpha + \dfrac{\beta_{vh}\alpha\vartheta P^*}{N_h^*} & 0 \\ 0 & 0 & 0 & 0 & 0 \\ 0 & \dfrac{\beta_{hv}\alpha N_v^*}{N_h^*} & 0 & 0 & 0 \\ 0 & 0 & 0 & 0 & 0 \\ 0 & 0 & 0 & 0 & 0 \end{bmatrix}.$$

Hence, the following result is established.

Lemma 3. *The disease-free equilibrium E_2, of the limiting system (12), given by (2) is locally asymptotically stable if $\mathcal{R}_e^{'} < 1$ and unstable if $\mathcal{R}_e^{'} > 1$.*

The feasible region can be defined as

$$\Omega_2 = \{(P, E, I, T, R, Y, Z) \in \Omega_1 : P \leq P^*\}.$$

Hence, the following result is claimed;

Theorem 2. *The disease-free equilibrium of system (12) given by (2), is globally asymptotically stable in Ω if $\mathcal{R}_e^{'} < 1$.*

Proof. It can be seen from the first equation of the limiting system (12) that,

$$\begin{aligned} \frac{dP}{dt} &= \tau\Lambda + g(N_h^* - P - E - I - T - R) + (1-\omega)R - \lambda_{hp}P - \mu_h P, \\ &\leq \tau\Lambda + gN_h^* - (g+\mu_h)P, \\ &= (g+\mu_h)\left(\frac{\tau\Lambda + gN_h^*}{g+\mu_h} - P\right), \\ &= (g+\mu_h)(P^* - P). \end{aligned}$$

Hence, $P(t) \leq P^* - (P^* - P_0)e^{-(g+\mu_h)t}$.
It follows that either $P(t)$ approaches P^* asymptotically or $P(t) \leq P^*$ in finite time $(note\ \ that\ \ \dfrac{dP}{dt} < 0\ \ if\ \ P(t) > P^*)$. Therefore, the set Ω is positively-invariant and attracting. Further, the equations for the infected components of (12) can be written as,

$$\begin{bmatrix} E^{'} \\ I^{'} \\ Y^{'} \\ Z^{'} \\ T^{'} \end{bmatrix} = (F_3 - V_2 - Q) \begin{bmatrix} E \\ I \\ Y \\ Z \\ T \end{bmatrix},$$

where, the matrices F_3 and V_2 are as defined above, and the matrix Q is given by

$$Q = \begin{bmatrix} \lambda_h & \lambda_h & 0 & \frac{\beta_{vh}\vartheta\alpha P^*}{N_h^*}\left(1-\frac{P}{P^*}\right) & 0 \\ 0 & 0 & 0 & 0 & 0 \\ 0 & 0 & \lambda_v & \lambda_v & 0 \\ 0 & 0 & 0 & 0 & 0 \\ 0 & 0 & 0 & 0 & 0 \end{bmatrix}.$$

Since $P \leq P^*$ $(for\ all\ t \geq 0)$ in Ω, it follows that, matrix Q is non-negative. Thus,

$$\begin{bmatrix} E' \\ I' \\ Y' \\ Z' \\ T' \end{bmatrix} = (F_3 - V_2)\begin{bmatrix} E \\ I \\ Y \\ Z \\ T \end{bmatrix}. \tag{14}$$

If $\mathcal{R}_e' < 1$, then $\rho(F_3V_2^{-1} < 1)$ (from the local stability result) which is equivalent to $F_3 - V_2$ [21]. It follows that, the linearized differential inequality system (14) is stable whenever $\mathcal{R}_e' < 1$. Consequently, by comparison theorem (Lakshmikantham et al., 1989), it follows that, $(E, I, Y, Z, T, R) \to (0, 0, 0, 0, 0, 0)$. Substituting $E = I = Y = Z = T = R = 0$ into the first equation of system (12) gives $P(t) \to P^*$ as $t \to \infty$. Further, since Ω is positively-invariant, and it follows that E_0 is globally asymptotically stable in Ω if $\mathcal{R}_e' < 1$.

The epidemiological implication of the above result is that malaria could be eliminated from the community if prevention and treatment can lead to a situation where $\mathcal{R}_e$ is less than unity. However, other factors have to be considered.

3. Numerical Simulations

We carry out numerical simulations using a fourth order Runge-Kutta scheme in Matlab. The main aim is to verify some of the analytical results on the stability of the system (1) and the effect of the interventions on the transmission dynamics of malaria in Malawi. Some of the parameter values were obtained from the National Statistical Office (NSO) in Zomba, Malawi [16], others were obtained from literature, and the rest were assumed for the purpose of illustration. For it malaria control programme, the Government of Malawi has put in place several strategies through the National Malaria Control Programme.

We simulate the model in the absence and presence of intervention strategies in order to find out the effects of varying each intervention parameter on the transmission dynamics of malaria. The figures are plotted using the parameter values in Table 1 and the initial conditions; $S_0 = 100$, $P_0 = 90$, $E_0 = 70$, $I_0 = 50$, $T_0 = 35$, $R_0 = 30$, $X_0 = 1000$, $Y = 900$, $Z_0 = 700$. These are estimated average values of the population which participated in the malaria survey which was conducted in Malawi.

In the absence of any intervention strategy, the basic reproduction number is $\mathcal{R}_0 = 1.3034$. Since this number is slightly greater than one, there are sufficient infective humans

Table 1. Parameter values of the model (Unit: *per day*)

Symbol	Value	Source
Λ	0.0015875	[18]
π	0.071	[18]
μ_h	$\frac{1}{(60 \times 365)}$	[3]
μ_v	1/15	[3]
ν	1/17	[3]
δ_h	0.02326	[16]
δ_v	0.07	Assumed
ϑ	0.475	Assumed
ω	$\frac{1}{(20 \times 365)}$	[3]
θ	0.0017	[15]
σ	0.01	[16]
α	0.86	[15]
β_{vh}	0.86	[18]
β_{hv}	0.083	[15]
τ	0.11	[15]
g	0.0833	[16]
ε	0.00722	[15]
γ	0.0035	[8]

and mosquitoes to continue generating new infection, whence the permanence/endemicity of the disease.

For the sub-model with interventions (ITNs and IRS), there is a steady decrease in the susceptible population and an increase in the protected population, highlighting the effectiveness of preventive measures in controlling malaria (Figure 2).

From Figure 2(b), the exposed population decreases steadily with time because of the preventive measure in place. The trend remained the same despite variation in the mosquito biting rate, α.

Treated individuals (Figure 2(d)) increase when there are more infectives and decrease following the reduction in the number of infected individuals. This increased trend is also seen in the recovered class when ITNs and IRS are adopted (Figure 3).

Treatment programs must be complemented with other interventions (such as vector reduction strategies and personal protection). This is further supported by varying the prevention rate g (Figure 4(b)) and the treatment rate σ (Figure 4(c)).

4. Discussion and Conclusion

We presented a deterministic model of the transmission dynamics of malaria. The model considered a varying population size that incorporated recruitment of new individuals into

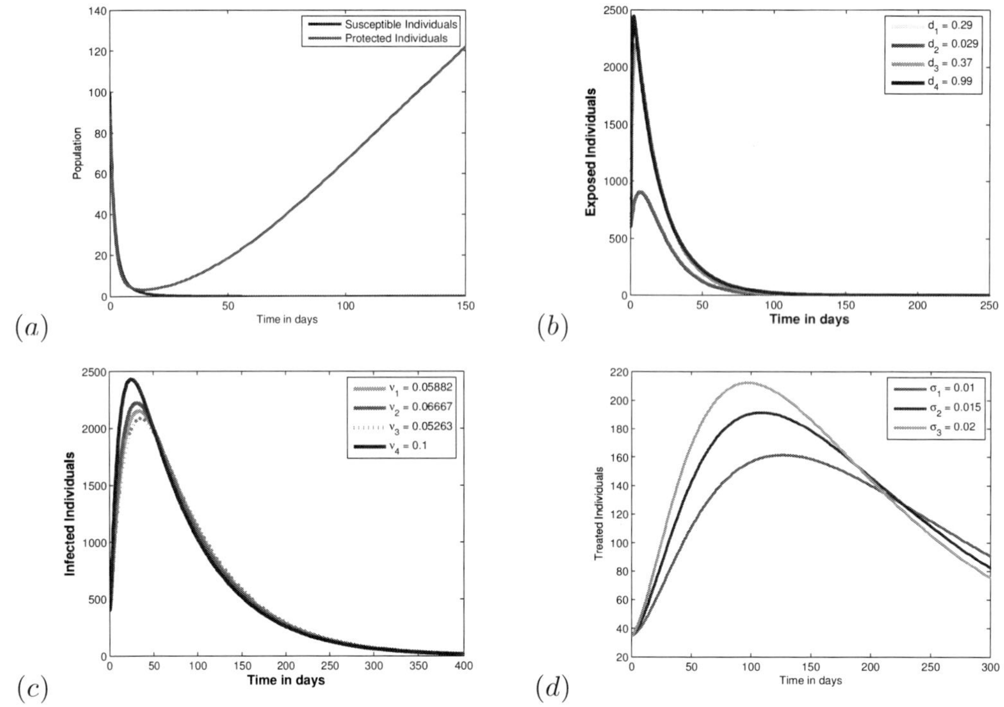

Figure 2. Time series of (a) Susceptible and protected individuals for the model with intervention - $\mathcal{R}_e = 0.0850$ (b) Exposed individuals (c) Infected individuals (d) Individuals in the Treated class

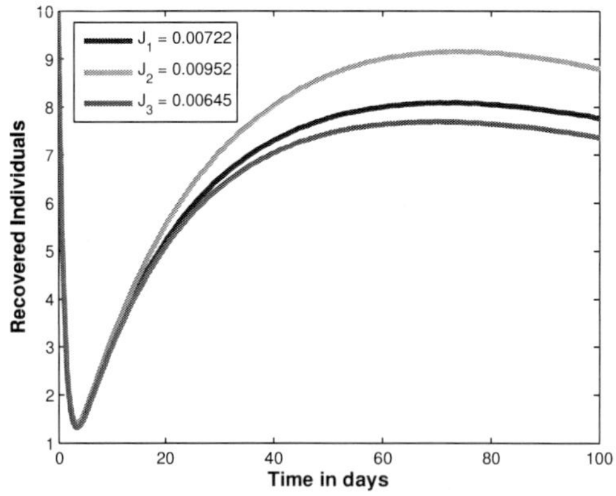

Figure 3. Recovered individuals for the malaria model with intervention strategies

the susceptible class through either birth or immigration. The model also had the protected and treated classes in order to assess the potential impact of protection and treatment strategies on the transmission dynamics of the disease.

Qualitative analysis of the model showed that there exists a domain where the model

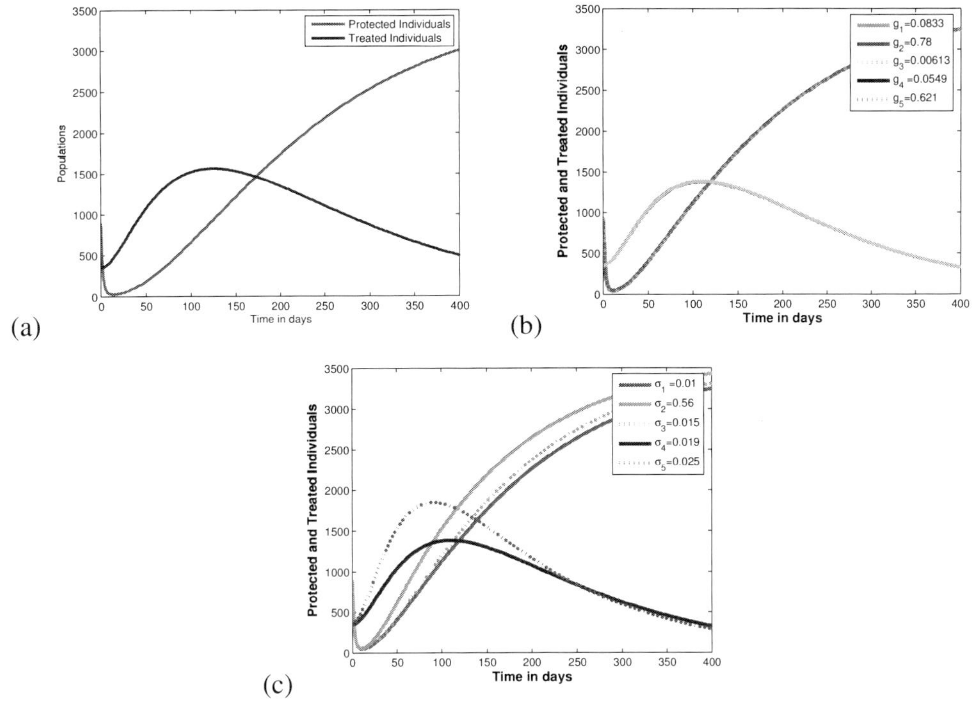

Figure 4. (a) Dynamics of protected and treated individuals (b) Protected and treated individuals as the protection rate g varies and (c) Protected and treated individuals as treatment rate σ varies

is epidemiologically and mathematically well-posed. The disease threshold parameter (or effective reproduction number) $\mathcal{R}_e$, was computed. The model was then qualitatively analyzed for the existence and stability of their associated equilibria. For $\mathcal{R}_e < 1$, the disease-free equilibrium E_0 is locally asymptotically stable, and for $\mathcal{R}_e > 1$ the endemic equilibrium E_1 exists. If $\mathcal{R}_e < 1$, the disease cannot persist, hence $\mathcal{R}_e$ is the useful indication of the effort required to eliminate/eradicate the disease. Also, if $\mathcal{R}_e \leq \mathcal{R}_0$, that is, increasing preventive and control measures have a positive effect on reduction of $\mathcal{R}_e$. Thus, malaria can be eradicated out of a community by deployment of a combination of strategies such as vector control via ITNs and IRS complemented with timely treatment of infected people.

Numerical analysis of the model suggested that effective control or eradication of malaria can be achieved by the combination of protection and treatment measures. We showed that at the time when infection is high in the community, treatment is highly required compared to prevention. Further, we found out that prevention measures were important to maintain the reduction or eradication of the disease transmission. This study concurs with the Chaves [6] suggestion that an intervention using insecticide-treated bed nets represents an excellent example of implementing an infectious disease control programme, and Smith's et al, study [19], which showed that both regular and non-fixed spraying resulted in a significant reduction in the overall number of mosquitoes, as well as the number of malaria cases in humans. Hence, the combination of these two findings, and treatment as an additional intervention strategy, showed a great impact in the reduction of the spread of malaria. Therefore, the combination of these interventions can play a bigger role in reducing or eradicating the transmission of the disease in the country.

As the resurgence of malaria continues to take its toll on individuals and communities in Malawi, the policy makers have to be informed about the research results. The following recommendations could be of interest to public health personnel and policy decision makers:

(i) Since most of the reductions in transmission come from the protection of a few humans, it is far more important to improve the killing effects by encouraging the use of ITNs and indoor residual spraying (IRS) around those who are mostly exposed to malaria; however, complete coverage and improved killing effects may be necessary to reach the control goals.

(ii) ITNs and IRS, with insecticidal and diversionary properties, would reduce the availability of hosts, and kill mosquitoes that are attempting to feed on human's blood, and reducing malaria transmission.

(iv) Because of the complications of measuring malaria at different transmission levels with different immunological status prevalent in different age and gender groups, and across different locations, some guidelines should be developed to give researchers and health professionals a more accurate foundation on which to select their indicators.

The proposed model can be extended to include the effects of the environment on the spread of malaria. Some parameters, such as the incubation period in mosquitoes and mosquito birth rate depend on seasonal environmental factors such as rainfall, temperature and humidity. We can include these effects by modeling these parameters as periodic functions of time. This would provide a more accurate picture of malaria transmission and prevalence than that obtained from models that use parameter values averaged over seasons. Addition of age, spatial variation on the implementation of insecticide mosquito treated nets (ITNs) and indoor residual spraying (IRS) as well as conducting a cost-effective analysis of thee interventions is viable.

Acknowledgments

PMM and MM acknowledge with thanks the financial support by the Norad's Masters programme in Mathematical Modeling (NOMA) at the University of Dar es Salaam, Tanzania.

References

[1] Aron, J.L. (1988), "Mathematical modelling of immunity to malaria, Nonlinearity in biology and medicine (Los Alamos, NM, 1987)", *Math. Biosci.*, Volume 90, pp 385-396.

[2] Birkhoff, G. and Rota, G.C. (1982), *Ordinary Differential Equations,* Ginn.

[3] Blayneh, K., Cao, Y., and Hee-Dae, K. (2009), "Optimal Control of Vector-borne Diseases: Treatment and Prevention", *Discrete and Continuous Dynamical Systems Series B*, Volume 11, No.3, pp: 1-xx.

[4] Bupa UK (2009), *Malaria symptoms, causes and treatment of disease,* http://hcd2.bupa.co.uk/fact sheets/html/malaria disease retrieved on Friday, 19th June, 2009.

[5] Carr, J. (1981), *Applications centre mainfold theory,* Springer-Verg, New York.

[6] Chaves, F.L., Kaneko, A., Taleo, G., Pascual, M. and Wilson, L.M. (2008), "Malaria transmission pattern resilience to variability is by insecticide-treated nets", *Biomed Central,* Volume 7, No. 100.

[7] Chavez-Castillo, C. and Song, B. (2004), "Dynamical models of tuberculosis and their applications", *Math.Biosci.Eng.,* No. 2, pp: 361-404.

[8] Chitnis, N., Hyman, J.M. and Cushing, J.M. (2008), "Determining important parameters in the spread of malaria through the sensitivity analysis of a mathematical model", *Math. Bio.,* Volume 70, pp: 1272-1296.

[9] Diekmann, O. and Heesterbeek, J.A.P. (1990), "On the definition and computation of the basic reproduction ratio R_0 in the model of infectious disease in heterogeneous populations", *J. Math. Biol.,* Volume 2, No. 1, pp: 265-382.

[10] Dushoff, J., Huang, W., and Castillo-Chavez, C. (1998), "Backwards bifurcations and castastrophe in simple models of fatal diseases", *J. Math. Biol.,* Volume 36, pp: 227-248.

[11] Garba, S.M., Gumel, A.B. and Abu Bakar, M.R. (2008), "Backward bifurcation in dengue transmission dynamics", *Math.Biosci.*, Volume 215, pp: 11-25.

[12] Hethcote, H.M. (2000), "The mathematics of infectious diseases", *SIAM Rev.*, Volume 42, No. 4, pp: 599-953.

[13] Kribs-Zaleta, C., and Valesco-Hernandez, J. (2000), "A simple vaccination model with multiple endemic state", *Math.Biosci.,* Volume 164, pp: 183-201.

[14] Lalloo, D.G., Shingadia, D., Pasvol, G., Chiodini, P.L., Whitty, C.J. and Beeching, N.J. (2007), "UK malaria treatment guidelines", *J. Infect 2007,* Volume 54, pp: 111-121.

[15] Miranda, I., Teboh-Ewungkem, Chandra, N. Podder and Gumel, A.B. (2009), "Mathematical Study of the Role of Gametocytes and an Imperfect Vaccine on Malaria Transmission Dynamics", *Bulletin of Mathematical Biology DOI 10,* Volume 1007, No. S11, pp: 538-009-9437-3.

[16] National Statistical Office (NSO) and UNICEF (2008), *Malawi Multiple Indicator Cluster Survey 2006, Final Report,* Lilongwe, Malawi.

[17] Ngwa, G.A. and Shu, W.S. (2000), "A mathematical model for endemic malaria with variable human and mosquito populations", *Math. Comp. Modeling,* Volume 32, pp: 747-763.

[18] Niger, A.M. and Gumel, A.B. (2008), "Mathematical Analysis of the Role of Repeated Exposure on Malaria Transmission Dynamics", *Differential Equations and Dynamical Systems,* Volume 16, No. 3, pp: 251-287.

[19] Smith, R.J. and Hove-Musekwa, D.S. (2008), "Determining effective spraying periods to control malaria via indoor residual spraying in sub-Saharan Africa", *Appl Math and Deci Sc.*, doi:10, Volume 1155, No. 745463, pp: 1-19.

[20] Tumwiine, J., Mugisha, J.Y.T. and Luboobi, L.S.(unpublished), "Threshold and stability results for a malaria model in a population with protective intervention among high-risk groups.

[21] Van den Driessche, P. and Watmough, J. (2002), "Reproduction numbers and subthreshold endemic equilibria for compartmental models of disease transmission", *Mathematical Bio-sciences*, Volume 180, pp: 29-48.

[22] Whitty, C.J.M., Lalloo, D. and Ustianowski, A. (2007), "Malaria: an update on treatment of adults in non-endemic countries", *BMJ.,* Volume 333, pp: 214-245.

[23] WHO, (2009), *malaria, Fact Sheets,* http://www.who.int/inf-fs/en/fact094.html retrieved on Wednesday, 17th June, 2009.

In: Dynamical Systems
Editor: Jean Michel Tchuenche, pp. 315-333
ISBN: 978-1-62808-001-8

Chapter 12

MODELLING THE DYNAMICS OF HIV-TB CO-INFECTION IN A VARIABLE SIZE POPULATION

Agraj Tripathi[1], Ram Naresh[2,*] and Dileep Sharma[3]

[1]Department of Mathematics, Bhabha Institute of Technology, Kanpur, India
[2]Department of Mathematics, Harcourt Butler Technological Institute, Kanpur, India
[3]Department of Mathematics, DBS(PG) College, Dehradun, India

Abstract

In this chapter, a nonlinear mathematical model is proposed to study the transmission dynamics of HIV and a curable TB pathogen within a population of varying size. The population, under consideration, is divided into five sub-classes, namely, the susceptibles, the TB infectives, the HIV infectives, the HIV-TB co-infectives and the AIDS patients. The model is analyzed qualitatively using stability theory of nonlinear differential equations.

The model exhibits five equilibria, namely, the disease- free, the TB only, the HIV only, the co-infection only, and the nontrivial endemic equilibrium. The basic reproduction number R_0 is the key factor of the model. It is observed that if $R_0 > 1$ then at least one of the diseases always persists in the population otherwise both the diseases die out. It is shown that the positive endemic equilibrium may be locally stable under certain conditions showing that the disease becomes endemic due to constant migration of the population into the habitat. A numerical study of the model is also carried out to investigate the influence of certain key parameters on the spread of the disease, and to support the analytical results.

Keywords: HIV/AIDS epidemic, TB pathogen, co-infection, equilibrium points, stability analysis, numerical simulation

* E-mail address: ramntripathi@yahoo.com (Corresponding author)

1. Introduction

Many infectious diseases which afflict human population around the world are spread by sexual contacts between susceptibles and infectives. The Human Immuno-deficiency Virus (HIV) infection which can lead to Acquired Immuno-deficiency Syndrome (AIDS), has become an important infectious disease in both the developed and developing nations. It is a fatal disease, which breaks down the body's immune system, leaving the victim vulnerable to a host of life threatening opportunistic infections, neurological disorders or unusual malignancies. It causes mortality of millions of people and expenditure of enormous amount of money in health care and disease control.

It is not a single epidemic but made up of distinct epidemics within the same state and continues to be driven strongly by heterosexual transmission. Mathematical models have been used extensively in research into the epidemiology of HIV/AIDS, to help improve our understanding of the major contributing factors in a given epidemic. Various refinements have been added into modeling frameworks, and specific issues have been addressed by researchers [see for instance 2, 8, 14 - 19, 22 - 23]. In particular, Busenberg et. al. [2] presented a model for the spread of HIV/AIDS in a community due to sexual interaction between a core group of female prostitutes and young unmarried males. Srinivasa Rao [22] presented a theoretical framework for transmission of HIV/AIDS epidemic in India. Tripathi et. al. [23] presented a nonlinear mathematical model to study the effect of screening of unaware infectives on the spread of HIV/AIDS in a homogeneous population with constant immigration of susceptibles.

Tuberculosis (TB) is a disease that affects both human and animal population. It was probably transmitted from animals to humans in areas where agriculture became dominant and animals were domesticated. The growth of human communities probably increased the recurrence of the TB epidemic leading to its currently overwhelmingly high levels of endemicity in some developing nations [3]. It is an airborne communicable disease caused by *Mycobacterium tuberculosis*, or the *tubercle bacillus*. The *tubercle bacilli* live in the lungs of infected hosts. They are spread in the air when infectious individuals sneeze, cough or speak. A susceptible individual may become infected with TB if he or she inhales bacilli from the air. The particles containing *M. tuberculosis* are so small that normal air current keeps them airborne and transport them throughout rooms or buildings [5, 24]. Infection begins with the multiplication of *tubercle bacilli* in alveolar macrophages, some of which spread through the bloodstream. TB is such a type of disease which also increases due to the environmental factors e.g. discharges of household wastes (garbage, trash etc.) in residential areas, open drainage of sewage in residential areas, open water storage tanks etc. Once infected with *M. tuberculosis*, a person stays infected for many years, probably for life.

The vast majority (90%) of people without HIV infection who are infected with *M. tuberculosis* do not develop tuberculosis disease. HIV is the most powerful risk factor for progression from TB infection to TB disease. An HIV positive person infected with *M. tuberculosis* has a 50% lifetime risk of developing TB whereas an HIV negative person infected with M. tuberculosis has only a 10% risk of developing TB. This is especially important in India where it is estimated that almost half of the adult population harbor *M. tuberculosis* [11, 12].

The enormous public health burden inflicted by these two diseases necessitates the use of mathematical modeling to gain insights into their transmission dynamics and to determine effective control strategies. Unfortunately, not much have been done in terms of modeling the dynamics of HIV-TB co-infection at a population level. The inextricably-linked pathogenesis and epidemiology of Mycobacterium tuberculosis (TB) and the Human Immuno-deficiency Virus (HIV) are well known. Some studies have been made by taking into account HIV-TB co-infection [3, 6, 8-10, 13, 20, 21, 25-26]. McLean and Nowak [10] proposed within host models for the dynamics between HIV and activated CD_4T cells specific to other pathogens. West and Thompson [25] developed models which reflect the transmission dynamics of both TB and HIV, and discussed the magnitude and duration of the effect that the HIV epidemic may have on TB. They found the effect that HIV will have on the general population, to be dependent on the contact structure between the general population and the HIV risk groups, as well as a possible shift in the dynamics associated with TB transmission. It is worth emphasizing that the two diseases differ in their modes of transmission. Whilst tuberculosis is an airborne disease, HIV is transmitted predominantly by sexual contacts or by needle sharing (particularly among IV drug users) [21]. Sharomi et. al. [21] proposed and analyzed a nonlinear mathematical model for HIV-TB co-infection and found that the disease-free equilibrium is locally asymptotically stable whenever the maximum of reproduction numbers is less than unity. The two diseases exhibit some sort of synergistic interaction, where each accelerates the progression of the other. The HIV/AIDS pandemic continues to play a major role in the resurgence of TB, resulting in increased morbidity and mortality worldwide. Additionally, HIV fuels progression to active disease in people infected with TB [7]. Rates of recurrence of TB, both due to endogenous reactivation and exogenous re-infection, are increased amongst people with HIV infection [6, 7].

In view of the above, in this chapter, a mathematical model is proposed to study the transmission of HIV/AIDS in presence of curable TB pathogen by considering a separate co-infection class of HIV-TB within a given population with varying size including demographic features. Qualitative and numerical analysis of the proposed model is carried out to understand the dynamics of HIV-TB co-infection. The purpose of the current study is to complement the aforementioned studies, by designing and qualitatively analyzing a new and more comprehensive deterministic model for gaining insights into the transmission dynamics and control of the two diseases in a population.

The outline of the chapter is as follows. In Section 2 we introduce an HIV-TB co-infection model. Section 3 deals with the stability analysis for the non-negative equillibria. Numerical simulations are carried out in Section 4 and Section 5 presents the conclusion.

2. Mathematical Model

We consider the population of size $N(t)$ at time t with constant immigration rate Q_0. The population size $N(t)$ is divided into five subclasses of susceptibles $S(t)$, tuberculosis infectives $I_1(t)$, HIV infectives $I_2(t)$ (also assumed to be infectious), HIV-TB co-infectives $I_3(t)$ and AIDS patients $A(t)$. The susceptibles become tuberculosis infected following contacts with the tuberculosis infectives I_1 at a rate β_1. The population in this class is diminished by HIV infection at a rate β_4.

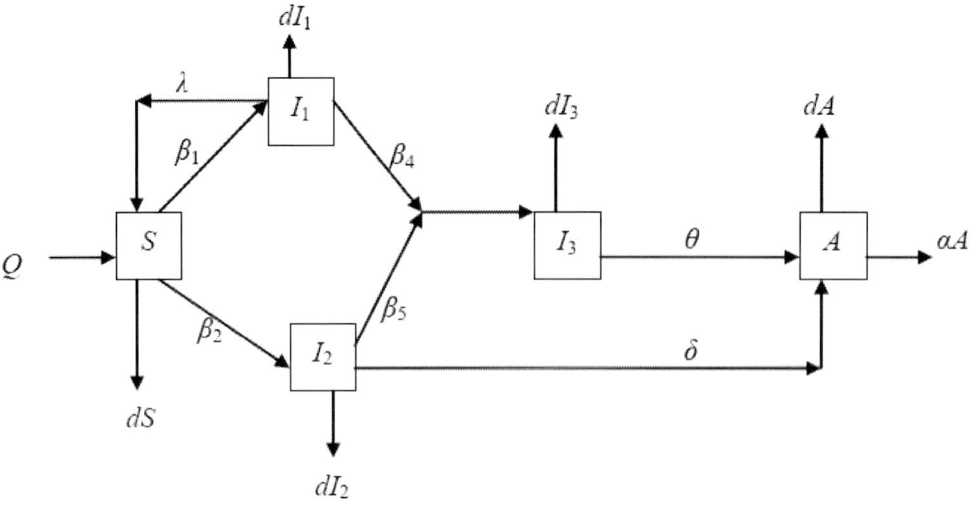

Figure 1. Transmission dynamics of HIV-TB co-infection.

Some members of this class are also cured and recovered at a rate λ and become susceptible again. The population in class I_2 is generated by the HIV infection of the susceptible individuals at a rate β_2. The population in this class is diminished by TB infection at a rate β_5. The population in class I_3 is also generated by the HIV-TB co-infection of the susceptibles at a rate β_3. Here we assume that the susceptibles, when interact sexually with the individuals of class I_3, develop both the tuberculosis and HIV infection to join the co-infectives class I_3. It is further assumed that the anti-HIV treatment is not available within the community and, therefore, individuals of the I_2 and I_3 classes are bound to develop 'full blown' AIDS with a rate δ and θ respectively, as shown in the transfer diagram above,

Thus, the spread of disease is assumed to be governed by the following system of nonlinear ordinary differential equations,

$$\frac{dS}{dt} = Q_0 - \frac{\beta_1 S I_1}{N} - \frac{\beta_2 S I_2}{N} - dS + \lambda I_1 \tag{2.1}$$

$$\frac{dI_1}{dt} = \frac{\beta_1 S I_1}{N} - \frac{\beta_4 I_1 I_2}{N} - (\lambda + d) I_1 \tag{2.2}$$

$$\frac{dI_2}{dt} = \frac{\beta_2 S I_2}{N} - \frac{\beta_5 I_1 I_2}{N} - (\delta + d) I_2 \tag{2.3}$$

$$\frac{dI_3}{dt} = \frac{\beta_3 S I_3}{N} + \frac{(\beta_4 + \beta_5) I_1 I_2}{N} - (\theta + d) I_3 \tag{2.4}$$

$$\frac{dA}{dt} = \delta I_2 + \theta I_3 - (\alpha + d) A \tag{2.5}$$

with

$$S(0) = S_0,\ I_1(0) = I_{10},\ I_2(0) = I_{20},\ I_3(0) = I_{30},\ A(0) = A_0,$$

where d is the natural mortality rate constant and α is the AIDS related death rate constant. It is assumed that all the dependent variables and parameters of the model are non-negative.

Since $N(t) = S(t) + I_1(t) + I_2(t) + I_3(t) + A(t)$, the above equations can now be written as,

$$\frac{dN}{dt} = Q_0 - dN - \alpha A \tag{2.6}$$

$$\frac{dI_1}{dt} = \frac{\beta_1(N - I_1 - I_2 - I_3 - A)I_1}{N} - \frac{\beta_4 I_1 I_2}{N} - (\lambda + d)I_1 \tag{2.7}$$

$$\frac{dI_2}{dt} = \frac{\beta_2(N - I_1 - I_2 - I_3 - A)I_2}{N} - \frac{\beta_5 I_1 I_2}{N} - (\delta + d)I_2 \tag{2.8}$$

$$\frac{dI_3}{dt} = \frac{\beta_3(N - I_1 - I_2 - I_3 - A)I_3}{N} + \frac{(\beta_4 + \beta_5)I_1 I_2}{N} - (\theta + d)I_3 \tag{2.9}$$

$$\frac{dA}{dt} = \delta I_2 + \theta I_3 - (\alpha + d)A \tag{2.10}$$

with $N(0) = N_0$, $I_1(0) = I_{10}$, $I_2(0) = I_{20}$, $I_3(0) = I_{30}$, $A(0) = A_0$.

From the model, it is noted that in the absence of deadly infection, the population size approaches the steady state value Q_0/d.

2.1. Invariant Region

Continuity of the right hand side of system (2.6) - (2.10) and its derivative imply that the model is well posed for $N > 0$, $\beta_1 > (\lambda + d)$ and $\beta_2 > (\delta + d)$. The invariant region where solution exists is obtained as follows:

$$\frac{Q_0}{(\alpha + d)} \leq \liminf N(t) \leq \limsup N(t) \leq \frac{Q_0}{d} \quad (\text{as } t \to \infty)$$

Since $N(t) > 0$ for all $t \geq 0$. Therefore, from eq. (2.6), $N(t)$ cannot blow up to infinity in finite time and consequently, the model system is dissipative (solutions are bounded) and the solutions exist globally for all $t > 0$ in the invariant and compact set,

$$\Omega = \left\{(N, I_1, I_2, I_3, A);\ 0 \leq I_1(t) + I_2(t) + I_3(t) + A(t) \leq N(t) \leq \frac{Q_0}{d} + \varepsilon\right\},$$

the region of attraction for any arbitrary small $\varepsilon > 0$.

As N tends to zero, S, I_i ($i = 1, 2, 3$) and A also tend to zero. Hence, each of these subpopulations, (S, I_i for i =1, 2, 3) tends to zero as N does. It is, therefore, natural to interpret these terms as zero at $N = 0$.

2.2. Positivity of Solutions

Lemma 1. Let the initial data be $N(0) = N_0 > 0$, $I_1(0) = I_{10} \geq 0$, $I_2(0) = I_{20} \geq 0$, $I_3(0) = I_{30} \geq 0$ and $A(0) = A_0 \geq 0$ for all $t \geq 0$. Then, the solution ($N(t)$, $I_1(t)$, $I_2(t)$, $I_3(t)$, $A(t)$) of the model remain positive for all time $t \geq 0$.

Proof. From eq. (2.6), we have

$$N'(t) \geq -(\alpha + d)N(t)$$

By applying a theorem on differential inequalities [1] we obtain

$$N(t) \geq c_1 e^{\{-(\alpha+d)\,t\}}$$

Here c_1 is a constant of integration. A similar reasoning on the remaining equations shows that they are always positive in Ω for $t > 0$

3. Equilibrium and Stability Analysis

The model (2.6) - (2.10) has five non-negative equillibria namely,

(i) E_0 (Q_0/d, 0, 0, 0, 0), the disease-free.

This equilibrium explains that in the absence of any of the diseases, the total population size approaches its equilibrium value Q_0/d.

(ii) $E_1\left(\frac{Q_0}{d}, \frac{Q_0[\beta_1 - (\lambda + d)]}{d\beta_1}, 0, 0, 0\right)$, the TB only, which exists if $\beta_1 > (\lambda + d)$.

This equilibrium explains that if only TB is present in the population then TB infectives will remain at its equilibrium $\frac{Q_0}{d}\left(1 - \frac{(\lambda + d)}{\beta_1}\right)$.

(iii) $E_2\left(\hat{N}, 0, \hat{I}_2, 0, \hat{A}\right)$, the HIV-only, which exists if $\beta_2 > (\delta + d)$.

This equilibrium implies that if only HIV is present in the population, the population size reduces to $\hat{N}$ from Q_0/d, with HIV-infectives maintained at its equilibrium $\hat{I}_2$ and AIDS population at $\hat{A}$, where

$$\hat{N} = \frac{\beta_2[\alpha + d + \delta]\,\hat{I}_2}{[\beta_2 - (\delta + d)](\alpha + d)},$$

$$\hat{I}_2 = \frac{Q_0(\alpha + d)[\beta_2 - (\delta + d)]}{[\beta_2 d(\alpha + \delta + d) + \alpha\delta\{\beta_2 - (\delta + d)\}]} \text{ and } \hat{A} = \frac{\delta\,\hat{I}_2}{\alpha + d}.$$

(iv) $E_3\left(\tilde{N}, 0, 0, \tilde{I}_3, \tilde{A}\right)$, the co-infection only, which exists if $\beta_3 > (\theta + d)$.

In this case, since no individual TB and HIV infectives are present, the susceptibles, interact directly with the co-infected individuals. As a result, the population of individuals with both the diseases increase which leads to generate the AIDS population.

where $\tilde{N} = \dfrac{\beta_3[\alpha + d + \theta]\,\tilde{I}_3}{[\beta_3 - (\theta + d)](\alpha + d)}$

$$\tilde{I}_3 = \frac{Q_0(\alpha + d)[\beta_3 - (\theta + d)]}{[\beta_3 d(\alpha + \theta + d) + \alpha\delta\{\beta_3 - (\theta + d)\}]} \text{ and } \tilde{A} = \frac{\theta\,\tilde{I}_3}{\alpha + d}$$

(v) $E^*\left(N^*, I_1^*, I_2^*, I_3^*, A^*\right)$, the nontrivial endemic equilibrium

where N^*, I_1^*, I_2^*, I_3^* and A^* is the positive solution of the model system (2.6) - (2.10). It is very cumbersome to obtain the values of the variables in closed form. However, the equilibrium exists and is shown numerically, (See Section 4).

Now we state a theorem for local stability of the above equilibrium points.

Theorem 1.

(i) The equilibrium point E_0 is locally asymptotically stable if $\dfrac{\beta_1}{\lambda + d} < 1$ $(i.e. R_1 < 1)$, $\dfrac{\beta_2}{\delta + d} < 1$ $(i.e. R_2 < 1)$ and $\dfrac{\beta_3}{\theta + d} < 1$ $(i.e. R_3 < 1)$.

(ii) The second equilibrium E_1 is also locally asymptotically stable.

(iii) The third equilibrium E_2 will be locally asymptotically stable if, a_1a_2-a_3 >0 and a_1, a_2, a_3 are all positive, the values of which are given in the proof.

(iv) The fourth equilibrium E_3 will be locally asymptotically stable if b_1b_2-b_3 > 0. The respective values of b_i (i = 1, 2, 3) are given in the proof and b_i should also be positive.

(v) The fifth equilibrium E^*, if it exists, is locally asymptotically stable provided c_1c_2-c_3 > 0 and $c_3(c_1c_2-c_3)-c_1^2c_4 > 0$ where c_1, c_2, c_3 and c_4 are given in the proof of the theorem.

Proof. The variational matrix corresponding to equilibrium point E_0 is given by,

$$M(E_0)=\begin{bmatrix} -d & 0 & 0 & 0 & -\alpha \\ 0 & [\beta_1-(\lambda+d)] & 0 & 0 & 0 \\ 0 & 0 & [\beta_2-(\delta+d)] & 0 & 0 \\ 0 & 0 & 0 & [\beta_3-(\theta+d)] & 0 \\ 0 & 0 & \delta & \theta & -(\alpha+d) \end{bmatrix}$$

From the matrix $M(E_0)$, it is clear that E_0 is locally asymptotically stable provided $R_1<1,\ R_2<1$ and $R_3<1$, where R_1, R_2 and R_3 are basic reproduction numbers associated with TB, HIV and HIV-TB co-infection, respectively defined as follows;

$$R_1=\frac{\beta_1}{\lambda+d},\ R_2=\frac{\beta_2}{\delta+d}\ \text{and}\ R_3=\frac{\beta_3}{\theta+d}$$

The eigenvalues of the matrix $M(E_0)$ are R_1, R_2 and R_3. The dominant eigenvalue, given by $R_0 = \max\{R_1, R_2, R_3\}$, is the basic reproductive number [4]. Thus, the disease-free equilibrium is locally asymptotically stable if $R_0 < 1$ and unstable if $R_0 >1$. If $R_1>1$, $R_2>1$ and $R_3>1$ the equilibrium point E_0 is a saddle point which is stable in *N-A* manifold and unstable in I_1-I_2 direction. In such a case, the infections are maintained in the population.

Now the variational matrix corresponding to the second equilibrium E_1 is given as

$$M(E_1)=\begin{bmatrix} -d & 0 & 0 & 0 & -\alpha \\ \beta_1 p^2 & -\beta_1 p & -(\beta_1+\beta_4)p & -\beta_1 p & -\beta_1 p \\ 0 & 0 & q & 0 & 0 \\ 0 & 0 & (\beta_4+\beta_5)p & r & 0 \\ 0 & 0 & \delta & \theta & -(\alpha+d) \end{bmatrix}$$

where, $p = [\beta_1-(\lambda+ d)] > 0$, $q = [\beta_2 - (\delta + d)]- (\beta_2 + \beta_5) [\beta_1 - (\lambda + d)] < 0$.

and $r = [\beta_3 - (\theta + d)]- \beta_3 [\beta_1 - (\lambda + d)] < 0$, since $\beta_1 > (\lambda + d)$, $\beta_2 < (\delta + d)$ and $\beta_3 < (\theta + d)$ in the equilibrium E_1.

The characteristic equation corresponding to matrix $M(E_1)$ is given as follows,

$$f(x)=(x+d)(x+\beta_1 p)(x+(\alpha+d))(x-q)(x-r)=0 \tag{3.1}$$

The matrix $M(E_1)$ contains all negative eigenvalues which makes it stable at this equilibrium point.

The variational matrix $M(E_2)$ corresponding to the third equilibrium E_2 is given as,

$$M(E_2) = \begin{bmatrix} -d & 0 & 0 & 0 & -\alpha \\ 0 & l & 0 & 0 & 0 \\ [\beta_2 - (\delta + d)]\frac{\hat{I}_2}{\hat{N}} & -(\beta_2 + \beta_5)\frac{\hat{I}_2}{\hat{N}} & m & -\frac{\beta_2 \hat{I}_2}{\hat{N}} & -\frac{\beta_2 \hat{I}_2}{\hat{N}} \\ 0 & (\beta_4 + \beta_5)\frac{\hat{I}_2}{\hat{N}} & 0 & n & 0 \\ 0 & 0 & \delta & \theta & -(\alpha + d) \end{bmatrix}$$

where, $l = [\beta_1\text{-}(\lambda + d)] - \frac{\beta_4 \hat{I}_2}{\hat{N}} - \frac{\beta_1 \hat{A}}{\hat{N}} < 0$, $m = [\beta_2 - (\delta + d)] - \frac{2\beta_2 \hat{I}_2}{\hat{N}} - \frac{\beta_2 \hat{A}}{\hat{N}}$

and $n = [\beta_3\text{-}(\theta + d)] - \frac{\beta_3 \hat{I}_2}{\hat{N}} - \frac{\beta_3 \hat{A}}{\hat{N}} < 0$ since $\beta_1 < (\lambda + d)$ and $\beta_3 < (\theta + d)$ in the equilibrium E_2.

The characteristic equation corresponding to matrix $M(E_2)$ is given by,

$$f(x) = (x - l)(x - n)\left(x^3 + a_1 x^2 + a_2 x + a_3\right) = 0 \tag{3.2}$$

where,

$$a_1 = \alpha + 2d + \beta_2\left[\frac{2\hat{I}_2}{\hat{N}} + \frac{\hat{A}}{\hat{N}}\right] - [\beta_2 - (\delta + d)]$$

$$a_2 = d(\alpha + d) + \frac{\beta_2 \delta \hat{I}_2}{\hat{N}} + (\alpha + 2d)\left[\beta_2\left(\frac{2\hat{I}_2}{\hat{N}} + \frac{\hat{A}}{\hat{N}}\right) - [\beta_2 - (\delta + d)]\right]$$

$$a_3 = \frac{\beta_2 \delta d \hat{I}_2}{\hat{N}} + d(\alpha + d)\beta_2\left[\frac{2\hat{I}_2}{\hat{N}} + \frac{\hat{A}}{\hat{N}}\right] + \left[\frac{\alpha \delta \hat{I}_2}{\hat{N}} - d(\alpha + d)\right][\beta_2 - (\delta + d)]$$

Here two eigenvalues are negative and the remaining two have negative real parts if $a_1 > 0$, $a_2 > 0$, $a_3 > 0$, and $a_1 a_2\text{-}a_3 > 0$ as from Routh-Hurwitz criteria.

Now the variational matrix $M(E_3)$ corresponding to the fourth equilibrium E_3 is given as,

$$M(E_3) = \begin{bmatrix} -d & 0 & 0 & 0 & -\alpha \\ 0 & u & 0 & 0 & 0 \\ 0 & 0 & v & 0 & 0 \\ [\beta_3 - (\theta + d)]\frac{\tilde{I}_3}{\tilde{N}} & -\frac{\beta_3 \tilde{I}_3}{\tilde{N}} & -\frac{\beta_3 \tilde{I}_3}{\tilde{N}} & w & -\frac{\beta_3 \tilde{I}_3}{\tilde{N}} \\ 0 & 0 & \delta & \theta & -(\alpha + d) \end{bmatrix}$$

where $u = [\beta_1-(\lambda+d)] - \frac{\beta_1 \tilde{I}_3}{\tilde{N}} - \frac{\beta_1 \tilde{A}}{\tilde{N}} < 0$, $v = [\beta_2-(\delta+d)] - \frac{2\beta_2 \tilde{I}_3}{\tilde{N}} - \frac{\beta_2 \tilde{A}}{\tilde{N}} < 0$

and $w = [\beta_3-(\theta+d)] - \frac{\beta_3 \tilde{I}_2}{\tilde{N}} - \frac{\beta_3 \tilde{A}}{\tilde{N}}$

The characteristic equation corresponding to matrix $M(E_3)$ is given by,

$$f(x) = (x-u)(x-v)\left(x^3 + b_1 x^2 + b_2 x + b_3\right) = 0 \tag{3.3}$$

where,

$$b_1 = \alpha + 2d + \beta_3\left[\frac{2\tilde{I}_3}{\tilde{N}} + \frac{\tilde{A}}{\tilde{N}}\right] - [\beta_3 - (\theta + d)]$$

$$b_2 = d(\alpha + d) + \frac{\beta_3 \tilde{I}_3}{\tilde{N}} + (\alpha + 2d)\left[\beta_3\left(\frac{2\tilde{I}_3}{\tilde{N}} + \frac{\tilde{A}}{\tilde{N}}\right) - [\beta_3 - (\theta + d)]\right]$$

$$b_3 = \frac{\beta_3 d \tilde{I}_3}{\tilde{N}} + d(\alpha + d)\beta_3\left[\frac{2\tilde{I}_3}{\tilde{N}} + \frac{\tilde{A}}{\tilde{N}}\right] + \left[\frac{\alpha\theta \tilde{I}_3}{\tilde{N}} - d(\alpha + d)\right][\beta_3 - (\theta + d)]$$

Here also two eigenvalues are negative and the other two have negative real parts if $b_1 > 0$, $b_2 > 0$, $b_3 > 0$, and $b_1 b_2 - b_3 > 0$ as from Routh-Hurwitz criteria.

The variational matrix corresponding to the nontrivial endemic equilibrium is given as,

$$M(E^*) = \begin{bmatrix} -d & 0 & 0 & 0 & -\alpha \\ [\beta_1 - (\lambda + d)]\frac{I_1^*}{N^*} & -\frac{\beta_1 I_1^*}{N^*} & -\frac{(\beta_1 + \beta_4) I_1^*}{N^*} & -\frac{\beta_1 I_1^*}{N^*} & -\frac{\beta_1 I_1^*}{N^*} \\ [\beta_2 - (\delta + d)]\frac{I_2^*}{N^*} & -\frac{(\beta_2 + \beta_5) I_2^*}{N^*} & -\frac{\beta_2 I_2^*}{N^*} & -\frac{\beta_2 I_2^*}{N^*} & -\frac{\beta_2 I_2^*}{N^*} \\ [\beta_3 - (\theta + d)]\frac{I_3^*}{N^*} & s_{42} & s_{43} & -s_{44} & -\frac{\beta_3 I_3^*}{N^*} \\ 0 & 0 & \delta & \theta & -(\alpha + d) \end{bmatrix}$$

where

$$s_{42} = -\frac{\beta_3 I_3^*}{N^*} + \frac{(\beta_4 + \beta_5) I_2^*}{N^*}, \quad s_{43} = -\frac{\beta_3 I_3^*}{N^*} + \frac{(\beta_4 + \beta_5) I_1^*}{N^*}$$

and

$$s_{44}=\frac{\beta_3 I_3^*}{N^*}+\frac{(\beta_4+\beta_5)I_1^* I_2^*}{N^* I_3^*}>0$$

The characteristic equation corresponding to $M(E^*)$ is given by

$$f(x)=(x^5+c_1x^4+c_2x^3+c_3x^2+c_4x+c_5)=0 \tag{3.4}$$

where, $c_1=\alpha+2d++s_{44}+\frac{\beta_1 I_1^*}{N^*}+\frac{\beta_2 I_2^*}{N^*}$

$$c_2=(\alpha+d)s_{44}+\left(\alpha+d+s_{42}+s_{44}+\frac{\beta_2 I_2^*}{N^*}\right)\frac{\beta_1 I_1^*}{N^*}+(\alpha+\delta+d+s_{43}+s_{44})\frac{\beta_2 I_2^*}{N^*}+\theta\frac{\beta_3 I_3^*}{N^*}$$

$$+d\left(\alpha+2d++s_{44}+\frac{\beta_1 I_1^*}{N^*}+\frac{\beta_2 I_2^*}{N^*}\right)-(\beta_1+\beta_4)(\beta_2+\beta_5)\frac{I_1^* I_2^*}{N^{*2}}$$

$$c_3=d(\alpha+d)s_{44}+d\left(\alpha+d+s_{42}+s_{44}+\frac{\beta_2 I_2^*}{N^*}\right)\frac{\beta_1 I_1^*}{N^*}+d(\alpha+\delta+d+s_{43}+s_{44})\frac{\beta_2 I_2^*}{N^*}$$

$$+d\theta\frac{\beta_3 I_3^*}{N^*}+d^2\left(\alpha+2d++s_{44}+\frac{\beta_1 I_1^*}{N^*}+\frac{\beta_2 I_2^*}{N^*}\right)-d(\beta_1+\beta_4)(\beta_2+\beta_5)\frac{I_1^* I_2^*}{N^{*2}}$$

$$+\left[(\alpha+\theta+d)s_{43}+(\alpha+\delta+d)s_{44}+(\theta-\delta)\frac{\beta_3 I_3^*}{N^*}+(\alpha+\delta+d+s_{42}+s_{43}+s_{44})\frac{\beta_1 I_1^*}{N^*}\right]\frac{\beta_2 I_2^*}{N^*}$$

$$+\left[(\alpha+d)s_{44}+\theta\frac{\beta_3 I_3^*}{N^*}+s_{42}(\alpha+\theta+d)-(s_{43}+\delta)\frac{(\beta_2+\beta_5)I_2^*}{N^*}\right]\frac{\beta_2 I_2^*}{N^*}$$

$$+\alpha\left[\delta[\beta_2-(\delta+d)]\frac{I_2^*}{N^*}+\theta[\beta_3-(\theta+d)]\frac{I_3^*}{N^*}\right]$$

$$-\frac{(\beta_1+\beta_4)I_1^*}{N^*}\left[(\alpha+d+s_{44})(\beta_2+\beta_5)\frac{I_2^*}{N^*}+s_{42}\frac{\beta_2 I_2^*}{N^*}\right]$$

$$c_4 = \left[(\alpha+\delta+d)s_{44} + s_{42}\delta + (\alpha+\theta+d)s_{43} + (\theta-\delta)\frac{\beta_3 I_3^*}{N^*} + d(\alpha+\delta+d+s_{42}+s_{43}+s_{44}) \right] \frac{\beta_1 I_1^* \beta_2 I_2^*}{N^{*2}}$$

$$-\frac{\beta_1 I_1^*}{N^*}(\beta_2+\beta_5)\left[s_{43}(\alpha+d) + \delta\frac{\beta_3 I_3^*}{N^*} \right]\frac{I_2^*}{N^*} + \alpha[\beta_2 - (\delta+d)(s_{43}\theta + s_{44}\delta)]\frac{I_2^*}{N^*}$$

$$+\alpha\frac{I_3^*}{N^*}[\beta_3 - (\theta+d)](\theta-\delta)\frac{\beta_2 I_2^*}{N^*} - (\beta_2+\beta_5)\alpha\delta[\beta_1 - (\lambda+d)]\frac{I_1^* I_2^*}{N^{*2}}$$

$$+\alpha\left[\delta[\beta_2 - (\delta+d)]\frac{I_2^*}{N^*} + \theta[\beta_3 - (\theta+d)]\frac{I_3^*}{N^*} \right]\frac{\beta_1 I_1^*}{N^*}$$

$$-\frac{(\beta_1+\beta_4)I_1^*}{N^*}\left[(\beta_2+\beta_5)\frac{I_2^*}{N^*}\left((\alpha+d)s_{44} + \theta\frac{\beta_3 I_3^*}{N^*} \right) + s_{42}(\alpha+\theta+d)\frac{\beta_2 I_2^*}{N^*} \right]$$

$$-\frac{(\beta_1+\beta_4)I_1^*}{N^*}\left[(\alpha+d+s_{44})(\beta_2+\beta_5)\frac{I_2^*}{N^*} + s_{42}\frac{\beta_2 I_2^*}{N^*} \right]d$$

$$+\frac{\beta_1 I_1^*}{N^*}\left[(\theta-\delta)s_{42}\frac{\beta_2 I_2^*}{N^*} - (s_{43}\theta + \delta s_{44})\frac{(\beta_2+\beta_5)I_2^*}{N^*} + d(\alpha+d)s_{44} + d\theta\frac{\beta_3 I_3^*}{N^*} \right]$$

$$+\frac{\beta_1 I_1^*}{N^*}d\left[s_{42}(\alpha+\theta+d) - (\delta+s_{43})\frac{(\beta_2+\beta_5)I_2^*}{N^*} \right]$$

$$+\left[(\alpha+\theta+d)s_{43} + (\alpha+\delta+d)s_{44} + (\theta-\delta)\frac{\beta_3 I_3^*}{N^*} \right]\frac{\beta_2 I_2^*}{N^*}$$

$$c_5 = d\left[(\alpha+\delta+d)s_{44} + s_{42}\delta + (\alpha+\theta+d)s_{43} + (\theta-\delta)\frac{\beta_3 I_3^*}{N^*} \right]\frac{\beta_1 I_1^* \beta_2 I_2^*}{N^{*2}}$$

$$-\frac{\beta_1 I_1^*}{N^*}d(\beta_2+\beta_5)\left[s_{43}(\alpha+d) + \delta\frac{\beta_3 I_3^*}{N^*} \right]\frac{I_2^*}{N^*}$$

$$-\frac{(\beta_1+\beta_4)I_1^*}{N^*}\left[(\beta_2+\beta_5)\frac{I_2^*}{N^*}\left((\alpha+d)s_{44}+\theta\frac{\beta_3 I_3^*}{N^*}\right)+s_{42}(\alpha+\theta+d)\frac{\beta_2 I_2^*}{N^*}\right]d$$

$$+\frac{\beta_1 I_1^*}{N^*}\left[(\theta-\delta)s_{42}\frac{\beta_2 I_2^*}{N^*}-(s_{43}\theta+\delta s_{44})\frac{(\beta_2+\beta_5)I_2^*}{N^*}\right]d$$

$$+\alpha\left[[\beta_2-(\delta+d)]\frac{I_2^*}{N^*}(s_{43}\theta+\delta s_{44})+\frac{\beta_2 I_2^*}{N^*}(\theta-\delta)[\beta_3-(\theta+d)]\frac{I_3^*}{N^*}\right]\frac{\beta_1 I_1^*}{N^*}$$

$$+\alpha[\beta_1-(\lambda+d)]\frac{I_1^*}{N^*}\left[\frac{\beta_2 I_2^*}{N^*}s_{42}(\theta-\delta)-(s_{43}\theta+\delta s_{44})(\beta_2+\beta_5)\frac{I_2^*}{N^*}\right]$$

$$-\alpha(\beta_1+\beta_4)\frac{I_1^*}{N^*}\left[[\beta_2-(\delta+d)]\frac{I_2^*}{N^*}s_{42}\theta+(\beta_2+\beta_5)\frac{I_2^*}{N^*}\theta[\beta_3-(\theta+d)]\frac{I_3^*}{N^*}\right]$$

$$+\alpha\frac{\beta_1 I_1^*}{N^*}\left[[\beta_2-(\delta+d)]\frac{I_2^*}{N^*}s_{42}\delta+(\beta_2+\beta_5)\frac{I_2^*}{N^*}\delta[\beta_3-(\theta+d)]\frac{I_3^*}{N^*}\right]$$

By Routh-Hurwitz criteria, E^* is locally asymptotically stable if,

$$c_i\,(i=1, 2, \ldots, 5) > 0, \begin{vmatrix} c_1 & 1 \\ c_3 & c_2 \end{vmatrix} > 0, \begin{vmatrix} c_1 & 1 & 0 \\ c_3 & c_2 & c_1 \\ 0 & 0 & c_3 \end{vmatrix} > 0, \begin{vmatrix} c_1 & 1 & 0 & 0 \\ c_3 & c_2 & c_1 & 1 \\ 0 & c_4 & c_3 & c_2 \\ 0 & 0 & 0 & c_4 \end{vmatrix} > 0$$

4. Numerical Simulations

In order to illustrate some of the properties of the HIV-TB co-infection model, numerical simulations are carried out using a Runge-Kutta numerical scheme coded in C++ programming language.

To see the dynamical behavior of the model system, we have used the following set of parameter values,

$$Q_0 = 2000, \beta_1 = 0.9, \beta_2 = 0.5, \beta_3 = 0.4, \beta_4 = 0.2, \beta_5 = 0.1$$

$$d = 0.02,\ \lambda = 0.196,\ \alpha = 1,\ \delta = 0.1,\ \theta = 0.076$$

with initial condition $S(0) = 20000$, $I_1(0) = 5000$, $I_2(0) = 3000$, $I_3(0) = 6000$, $A(0) = 500$.

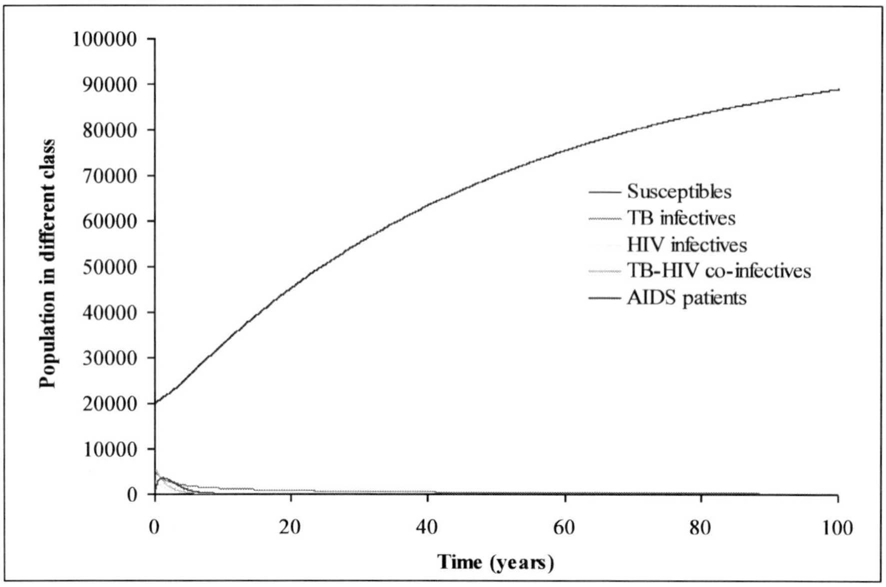

Figure 2. Numerical Simulations for R_1=0.978>R_2=0.543>R_3=0.435<1 with Q_0 = 2000, β_1 = 0.9, β_2 = 0.5, β_3 = 0.4, β_4 = 0.2, β_5 =0.1, d =0.02, λ = 0.9, α =1, δ = 0.9, θ = 0.9 and initial condition S (0) = 20000, $I_1(0)$ = 5000, $I_2(0)$ = 3000, $I_3(0)$ = 6000, $A(0)$ = 500.

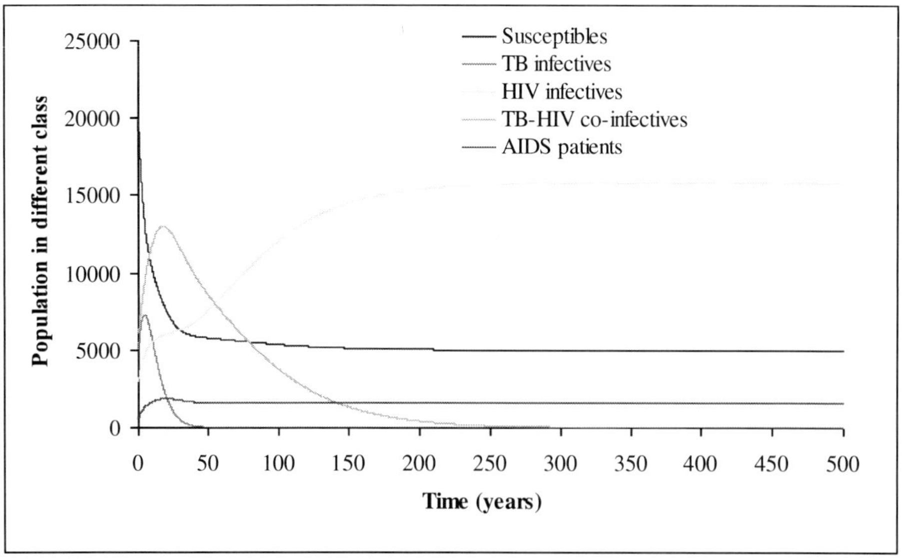

Figure 3. Numerical Simulations for R_1= 2.813<R_2= 4.167>R_3= 3.333>1 with Q_0 = 2000, β_1 = 0.9, β_2 = 0.5, β_3 = 0.4, β_4 = 0.2, β_5 =0.1, d =0.02, λ = 0.3, α =1, δ = 0.1, θ = 0.1 and initial condition S (0) = 20000, $I_1(0)$ = 5000, $I_2(0)$ = 3000, $I_3(0)$ = 6000, $A(0)$ = 500.

The equilibrium values are obtained as,

$$S^* = 6454.866,\ I_1^* = 1223.099,\ I_2^* = 67.218,\ I_3^* = 19155.88,\ A^* = 1433.88$$

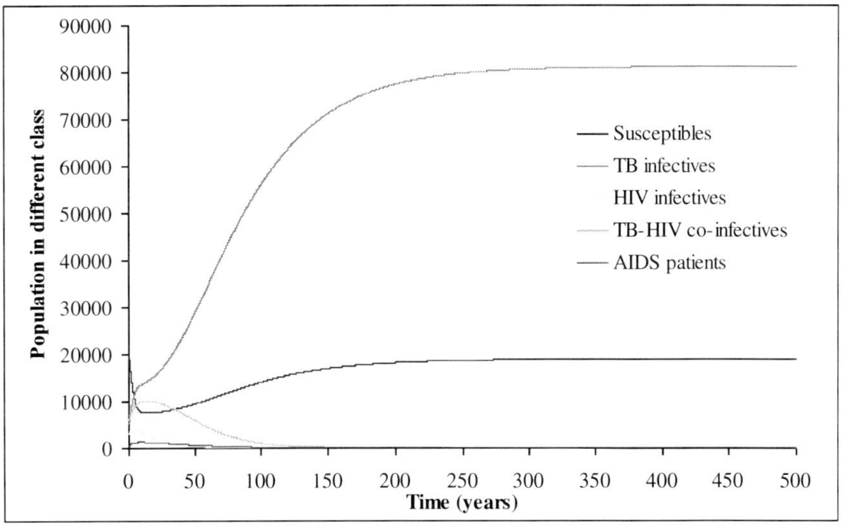

Figure 4. Numerical Simulations for R_1= 5.294>R_2= 4.167>R_3= 3.333>1 with Q_0 = 2000, β_1 = 0.9, β_2 = 0.5, β_3 = 0.4, β_4 = 0.2, β_5 =0.1, d =0.02, λ = 0.15, α =1, δ = 0.1, θ = 0.1 and initial condition S (0) = 20000, $I_1(0)$ = 5000, $I_2(0)$ = 3000, $I_3(0)$ = 6000, $A(0)$ = 500.

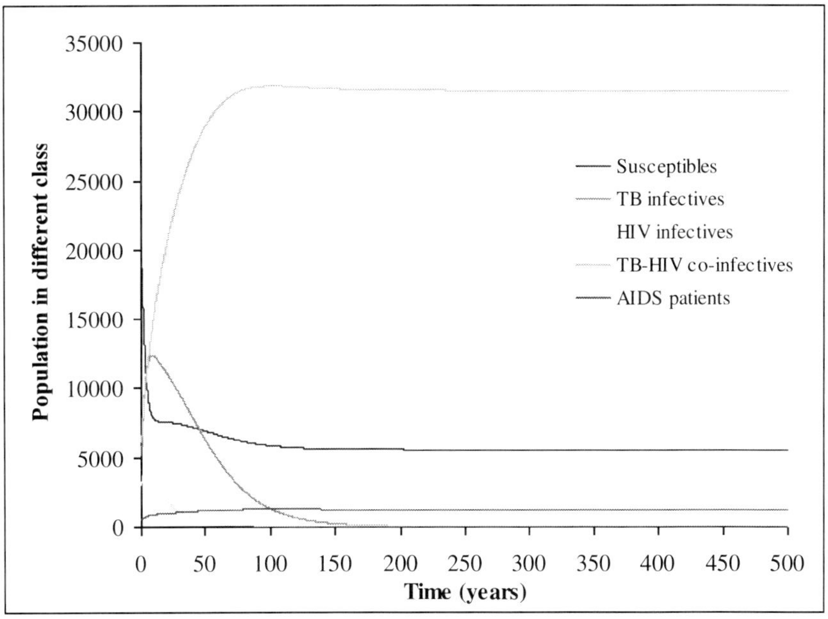

Figure 5. Numerical Simulations for R_1= 5.294>R_2= 4.167<R_3= 6.667>1 with Q_0 = 2000, β_1 = 0.9, β_2 = 0.5, β_3 = 0.4, β_4 = 0.2, β_5 =0.1, d =0.02, λ = 0.15, α =1, δ = 0.1, θ = 0.04 and initial condition S(0) = 20000, $I_1(0)$ = 5000, $I_2(0)$ = 3000, $I_3(0)$ = 6000, $A(0)$ = 500.

Figures 2-9 illustrate the solution profiles of the susceptibles $S(t)$, the tuberculosis-only infectives $I_1(t)$, the HIV-only infectives $I_2(t)$, the HIV-TB co-infectives $I_3(t)$ and the AIDS patients $A(t)$.

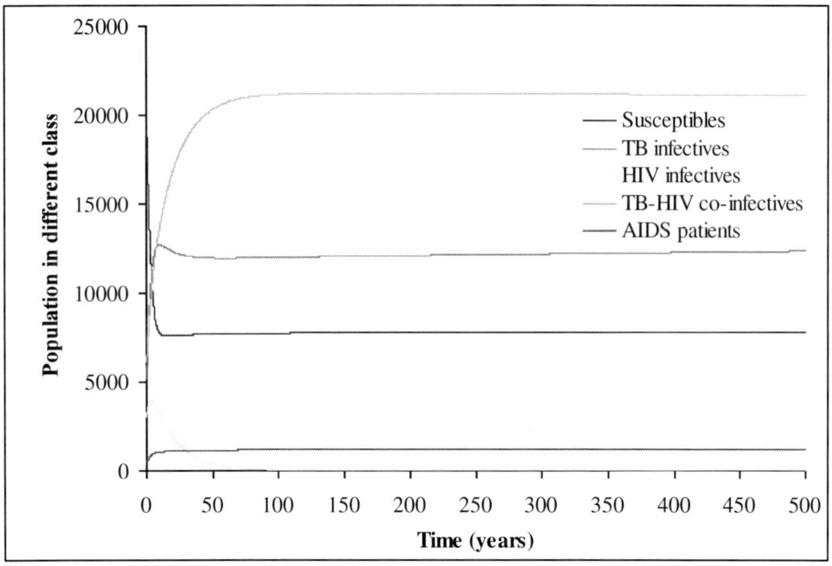

Figure 6. Numerical Simulations for R_2= 4.167< R_1= R_3= 5.29>1 with Q_0 = 2000, β_1 = 0.9, β_2 = 0.5, β_3 = 0.4, β_4 = 0.2, β_5 =0.1, d =0.02, λ = 0.15, α =1, δ = 0.1, θ = 0.0556 and initial condition S(0) = 20000, $I_1(0)$ = 5000, $I_2(0)$ = 3000, $I_3(0)$ = 6000, $A(0)$ = 500.

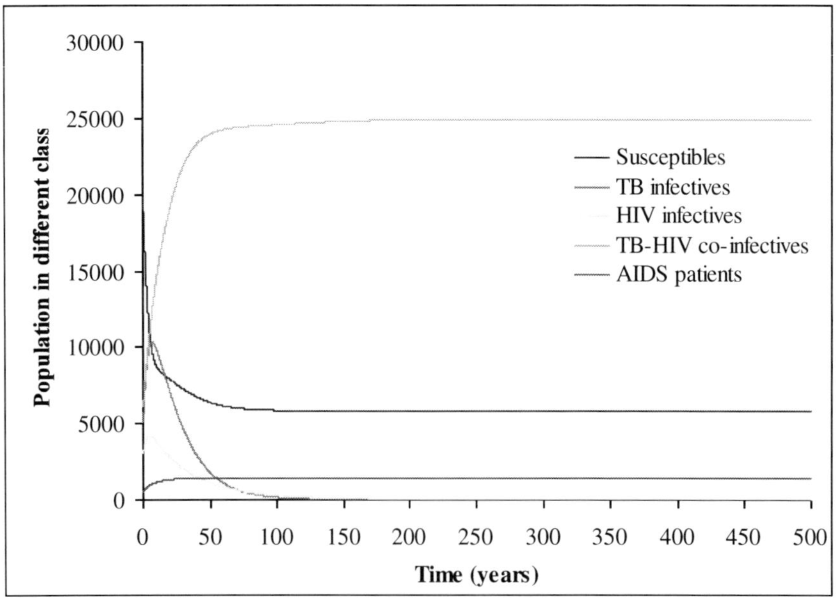

Figure 7. Numerical Simulations for R_1= R_2= 4.167<R_3= 5.29>1 with Q_0 = 2000, β_1 = 0.9, β_2 = 0.5, β_3 = 0.4, β_4 = 0.2, β_5 =0.1, d =0.02, λ = 0.196, α =1, δ = 0.1, θ = 0.0556 and initial condition S(0) = 20000, $I_1(0)$ = 5000, $I_2(0)$ = 3000, $I_3(0)$ = 6000, $A(0)$ = 500.

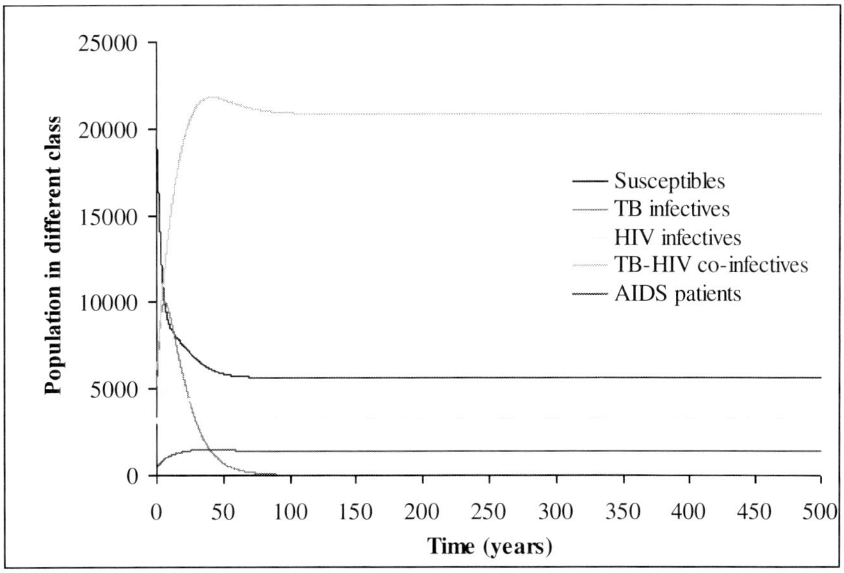

Figure 8. Numerical Simulations for $R_1= 4.167<R_2= R_3= 5.29>1$ with $Q_0 = 2000$, $\beta_1 = 0.9$, $\beta_2 = 0.5$, $\beta_3 = 0.4$, $\beta_4 = 0.2$, $\beta_5 = 0.1$, $d = 0.02$, $\lambda = 0.196$, $\alpha = 1$, $\delta = 0.0745$, $\theta = 0.0556$ and initial condition $S(0) = 20000$, $I_1(0) = 5000$, $I_2(0) = 3000$, $I_3(0) = 6000$, $A(0) = 500$.

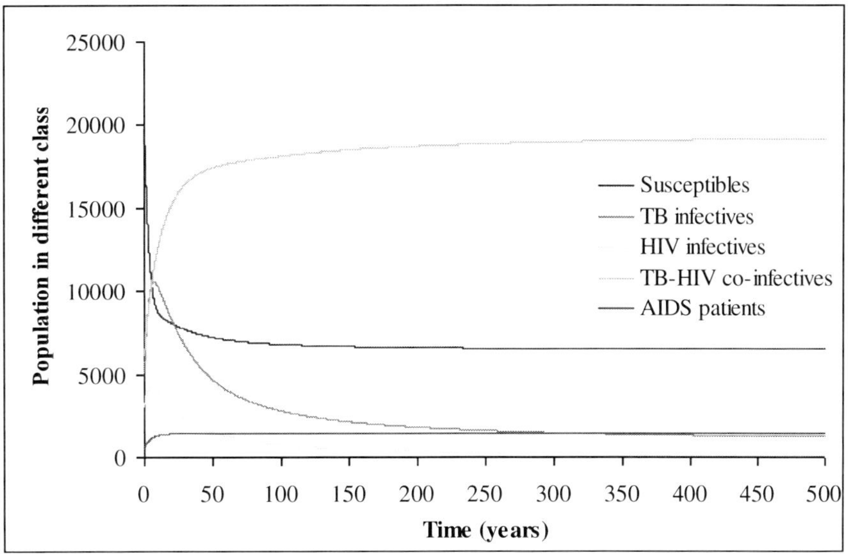

Figure 9. Numerical Simulations for $R_1= R_2= R_3= 4.167>1$ with $Q_0 = 2000$, $\beta_1 = 0.9$, $\beta_2 = 0.5$, $\beta_3 = 0.4$, $\beta_4 = 0.2$, $\beta_5 = 0.1$, $d = 0.02$, $\lambda = 0.196$, $\alpha = 1$, $\delta = 0.1$, $\theta = 0.076$ and initial condition $S(0) = 20000$, $I_1(0) = 5000$, $I_2(0) = 3000$, $I_3(0) = 6000$, $A(0) = 500$.

Simulation results for the co-infection model in Figure 2 illustrate that when the partial reproductive numbers are less than unity ($R_1 < 1$, $R_2 < 1$, and $R_3 < 1$; $R_0 < 1$), disease prevalence shows convergence to the disease-free equilibrium which is in line with the results in Theorem 1. Simulation results in Figures 2 - 8 for the cases, (i) $R_1 = 0.978 > R_2 = 0.543 >$

$R_3 = 0.435 < 1$ (ii) $R_1 = 2.813 < R_2 = 4.167 > R_3 = 3.333 > 1$ (iii) $R_1 = 5.294 > R_2 = 4.167 > R_3 = 3.333 > 1$ (iv) $R_1 = 5.294 > R_2 = 4.167 < R_3 = 6.667 > 1$ (v) $R_2 = 4.167 < R_1 = R_3 = 5.29 > 1$ (vi) $R_1 = R_2 = 4.167 < R_3 = 5.29 > 1$ (vii) $R_1 = 4.167 < R_2 = R_3 = 5.29 > 1$ illustrate that for the basic reproductive number greater than unity $R_0 > 1$, the endemic equilibrium exists and the disease will prevail in the population group with the highest partial reproductive number irrespective of the value of the other partial reproductive numbers being greater than unity, thus in this case, the model shows competitive exclusion. Figure 9 illustrates that the three diseases can co-exist in the population when their partial reproductive numbers are equal i.e. when $R_1 = R_2 = R_3 = 4.167 > 1$ and all diseases are endemic in this case.

Conclusion

In this chapter, a nonlinear mathematical model is proposed and analyzed to study the transmission of HIV/AIDS and a curable TB pathogen within a population of variable size structure. It is shown that there exist threshold parameters R_1, R_2 and R_3 for each disease (i.e. TB, HIV and HIV-TB co-infection respectively) and R_0, for the model in general. It is noted that when $R_0 < 1$ (i.e. $R_1 < 1$, $R_2 < 1$ and $R_3 < 1$) then both the infections die out and when $R_0 > 1$ then at least one disease will be present in the population.

The model has five non-negative equilibria namely E_0, the disease free, E_1 the TB only, E_2 the HIV only, E_3 the co-infection only (only HIV-TB co-infected population) and E^* the non-trivial endemic equilibrium. The stability results for equilibria are formulated in Theorem 1. The endemic equilibrium E^* is locally asymptotically stable under certain conditions. It is also found that these infectious diseases become more endemic due to immigration. Thus, if the immigration of the population into susceptible community is restricted, the spread of the disease can be kept under control. The number of HIV infected individuals in population increases rapidly due to presence of another disease that is tuberculosis. Thus, if tuberculosis in the population is effectively treated, the spread of HIV can be slowed down.

References

[1] G. Birkhoff, G.C. Rota, Ordinary Differential Equations, *Ginn*, 1982.

[2] K. Busenberg, K. Cooke and H. Ying-Hen, A model for HIV in Asia, *Math. Biosc.* 128, 185-210, 1995.

[3] C. Castillo Chavez and B. Song, Dynamical models of tuberculosis and other applications, *Math. Biosc. Engg.* 1, 361-404, 2004.

[4] O. Diekmann, P. J. A. Heesterbeek and J. A. J. Metz, On the definition and the basic reproduction ratio R_0 in models for infectious diseases in heterogeneous populations, *J. Math. Biol.* 28, 365-382, 1990.

[5] L. Esteva and M. Matias, A model for vector transmitted disease with saturation incidence, *J. Biol. Sys.* 9, 235-245, 2001.

[6] HIV-TB co-infection- A guide for medical officers, National AIDS control organization, Ministry of Health and family welfare, Government of India, 2006. www.nacoonline.org/publication.

[7] D. Kirschner, Using mathematics to understanding HIV immune dynamics, *Notices of the AMS* 43, 191-202, 1996.

[8] D. Kirschner, Dynamics of co-infection with *M. tuberculosis* and HIV-1, *Theor. Popu. Biol.* 55, 94-109, 1999.

[9] E. Massad, M. N. Burattini and, F. A. B. Coutinho *et al.*, Modelling the interaction between AIDS and tuberculosis, *Math. Comp. Model.* 17, 7-21, 1993.

[10] R. McLean and M. A. Nowak, Models of interactions between HIV and other pathogens, *J. Theoret. Biol.* 155, 69-86, 1992.

[11] S. M. Moghadas and M. E. Alexander, Exogenous reinfection and resurgence of tuberculosis: A Theoretical Framework, *J. Biol. Sys.* 12, 231-247, 2004.

[12] B.M. Murphy, B. H. Singer and D. Kirschner, On treatment of tuberculosis in heterogeneous populations, *J. Theoret. Biol.* 223, 391-404, 2003.

[13] R. Naresh and A. Tripathi, Modelling and analysis of HIV-TB co-infection in a variable size population, *Math. Model. Anal.* 10, 275-286, 2005.

[14] R. Naresh, A. Tripathi and S. Omar, Modelling the spread of AIDS epidemic with vertical transmission, *Appl. Math. Comp.* 178, 262-272, 2006.

[15] R. Naresh, A. Tripathi and D. Sharma, Modelling and analysis of the spread of AIDS epidemic with immigration of HIV infectives, *Math. Comp. Model.*, 49, 880-892, 2009.

[16] R. Naresh, A. Tripathi and D. Sharma, A nonlinear AIDS epidemic model with screening and time delay, *Appl. Math. Comp.,* 217(9), 4416-4426, 2011.

[17] R. Naresh, A. Tripathi and D. Sharma, A nonlinear HIV/AIDS model with contact tracing, *Appl. Math. Comp.*, 217, 9575-9591, 2011.

[18] R. Naresh and D. Sharma, An HIV/AIDS model with vertical transmission and time delay, *World Journal of Modelling and Simulation*, 7(3), 230-240, 2011.

[19] A. S. Perelson and P. W. Nelson, Mathematical analysis of HIV-1 dynamics in vivo, *SIAM Rev.* 41, 3-44, 1999.

[20] R. B. Schinazi, Can HIV invade a population which is already sick? *Bull. Braz. Math. Soc.* (N.S.) 34, 479-488, 2003.

[21] O. Sharomi, C. N. Podder, A. B. Gumel and B. Song, Mathematical analysis of the transmission dynamics of HIV/TB co-infection in the presence of treatment, *Math. Biosc. Engg.* 5, 145-174, 2008.

[22] A. S. R. Srinivasa Rao, Mathematical modeling of AIDS epidemic in India, *Current Science* 84, 1192-1197, 2003.

[23] A. Tripathi, R. Naresh and D. Sharma, Modelling the screening of unaware infectives on the spread of HIV infection, *Appl. Math. Comp.* 183, 1053-1068, 2007.

[24] W. F. Wells, Aerodynamics of droplet nuclei, airborne contagion, and air hygiene, Harvard University Press, Cambridge, 1995.

[25] R.W. West and A. R. Thompson, Modeling the impact of HIV on the spread of tuberculosis in the United States, *Bull. Math. Biol.* 59, 35-60, 1996.

[26] B.G. Williams and C. Dye, Antiretroviral drugs for tuberculosis control in the era of HIV/AIDS, *Science* 301, 1535-1537, 2003.

In: Dynamical Systems
Editor: Jean Michel Tchuenche, pp. 335-348
ISBN: 978-1-62808-001-8

Chapter 13

CASE DETECTION ON TUBERCULOSIS DYNAMICS WITH ISOLATION IN A SANATORIUM

*Nyimvua Shaban**
Department of Mathematics, University of Dar es Salaam,
Dar es Salaam, Tanzania

Abstract

This chapter presents a deterministic model for assessing the impact of early detection on the dynamics of tuberculosis (TB) epidemic. Qualitative analysis of the model reveals conditions under which the model can exhibit backward bifurcation where a stable disease-free equilibrium co-exists with a stable endemic equilibrium when the effective reproduction number is less than unity. Furthermore, an explicit expression for the effective reproduction number was derived, above which the disease will persist and below which TB could be eliminated in community. Graphical representations are provided to support the analytical results. The higher the detection rate the better the control of the disease.

Keywords: Disease detection, reproduction number, sanatorium, treatment

1. Introduction

Tuberculosis (TB) is an infectious disease caused by bacillus Mycobacterium tuberculosis. Typically, it affects the lungs (pulmonary TB) but can affect other parts of the body as well (extra pulmonary TB) (CDC 2005). Despite the availability of short course chemotherapy as the most effective treatment for infected individuals, TB remains a major health problem (Dye et al. 1998, Ssematimba et al. 2005, WHO 2011). The current observed patterns of TB can be attributed to its slow dynamics and the emergence of multi-drug resistance TB strains; two processes which take place over distinct spatial and temporal scales (Aparicio and Castillo-Chaves 2009). Various dynamics model the transmission of TB have been carried out (Anderson and May 1991, Murray and Solomon 1998, Dye et al. 1998, Feng and Castillo-Chavez 2000, Elad et al. 2001). Contact tracing of new cases of active TB

*E-mail address: shabanmbare@yahoo.com

have been investigated by Aparicio and Hernandez (2006) and case detection under direct observation therapy strategy by Korobeinikov and Oknonghae (2007). The identification of infectious individuals has attracted attention as a disease control that seeks to uncover their contacts and prevent further spread of the disease (Hughes et al. 2006, Murray and Solomon 1998).

In this chapter, we investigate the potential impact of early detection and isolation in a sanatorium for supervised treatment. A sanatorium is a medical facility for long term illness, most typically associated with treatment of TB. A sanatorium can as well be used as an educational centre where TB patients receive both didactic and practical instructions (ACCP 1940). Transmission of TB in a sanatorium is assumed to be negligible, since isolated infectious individuals cannot interact with the susceptible except the health workers. We assume that individuals mix homogeneous. We further assume that all immigrants and newborns are uninfected and therefore join the susceptible class.

We derive the reproduction number R_0, defined as the average number of new cases generated by a typical infectious individual in a completely susceptible population during the epidemic, for an epidemic without intervention. The quantity R_0 is of fundamental importance to the infections, since a major outbreak is possible if and only if $R_0 > 1$ (Keeling and Rohani 2008). We also derive the effective reproduction number R_e when control measures (detection and isolation) are implemented. Using the simulation results we illustrate how effective the intervention model is, and show that the number of undetected infectious individuals decreases exponentially as the detection rate increases.

The rest of the chapter is organized as follows. The model and parameters used are defined in Section 2. In Section 3, the model is analyzed and the reproduction numbers with and without detection are computed. Conditions for the existence of backward bifurcation are derived in Section 4. Graphical representations are provided in Section 5. Section 6 concludes the chapter.

2. The Model

Let the total population at time t be $N(t)$. During the epidemic, individuals can be in one of the following sub-populations: susceptible $(S(t))$, latent $(E(t))$, undetected infectious individuals $(I(t))$, detected infectious individuals who are sent to the sanatorium for treatment $(T(t))$, and recovered individuals $(R(t))$. Here, the use of a sanatorium includes supervised treatment and complemented with health education or counseling against practices that may result in TB infection. The model takes the form of the following deterministic system of non-linear differential equations.

$$
\begin{aligned}
\frac{dS}{dt} &= \Lambda - (\lambda + \mu)S \\
\frac{dE}{dt} &= \lambda(1-p)S + \tau\lambda R + rI - (k + \alpha I + \mu)E \\
\frac{dI}{dt} &= p(1-\omega)\lambda S + (1-\omega)(k+\alpha I)E - (\mu + d_1 + r)I \qquad (1) \\
\frac{dT}{dt} &= p\omega\lambda S + \omega(k + \alpha I)E - (\mu + d_2 + \gamma)T \\
\frac{dR}{dt} &= \gamma T - (\tau\lambda + \mu)R,
\end{aligned}
$$

where $\lambda = \dfrac{\beta I}{N}$ and $N = S + E + I + T + R$.

The rate λ above is the force of infection associated with TB transmission, and the parameter β is the effective contact rate (i.e contact that may result in TB infection). Recruitment into susceptible population occurs at a constant rate Λ (this may include newborns and immigrants), and susceptible people acquire infection following effective contact with infected individuals at the rate λ. A fraction p of infected individuals progress faster to active TB, and k is the progression rate of infected individuals from latent class into infectious class. A proportion ω of individuals are detected and taken to the sanatorium for treatment, while r is the rate at which undetected infectious persons get self-recovery and revert to latent stage. It is assumed that no treatment failure in the sanatorium, implying that detected infectious individuals will either recover or die. Individuals who recover from TB may progress to the latent class at the rate $\tau\lambda$, where $\tau < 1$ accounts for the reduced infection due to acquired partial immunity. Individuals in all sub-populations suffer natural death at a rate μ. Additionally, TB infectious individuals die at rate d_1 (for undetected class) or d_2 (for detected class). A schematic diagram of the model is shown in Figure 1, and the associated variables and parameters are described in Table 1.

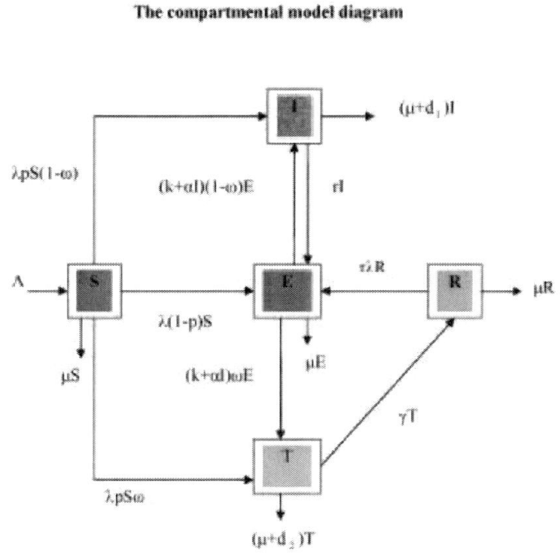

Figure 1. Schematic diagram showing movements of individuals between compartments.

3. Analysis of the Model

The properties of the model (1) are discussed and analyzed in a suitable feasible region Φ of biological interest. We begin by showing that the system is dissipative, that is, all feasible solutions are uniformly-bounded in a proper subset $\Phi \in \mathbb{R}^5_+$. We consider the feasible region Φ with,

Table 1. Description of variables and parameters of the model

Variable	Description
S	The number of susceptible individuals at time
E	The number of latently infected individuals at time
I	The number of undetected infectious individuals at time
T	The number of detected infectious individuals undergoing treatment at time
R	The number of recovered individuals at time
Parameters	**Description**
Λ	Recruitment rate
μ	Natural death rate
d_1	TB induced death rate for undetected TB cases
d_2	TB induced death rate for detected TB cases undergoing treatment
k	The rate of progression of individuals in the latent state to active TB
ω	Fraction of those who are detected and undergoing treatment in the sanatorium
p	Fraction of fast progressors to the active disease
r	Rate at which undetected individuals get self-cured and revert to latent state
α	Effective transmission rate for the latent class due to exogenous reinfection
β	Transmission rate for the susceptible class to the infective class
τ	Rate that accounts for reduced infection in recovery class
γ	Recovery /treatment rate for the detected infectious individuals in the sanatorium

$$\Phi = \left\{(S, E, I, T, R) \in \mathbb{R}^5_+ : N \leq \frac{\Lambda}{\mu}\right\}.$$

Adding all the equations of the model (1) yields

$$\begin{aligned}\frac{dN(t)}{dt} &= \Lambda - \mu N(t) - d_1 I - d_2 T \\ &< \Lambda - \mu N(t).\end{aligned} \tag{2}$$

Applying Birkhoff and Rota's theorem (1989) on differential inequalities we obtain

$$N(t) \leq N(0)e^{-\mu t} + \tfrac{\Lambda}{\mu}(1 - e^{-\mu t}),$$

where $N(0)$ represents the initial values of the respective variables. Then $0 \leq N \leq \frac{\Lambda}{\mu}$ as $t \to \infty$. Therefore, $\frac{\Lambda}{\mu}$ is an upper bound of $N(t)$ provided $N(0) \leq \frac{\Lambda}{\mu}$. Furthermore if $N(t) > \frac{\Lambda}{\mu}$, then either the solution enters Φ in finite time or $N(t)$ approaches $\frac{\Lambda}{\mu}$. Thus, all feasible solutions of the system eventually approach, enter or stay in Φ. Hence, the region of biological interest Φ is positively invariant under the flow induced in the system of model (1). Therefore, the dynamics of model (1) can be considered in Φ and it can be shown, using the theory of permanence in Huston and Schmitt (1992), that all solutions on the boundary of Φ eventually enter the interior of Φ. Furthermore, in Φ, the usual existence, uniqueness and continuity of results hold for the system of equations of the model (1), making the system well-posed. Thus, we summarize the result in the following lemma.

Lemma 3.1. *The closed set* $\left\{(S, E, I, T, R) \in \mathbb{R}^5_+ : N \leq \frac{\Lambda}{\mu}\right\}$ *is positively invariant and attracting with respect to the model (1).*

3.1. Local Stability of Disease-Free Equilibrium

This is the situation in which the population is free from the disease, implying that both classes E and I are zero. The model (1) has a unique disease-free equilibrium obtained by setting the right-hand sides of the equations of the model to zero, given by

$$X_0 = \left(S^*, E^*, I^*, T^*, R^*\right) = \left(\frac{\Lambda}{\mu}, 0, 0, 0, 0\right) \tag{3}$$

We define $S_0 = \frac{\Lambda}{\mu}$ as the asymptotic carrying capacity of the population. It can be shown that X_0 attracts the region (the stable manifold of X_0)

$$\Phi_{X_0} = \{(S, E, I, T, R) \in \Phi : E = I = T = R = 0\}.$$

Using the next generation matrix method (as in Van Den Driessche and Watmough 2002), the associated matrices **F** for the new infection terms and **V** for the transition terms at the disease-free equilibrium are respectively given by

$$\mathbf{F} = \begin{bmatrix} \dfrac{\beta(1-p)SI + \tau\beta RI}{N} \\ \dfrac{\beta p(1-\omega)SI}{N} \\ \dfrac{\beta p\omega SI}{N} \end{bmatrix}$$

and

$$\mathbf{V} = \begin{bmatrix} (k + \alpha I + \mu)E - rI \\ (r + d_1 + \mu)I - (k + \alpha I)(1-\omega)E \\ (\gamma + d_2 + \mu)T - \omega(k + \alpha I)E \end{bmatrix}$$

It follows that the effective reproduction number denoted by R_e is given by

$$R_e = \rho(FV^{-1}) = \frac{\beta(p\mu + k)(1-\omega)}{(k+\mu)(d_1+\mu) + r\mu - rk\omega}, \tag{4}$$

where ρ is the spectral radius (i.e. the dominant eigenvalue) of the matrix FV^{-1}.
The effective reproduction number measures the average number of new infections generated by a single TB infected individual in a community where detection with treatment in

a sanatorium is used as a control strategy. Moreover, in the absence of detective strategy ($\omega = 0$) the quantity in (4) becomes

$$R_e = \frac{\beta(p\mu + k)}{(k+\mu)(d_1+\mu)+r\mu} = R_0, \tag{5}$$

where R_0 is the basic reproduction number which represents the average number of new cases generated by a single TB infected individual in a completely susceptible population. The basic reproduction number measures the power of a disease to invade a population under conditions that facilitates maximal growth. From Theorem 2 of Van Den Driessch and Watmough (2002) we establish the following result:

Theorem 3.2. *The model (1) always has the disease-free equilibrium X_0. If $R_e < 1$ the disease-free equilibrium is locally asymptotically stable in Φ, and unstable otherwise.*

Epidemiologically, Theorem (3.2.) means that TB can be eliminated from the community when $R_e < 1$, provided that the initial sizes of the sub-populations of the model (1) are within the domain of attraction of X_0.

The performance of the detective intervention can be assessed by re-writing R_e as a function of R_0. Using equation (5) and some algebra, Equation (4) can be simplified to yield the representation of R_e in terms of the basic reproduction number as

$$R_e = \frac{\Big[(k+\mu)(d_1+\mu)+r\mu\Big](1-\omega)}{(k+\mu)(d_1+\mu)+r\mu-rk\omega}R_0.$$

Hence, R_e varies linearly with the basic reproduction number R_0. In order to surely prevent an outbreak of TB when this intervention is implemented, we must have that $R_e \leq 1$, or equivalently

$$\frac{\Big[(k+\mu)(d_1+\mu)+r\mu\Big](1-\omega)}{(k+\mu)(d_1+\mu)+r\mu-rk\omega}R_0 \leq 1,$$

implying that

$$\omega \geq \frac{\Big[(k+\mu)(d_1+\mu)+\mu r\Big](R_0-1)}{\Big[(k+\mu)(d_1+\mu)+\mu r\Big]R_0 - rk}.$$

The inequality above indicates that the larger the proportion of detected infectious individual the better, because undetected or late detected infectious persons are responsible for further spread of TB infections.

4. Existence of Backward Bifurcation

We need to investigate whether the disease-free equilibrium is globally asymptotically stable. Backward bifurcation phenomenon in epidemic models is usually associated with the co-existence of disease-free equilibrium and endemic equilibria when the basic reproduction number (R_0) is less than unity (Hadeler and Van Den Driessche, 1997; Kribs-Zaleta and Valesco-Hernandez, 2000; Elibasha and Gumel, 2006). This phenomenon implies that

the requirement of having the reproduction number less than unity is necessary but not a sufficient condition for control of the disease (Hussaini et al., 2010).
In the presence of infection (i.e $E \neq 0$ and $I \neq 0$) the model(1) has a non-trivial equilibrium point denoted by

$$X_1 = \left(S^{**}, E^{**}, I^{**}, T^{**}, R^{**}\right).$$

Now let

$$\lambda^{**} = \frac{\beta I^{**}}{N^{**}} \tag{6}$$

be the force of infection at an arbitrary equilibrium point of model (1). The variables of the model can be re-written in terms of λ^{**} as follows.

$$\begin{aligned}
S^{**} &= \frac{\Lambda}{\lambda^{**} + \mu} \\
E^{**} &= \frac{(\lambda^{**} + \mu)(r + d_1 + \mu)I^{**} - p\Lambda(1 - \omega)\lambda^{**}}{(k + \alpha\lambda^{**})(\lambda^{**} + \mu)(1 - \omega)} \\
I^{**} &= \frac{\lambda^{**}\Lambda p(1 - \omega) + (k + \alpha\lambda^{**})(1 - \omega)(\lambda^{**} + \mu)E^{**}}{(\lambda^{**} + \mu)(r + d_1 + \mu)} \\
T^{**} &= \frac{\omega(r + d_1 + \mu)I^{**}}{(\gamma + d_2 + \mu)(1 - \omega)} \\
R^{**} &= \frac{\gamma\omega(r + d_1 + \mu)I^{**}}{(\tau\lambda^{**} + \mu)(1 - \omega)(\gamma + d_2 + \mu)}
\end{aligned} \tag{7}$$

We now substitute (7) into (6) and simplify to get $\lambda^{**} = 0$ (which corresponds to disease-free equilibrium, X_0) and the following quadratic equation in terms of λ^{**}:

$$a_1(\lambda^{**})^2 + a_2(\lambda^{**}) + a_3 = 0, \tag{8}$$

where

$$\begin{aligned}
a_1 &= (\gamma + d_2 + \mu)(1 - \omega)\alpha\tau\Lambda \\
a_2 &= (\gamma + d_2 + \mu)(k\tau + p\mu\tau + \mu\alpha)(1 - \omega)\Lambda \\
a_3 &= (\gamma + d_2 + \mu)(1 - \omega)pk\mu^2\Lambda
\end{aligned}$$

Since all the model parameters are non-negative, it follows that a_1, a_2 and a_3 are non-negative as well. Suppose that $\omega = 1$, so that the detective strategy is completely effective. Then we have $a_1 = a_2 = a_3 = 0$, corresponding to zero prevalence and elimination of the TB epidemic when $R_e < 1$. This precludes the possibility of backward bifurcation in this case.

However, using the centre manifold theorem (as in Castillo-Chavez and Song, 2004) the following theorem gives the conditions for the existence of backward bifurcation of the model. The proof of the following result is given in the Appendix.

Theorem 4.1. *If (A.4.1) holds, then model (1) has backward bifurcation at* $R_e = 1$ *and the bifurcating branch is unstable near* $R_e = 1$.

We note from the proof of the theorem above that the model (1) cannot undergo a backward bifurcation if $\alpha = 0$, implying that exogenous reinfection is necessary for a bifurcation to occur (Gumel and Song, 2008).

5. Numerical Simulations

We have shown how model (1) works by simulations for different parameter values and using the same sub-populations. Parameter values used in Table 2 were obtained from various literatures and those not available were estimated or assumed. The initial conditions used for the sub-populations are as follows: $S(0) = 95,000$, $E(0) = 25,000$, $I(0) = 19500$, $T(0) = 15500$ and $R(0) = 13,500$. These initial numbers for the sub-populations are just a representation of many other values which can be chosen for the same purpose. In order to illustrate the impact of the detection rate in preventing further TB infection, we have chosen the values of ω as 0.51, 0.65 and 0.85 with reference to simulation results.

Figure 2 illustrates change in size with time of each sub-population from the simulation results. Figure 2(a) shows the decrease of the number of exposed individuals after the implementation of the strategy. The detection strategy has reduced the number of those who would enter the exposed class through infection and many individuals from this class join the infectious class. Few recovered individuals join the exposed class due to reduced infection rate. In Figure 2(b) the number of undetected infectious individuals has risen initially, implying that some infected individuals progress faster infectious class due to high infection rate before the implementation of the detective strategy. Also, it is the same time interval that we observed the decrease of the number of susceptible population. In Figure 2(c) it is shown that the number of detected individuals admitted for treatment decreases as meaning that many people are recovering from TB disease.

The effect of detection rate to the population of undetected infectious individuals has been illustrated in Figure 3. We observed that the number of undetected infectious individuals decreases exponentially as detection rate is increased. This decrease is quite significant when the detection rate is higher than 0.65, which is in agreement with results of other studies (e.g see Okuonghae and Aihie, 2008, Korobeinikov and Okuonghae, 2007 and the refernce therein).

Table 2. Epidemiological parameter values

Parameter	Value	Source
μ	0.02041 per year	Okuonghae and Aihie, 2008
Λ	2041 per year	Korobeinikov and Okuonghae, 2007
r	0.2 per year	Agusto, 2009
k	0.05 per year	Okuonghae and Aihie, 2008
ω	0.22 per year	Estimated
d_1	0.365 per year	Adetunde, 2007
d_2	0.3 per year	Estimated
p	0.35 per year	Estimated
α	0.000125 per year	Estimated
β	0.35 per year	Agusto, 2009
γ	1.5 per year	Estimated
τ	0.9 per year	Feng, 2000

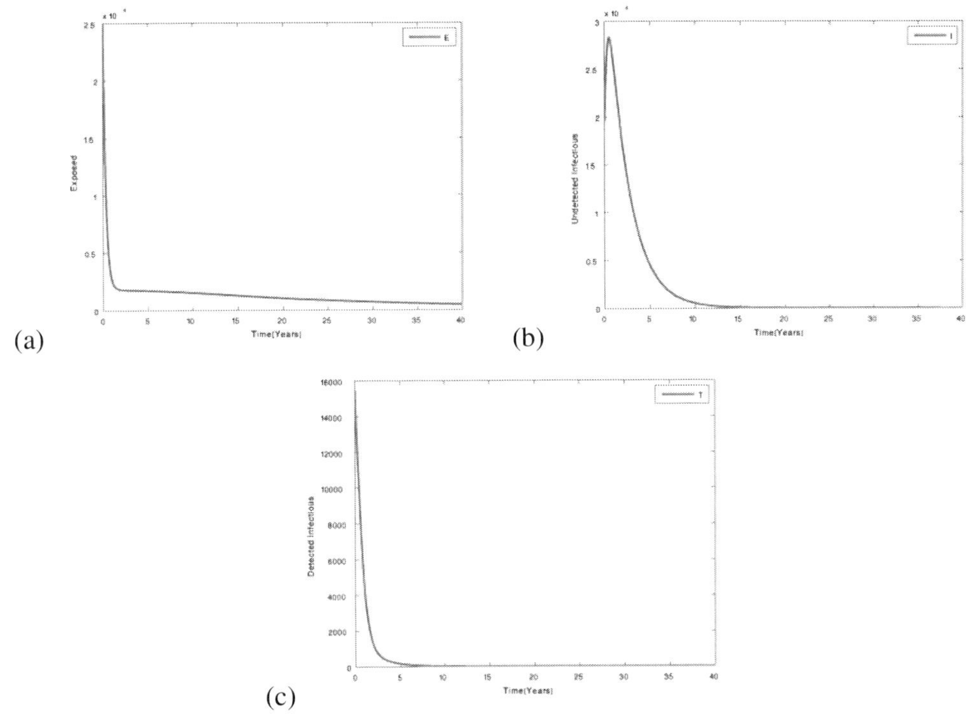

Figure 2. Simulations of model (1) showing the variations of population size of each epidemiological class.

6. Conclusion

We formulated a deterministic model and investigated the potential impact of early detection with isolation in a sanatorium on the transmission dynamics of TB in a community. We derived an expression for the reproduction numbers when there are no detection and isolation strategies in place (i.e R_0) and when these control strategies are implemented (i.e R_e). Qualitative analysis of the model has shown that disease-free equilibrium point exists and is asymptotically stable if the effective reproduction number is less than unity, and unstable otherwise. This implies that TB can be eradicated from the population when $R_e < 1$ provided the initial sizes of the sub-populations lie within the domain of the disease-free attraction.

The effectiveness of the detection strategy was assessed through numerical simulations. It was shown that the number of undetected TB infectious individuals decreases when the rate of detection is increased. Furthermore, this decrease is significant when the detection rate ω is greater than 0.65, which is in agreement with the results obtained by Korobeinikov and Okuonghae (2007) and Okuonghae and Aihie (2008). That is, to bring down the incidence of TB requires a case detection of about 70%. However, this rate may not be feasible in most limited resource countries and consequently, early detection and isolation alone are insufficient to combat the disease and these should be combined with other control strategies

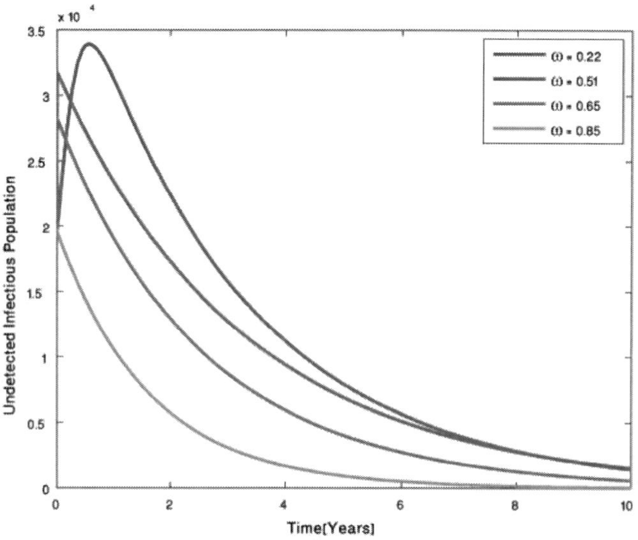

Figure 3. The diagram showing the effect of detection rate to the population of undetected infectious individuals.

such as treatment and vaccination. The model could be extended to include heterogeneous contact rates, susceptibility and infectivity.

Acknowledgments

The author gratefully acknowledges financial support from the International Science Programme (ISP) from Uppsala University, Sweden.

Appendix A. Proof of Theorem 4.1

We use the Center Manifold Theorem as in Castillo-Chavez and Song (2004) to show the possibility of the existence of backwawrd bifurcation in model (1) when $R_e = 1$. For convenience, we re-label the state variables as $S = x_1, E = x_2, I = x_3, T = x_4, R = x_5$ so that $N = x_1 + x_2 + x_3 + x_4 + x_5$. The model (1) can be re-written as follows:

$$\begin{aligned}
\frac{dx_1}{dt} &= f_1(X) = \Lambda - \frac{\beta x_1 x_3}{N} - \mu x_1 \\
\frac{dx_2}{dt} &= f_2(X) = \frac{\beta(1-p)x_1x_3}{N} + \frac{\tau\beta x_3 x_5}{N} + r x_3 - \rho_3 x_2 - \alpha x_2 x_3 \\
\frac{dx_3}{dt} &= f_3(X) = \frac{\beta p(1-\omega)x_1x_3}{N} + k(1-\omega)x_2 + \alpha(1-\omega)x_2x_3 - \rho_1 x_3 \\
\frac{dx_4}{dt} &= f_4(X) = \frac{\beta p\omega x_1 x_3}{N} + \omega k x_2 + \alpha\omega x_2 x_3 - \rho_2 x_4 \\
\frac{dx_5}{dt} &= f_5(X) = \gamma x_4 - \beta\tau x_3 x_5 - \mu x_5.
\end{aligned} \tag{9}$$

where $X = (x_1, x_2, ..., x_5)^T$ and $F = (f_1, f_2, ..., f_5)^T$.

The Jacobian of the system (9) at disease-free equilibrium denoted by J_β is given by

$$\mathbf{J}_\beta = \begin{bmatrix} -\mu & 0 & 0 & 0 & 0 \\ 0 & -\rho_3 & \beta(1-p)+r & 0 & 0 \\ 0 & k(1-\omega) & -\rho_1 & 0 & 0 \\ 0 & \omega k & \beta p\omega & -\rho_2 & 0 \\ 0 & 0 & 0 & \gamma & -\mu \end{bmatrix} \tag{10}$$

where $\rho_1 = r + d_1 + \mu$, $\rho_2 = \gamma + d_2 + \mu$ and $\rho_3 = k + \mu$.
Suppose β is chosen as a bifurcation parameter. Solving (9) for β when $R_e = 1$ gives

$$\beta = \beta^* = \frac{\rho_1\rho_2 - rk(1-\omega)}{p(1-\omega)\rho_3 + k(1-\omega)(1-\omega)}.$$

The right and left eigenvectors associated with the zero eigenvalue of the Jacobian J_β evaluated at β^* are given, respectively by w= $[w_1, w_2, w_3, w_4, w_5]$ and v= $[v_1, v_2, v_3, v_4, v_5]$, where

$$\begin{aligned} w_1 &= 0,\ w_2 = \sigma_2 w_3,\ \ w_3 > 0 \\ w_4 &= \frac{\mu}{\gamma}\sigma_1 w_3,\ \ w_5 = \sigma_1 w_3 \end{aligned}$$

and $\sigma_1 = \dfrac{\gamma\omega(\beta^* p(1-\omega) + \rho_1)}{\mu(1-\omega)\rho_2}$ $\quad \sigma_2 = \dfrac{\mu\rho_2 - \beta^* p\gamma\omega}{\gamma\omega k}.$

Also

$$\begin{aligned} v_1 &= v_4 = v_5 = 0, \\ v_2 &= k(1-\omega)v_3,\ \ v_3 > 0\,. \end{aligned}$$

To determine the direction of bifurcation, we find the signs of a and b where

$$a = v_2 \sum_{i,j,k=1}^{5} w_i w_j \frac{\partial^2 f_2(0,0)}{\partial x_i \partial x_j} + v_3 \sum_{k,i,j=1}^{5} w_i w_j \frac{\partial^2 f_3(0,0)}{\partial x_i \partial x_j}$$

and

$$b = \sum_{k,i=1}^{5} v_k w_i \frac{\partial^2 f_k}{\partial x_i \partial \varphi}(0,0).$$

It can be shown, after using the associated non-zero partial derivatives of F at the disease-free equilibrium that

$$\begin{aligned} \frac{a}{2v_2 w_3} &= -\Big(\frac{(w_3 + 2w_4)(\beta^*(1-p)\mu)}{\Lambda} + \frac{(w_2 + w_5)(\beta^*(1-p)\mu)}{\Lambda} \\ &\quad + \frac{(\beta^* p\mu)(w_2 + w_4 + w_5)}{k\Lambda} - \frac{\beta^*\tau\mu}{\Lambda} w_5\Big) + w_2\alpha\Big(\frac{1-k}{k}\Big). \end{aligned}$$

Thus, $a > 0$ iff

$$\alpha > \beta^* \frac{k\mu}{(1-k)\Lambda\sigma_2}((1-p) + \frac{\mu}{\gamma}\sigma_1(2-p) + \sigma_2 + \sigma_1(1-\tau)).$$

For the sign of b, we substitute vectors v and w and the respective associated non-zero partial derivatives of F at the disease-free equilibrium into

$$b = \sum_{k,i=1}^{5} v_k w_i \frac{\partial^2 f_k}{\partial x_i \partial \varphi}(0,0),$$

which yields

$$b = v_2 w_3 (1-p) + v_3 w_3 p = \frac{k(1-p)+p}{k} v_2 w_3 \ > \ 0.$$

References

[1] ACCP (1940): Sanatorium treatment of tuberculosis, *Chest*, 6: 278-280.

[2] Adetunde I. A. (2007): The Mathematical Models of the Dynamical Behaviour of Tuberculosis Disease in the Upper East Region of the Northern Part of Ghana. A case study of Bawku, *Research Journal of Applied Sciences*, 2(9):943-946.

[3] Agusto, F. B. (2009): Optimal Chemoprophylaxis and treatment control Strategies of Tuberculosis Transmission Model", *World Journal of Modelling and Simulation* , 5(3):163-173.

[4] Anderson, R. M. and May, R. M. (1991): Infectious Diseases of Humans. Oxford University Press

[5] Aparicio, J. P., and Castillo-Chavez, C.(2009): Mathematical Modelling of Tuberculosis Epidemics", *Mathematical Biosciences and Engineering*, 6(2): 209-237.

[6] Aparicio, J. P. and Hernandez, J. C. (2006): Preventive Treatment of Tuberculosis Through Contact tracing", *Contemporary Mathematics*, 410: 17-29.

[7] Birkhoff, G. and Rota, G.C.(1989): Ordinary Differential Equations, John Willy and sons, New York.

[8] Castillo-Chavez, C. and Song, B. (2004): Dynamical models of tuberculosis and their applications, *Mathematical Biosciense*s *and Engineering*, 1(2), 361-404.

[9] CDC (2005): Tuberculosis Cases and Percentages by Pulmonary and Extra-pulmonary Disease.Atlanta, G A, *States, US Department of Health and Human Services.*

[10] Dye, C., Garnett, G.P., Sleeman, K. and Williams, B.G. (1998): Prospects for world-wide tuberculosis control under the WHO DOTS strategy. Directly Observed Short-course therapy", *Lancet*, 352(9144):1886-1891.

[11] Elad, Z., Daley, C.L. and Blower, S.M. (2001): Early therapy for latent tuberculosis infection, *Am. J. Epidemiol.*, 153(4): 381-385.

[12] Elibasha, E.H. and Gumel, A.B. (2006): Theoretical assessment of public health impact of imperfect prophylactic HIV-1 vaccines with therapeutic benefits. *Bull. Math. Biol.,* 68, 577-614.

[13] Feng, Z. and Castillo-Chavez, C. and Capurro, A.F. (2000): A model for tuberculosis with exogeneous reinfection, *Theor. Pop. Biol.* 36: 227-248.

[14] Gumel, A. B. and Song, B. (2008): Existence of Multiple-stable Equilibria for a multi-drug resistance model of Mycobacterium Tuberculosis, *Mathematical Biosciences and Engineering*, 5(3), 437-455.

[15] Hadeler, K.P. and Van Den Driessche, P. (1997): Backward bifurcation in epidemic control, *Math. Biosc.* 146, 15-35.

[16] Hughes, G.R., Currie, C. S. M. and Corbett, E. L.(2006). Modelling Tuberculosis in areas of High HIV prevalence. *Proceedings of 2006 Winter simulation conference*, University of Southampton, UK.

[17] Hussaini, N., Winter, M. and Gumel, A.B. (2010): Qualitative assessment of the role of public health education program on HIV transmission dynamics, *Mathematical Medicine and Biology*, doi:10.1093/imammb/dqq009, 1-26.

[18] Hutson, V. and Schmitt, K. (1992): Permanence and the Dynamics of Biological Systems, *Math Biosci*, 111: 1-71.

[19] Keeling, M.J. and Rohani, P. (2008): Modeling Infectious Diseases in Humans and Animals, Princeton University Press.

[20] Korobeinikov, A. and Okuonghae, D. (2007): Dynamics of Tuberculosis: The effect of Direct Observation Therapy Strategy (DOTS) in Nigeria", *Mathematical Modeling of Natural Phenomena*, 2(1):101-113.

[21] Kribs-Zaleta, C. and Valesco-Hernandez, J. (2000): A simple vaccination model with multiple endemic states, *Math. Biosci.* 164, 183-201.

[22] Murray, C.J.L. and Salomon, J. A. (1998): Modeling the impact of global tuberculosis control strategies", *PNAS*, 95(23): 13881-13886.

[23] Okuonghae, D. and Aihie, V. (2008): Case detection and Direct Observation Theraphy Strategy (DOTS)in Nigeria: Its Efects on TB Dynamics", *Journal of Biologial Systems*, 16(1): 1-31.

[24] Ssematimba A., Mugisha J.Y.T. and Luboobi L.S. (2005): Mathematical Models for the Fynamics in Density-dependent Populations: The Case of Internally Displaced Peoples' Camps (IDPCs) in Uganda, *Journal of Mathematics and Statistics* 1 (3): 217-224.

[25] van den Driessche, P. and Watmough, P. (2002): Reproduction numbers and sub-threshold endemic equilibria for compartmental models of disease transmission, *Math Biosci*, 180, 29-48.

[26] WHO (2011): Global Tuberculosis Control, WHO Report 2011.

INDEX

A

B

C

D

E

F

G

J

K

L

M

N

S

T

U

V

W

Y

Z